Network Processors

The Morgan Kaufmann Series in Systems on Silicon
Series Editor: Wayne Wolf, Georgia Institute of Technology

Network Processors
Architecture, Programming, and Implementation

Ran Giladi

Ben-Gurion University of the Negev
and EZchip Technologies Ltd.

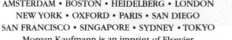

AMSTERDAM • BOSTON • HEIDELBERG • LONDON
NEW YORK • OXFORD • PARIS • SAN DIEGO
SAN FRANCISCO • SINGAPORE • SYDNEY • TOKYO
Morgan Kaufmann is an imprint of Elsevier

MORGAN KAUFMANN PUBLISHERS

Morgan Kaufmann Publishers is an imprint of Elsevier.
30 Corporate Drive, Suite 400, Burlington, MA 01803

This book is printed on acid-free paper. ∞

Library of Congress Cataloging-in-Publication Data
Giladi, Ran.
    Network processors: architecture, programming, and implementation/Ran Giladi.
        p. cm.—(The Morgan Kaufmann systems on silicon series)
Includes bibliographical references and index.
ISBN 978-0-12-370891-5 (alk. paper)
 1. Network processors. 2. Routing (Computer network management)—Equipment and supplies.
 3. Packet switching (Data transmission)—Equipment and supplies. I. Title.
TK5105.543.G55 2008
621.382'1–dc22

                                                                                              2008024883

For information on all Morgan Kaufmann publications,
visit our Website at *www.mkp.com* or *www.books.elsevier.com*

Printed and bound by CPI Group (UK) Ltd, Croydon, CR0 4YY

Transferred to Digital Print 2011

In memory of my father Benjamin (Sontag) Z"L
To my mother Rita (Aaron)
To my beloved wife Kora
To Ornit, Niv, and Itamar, our wonderful children,
and to my dear brother Eival

Contents

Preface

Network Processor Units (NPUs) are designed for use in high-speed, complex, and flexible networking applications, and they have very unique architectures and software models. The first generation of NPUs appeared in about 2000, followed by a second generation of specialized NPUs for various segments of networking applications, usually equipped with the traffic managers required by contemporary networks. NPUs can be compared to Digital Signal Processors (DPSs) that are targeted for high-speed, complex, and flexible signal processing, or to Graphical Processing Units (GPUs) that are required in demanding video processing.

Using NPUs is not trivial, and professional engineers have to learn how to use and program these processors. System architects and system engineers must be aware of the capabilities and advantages that NPUs can provide them in designing network devices. Since data communications and telecommunications industries are bound to use these devices, I found it necessary to include a basic course on NPUs in my department of Communication Systems Engineering at Ben-Gurion University. I have been teaching the subject of NPUs since 2003, and during this period an increasing number of similar courses have been given by other electrical engineering, computer science, and communication systems engineering departments worldwide. In addition, many of my graduate students (now engineers) have called specifically to tell me how important this course has been for their professional careers, helping them to use any NPU, or even developing and using a special-purpose ASIC, for networking applications.

After teaching this subject for several years, I was asked to write a comprehensive book to cover all aspects of NPUs, from network technologies, computer architecture, and software, to specific network processing functions and traffic engineering. Taking advantage of my long association with EZchip, one of the leading NPU vendors, what I wrote emphasizes what practitioners require while studying NPUs for their projects. The result is this book, which took me almost two years to complete.

Network Processors: Architecture, Programming, and Implementation is organized in three parts: (1) Networks—from fundamentals, to converged core and metro networks, to access networks; (2) processing—from network processing and algorithms, to network processor architectures and software modeling; and (3) an in—depth example of a network processor (EZchip), including hardware, interfaces, programming and applications.

The book's target audience is practitioners and students alike. Practitioners include professionals—system architects and system engineers who plan their next-generation equipment and require more in-depth knowledge for choosing and using NPUs; network engineers and programmers who manage projects or write applications for high-speed networking, and need the knowledge and the terminology used in networks, network algorithms and software modeling; and possibly product managers and other professionals who want cutting-edge knowledge in order to design products or to understand them. Research and graduate students, or students in the last phase of their undergraduate studies, may need this book for projects, research on high-speed network applications, or simply a deeper understanding of networking; it will prepare them well for their professional careers.

Acknowledgments

While writing, many assisted me in various ways—too many to list here, and I thank them all. However, I must thank first and foremost a dear friend, Eli Fruchter, who I have known for the last 25 years since we started working together as engineers. Eli founded EZchip and has headed the company from the beginning, and he agreed to disclose EZchip's architecture and have it included in this book. Amir Eyal, EZchip's vice president of business development, who knows the NPU market and technology inside out, provided me with his vision and insight into this field. Eyal Choresh, vice president of software at EZchip, whom I also have known for many years, assisted me in explaining the NP architecture's details, as well as providing other clever ideas on software issues. Alex Tal, EZchip's first chief technical officer, gave me my first insight into how this book should be structured and provided me with his perspectives on the architecture of NPUs. Thanks go to all the EZchip family for various kinds of assistance, including, EZchip's current CTO Guy Koren, Nimrod Muller, Aweas Rammal, Daureen Green, and Anna Gissin.

Many of my students and colleagues were very helpful in writing and checking software, reviewing and commenting; they include Mark Mirochnik, Dr. Nathalie Yarkoni, Dr. Iztik Kitroser, Micahel Borokhovich, Kfir Damari, Arthur Ilgiaev, Dr. Chen Avin, Dr. Zvi Lotker, Prof. Michael Segal, and Dr. Nissan Lev-Tov.

Among the many friends and colleagues who pushed and supported me during the long period of writing, I must thank Prof. Gad Rabinowitch, Prof. Gabi Ben-Dor, Shai Saul, Beni Hanigal, Prof. Yair Liel, and Moshe Leshem.

Special thanks go to Dr. Tresa Grauer who worked with me on editing everything I wrote, watching me to make sure that everything I wrote was clear, who made me rewrite and reedit until she was satisfied. Her devoted efforts have been remarkable. I am grateful to Chuck Glaser, Greg Chalson, Marilyn Rash, and the editorial and production staff at Morgan Kaufmann and Elsevier for their wonderful work, patience, encouragement, and professionalism.

I want to take the opportunity to thank my parents who form the basis of everything I've got, each of them in a special way—the endless support and warmth of my mother and the drive for knowledge and accomplishments from my father. I owe them the many things I have achieved, this book included. I regret that my father's wish to see it completed was unfulfilled.

Last, but not least: Kora, my wife and lifetime friend, and our three children, Ornit, Niv, and Itamar. All were so tolerant of me and my absence while working on this book (and on other projects, in their turn). They assisted with drawings, and Itamar even arranged all the acronyms. Their support and love allowed me to devote time to read many hundreds of papers, learning and thinking, designing, and writing this book.

Introduction and Motivation

Network processors (NPs) are chips—programmable devices that can process network packets (up to hundreds of millions of them per second), at wire-speeds of multi-Gbps. Their ability to perform complex and flexible processing on each packet, as well as the fact that they can be programmed and reprogrammed as required, make them a perfect and easy solution for network systems vendors developing packet processing equipment.

Network processors are about a decade old now and they have become a fundamental and critical component in many high-end network systems and demanding network processing environments.

This chapter introduces this relatively new processing paradigm, and provide a high-level perspective of what NPs are, where to use them and why, and conclude with a brief description of the contents of the rest of the book.

1.1 NETWORK PROCESSORS ECOSYSTEM

Telecommunications and data networks have become essential to everything that we do, to our well-being and to all of our requirements. The prevalence of Internet technology, cable TV (CATV), satellite broadcasting, as well as fixed and cellular mobile telephony, tie many and expanding services to a very large population that is growing exponentially. The increasing speed of the communication links has triggered a wide range of high-speed networks, followed by an increasingly broad spectrum of services and applications.

We are witnessing this dramatic growth in communication networks and services; just think of the changes that networks and services have undergone in the past 10 years in the areas of mobile, video, Internet, information availability, TV, automation, multimedia, entertainment, online services, shopping, and multiplayer games. You can safely assume that an equivalent jump in technology and services will happen again in the next 5 years or so.

Networks and infrastructures have enabled it all. Most houses, vehicles, pieces of equipment, and people, maintain a communication link to the "network," a giant octopus with zillions of arms. And the oxygen that runs in its veins, pipes, and trunks are packets and cells; zillions of zillions of them are flying around us and surrounding us at every moment. These packet flows undergo various treatments, processing, and forwarding, in many kinds of network devices. These network devices are systems by themselves, and they keep the octopus, or the "network," alive. Such network systems include switches, adapters, routers, firewalls, and so on.

Network systems, therefore, face an ever-increasing magnitude of packets they have to handle, while at the same time, the processing of these packets becomes more and more complex. This creates a gigantic performance problem. In order to cope with it, vendors have replaced the traditional general purpose Central Processing Unit (CPU) in the network systems with Application Specific Integrated Circuits (ASICs), which are hardwired processing devices. However, as vendors also face rapid changes in technology, dynamic customer requirements, and a pressure for time-to-market, short developing cycles and lots of revisions have become necessary. All of this has required an innovative approach toward network systems architecture and components.

This is the foundation on which NPs flourish. NPs enjoy the advantages of two worlds—they have the performance capabilities of an ASIC, and the fast, flexible, and easy programmability of a general-purpose CPU. NPs constitute the only option for network systems developers to implement dynamic standardized protocols in performance-demanding environments.

1.2 COMMUNICATION SYSTEMS AND APPLICATIONS

Communication systems are composed of networks and network devices (which are sometimes referred to as network elements, or network systems, as we called them above, since they are computerized, special purpose systems that include both software and hardware).

Communication networks can be separated into three main categories: the core, the aggregation (or metro), and the access networks. Each of these network categories are characterized by different requirements, technologies, and equipments.

In addition, there are traditionally two communication systems we use: telecommunications (telecom) and data communications (datacom). The more established and older network, the telecommunication network, is based on "circuits," or channels of continuous bit streams, which grew out of its original application to telephony services. The more recent networks, the data communications networks, were initially based on packets of data to carry information among computers. This resulted in two paradigms of network technologies—*circuit switching* and *packet switching*.

The main technology in telecommunication systems is the Synchronized Optical Networks/Synchronous Digital Hierarchy, which is used in both the core and the aggregation networks. Radio Access Networks through which most mobile telephony is conducted, as well as the wireline telephony access networks, are usually attached to these networks, using their own technologies. CATV is a parallel network that is traditionally used for TV services.

The main technologies in the datacom systems are the Ethernet and the Internet Protocol (IP). For the last two decades, converged telecom–datacom networks have been subject to vast research, industry implementation trials, and services. This convergence happens in the technology plane as well as in services: recent Telecom networks' cores are implemented using datacom technologies (such as Ethernet, Multiprotocol Label Switching [MPLS] and IP). Recent trends in converged services, starting with "triple-play" service (or "quadruple-play" for Internet, TV, telephony, and data-oriented services) to Voice over Internet Protocol telephone services and TV over IP, are just a few examples of the transition to a unified, converged network.

The result is that packet networks have become the prevalent technology for communication systems. Network systems are making the transition from circuit-switched based technologies (multiplexers, cross connects, branch exchanges, etc.) to packet-switched based technologies (such as bridges and routers).

Services are also developing in scale and complexity; from plain telephony, TV, and Internet surfing and e-mails, we are now facing High Definition TV, new generations of web and web services, digital libraries, and information availability, gaming, and mobile 3G and 4G services that include information, multimedia, video, and TV, and many other demanding applications. Other adjunct services, such as security, provisioning, reliability, and accounting must be supported by network systems, which may impose additional significant load on them.

As networks become the infrastructure for information, interactive data, real-time data, huge multimedia content transport, and many other services described previously, the technology of networks must cope with various requirements, but primarily that of speed. High-speed networking refers to two aspects of speed: the links' transmission rates—from multi Mbps (10^6 bits per second) to multi Gbps (10^9 bits per second)—and the complexity and speed of the required processing due to the number of networks, addresses, services, traffic flows, and so on.

If we examine the speed of network links over the years, we find a similar but higher growth pattern than that of processing capabilities. In computing, this exponential growth is modeled as doubling roughly twice every 2 years (after Moore's law);[1] however, in the last decade this growth rate has shrunk, and is roughly at 41% annually (with clock speedups increasing only 29% annually). [116] If we look, for example, at Ethernet bandwidths, we find a $\times 10^4$ speedup in 27 years (from 10 Mbps approved in 1983 to 100 Gbps expected to be approved

---

[1]Moore's law is an interpretation of the 1965 statement of Gordon Moore of Fairchild, who was a founder of Intel. The law refers to the doubling of the number of transistors on a chip, or the chip performance, every 18 to 24 months.

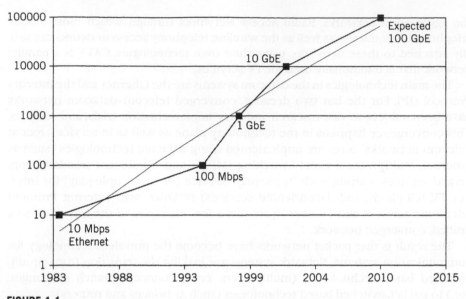

FIGURE 1.1

Ethernet-approved interface speeds

in 2010, as shown in Figure 1.1), which is doubling the bandwidth every 24 months. However, if we examine the increase from 100 Mbps (approved in 1995) to 100 Gbps ($\times 10^3$), it is doubling the bandwidth every 18 months. This pattern of growth is similar to that of telecom links (Optical Carrier—availability during these years).

Ethernet bandwidth growth is in-line with Moore's law, although a bit faster. Add to that the increase in network utilization and the increased number of networks, which are doubling roughly every four years, as the number of BGP[2] entries indicate, depicted[3] in Figure 1.2; and we have aggregated traffic that is doubling approximately once a year [85].

This traffic increase is twice as fast as Moore's law, or twice as fast as computing processing is capable of meeting, including network systems (such as routers) that are based on the same computing paradigm. The many different applications used on the Internet, with varying demands that are changing constantly, add another dimension of difficulty for network systems. Figure 1.3 shows a snapshot demonstrating traffic growth and application mix in a major Asian-Pacific ISP[4] from July 1, 2007, to April 1, 2008.

---

[2]Border Gateway Protocol (BGP) routers possess entries that indicate the number of IP networks.
[3]The number of active BGP entries are from *http://bgp.potaroo.net/as2.0/bgp-active.txt*.
[4]The data are sampled on the first of every month, at 07:00 hours.

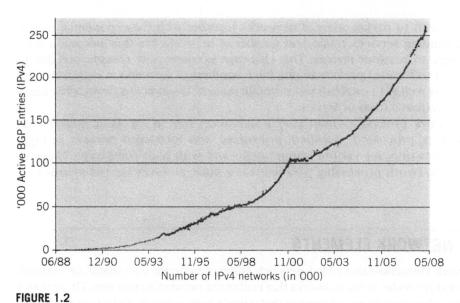

FIGURE 1.2

Number of IP networks known to BGP routers

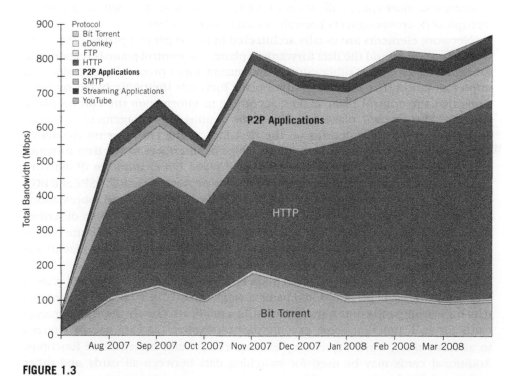

FIGURE 1.3

Application mix in internet (Courtesy of Allot Communications.)

The shift to packet-oriented networks, together with the exponential growth of demanding services, traffic, and number of networks has thus become a great challenge to network systems. This challenge is enormously complicated, due to the need to forward packets among huge numbers of networks at extremely high speeds, as well as to establish and maintain tunnels, sessions, and flows, while keeping the required Class of Service.

Network processors must play a significant role in inputting packets and outputting processed, classified, prioritized, and forwarded packets, in high-speed networks, for various requirements and with high complexity. The shift to new network processing paradigms is a must, and NPs are the vehicles for doing it.

1.3 NETWORK ELEMENTS

Network elements (also called network systems, network equipment, or network devices) are nodes in the networks that enable the network to function. These nodes are complex computerized systems that contain both software and hardware (sometimes also operating systems), and are special purpose systems that are used for networking and, more specifically, for networking functions such as switching, routing, multiplexers, cross-connects, firewalls, or load balancers [380].

Network elements are usually architected in two separated planes [255, 435]: the control plane and the data forwarding plane. The control plane is responsible for signaling as well as other control and management protocol processing and implementation. The data forwarding plane forwards the traffic based on decisions that the control plane makes, according to information the control plane collects. The control plane can use some routing or management protocol, according to which it decides on the best forwarding tables, or on the active interfaces, and it can manipulate these tables or interfaces in the data forwarding plane so that the "right" actions will take place. For example, in IP network elements, the control plane may execute routing protocols like RIP, OSPF, and BGP, or control and signaling protocols such as RSVP or LDP. (All these abbreviations indicating various protocols are covered in Part 1 of the book, which describes networks.) The data forwarding in these IP network systems may execute packet processing, address searches, address prefix matching for forwarding, classifying, traffic shaping, and metering, network address translation, and so on.

A typical system can be either a "pizza box" (Figure 1.4) or a multicard chassis (see Figure 1.5). A "pizza box" contains a single board, on which all processing, data forwarding, and control are executed. In a multicard chassis, there are separate cards for each function; some execute data forwarding (and are usually referred to as *line-cards*), while others may execute control and management functions. Additional cards may be used for switching data between all cards, and other utilities of the chassis.

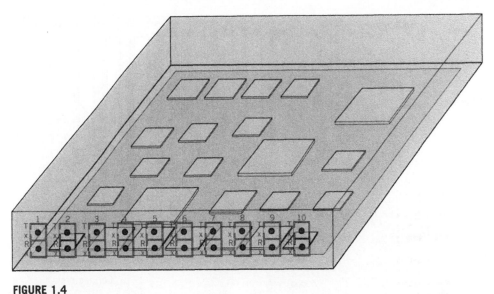

FIGURE 1.4

Stand-alone, "pizza box" system

Since the data forwarding plane is responsible for per-packet processing and handling, it is executed by high-speed devices such as ASICs or NPs, as described before. Control plane processing is usually executed by general-purpose processors, since the load on the control plane is significantly less than that of the data forwarding plane.

There are several alternatives for components in the data forwarding plane that do the processing for various functions; these alternatives include switching chip sets, programmable communication components, Application Specific Standard Products (ASSPs), configurable processors, Application Specific Instruction Processors (ASIPs), Field Programmable Gate Arrays (FPGAs), NPs, and ASICs [176]. Among these options, ASICs and NPs are the two real options for the main processing units of network systems. These alternatives offer processing power and best fit to the required applications.

It is worth noting here, that ASIC was perceived as a better solution, commercially, although it suffers from a very high development cycle and cost, and is absolutely inflexible. NPs, on the other hand, used to be more expensive, but had a short and low cost development cycle. However, as technological advantages made the size and cost of NPs and ASICs about equal, ASICs lost their advantage. Indeed, at the beginning of 2008, leading networking vendors announced that they had developed NPs in-house in order to support their network systems, in addition to their continued use of NPs from various external vendors.

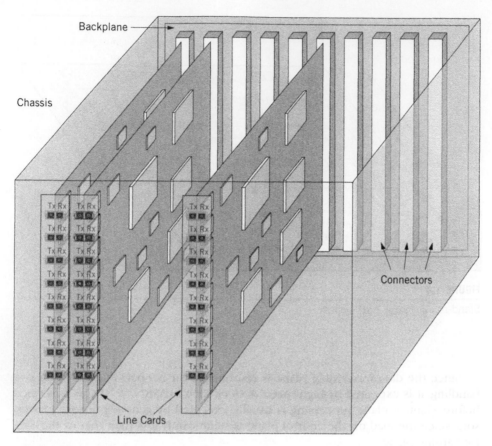

FIGURE 1.5

Multicard chassis system

1.4 NETWORK PROCESSORS

Network processors are categorized into three groups according to their use and the way they have evolved: entry-level or access NPs, mid-level processors (legacy and multiservice NPs), and high-end NPs that are targeted for core and metro networking equipment, usually on line cards.

Entry-level NPs, or access NPs, process streams of up to 1 to 2 Gbps packets, and are sometimes used for enterprise equipment. Applications for such access NPs include telephony systems (e.g., voice gateways and soft switches), Digital Subscriber Loop (xDSL) access, cable modems, RANs, and optical networks (e.g., Fiber to the Home, Passive Optical Networks, etc.). A few examples of such NPs are EZchip's NPA, Wintegra's WinPath, Agere, PMC Sierra, and Intel's IXP2300.

Mid-level NPs (2–5 Gbps) contain two subgroups of NPs: legacy NPs and multiservice NPs, which are usually used for service cards of communication equipment, data center equipment, and multiservice platforms, in Deep Packet Inspection and Layer 7 applications (security, storage, compression, etc.). In the legacy subgroup, one can include the classical, multipurpose NPs like AMCC, Intel's IPX, C-port, Agere, Vittese, and IBM's NP (which was sold to Hifn). Examples of multiservice and application (Layer 7) NPs are Cisco's QFP, Cavium, RMI, Broadcom, Silverback, and Chelsio.

High-end NPs (10–100 Gbps) are used mainly for core and metro networks, usually on the line cards of the equipment. These NPs can process hundreds of millions of packets per second, at wire-speed. Examples of such NPs are EZchip's NPs, Xelerated, Sandburst (which was bought by Broadcom), Bay Microsystems, or the in-house Alcatel-Lucent SP2.

The choice of which category can significantly impact the architecture of the NPs. In any case, the architecture and programming models of NPs are completely different from general purpose processors. NPs' architecture can be characterized as being based on parallel processing and pipelining. Most NPs include integrated hardware assists such as integrated traffic management and search engines, and also have high-speed memory and packet I/O interfaces. The main processors of NPs can be based on MIPS RISC processors, VLIW or reconfigurable processors, or special packet processors (generic RISC, also called micro engine, pico engine, or Task-Optimized Processors [TOPs], etc.).

Network processors are sometimes distinguished from packet processors by being programmable, whereas packet processors are just configurable. The programming paradigms, models, styles, and languages that are used for NPs are also very different from those used for applications running on general purpose processors. No interrupts are used in NPs, and the main principle is to use an event-driven control program, in which packets dictate how the software runs. Network processors perform several key functions, including:

- Parsing the incoming frames in order to understand what they are, and where to find the relevant information that is required for processing.
- Retrieving the relevant information from the frames, which may be complicated as encapsulation, variable fields length, and various levels of protocols may be involved.
- Deep packet analysis when required, such as understanding HTTP names, identification of XML protocols, and so on. This may be required for priority assignments for various kinds of traffic flows.
- Searching for related information in repositories; for example, routing tables, access lists, and so on.
- Classifying the frames according to the various forwarding and processing schemes that the frames should undergo.
- Modifying the frame's contents or headers, possibly adding tags, changing addresses, or altering the contents.
- Forwarding, which may potentially be coupled with various traffic management tasks such as metering, policing, shaping, queueing, scheduling, and statistics.

Network processors appeared in the late 1990s, and flourished as major processor and network vendors led the NPs market (companies like Intel, IBM, Motorola, and Agere). Since the concept was new, it created a lot of enthusiasm, and caused a wave of established companies as well as many newcomers to invest and innovate. Then the market suffered a slowdown, and most players abandoned their NPs, or sold them. Some say this was a consequence of the "bubble" phenomena of the early 2000s that hit the telecom markets the most (for which NPs are targeted); others say it was a normal market reaction to any new revolutionary technology—the phase of absorbing, understanding, and applying the rules of natural selection ultimately narrows the field until only the best remain.

Whatever the case may be, those who survived established themselves and their NPs as solid solutions for network systems, and the market flourished again. As of the beginning of 2008, major network vendors announced new generations of NPs that are the only way for them to compete and introduce network devices that can sustain network demands. Total investment in NPs development has reached approximately 1 billion U.S. dollars, as of 2008.

Companies that introduced various levels of NPs include:

Network and packet processors: Agere, Applied Micro Circuits Corp (AMCC, which bought MMC Networks), Bay Microsystems, C-Port (acquired by Motorola, which is now FreeScale), Cognigine, ClearSpeed Technology, Clearwater Networks (formerly XStream Logic), Entridia, Ethernity Networks, EZchip Technologies, Fast-Chip, Greenfield Networks (acquired by Cisco), Hyperchip, IBM (sold its NP line to Hifn), Intel, Internet Machines (changed to IMC semiconductor and a new business), Paion, PMC-Sierra, Sandburst (acquired by Broadcom), Sand-Craft, Silicon Access Networks, Sitera (acquired by Vitesse), Teradiant (formerly SiOptics), Terago Communications, Xelerated, Zettacom (acquired by Integrated Device Technology).

Access processors: Agere, AMCC, Audiocodes, Broadcom, Centillium, Conexant, Ethernity, EZchip Technologies, Freescale, Infineon, Intel, Mindspeed, Octasic, Texas Instruments, TranSwitch, Wintegra.

1.5 STRUCTURE OF THIS BOOK

Part 1 is concerned with the first part of the phrase *Network Processors*, the networks. This part contains a brief summary of networks' technologies, standards and protocols. It begins in Chapter 2 with fundamentals, discussing data and tele-communication network technologies, then goes on to provide more in-depth description of contemporary converged networks, and ends with a description of access networks and home networking.

Chapter 2 describes network models and architectures, and then data networks—namely, Ethernet and Internet Protocol networks. The basics of telecommunications networks are also described in this chapter (e.g., PDH and SDH/SONET networks), with an emphasis on relevant technologies for data applications.

Chapter 3 focuses on converged networks. In these networks, data and telecommunications applications interface and mix, despite their very different natures. Data networks are bursty, and oriented toward connectionless traffic, whereas telecommunications networks are streamed, and connection-oriented. This chapter covers Asynchronous Transfer Protocol (ATM), Multiprotocol Label Switching (MPLS), Layer 2 and Layer 3 Virtual Private networks (VPNs), Carrier Ethernet, and Next Generation SONET/SDH technologies.

Chapter 4 discusses access and home networks, or customer premises networks. These networks also use converged technologies, and equipment in these networks is increasingly based on network processors, as bandwidth and complexity reach a degree that justifies them.

Part 2 is concerned with the second part of the phrase *Network Processors*, the processing and processors. It discusses the theory behind network processors, starting with frame and packet processing, the algorithms used, data structures, and the relevant networking schemes that are required for packet processing. Then it describes the theory behind the processors themselves, beginning with hardware (architecture), moving on to software (programming models, languages, and development platforms), and concluding with network processors' peripherals.

Chapter 5 describes packet structure and all the processing functions that the packet must go through (i.e., framing, parsing, classifying, searching, and modifying). Searching (lookups) and classification are treated in a detailed manner, since they are the most important and demanding tasks.

Chapter 6 addresses various aspects of packet flows, traffic management, and buffers queuing. The chapter deals with Quality of Service (QoS) and related definitions, and QoS control mechanisms, algorithms, and methods.

Chapter 7 describes the basic architectures and definitions of network processors. It covers various computation schemes, as well as network processing architectures in general. In particular, parallelism, threading, and pipelining are described. Other architectural components (e.g., I/O and memories) as well as interface standards are also described, in order to provide a comprehensive understanding of network processors' design and interface, both at the system level (board and equipment), and at the networking level.

Chapter 8 describes programming models of network processors, as well as some important principles that are relevant to their programming, and concludes by describing the typical programming environment of network processors. Programming a network processor is very different from programming any other processor. Parallel and pipelining processing and programming are covered.

Chapter 9 concludes the second part of the book with a description of two important network processors' peripherals: switch fabrics (the interconnection functions), and coprocessors (for adjunct functions).

Part 3 examines the two subjects of networks and processing together with a concrete example of a network processor. It provides an in-depth description of EZchip's network processor, which dominates the metro networks markets. It begins with a description of the hardware architecture, and then continues with the software architecture and programming. Following these chapters, each of the processors and the functional units of the NP is described in a separate chapter. This part concludes with a comprehensive example of writing a program and using the network processor.

Chapter 10 describes the general architecture of EZchip's NP-1 network processor, its heterogeneous pipeline of parallel processors, the TOPs, the interfaces, and the data flow of a packet inside the NP-1.

Chapter 11 explains how to program the NP, how to use the development environment of the network processor (including compiling, debugging, simulating, etc.), and how to run the programs (i.e., initializing the NP, downloading it, and working with the attached external host processor). It covers pipelining programming in depth, as well as the NP-1 Assembly.

Chapters 12, 14, and 15 outline the TOPparse, TOPresolve, and TOPmodify architectures, respectively, as well as their internal blocks and their instruction sets. Simple examples are also given to demonstrate the use of these TOP engines.

Chapter 13 describes TOPsearch I very briefly, and explains how to carry out various simple lookup operations with the TOPsearch engine by providing search keys, and getting results that match these keys.

Chapter 16 describes how to use the EZchip development system based on the example given in previous chapters; that is, how we load it, compile it, and debug it. In addition, the chapter provides a quick and basic review of how to define frames (which the simulator will use) and how to build the search database (structures that also can be used during the debugging phase).

Chapter 17 concludes the third part of the book by demonstrating how to use the EZchip NP, with a high-speed network application, a multi-Gbps routing, and an "on the fly" screening filter for prescribed words that are to be identified and masked. This chapter shows how to design, write, run, debug, and simulate an application with the Microcode Development Environment (MDE).

The EZmde demo program can be used to write a code, debug it, and simulate an NP with this code (with debugging features turned on). The EZmde can be downloaded from *http://www.cse.bgu.ac.il/npbook* (access code: CSE37421), as well as the EZmde design manual.

1.6 SUMMARY

Network processors became an essential component for network system vendors. This chapter outlined the ecosystem of NPs, the reasons for their importance, their growth, and their capabilities.

This subject is very dynamic; the interested reader may use periodicals, books [86, 92, 134, 135, 278], and Internet sources for remaining updated in the area of NPs. Some excellent Internet sources for example are Light Reading [10] and the Linley Group [286].

Networks

1

This part of the book, *Network Processors: Architecture, Programming, and Implementation,* is concerned with the first part of its title, Networks.

Since network processing is about processing packets and frames in networks, according to network protocols, demands, and behavior, it is essential to have a thorough knowledge of them before network processing is possible. For example, even analyzing a packet or a frame cannot be done without this understanding because the frame is structured according to the network protocol used, among many other issues.

To provide unified terminology for this book, as well as some background on the relevant networking concepts, this part contains a brief summary of networks' technologies, standards, and protocols. It begins with fundamentals, discussing data, and telecommunication network technologies, with an emphasis on the common networks that are most likely to be of interest to users who want to learn about network processing and processors. It then goes on to provide more in-depth descriptions of contemporary converged networks, as well as some good background on the technologies required for the reader who is interested in implementing network processors into metro or core networks. This part ends with a description of access networks, both wireline, and Radio Access Networks (RANs), and home networking.

As this is not a general textbook on networking, the descriptions provided in this part do not cover networking comprehensively; rather, they provide the necessary information required for and relevant to network processors. It contains the following chapters:

- Chapter 2—Networking Fundamentals
- Chapter 3—Networks Convergence
- Chapter 4—Access to Home Networks

Part 2 deals with the second part of this book's title—processors and processing.

Networking Fundamentals

The previous chapter introduced the huge field of networks—what a network processor is and how to use it, as well as services that are relevant and the challenges that make network processors so important. Before we jump into a discussion of the requirements, roles, and benefits of network processors, however, it is first essential to be familiar with networking principles and technologies. Because we assume that most of the readers have at least some background knowledge of networking, our overview of the fundamentals of networking is quite general, and our descriptions are provided primarily in order to establish our terminology. In the next chapter, we move on to discuss more advanced, converged, and contemporary networking technologies.

When talking about networks in the context of network processors, it is important to remember that networks can be found in many places and in many shapes, on many kinds of media and serving many purposes. We focus on a small but important segment of networks, where network processors are used in network nodes to carry the networking functions. We are mainly concerned here with data networks and telecommunications networks. In this chapter, we begin by describing network models and architectures. Then we describe data networks—namely, Ethernet and Internet Protocol (IP) networks. Ethernet technology is used primarily and traditionally for data communications in local area networks (LANs), and more recently in metropolitan area networks (MANs), or Metro networks. IP, which is a network of networks, is used as the underlying technology for the Internet and most wide area networks (WANs), including enterprise and campus networks. The basics of telecommunications networks are also described in this chapter, with an emphasis on relevant technologies for data applications.

In the next chapter, we describe contemporary networks—networks that combine data applications and services with telecommunication (or data networks, telecommunication networks, and what is in between them)—and elaborate on our discussion of converged data and telecommunication networks such as Multiprotocol Label Switching (MPLS) and metro networks technologies.

A disclaimer: As mentioned above, we describe and generalize network concepts with a bias toward issues that are relevant to network processors. It is not the goal of this chapter or the next one to work systematically through an explanation of networking itself.

2.1 INTRODUCTION

In recent years, telecommunications, computers, networks, contents, and applications concepts have been combined and reshaped into new paradigms of infrastructure (networks and equipment), services (applications), and information (content). We introduce networks in this chapter from the perspective of infrastructure, that is, starting from the hosts, communication links and network equipment. We then go on to describe algorithms, protocols, and data structures that are part of the communication system.

Communication networks started by creating physical connections between peers, first carrying analog-streamed data and then evolving to connections carrying digital-streamed data. With the ability to carry digital information came the possibility of organizing and packaging information in packets for networking. Data communications networks were initially based on packets to carry data, whereas telecommunication networks were based on "circuits," or channels of continuous bit streams. This was reflected in two communication network paradigms—*circuit switching* and *packet switching*. In circuit switching, communication channels are dedicated to the communicating peers throughout the communication session, whereas in packet switching, the physical channels are shared by many communication sessions simultaneously, and packets are routed and switched between the communicating peers.

Circuit switching in its original form has been mostly replaced by packet-switching technologies that emulate the circuit-switching paradigm, providing virtual circuits between the communicating peers. Packet networks are subdivided into *connection-oriented networks*, and *connectionless-oriented networks*. The first category describes "ordered" and reliable communications procedures like the telephone system (e.g., call set-up, transmission acknowledgment and verification, and call termination), while the second describes "lighter" procedures and requirements that simplify the communication procedure (e.g., a mail or messaging system that does not require call set-up or the other procedures).

Today, almost the entire world of communication networks uses packets, and we are surrounded by packets flying and flowing around us, both in wires and wireless. The rapid growth in networking and communications applications has been accompanied by exponential growth in the number and rate of packets flowing around—packets that have to be analyzed, treated, processed, routed, and accounted for. This is the ground on which network processors emerged.

This chapter begins with a networks primer and some functional, physical, and architectural models. Then we describe networks according to their classification as either Telecom or Datacom:

- Pure Telecom networks (global, state, regional, or public networks):
  - Plesiosynchronous Digital Hierarchy (PDH).
  - Synchronized Optical Networks/Synchronous Digital Hierarchy (SONET/SDH).
  - Optical networks.
- Pure Datacom networks (office or campus-wide private networks):
  - Enterprise Ethernet.
  - Internet Protocol.

The network technologies that are the most likely to be used by Network Processors applications are Ethernet (enterprise and carrier class), MPLS, and IP. Therefore, most of the description in this chapter is focused on these networks, with discussion of the other networks provided to fill out the general overview and to provide the reader with a framework of how networks are combined, relate to each other, and develop.

2.2 NETWORKS PRIMER

Networks exist with wide ranges of functions, speeds and distances—from a scale of millimeters to global span (including satellite networks). Networks are even used in chips (called Networks on Chips, NoCs). Networks' speeds vary from very low bps to as many as thousands of Giga bps (Gbps, ending up in the Tera bps, Tbps range). As a rule of thumb, the smaller the span, the faster the network (Figure 2.1), and technology pushes the speed-span curve to ever higher speeds at larger spans.

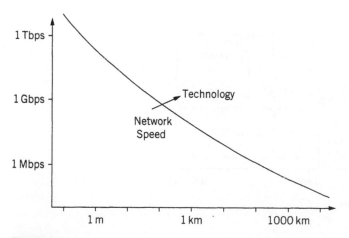

FIGURE 2.1

Network span and speed

Network processing covers several segments of networks and technologies, as described in this subsection, and shown in Figure 2.2. Computer-Peripheral Networks (CNs) are entirely beyond the scope of this book, although, as we shall see in the following chapters, some switching systems are offered that are based on these kinds of networks. Personal area networks (PANs) are also beyond the scope of this book, although as pervasive and ubiquitous computing grows in popularity (and might be the "killer application" for PAN), it might overload the other kinds of networks that we are describing.

Local area networks are definitely relevant to network processors, and many applications of network processors are frequently used in these kinds of networks. Since it is important for writing network processors applications to absolutely understand how LANs operate (mainly Ethernet, the dominant technology), Ethernet will be covered here in some detail.

Wide area networks are also very relevant to network processors, and the dominant internetworking technologies are also described here in detail. WAN is overlaid on metro or other telecom and core networks for providing data applications services.

Data networks are used mainly for computer data transfers, and telecommunications networks are used for streaming services (voice, video), as well as for large trunks of data channels to interconnect data networks. In Figure 2.2, for example, data networks are LANs and WANs, whereas Core networks are considered to be telecommunications networks. Metro networks can be either telecommunications networks or data networks, depending on the technology, the definition, and the current trend.

Despite the fact that the "convergence" trend of data, voice, and video is both exciting and long-awaited, there is (still) a separation between data networks and

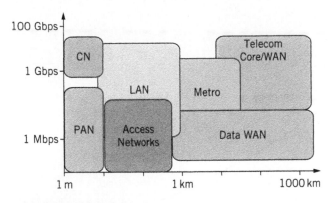

CN—Computer Peripherals Network
PAN—Personal Area Network
LAN—Local Area Network
WAN—Wide Area Network

FIGURE 2.2

Relevant networks

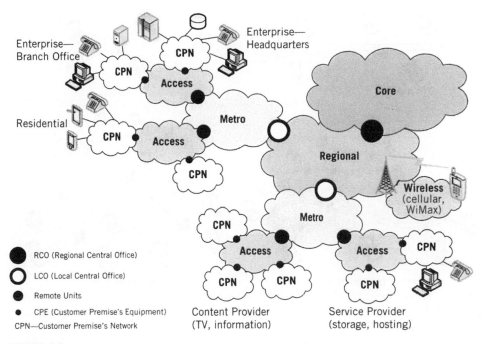

FIGURE 2.3

Network hierarchy

telecommunications networks. Convergence is happening though, and buzzwords like triple-play or available services that are data-voice-video converged are here to stay (e.g., Voice over IP, VoIP; or TV over IP, IPTV).

Another way to categorize networks is based on their functions and the relationships between them (from which the span and the speed are derived). Schematically, at one extreme we have PAN, LAN, home networking, data-centers, and enterprise networking (Customer Premise's Networks, CPN), with core networks at the other extreme (as shown in Figure 2.3). As mentioned above, WANs are overlaid on the access, metro, regional, and core networks.

Networks have basic hierarchies, as can be seen in Figure 2.3. Core networks (sometimes referred to as long-haul, or backbone networks) are networks that span globally, nationwide, and long distance (hundreds and even thousands of miles), carrying 10 Gbps and more, to which regional networks are connected through a Regional Central Office (RCO).[1] The regional networks (sometimes referred to as core-metro) are many tens of miles in span, carrying 10 Gbps, and

---

[1]Regional Central Office (RCO, also called Toll-Center, TC) uses interfacing and switching equipment that is analogous to the class 4 telecommunications switches, or tandem switches, used in the traditional telephone network hierarchy.

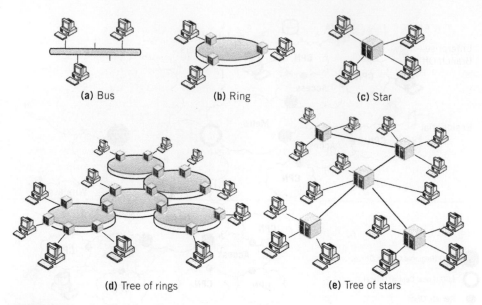

(a) Bus   (b) Ring   (c) Star

(d) Tree of rings   (e) Tree of stars

FIGURE 2.4

Network topologies

connecting metro networks through a Local Central Office (LCO).[2] The metro networks (sometimes referred to as metro edge, metro access, or aggregation networks) run at 2.5 to 10 Gbps, are up to a few tens of miles, and are connected to the access networks through remote units. The access networks have a span of about a mile, run up to 2.5 Gbps, and connect the CPN to the entire network through the customer premise's equipment. At the customer premises, there are many types of networks, depending on the type of customer—residential, enterprise, content or service provider, and so on. Wireless networks are also part of this setup, where public wireless networks such as cellular and WiMAX are connected like the metro networks (sometimes through the metro networks), and private networks (i.e., residential or enterprise) are within the customer premises (e.g., Wi-Fi and all PAN networks).

Although most readers are familiar with networks topologies (Figure 2.4), it is worth noting that most networks today are either based on a star topology (like most enterprise networks), or a ring topology (like most regional and metro networks).

---

[2] Local Central Office (LCO, also called End-Office, EO) uses interfacing and switching equipment that is analogous to the class 5 telecommunications switches, or telephone exchange, used in the traditional telephone network hierarchy. Actual telephones were connected to this equipment, whereas today metro or access trunks are connected to the LCO.

2.3 DATA NETWORKING MODELS

Network modeling can be done in any one of the following two ways: either by modeling the data and the communications' protocols between the communicators, or by modeling the physical components of the network and their interconnections. Eventually, the two models converge into one representation of network modeling.

Communication between two nodes can be done by a program that handles everything from taking care of bit and byte ordering and transmission or receiving to inter-application inputs and outputs. Such a program also handles all aspects of networking (routing and forwarding), error recovery, handshakes between the applications, data presentation, and security. These programs existed in the early days of data communications; however, in modern, sophisticated networks, it is now impractical not only to handle communications programs in this way, but also to maintain them or to reuse parts of them when required.

As data communications and telecommunications programming, interfaces, and equipment grew more sophisticated, the International Standard Organization (ISO) suggested a structured, layered architecture of networking called Open System Interconnect (ISO/OSI). The ISO/OSI is an abstract reference model of layered entities (protocols, schemes), depicting how each entity interfaces with the entities that reside in the layers directly above and below it (except for the lowest layer, which communicates only with peer entities). At about the same time, the U.S. Department of Defense (DoD) offered another layered model that concentrated on data-network modeling. These two models provide fundamental concepts in communications, and most systems and definitions use their language.

According to the ISO/OSI model, which is also called the seven-layer model, any two peered layers interact logically, carrying the relevant data and parameters, and executing the functionality of that layer. These layers actually interface with the layers above or below them (i.e., they hand them the data and parameters). The seven layers are shown in Figure 2.5.

The physical layer handles bits and the physical transmission of bits across the communication channel through some sort of medium (whether it be a kind of wire, fiber, radio-waves, or light). The second layer (referred to as L2) is the data-link layer, which takes care of framing bytes or a block of bytes, and handles the integrity and error recovery of the data transmitted at this level between two nodes connected by a physical channel. The third layer (referred to as L3) is the network layer, which is responsible for carrying blocks of data between two or more nodes across a network that is composed of multiple nodes and data-links; L3 responsibilities include the required addressing, routing, and so on. The transport layer is the lowest application (or host) layer that carries data between applications (or hosts), independently and regardless of the networks used. It is responsible for the end-to-end data integrity and reliability, and it works through either *connection* or

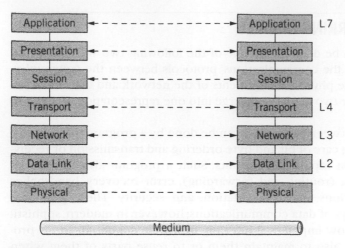

FIGURE 2.5

ISO/OSI seven layers model

connectionless transport mechanisms.[3] The session layer controls the session (e.g., determining whether the relationship between the nodes is peered or master/slave; establishing, maintaining, and terminating a session). The presentation layer determines such things as the format, encryption, compression, structure, and encoding of the application data. The application layer determines the way the application uses the communication facilities, that is, e-mailing, file transfer, and so on.

The upper four layers (the transport, session, presentation, and application) are considered the host layers, while the lower three (the physical, data-link, and network) are the network layers. The network layers are considered the most important in network processing; nevertheless, many networking decisions are made based on the upper four layers, such as priority, routing, addressing, and so on.

The equivalent data-networking model of the DoD (often called the Internet model, or more commonly, the TCP/IP model), is simpler, and contains fewer layers. (It originally had only four layers, without the physical layer; see Figure 2.6.)

The ISO/OSI model layers are not mapped exactly onto the TCP/IP model layers; however, roughly speaking, the TCP/IP model shrinks all host layers into the host-to-host (transport) layer (L4), and adds a new, internetworking layer that is composed mainly of the ISO/OSI network layer (L3). TCP/IP's network layer (L2) is composed mainly of the functionalities of ISO/OSI's data link layer and some of its network layer. Recently, this model has been amended by a "half" (or a "shim")

---

[3]Connection and connectionless communication interfaces are fundamental in networking, and the concepts are briefly described in the introduction. In connection-oriented communication, one node asks the other to establish a link ("call setup"), and once allowed, uses this link until it "hangs up" (like a telephone conversation). In connectionless communications, the originating node simply "throws" data to the network (like letters or e-mails).

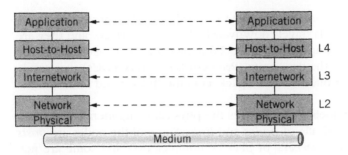

FIGURE 2.6

TCP/IP mode

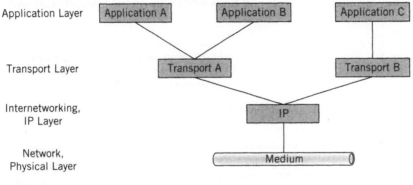

FIGURE 2.7

Using the TCP/IP mode

layer as new technologies have been introduced, so this model can better fit systems more precisely. These new technologies, which appear mainly in the "2.5" layer, are described later, and are actually extensions to the network layer in the TCP/IP model.

Application entities, as well as the other entities in the various layers, can interface with other entities up or down the communication model to allow for specific use of protocols and interfaces. Figure 2.7 demonstrates how three application entities use two transport entities (which manifest themselves as protocols when interacting with peer entities in the other host).

The way the applications and other entities are multiplexed in one host and demultiplexed in the target hosts is by using headers in each layer, and data encapsulation. In ISO/OSI and TCP/IP models, each layer's entity interfaces with the entities in the layer underneath it by adding a layer-header describing the parameters required for the entity in the layer underneath to function properly. This header also lets the peered entity in the same layer in the other host to have the required information about how to process the data (according to the protocol). These headers are added as the data travels downwards through the layers, and they are removed as the data travels upwards. Figure 2.8 depicts data encapsulation in the TCP/IP model, and names the data units (Protocol Data Units, PDUs) that result; that is, datagrams in the application layer, packets in the IP layer, and frames in the physical layer.

The other network modeling emphasizes the physical components of the network, and is composed basically of *nodes* and *links*. Some of the nodes are *hosts* or *end systems* (*clients* and *servers*, or *peer* communicators), and some are network *edge devices* or network *core devices*, as shown in Figure 2.9. Edge and core devices are *gateways, routers, switches, bridges, hubs,* and *repeaters*.

Network devices work up to the third layer (with the exception of gateways that work on all layers, and connect very different types of networks and applications). Generally speaking, repeaters and hubs work only on the first, physical layer,

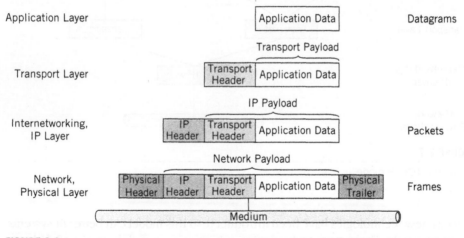

FIGURE 2.8

Data (payload) encapsulation

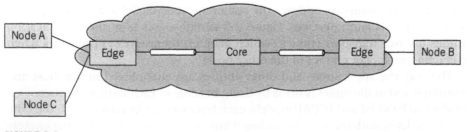

FIGURE 2.9

Network model

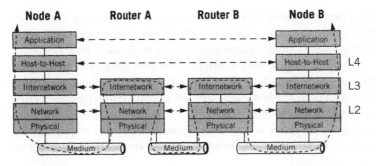

FIGURE 2.10

Router layers

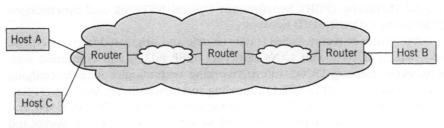

FIGURE 2.11

Internet model

bridges, and switches work on the second layer, and routers work on the third layer. Thus, for example, routers interconnect networks and links by analyzing and forwarding packets based on the headers of the third layer, as shown in Figure 2.10, where node A and router A are connected by one link (the left medium), and node B and router B are connected by another link (the right medium). The routers are connected by a third link (the middle medium).

Links can be replaced, or generalized, by networks, such that, for example, the routers are connected by networks in between them, as shown in the Internet model in Figure 2.11.

2.4 BASIC NETWORK TECHNOLOGIES

Network technologies can be grouped according to different dimensions; we already saw the layered approach, in which various levels are defined for some function (e.g., data link, internetworking, or application). There are network technologies that are specific to each of these layers, such as forwarding, routing, encryption, and data-link control. In this subsection, we use other two major dimensions that are imperative to network processing.

Data plane and *control plane* describe two basic processing levels in networks as described in the previous chapter.[4] Data plane processing deals with *actual data forwarding*, while control plane processing deals with *how the data is forwarded*; that is, mainly builds and maintains various data structures and repositories, or databases of forwarding rules. In network processing, we relate these two terms to the data path (or fast path) and the control path (or slow path), as described in more detail in Chapter 5.

The second grouping of network technologies results from the requirements of the networks for which these technologies are used, the applications they serve and the traffic patterns they meet.

One group of technologies that originated from the telecom networks is that of voice-centric transport networks that emerged from telephony services. These are connection-oriented, streaming, synchronous networks, and include Plesiochronous Digital Hierarchy (PDH), Synchronous Optical NETwork and Synchronous Digital Hierarchy (SONET/SDH) networks.

The other group of technologies is data-centric and originated from enterprise networks, which started with LANs (Ethernet, IEEE 802.3) and continued with internetworking (among LANs). Internetworking technologies include bridging and switching in layer 2 (IEEE 802.1), tunneling and forwarding in layer 2.5 (MPLS), and routing in layer 3 (IP), all of which are the infrastructure of WANs, MANs, and core networks. These are usually connectionless-oriented, packet-based, bursty, and asynchronous networks.

A third group of technologies are hybrid technologies, that is, technologies that aim to bridge these two groups of technologies (traditional telecom and data-centric). There are two approaches to provide such multiservice networks that reflect current traffic patterns on the core networks (which consists mostly of data): first, taking data networks to the telecom, and second, adapting telecom networks to data traffic. The remainder of this chapter describes the telecom networks and the data-centric networks, and the next chapter deals with the converged networks.

2.5 TELECOM NETWORKS

Telecommunication networks are used mainly for traditional telecommunications services such as telephone and TV services, as well as some data (point-to-point connectivity) and video services. We describe here the main telecommunication networks that are used for telephony and data services. Other telecommunications networks (e.g., cellular networks, or broadcast networks such as CATV) are briefly discussed in Chapter 4, since they usually do not use network processors.

---

[4]A third plane, the *management plane*, deals with configuration and monitoring of network elements. Network processors function mainly in the data plane, less in the control plane, and almost not at all in processing of the management plane.

The main service of traditional, pure telecommunications networks is telephony. Thus, telecommunications networks technologies mainly stream data of digitized voice; they are usually circuit switched, the duration of sessions is measured in seconds and minutes, and they have clear call set-up and termination procedures. Telecommunications links and trunks convey multiple simultaneous voice channels through Time Division Multiplexing (TDM); hence, in many network infrastructures (and not only in telecom), TDM services has come to refer to those services that are supported by these telecommunication networks (such as voice channels, video, and other real-time, streaming services).

Optical networks, which emerged mainly from and for the telecom industry, are based on Wave Division Multiplexing (WDM), which allows large trunks to be carried for distances and to be switched and routed much like the TDM networks.

It is important to note here that the transport networks in the telecommunications industry were voice-centric (traditionally designed for voice traffic); however, during the last decade or two, traffic on these transport networks has become data-centric, as Internet and data applications has become the main communications means for entertainment, business, news, peer-to-peer, and other applications (and this is before TV services shifted to IPTV). In fact, for some time now, the volume of data traffic has surpassed that of voice, and the ratio grows rapidly in favor of data traffic. The burstiness and noncontinuous, unpredictable nature of data traffic requires short-lived high bandwidth connections, and dynamic, flexible, and on-demand transport services. Therefore, the telecom industry, vendors, and operators, are amending and modifying their telecommunications transport networks, which are described in the next chapter.

2.5.1 PDH

Plesiochronous Digital Hierarchy [194] (also called T-Carrier in North America) was the main digital transmission mechanism used in telecommunications trunks, and was the foundation for later technologies. In Greek, "plesio" means *almost* and "chronos" refers to *time*, and the etymology of the name describes the technology fairly well; that is, "plesiochronous" refers to almost-synchronous data transmission in which the digital data streams are clocked at about the same speed. "Digital hierarchy" means that PDH networks multiplex voice channels into a higher hierarchy of trunks' capacities, and demultiplex them into decreasing capacities. In other words, several digital channels are combined (multiplexed) to create a higher hierarchy digital channel in the PDH network.

Since the roots of PDH are in the telephony system, the basic channel of the digital hierarchy is a voice telephone channel. Each voice channel is 64 Kbps and is an uncompressed digitized voice channel,[5] called DS0 for Digital Signal level 0 (sometimes also called Digital Service level zero).

---

[5]The basic digitized voice is represented by Pulse Code Modulation (PCM), 8000 8-bit samples per second, which are enough to carry the 4 KHz analog line—the bandwidth of the telephone voice channel.

In North America, the basic T-Carrier link is called DS1 for Digital Signal Level One, and is 1.544 MHz, multiplexing 24 DS0 channels. The multiplexing of the voice channels into DS1 is done according to time division; that is, 1 byte from each channel is picked up in turn, cyclically. The DS1 frame[6] is composed of 24 bytes (from all channels) and one control bit (total of 193 bits). There are 8000 such frames per second, or 125 microseconds for frame processing. The resulting control channel is 8 Kbps wide (from the one extra bit per frame). The DS1 signal is carried in the T1 channel (J1 in Japan), which defines the framing, the physical layer, and the line interfaces, and is also used to define any 1.544 Mbps service.

T1 was a very popular service, and it was used quite extensively to connect private (enterprise) telephone exchanges to the telephony system. In addition, and although PDH was designed primarily for telephone services, PDH links were also used very often for data networking applications, where connections of two networks or data equipment was done through a point-to-point PDH link (e.g., T1 service).

Higher hierarchy links are DS2 (which multiplexes four DS1 channels, or 96 DS0 channels, into a 6.321 Mbps channel), and DS3 (which multiplexes 28 DS1 trunks into 44.736 Mbps). The same applies to T2 and T3, which multiplex four and 28 T1 links, respectively. In Japan, J2 is similar to T2, but J3 multiplexes just five J2 lines (and therefore is 32 Mbps), and J4 multiplexes three J3 lines (and is 98 Mbps wide).

The basic PDH link that is used in Europe, called E1, is 2 Mbps wide, and it multiplexes 32 channels—two for control purposes, and 30 voice channels. E1's frame is 256-bit long (1 byte from each of the 32 channels), and also lasts 125 microseconds. Higher hierarchies of SDH are E2 (which multiplexes four E1 lines into 8 Mbps), E3 (which multiplexes 16 E1 lines, or four E2 lines, into 34 Mbps), and E4 (which multiplexes 64 E1 lines, or 16 E2 lines, or four E3 lines, into a 140 Mbps trunk).

Plesiochronous Digital Hierarchy links were used for trunks that connected telecommunication equipment, as well as to provide point-to-point digital links for enterprises and other wide access network customers; for example, intra-network communication equipment in central offices. A summary of PDH links is provided in Table 2.1 as follows. These lines used coax cables, twisted pairs cables, or fiber-optic links.

Plesiochronous Digital Hierarchy lines were used from the 1960s and were gradually replaced by a fully synchronous network, SDH, based on optical networking, as described in the next subsection. The major reason for their replacement is that PDH was basically a point-to-point connection, and in order to connect various PDH hierarchies, or any other required manipulation of channels, the multiplexed PDH line had to be decoded, demultiplexed at every end-point, and multiplexed again as required. This procedure was necessary because PDH was not fully synchronized and some buffering was required, and also because the multiplexing involves inter-channel dependencies. Moreover, PDH links were not designed for the high-rate optical networks that became available.

---

[6]Please note that "frame" in the context of PDH is meant to describe the repetitive structure of TDM signals and is not a frame of the packet-oriented networks, specifically not a Layer 2 frame.

Table 2.1 PDH Lines

Link	# of DS0	Frame size	Bandwidth	Frames per second
DS0	1		64 Kbps	
DS1, T1, J1	24	193 bits	1.544 Mbps	8000
E1	32	256 bits	2.048 Mbps	8000
DS2, T2, J2	96	1176 bits	6.231 Mbps	5300
E2	128	848 bits	8.448 Mbps	9962
J3	480		32.064 Mbps	
E3	512	1536 bits	34.368 Mbps	22,375
DS3, T3	672	4760 bits	44.736 Mbps	9400
J4	1440		97.728 Mbps	
E4	2048	2928 bits	139.264 Mbps	47,563
DS4, T4	4032		274.176 Mbps	

2.5.2 Synchronized Optical Networks/Synchronous Digital Hierarchy

The next phase of telecommunications networks was to create a fully synchronized, end-to-end, data transmission service, with trunking hierarchy that enabled flexibility and scalability on optical networks. The Synchronous Digital Hierarchy telecommunication networks is the international standard [197], while in North America the SDH variant is called SONET, for Synchronous Optical NETwork.

Synchronous optical networks and synchronous digital hierarchy networks were also voice-centric, working through streams of digitized 64 Kbps telephone voice channels. However, SONET/SDH provides the means to support other data communications requirements, such as packet over SONET/SDH (POS) and Asynchronous Transfer Mode (ATM) services, as described in the next chapter. Next Generation SONET/SDH (NG-SDH/SONET) is a collection of several attempts that aim towards the inclusion of data networking in the SONET/SDH framework, mainly bursty, noncontinuous data streams (such as Ethernet), which, by nature, is opposed to the synchronous, streamed data bits of TDM applications (mainly voice). NG-SONET/SDH is described in the next chapter.

SDH networks can be built in ring, point-to-point, linear add/drop, or mesh configurations, but they are usually built in rings. The rings are actually double rings; that is, two optical fibers carry data in opposite directions, backing each other, thus resulting in resilience (see Figure 2.12). The active fibers (or working fibers) use the backup fibers (or protection fibers) for 50 ms protection; that is, in case of a working fiber fault, data flow will resume in <50 ms.

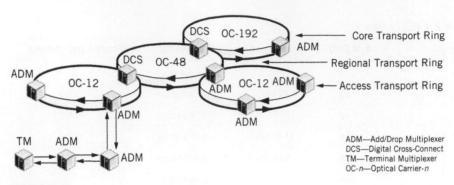

ADM—Add/Drop Multiplexer
DCS—Digital Cross-Connect
TM—Terminal Multiplexer
OC-*n*—Optical Carrier-*n*

FIGURE 2.12

SDH network

Table 2.2 SDH Lines

SONET signal	Optical carrier	SDH signal	Bit rate (Mbps)	DSO capacity
STS-1		STM-0	51.84	783[7]
STS-3	OC-3	STM-1	155.52	2349
STS-9	OC-9	STM-3	466.56	7047
STS-12	OC-12	STM-4	622.08	9396
STS-48	OC-48	STM-16	2488.32	37,584
STS-192	OC-192	STM-64	9953.28	56,376
STS-768	OC-768	STM-256	39,813.12	601,344
STS-1536	OC-1536	STM-512	79,626.24	1,202,688
STS-3072	OC-3072	STM-1024	159,252.48	2,405,376

SDH and SONET are technically almost identical, but there are several differences, mainly in terminology and the basic transmission rates. Hierarchy in signal bandwidth of SONET starts with a basic Synchronous Transport Signal 1 (STS-1), which is 51.84 Mbps, carried in an Optical Carrier 1 (OC-1) link. SDH starts its hierarchy with a basic Synchronous Transport Module 1 (STM-1), which is 155.52 Mbps, designed for one E4 link, and is equivalent to OC-3.[8] A summary of SONET/SDH links is provided in Table 2.2.

---

[7]Due to practical multiplexing considerations and techniques, STS-1 can carry a DS-3 signal, or 28 DS-1 signals, which are less than this theoretical capacity.

[8]STM-*n* interfaces are defined as logical and electrical signals and are also defined for optical transmission, whereas STS-*n* is the electrical signal of SONET, and OC-*n* is the corresponding optical signal for SONET.

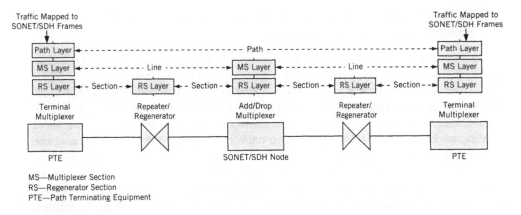

FIGURE 2.13

SONET/SDH network elements

The multiplexing technique in SONET/SDH is much more sophisticated than it is in PDH (in which the demultiplexing of the entire channel is forced down to its lowest rate channels in order to isolate one channel). Here, all frames[9] are transmitted 8000 times per second, or, all frames last $125\,\mu s$. Higher hierarchy SDH signals are obtained by interleaving the bytes of the lower hierarchy aligned frames, so frames are essentially getting bigger and contain all lower hierarchies' frames. The frame structure is based on pointers that specify the location of the multiplexed payload data in the frame.

Before going into the structure of SONET/SDH, there are some definitions of the SONET/SDH physical network entities that should be familiar in order to understand the frame's overhead. There are two types of network elements in SONET/SDH: Digital Cross Connect (DCS), which is used to add, drop, multiplex, demultiplex, and switch streams from various inputs to required outputs, and Path Terminating Equipment (PTE). PTE includes Add/Drop Multiplexers (ADM), which add and drop lower rate data streams onto or from the multiplexed stream, and Terminal Multiplexers (TM), which multiplex and demultiplex PDH and SDH links to or from the SDH network. This structure can be seen in Figure 2.12, where three rings of different hierarchies that are used for different purposes (access networks, regional networks and core networks) are interconnected by, and use, various equipment types. Communications between SONET/SDH physical network elements are defined in three layers, as can be seen in Figure 2.13:

- The path layer, which does the end-to-end connectivity and maps data streams onto the SONET/SDH payload.
- The line layer (also called the Multiplexer section, MS), which multiplexes the path layer frames on a single line, synchronizes the frames, and activates protection switching when required.

---

[9]As in PDH, "frame" in the context of SDH is meant to describe the repetitive structure of TDM signals and not a frame of the packet-oriented networks, specifically not a Layer 2 frame.

- The section layer (also called the Regenerator section, RS), which does the framing using the physical interface.

Each SONET frame contains overhead (called Transport Overhead, TOH), and payload (called Synchronous Payload Envelope, SPE). The basic SONET's STS-1 (OC-1) frame is composed of 810 bytes, organized in nine rows of 90 bytes each. The frame is transmitted row by row, starting from the top row, and each row is transmitted from left to right (i.e., the most significant byte first, and the most significant bit first). The first 3 bytes in each row are used for the TOH, and the rest of the 87 bytes are used for the SPE (see Figure 2.14). This gives a total of 50,112 Kbps payload rate (87 bytes × 9 rows × 8 bits × 8000 frames per second), which is exactly the right rate for 783 64 Kbps DS0 voice channels.[10] Higher hierarchies of SONET frames are built by byte interleaving of the lower hierarchy frames, so, for example, an STS-3 frame would use 9 bytes in each row for TOH, and 261 (87 × 3) bytes for the SPE in each row.

The TOH, which is 27 bytes per frame or a 217 Kbps link in STS-1, is further divided into Section overhead and Line Overhead (LOH) that are used by the corresponding network layers; that is, the section endpoint equipment and the line endpoint equipment. The Path Overhead (POH) is the first byte of every row in the payload (SPE), so in STS-1 there are 9 bytes for POH at each frame (or 72 Kbps link).

The basic SDH's STM-1[11] frame is very similar to the SONET frame structure (STS-3 frame, see Figure 2.14), and is composed of 9 rows, each of 270 bytes, for a total of 2340 bytes [195]. The first 9 bytes of each row are used for Section Overhead (SOH, which is similar to the TOH of SONET), and the rest of the 261 bytes are used for the payload. The SOH is divided, like the TOH, into a Regenerator SOH (RSOH), equivalent to SONET's SOH, a pointer to the payload, and a Multiplex SOH (MSOH) that is almost equivalent to SONET's LOH (the LOH contains this pointer).

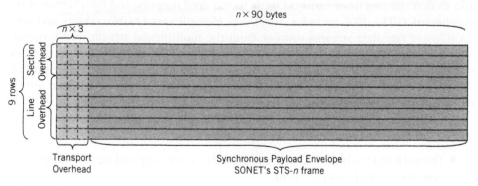

FIGURE 2.14

SONET/SDH frame

---

[10]See footnote 7.
[11]STM-0 was defined later to match SONET's basic frame STS-1.

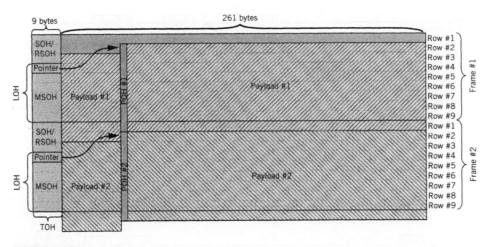

FIGURE 2.15

STM-1/STS-3 payload pointing

As noted before, the first row in the LOH is used as a pointer to the first byte of the payload in each of the frames. For example, in the STS-1 frame, the pointer is located in the three bytes starting at an offset of 270 bytes, which begins the fourth row of the frame. This pointer is defined in SDH's STM-1 frame at exactly the same place (i.e., the beginning of the fourth row, at an offset of 810 bytes). An example of payload pointing is shown in Figure 2.15.

This pointer mechanism allows SDH to accommodate payloads that originate with some differences in speed between the SDH frames and the payload. It can also be used for pointing to several payloads that are made up of smaller capacities than the SPE. The last issue in SONET/SDH that we deal with here, is how to map various sources of data streams onto the SONET/SDH frames (or actually, onto the payload).

In SONET, the basic sub-SPE synchronous signal is called Virtual Tributary (VT). There are several VT definitions for 3, 4, 6, or 12 columns of the SPE. For example, VT-1.5 occupies three columns; it therefore has 27 bytes per frame, and is 1.728 Mbps in bandwidth (8000 frames per second). VT-2 occupies four columns, or 36 bytes per frame, and thus has 2.304 Mbps and so forth. VT-1.5 can contain T1 (DS-1 signal), and VT-2 can contain an E1 signal. Similar VTs are grouped into one block of 12 columns of the SPE (i.e., 108 bytes); seven VT groups can fit into one STS-1 SPE,[12] where the seven VT groups can be of different VT types (e.g., three VT groups of VT-1.5, and four VT groups of VT-2). The multiplexing scheme of SONET is depicted in Figure 2.16 on next page.

---

[12]It occupies 84 of the 87 columns of the SPE; one of the remaining three is used for the POH (column 1), and the other two (columns 30 and 59) are fixed.

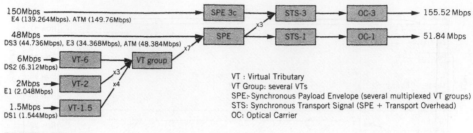

FIGURE 2.16

SONET multiplexing scheme

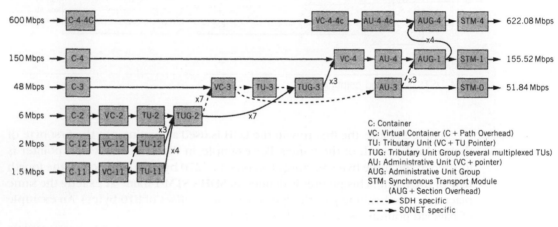

FIGURE 2.17

SDH multiplexing scheme

A similar scheme, though a bit more complicated, is used for SDH, as shown in Figure 2.17. The synchronous payload is packaged in a container.[13] As noted above, STM-1 contains 270 columns (bytes) of nine rows, of which 260 columns (bytes) are used for payload and are defined as Container-4, or C-4. Smaller containers, which are termed low-order containers, are used for lower bandwidth streams and are multiplexed to become a STM-1 frame eventually. Higher data rates are organized by associating several containers together through a procedure called concatenation, which results in a combined capacity that can be used as a single container across the network. In other words, concatenation, according to ITU-T G.707 [195], is a process of summing the bandwidth of a number of smaller containers into a larger bandwidth container. The concatenated containers are marked by the number of C-4 containers multiplexed together; for example, C-4-4c means four contiguously concatenated C-4 containers, resulting eventually in 622 Mbps STM-4, and so forth.

---

[13]Container is defined in ITU-T G.707 [195] as the information structure that forms the network's synchronous information payload.

Table 2.3 VC Types and Capacities

VC type	VC bandwidth (Mbps)	VC payload (Mbps)	SHD frame	SDH frame rate (Mbps)
VC-11	1.644	1.6		
VC-12	2.240	2.176		
VC-2	6.848	6.784		
VC-3	48.960	48.384	STM-0	51.84
VC-4	150.336	149.76	STM-1	155.52
VC-4-4c	601.344	599.04	STM-4	622.08
VC-4-16c	2405.376	2396.16	STM-16	2488.32
VC-4-64c	9621.504	9584.64	STM-64	9953.28
VC-4-256c	38,486.016	38,338.56	STM-256	39,813.12

Each Container, together with its POH, is called a Virtual Container (VC), as it can be placed anywhere in the payload, using two levels of pointers. SDH's Virtual Container (VC) is thus equivalent to SONET's SPE. The relation between the information rates and the multiplexed and overhead frame rates are shown in Table 2.3.[14]

Small VCs (sub-STM-1 frame in size, which SDH calls "low order VCs") are multiplexed to become a Tributary Unit Group, after each VC is aligned with a Tributary Unit (TU) pointer. The bigger VCs (high order VCs) become Administrative Units (AU) after the frame's pointer is attached to them. Several AUs can be multiplexed into an Administrative Unit Group (AUG) for higher STM-n signals. The last operation is to add the Section Overhead (SONET's equivalent of TOH) to the AUG, and to make it a STM frame. Figure 2.17 summarizes this multiplexing structure, where formations from containers to VCs are done by mapping, from VCs to TUs or AUs by aligning (both TUs and AUs are themselves pointer processing), and all the rest are done by multiplexing.

Several enhancements to SONET/SDH were standardized to handle data traffic. These include Virtual Concatenation (VCAT), Link Capacity Adjustment Scheme (LCAS), and Generic Framing Procedure (GFP). These techniques are described in the next chapter, where data convergence is explained.

---

[14] The calculations for STM-1, for example, are based on a frame of 270 columns (bytes) by nine rows, 8000 frames per second, and 8 bits per byte, which yields a 155.52 Mbps SDH frame rate; when 261 bytes are considered (which is the frame without the nine-column overhead), the VC bandwidth comes to 150.336 Mbps, and when just 260 bytes are considered (without the nine-column frame overhead and the one column path overhead), the net VC payload comes to 149.76 Mbps.

2.5.3 Optical Networks

The underlying technology of SDH is based on optical links (fibers). Optical networking evolved from plain optical links—from point-to-point connections that mainly supported the long-haul telecom networks—to become an important telecommunication network infrastructure in itself. In addition, multiservice provisioning,[15] or networks used not only for traditional telecommunication applications, but also for a variety of services such as telephony, TV, Internet, and data applications, pushed optical infrastructure to support Multiservice Provisioning Platforms (MSPP) that include direct data communications interfaces.

The main advantages of optical networks are their huge capacity, solid reliability, and low price. Although an "all-optical-network" is desirable, optical networks are usually hybrid in terms of electronics, and optical–electrical conversions are often required in optical networks. The principal technologies that enable optical networking are optical switching and WDM.

Optical switching can be done through several different means: electro-optical (e.g., liquid crystals or semiconductors), electro-mechanical (e.g., Micro Electro Mechanical Systems or micro-optoelectromechanical systems), or thermo-optical. Optical switching is used for Optical Cross-Connects and for other optical devices and network elements. Switching is done either for Optical Circuit Switching, for Packet Switching (OPS), or for Burst Switching, which is used for aggregated packets [339, 438]. OPS requires that packet headers be processed in a matter of nanoseconds, and network processors must be used for this.

Wave Division Multiplexing means simultaneous transmission of several optical carriers, each with a different wave-length (color), on a single fiber.[16] Each of the "colors" of the lights represents a different channel, and filters (or different light color pairs of sources and receivers) at the fiber-ends separate and distinguish between these channels. There are two basic approaches to WDM: Coarse WDM (CWDM), which handles several channels simultaneously, and Dense WDM (DWDM), which handles tens of channels. WDM enables a single optical fiber to carry as many as several Tbps.

Various optical devices and network elements are used in WDM networks for multiplexing, routing, and switching the light paths from sources to destinations. The simplest device is a passive multiplexer, which combines several fibers carrying different colors. A demultiplexer can be composed of a passive splitter, coupled with an optical band-pass filter. An optical add/drop multiplexer (OADM) adds and drops channels (colors) from the fibers, analogously to the ADM of the SDH networks. Attaching optical switches to an OADM results in a reconfigurable OADM (ROADM), which adds flexibility in setting up or breaking light-paths in the network,

---

[15]An often used buzzword that describes multiservice provision by operators is "triple play" (for telephony, TV, and Internet) or "quadruple play" (which also includes data communications; for example, Storage Are Networks, SAN).

[16]Wave-length is related to frequency, and the same WDM principle is actually used in wireless and other radio systems, where it is called Frequency Division Multiplexing (FDM).

and even makes it possible to route light paths dynamically, on the fly (again, usually controlled by a network processor).

In recent years, an Optical Transport Network (OTN) was standardized by the ITU-T [196, 200, 201] that unifies SONET, SDH, and other packet-based services, and is based on WDM networks. It offers data rates that are several orders of magnitude higher than current SDH/SONET rates, but in practice, OTN was not massively adopted because of SDH/SONET improvements, the high cost of replacing the installed base, and the potential leap to Ethernet-based transport networks.

Optical Transport Network provides separated and transparent digital multiservices that go beyond those of the analog WDM networks, including error management and quality of service provisioning at the physical, optical layer. It specifies a new optical channel frame,[17] the Optical channel Transport Unit (OTU), which is transmitted in one of three bit rate classes, called k. The frame is fixed in size for all rates, in contrast to SONET/SDH, and structured as four rows of 4080 bytes each. The first class, OTU1, is 2.67 Gbps, and the frame transmit time is 48 µs, as opposed to SONET/SDH in which all frame periods are always 125 µs. The second rate, OTU2, is 10.7 Gbps, and the frame duration is 12 µs. Currently, the highest rate is OTU3, 43 Gbps, for a frame time of 3 µs. The Optical channel Payload Unit (OPU) has 3808 bytes in each row, and has an additional two columns overhead. The Optical channel Data Unit (ODU) encapsulates the OPU, and has 3824 columns, of which 14 columns are overhead that include frame alignment and maintenance signals, path and tandem connection monitoring, general communication, and protection control Channels. A 256-column forward-error-correction code is added to the ODU to create the OTU.[18] This entire structure is shown in Figure 2.18. Low-order OPU payload and ODU frames can be multiplexed by interleaving to

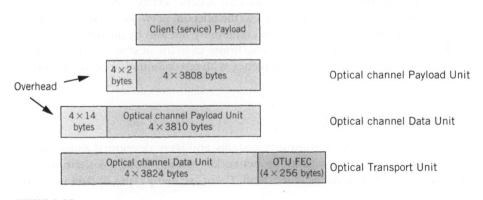

FIGURE 2.18

Optical transport unit frame

---

[17]As with the case of SDH and PDH, OTN's frame describes the repetitive structure of TDM signals, and is not a frame of the packet-oriented networks (specifically not a Layer 2 frame).
[18]ODU's data rate is lower than OUT's data rate: ODU1 is then 2.5 Gbps, ODU2 is 10 Gbps, and ODU3 is 40 Gbps.

create a higher order OPU payload and ODU frame; for example, four OPU1/ODU1 frames create one OPU2/ODU2 frame.

Optical channel Transport Unit (OTU) frames are electrical signals representing the frames, and they are placed on the optical fiber by converting them first to the optical channel layer, which is the first of the optical layer hierarchy. (The optical layer hierarchy is a physical layering system that we are not discussing here.)

SDH, SONET, and ATM payloads are mapped directly into the OPU. Other packet-type payloads such as IP and Ethernet are mapped by various methods into the OPU for transport services.

The control plane is an important layer in optical networks, and uses call and connection functions in real time, which enables efficient use of the underlying optical networks. It is becoming increasingly important, as switching and routing decisions are dynamic, flexible, fast, and interoperable in a data-centric environment. Toward this end, the standard of Automatically Switched Optical Networks (ASON) was established by the ITU-T [212, 213] to be used by SONET/SDH and OTN.[19] ASON includes not just call and connection functions, but also path control (i.e., routing, protection, and restoration) and discovery services for self-configuration. ASON's signaling protocols, used to transport messages between all communicating entities of the control plane, are based on a variety of signaling protocols, for interoperability of multi-domain networks. Although ASON is a protocol-neutral framework, it uses several signaling protocols,[20] and is likely to implement Generalized Multi-Protocol Label Switching (GMPLS)-based signaling protocols [234]. GMPLS by itself offers an alternative approach for the control plane implementation [297], which evolved from data networks.

All of the optical networks described so far are used primarily for long hauls, metro networks, and core networks. In the next chapter we will briefly describe Passive Optical Networks (PONs), which are used for access networks or aggregation networks that connect customers' premises to the telecom network. PONs are standardized not just by the telecommunication community, but also by the data community as well, that is, the "first mile" Ethernet. This brings us to the next section, in which we discuss data networks and their convergence with telecommunication networks.

2.6 DATA NETWORKS

Data networks are as old as computer systems, and were developed alongside the already existing telecommunications networks—some relying on their services and some completely separate from them. There are many types of data networks

---

[19]ASON is a framework that continues, extends, and eventually replaces its root standard, Automatic Switched Transport Network (ASTN) [199], since ASON is general enough to handle nonoptical transport networks as well.

[20]Such protocols include Private Network to Network Interface (PNNI) [209], Generalized Multi-Protocol Label Switching—Reservation Protocol Traffic Engineering (GMPLS RSVP-TE) [210] and GMPLS Constraint Routing Label Distribution Protocol (GMPLS CR-LDP) [211].

and many typologies for categorizing them, and they use many different technologies. It is not our intention here to cover data networks generally—there are numerous books on this subject—but instead to focus on the two dominant technologies: Ethernet and Internet.

2.6.1 Enterprise Ethernet

The Ethernet is basically an L2 protocol that dominates most LANs without any real competition from any other technology (at least when it comes to wired networks, as wireless networks have their own standards). The original version of the Ethernet was proposed in the 1970s by Xerox, and standardized in the 1980s by DEC, Intel, and Xerox to what is known as Ethernet DIX, or Ethernet II. Basically, the Ethernet offered a communication protocol and interface between several hosts ("stations," in the original notation) by using a shared medium. The Ethernet was based on the Carrier Sense Multiple Access/Collision Detection (CSMA/CD)[21] protocol. At the same time IEEE standardized all LAN technologies in the 802 standards committee,[22] the Ethernet included, and the Ethernet was slightly modified to become what was known as Ethernet 802.3, after the IEEE 802.3 standard [186]. Despite the passage of time and many proposed LAN alternatives, technologies, and usages, the Ethernet survived and evolved to become the major LAN technology due to its simplicity and price. (It is in use in over 85% of all LANs.) This technology now expands the scope of LANs and is proposed as an infrastructure to MANs (pure Ethernet metro), as described in the following.

The Ethernet now refers to a family of LAN standards that cover many media, speeds, and interfaces, all still sharing the CSMA/CD protocol, and that have similar (but not identical) frame formats. The Ethernet specification relates to the two lower layers of the ISO OSI model; that is, the data-link and the physical layers [178]. The data link layer in the Ethernet is composed of the Media Access Control (MAC) sublayer, and the Logical Link Control (LLC) sublayer. The MAC layer controls the node's access to the media, and describes the data structure and frame format, defined by IEEE802.3, as described in the following.

---

[21]*Carrier sense* means that before attempting any transmission, the host listens to the medium to find a gap between transmitted frames for its own transmission. It is *multiple access* in the sense that many hosts can do such listening, but there is still a chance of collision when two distant hosts sense a clear medium and start their transmission at about the same time. In this case, *collision detection* happens, when each host checks, while transmitting, that its transmitted frame is uninterrupted, that is, no "foreign" bits interfered with its own bits. If there is a collision, the host stops transmitting, waits some random amount of time, and retransmits its frame again.

[22]IEEE 802 standards committee started as an initiative of the IEEE Computer Society called "Local Network Standards Committee," project 802, in February 1980. The name "802" has no special meaning; it is just a number in the sequence of the IEEE numbers issued for standards projects. Later on, this standards committee was named "LAN/MAN Standards Committee (LMSC)" after expanding from issues of LANs to MANs.

The LLC interfaces between the Ethernet MAC and the upper protocol layers, and is defined in the IEEE802.2 standards [185]. The physical layer defines bit rate, signal encoding, and media interface, and is also described in IEEE802.3. There are some differences in the physical layer of the different Ethernet implementations, as can be seen in the architectural positioning of the Ethernet (Figure 2.19). Above the Ethernet, there are higher layers, described by other protocols and standards; however, IEEE802.1 defines a framework of using the Ethernet and offers some networking capabilities that go beyond using a shared media, as IEEE802.3 does. This is described in later subsections.

2.6.1.1 Ethernet Physical Layer

The reconciliation sublayer maps the signals between the medium interface and the MAC layer. As can be seen in Figure 2.19, the physical layer defines several alterative

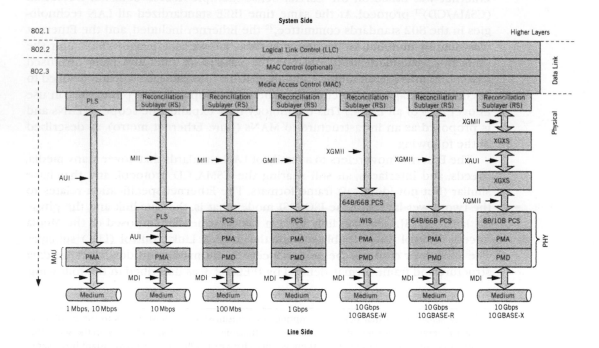

AUI—Attachment Unit Interface
MDI—Medium Dependent Interface
MAU—Medium Attachment Unit
PHY—Physical layer device
PMA—Physical Medium Attachment
WIS—WANInterface Sublayer
XGMII—10 Gigabit MII

GMII—Gigabit MII
MII—Medium Independent Interface
PCS—Physical Coding Sublayer
PLS—Physical Layer Signaling
PMD—Physical Medium Dependent
XAUI—10 Gigabit AUI
XGXS—XGMIIExtended Sublayer

FIGURE 2.19

Ethernet architectural positioning

and compatible interfaces (which are described in more detail in Chapter 7 in the subsection that describes interfaces):

- MII/GMII/XGMII
- MDI
- AUI/XAUI

The MII, GMII, and the XGMII are optional, and are designed to connect the MAC to remote physical devices, to different medium dependent physical devices, or to be used in chip-to-chip interfaces; thus, they can be used as compatible interfaces. The MDI is independent of the upper layers, defining the physical cable interface (e.g., media signals), and relating to the bit interface. AUI and XAUI are very different interfaces, but have functions similar to the MII/XGMII; they extend the connection between the MAC and the physical device. In the beginning of the Ethernet, AUI interface was very common, used by a cable that connected the host (with the MAC at its lowest communication stack) to a nearby MAU (through some required circuitry attached directly to the LAN cable). The XAUI is used mainly for chip-to-chip interface, and is optional. The physical interfaces, or cable schemes, are marked by the IEEE as this table shows:

1BASE5	1 Mb/s over two pairs of twisted-pair telephone wire
2BASE-TL	Up to 5.696 Mb/s point-to-point link over single copper wire pair
10BASE2	10 Mb/s over RG 58 coaxial cable (up to 200 m)
10BASE5	10 Mb/s over coaxial cable (i.e., thicknet, up to 500 m)
10BASE-F	10 Mb/s over fiber-optic cable
10BASE-T	10 Mb/s over two pairs of twisted-pair telephone wire (up to 100 m)
10BROAD36	10 Mb/s over single broadband cable
100BASE-X	100 Mb/s that uses FDDI encoding on two strands of fiber or two Category 5 UTP wires
100BASE-T	100 Mb/s (up to 100 m) that uses two or four copper pairs in many encoding schemes
100BASE-T2	100 Mb/s over two pairs of Category 3 or better balanced cabling
100BASE-T4	100 Mb/s over four pairs of Category 3, 4, and 5 UTP wire
100BASE-TX	100 Mb/s over two pairs of Category 5 UTP or STP wire
100BASE-FX	100 Mb/s over two multimode optical fibers (up to 2 km)
100BASE-BX10	100 Mb/s point-to-point link over one single-mode fiber
100BASE-LX10	100 Mb/s point-to-point link over two single-mode fibers
10PASS-TS	100 Mb/s point-to-point link over single copper wire pair
1000BASE-X	1 Gb/s using two strands of fiber or two STP copper wire pairs

(continued)

1000BASE-T	1 Gb/s using four pairs of Category 5 balanced UTP copper cabling (up to 75 m)
1000BASE-CX	1000BASE-X over specialty shielded balanced copper jumper cable assemblies
1000BASE-SX	1000BASE-X using short wavelength laser devices over multimode fiber (<500 m)
1000BASE-LX	1000BASE-X using long wavelength over multimode and single-mode fiber (<5 km)
1000BASE-LX10	1 Gb/s point-to-point link over two single-mode or multimode optical fibers
1000BASE-PX10	1 Gb/s point to multipoint link over one single-mode fiber, with a reach of up to 10 km
1000BASE-PX20	1 Gb/s point to multipoint link over one single-mode fiber, with a reach of up to 20 km
1000BASE-BX10	1 Gb/s point-to-point link over one single-mode optical fiber
10GBASE-X	Physical coding sublayer for 10 Gb/s operation over XAUI and four lane PMDs
10GBASE-T	10 Gb/s over UTP
10GBASE-CX4	10GBASE-X encoding over four lanes over shielded balanced copper cabling
10GBASE-R	Physical coding sublayer for serial 10 Gb/s operation
10GBASE-W	Physical coding sublayer for serial 10 Gb/s operation that is data-rate and format compatible with SONET STS-192c
10GBASE-S	PMD specifications for 10 Gb/s serial transmission using short wavelength
10GBASE-SR	10GBASE-R encoding and 10GBASE-S optics (up to 300 m)
10GBASE-SW	10GBASE-W encoding and 10GBASE-S optics
10GBASE-L	PMD specifications for 10 Gb/s serial transmission using long wavelength
10GBASE-LR	10GBASE-R encoding and 10GBASE-L optics (up to 10 km)
10GBASE-LW	10GBASE-W encoding and 10GBASE-L optics
10GBASE-LX4	10GBASE-X encoding over four WWDM lanes over multimode fiber
10GBASE-E	PMD specifications for 10 Gb/s serial transmission using extra long wavelength
10GBASE-ER	10GBASE-R encoding and 10GBASE-E optics
10GBASE-EW	10GBASE-W encoding and 10GBASE-E optics

There are some converters and other interfaces such as the Gigabit Interface Converter (GBIC) that allow configuration of each gigabit port for short-wave (SX), long-wave (LX), long-haul (LH), and copper physical (CX) interfaces.

2.6.1.2 Ethernet MAC Layer

The Ethernet's MAC layer provides services to its clients, that is, the sublayers above it—the LLC sublayer, bridge relay entity (as described later), or other users of the MAC layer. Basically, these services include sending and receiving frames to allow the LLC sublayer to exchange data with a peer LLC sublayer. MAC services may provide support for resetting the MAC sublayer to some

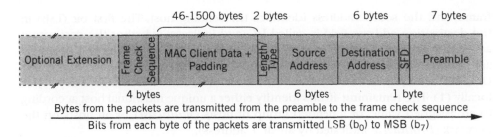

FIGURE 2.20

Ethernet MAC frame format (from right to left)

state, and an optional MAC control sublayer may add control services to the MAC sublayer, such as flow control between the MAC clients. From a frame processing perspective, the main issue of Ethernet MAC is Ethernet frame types and formats, and it is important to know them since manipulations of these frames might be required in network processing. The Ethernet, however, has many types and formats of frames, depending on the standards, and the evolution of the Ethernet.

The original Ethernet II frame contained addresses (sender and receiver), the type of the payload, and some header bits. Although the length is significant in the Ethernet protocol, the length was supposed to be defined in the payload; this was unacceptable to the IEEE802.3 standards committee due to the independency principle required in each of the layers.[23] So, the defined 802.3 frame is a bit different from that of the Ethernet II, and length field is used instead of the type field as described in Figure 2.20.[24] Nevertheless, as can be seen, the format of the Ethernet II frame is supported by including the option of having the type of the MAC clients (commonly known as the *Ethertype*) in the Length/Type field. As a matter of fact, most Ethernet frames still contain Ethertype rather than frame Length.

The preamble is used for synchronization, and is composed of alternating bits of 1 and 0, forming seven bytes of 0xAA. The preamble is terminated by the Start of Frame Delimiter (SFD) field, which "breaks" the alternating bits by one byte of 0xAB. Following this header, the destination and source address fields are defined, where the destination address identifies the host(s) that should receive the

---

[23]Each layer has to function independently, without using fields that contain information managed by other layers. Length is an outcome of a specific layer; in our case, of the MAC. The MAC layer should not expect another layer to put length somewhere in order to function properly, since it breaks the independency principle. And, if length is not there (that is, not provided by another layer, as Ethernet II MAC expects), the MAC layer will not work.

[24]Please note that the order of the bytes in Figure 2.20, moves from right to left, and not as is customary in similar figures. But, as Figure 2.20 indicates, the *flow* of the frame moves from left-to-right; that is, transmitting bytes into the imaginary medium to the right of the frame.

frame, and the source address identifies the sending host. The first bit (LSB) in the destination address field is called I/G, and indicates whether the destination address is an individual (0) or a group address (1). The source address is always an individual address; hence, its leftmost bit is always 0. The second bit is called U/L, and distinguishes between addresses that are administered universally (0) or locally (1). The remaining 46 bits identify either a unique individual host according to an assigned value, or a defined group of hosts (*multicast*), or all hosts on the network (*broadcast*, indicated by an all '1' address).

Addresses are known as MAC addresses, and are usually marked as six bytes in the notation XX:XX:XX:XX:XX or XX-XX-XX-XX-XX. In order to approach a specific host, its Ethernet MAC address must be known to the sending host. If there are several Ethernet interfaces to a host, it can be approached by any of the various interfaces, using their specific MAC addresses. This poses the problem of having to maintain a dictionary or an address book in order to communicate between peer hosts, if broadcast or multicast addresses are not to be used.

The next two bytes have two possible meanings. Since frame length is limited to 1536 bytes (which equals 0x0600, in hexadecimal notation), then a frame containing any value greater than or equal to 0x0600 indicates that this field contains the *type* (or nature) of the MAC client; that is, this field should be interpreted as the Ethertype, and the frame is an Ethernet II frame. This frame format, Ethernet II, that uses the Length/Type field for Ethertype, is the most frequent format used. If, on the other hand, the value is less than 1536 (or 0x0600), this field contains the *length* of the MAC client data, and the frame is called an 802.3 frame.

All 802.3 frames must contain a LLC header at the beginning of their payload. Ethertype is conveyed in 802.3 frames by a specific protocol and header, the Subnetwork Access Protocol (SNAP), which has an encapsulated LLC/SNAP header, as described in the following. Some common Ethertypes are listed in the table below.[25]

The payload—the MAC client data—comes after the Length/Type field. The data field must be of some minimum size for correct CSMA/CD protocol operation; these minimums are described in the following. If it does not meet the minimum, the data field is extended by adding extra bytes to pad it.

Minimal frame size is necessary, since the collision detection mechanism requires that the worst round trip time of a transmitted frame in the LAN segment will be at least the same as the frame transmission time. This ensures that a

0x0800	Internet Protocol (IPv4)	0x0806	Address Resolution Protocol
0x8035	Revered ARP (ARP)	0x809B	Ethertalk (AppleTalk)
0x8100	802.1Q tagged frame	0x86DD	Internet Protocol (IPv6)
0x8847	MPLS unicast	0x8848	MPLS multicast

---

[25]The entire list is at *http://standards.ieee.org/regauth/ethertype/eth.txt*

collision, if it happens, will be detected by the transmitting host. In the Ethernet, a round trip is limited to 50 μs (equivalent to 2500 m propagation of the bits in a copper transmission channel). This means, for example, that at a transmit rate of 10 Mbps, 500 bits are transmitted in this time.

This round-trip propagation is called the *slot time* and is expressed in bits time. At 8 bits per byte, the slot time of 10 Mbps is slightly <64 bytes, which determines the minimum frame size of 10 Mbps Ethernet. The same minimal frame size is used for the 100 Mbps Ethernet, which means that the network span (diameter) is reduced by a factor of 10; that is, to about 200 m. In 1 Gbps Ethernet, it became impractical to scale the network diameter down again by a factor of 10 just to maintain frame compatibility (i.e., a minimal frame size of 64 bytes), so the solution was to extend the frame by appending bytes *after* the FCS field, just to have the collision detection mechanism work. (The receiver automatically ignores this extension field.) In this case, the 200-meter network diameter remains, as well as the minimal 64 bytes frame, and the slot time becomes 4096 bits time. This means that the minimum frame, with the extension field, should be 512 bytes. IEEE802.3 standard defined a minimum frame size of 520 bytes for 1000BASE-T, and 416 bytes for 1000BASE-X (because it uses 8B/10B encoding, which requires 10 bits transmission or receiving for each byte).

The last field is the Frame Check Sequence, which contains the calculated Cyclic Redundancy Check (CRC) of the main frame-fields (addresses, type/length and data fields), for validating the frame by the receiving host.[26] This field is transmitted from the most significant bit (MSB) of the entire 4 bytes field to its least significant bit (LSB) (i.e., b_{31} to b_0). This is in contrast to all other fields of the frame, which are sent byte after byte, with their least significant bit first.

An additional inter-frame gap of 96 bits was defined by the IEEE802 for all Ethernet rates in order to separate frames and allow the correct CSMA/CD algorithm to work. In the case of 1 Gbps Ethernet, a host can initiate a transmission of a sequence of frames, known as *burst mode* transmission, without contending for the medium as required in a normal CSMA/CD operation. The maximum burst size is 8K bytes, composed of a sequence of regular frames (of which only the first requires the extended field), with an inter-frame gap of 96 bits in between them. This is not to be confused with a nonstandard industry practice of using *Jumbo Frames* of size 9216 bytes (which is large enough to carry 8K bytes and small enough to maintain the CRC effectiveness).

Subnetwork Access Protocol is used to identify the Ethertype when the IEEE802.3 frame is used, as described above (Type/Length of the frame is its length, and is <0x0600) [186]. A SNAP header is preceded by the LLC header, and both are encapsulated in the 802.3 frame (since LLC is described in IEEE802.2 [185], this type of frame is often called IEEE802.2/802.3 Encapsulation [358]). Basically, LLC provides mechanisms for connection and connectionless modes of data link services, with optional flow control and retransmission based on

---

[26]More details are provided in Chapter 5.

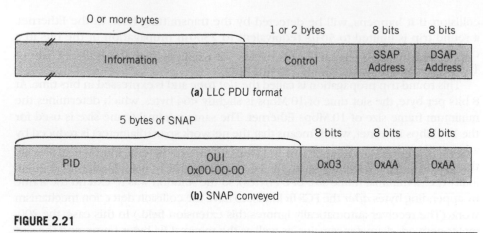

(a) LLC PDU format

(b) SNAP conveyed

FIGURE 2.21

IEEE802.2 LLC/SNAP encapsulation (from right to left)

sliding window protocol (e.g., High-level Data Link Control, HDLC [193]), but it is out of the scope and interest of this book. For our purposes, for providing Ethertype, LLC uses an "unnumbered information" (UI) frame, carrying the SNAP PDU, which contain the Ethertype (called Protocol IDentifier, PID), as shown in Figure 2.21.

Logical Link Control uses specific addresses, called Service Access Points (SAP, an OSI terminology) to indicate the services in the higher layers that use the LLC. The Destination Service Access Point (DSAP) is the address of the frame's intended destination, and the Source Service Access Point (SSAP) is the address of the source that initiated the frame; in the case of SNAP, both addresses are 0xAA. The control Field contains either commands or responses according to the LLC protocol and the state of the protocol, and for SNAP it contains 0x03, which means Unnumbered Information, UI. The information field contains the required information according to the LLC protocol and its state, and in the SNAP case, it contains three bytes of the Organizationally Unique Identifier (OUI) (which is always 0x00-00-00), followed eventually by the two bytes Protocol Identifier (PID). For instance, the Ethernet 802.2/802.3 encapsulated frame that conveys an IP datagram looks like the frame shown in Figure 2.22.

2.6.1.3 Ethernet Evolution

There has been continuous effort by the IEEE802.3 standards committee to keep updating the Ethernet protocol suite. As of 2008, the following standards are completed; some are integrated into the current IEEE802.3-2005 standard:

IEEE Std 802.3z-1998, Gigabit Ethernet (1000BASE-X).
IEEE Std 802.3ab-1999, 1000BASE-T (1GbE over 4 pairs of category 5 copper wire).
IEEE Std 802.3ac-1998, Frame extensions for Virtual LAN (VLAN) tagging.

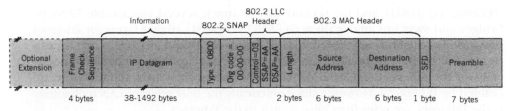

FIGURE 2.22

IP datagram in 802.2/802.3 frame (from right to left)

IEEE Std 802.3ad-2000, Aggregation of multiple link segments.
IEEE Std 802.3ae-2002, 10Gb/s Ethernet (over fiber; 10GBASE-S/L/E R/W).
IEEE Std 802.3af-2003, DTE Power via MDI.
IEEE Std 802.3ah-2004, Ethernet in the First Mile (subscriber access network).
IEEE Std 802.3ak-2004, 10GBASE-CX4 (over twinax).
IEEE Std 802.3an-2006, 10GBASE-T (over UTP).
IEEE Std 802.3ap-2007, Backplane Ethernet (for printed circuit boards).
IEEE Std 802.3aq-2006, 10GBASE-LRM (over multimode fiber).
IEEE Std 802.3as-2006, Frame Expansion.
IEEE Std 802.3-2005/Cor 1-2006 (IEEE 802.3au) DTE Power Isolation Corrigendum.
IEEE Std 802.3-2005/Cor 2-2007 (IEEE 802.3aw), 10GBASE-T Corrigendum.

As of 2008, there are still several study groups working on various issues, and drafts that are in process, as follows:

IEEE P802.3ar, Congestion Management Task Force.
IEEE P802.3at, DTE Power Enhancements Task Force.
IEEE P802.3av, 10Gb/s PHY for EPON Task Force.
IEEE 802.3ax, (IEEE P802.1AX) Link Aggregation Task Force.
IEEE 802.3az Energy-Efficient Ethernet Task Force.
IEEE P802.3ba 40Gb/s and 100Gb/s Ethernet Task Force (previously Higher Speed Study Group—HSSG).

2.6.1.4 Ethernet Networking

The Ethernet started as a communication means for many hosts, sharing one medium, as one LAN. It was possible to extend the diameter of this LAN by using repeaters and hubs functioning at the physical layer, as described above. It created a broadcast domain; that is, a domain in which every frame reaches all the hosts on the network. However, scalability became impractical when many hosts competed for a clear gap in which to transmit, and when all collisions had to be detected throughout the entire network.

Two main directions were taken to allow Ethernet networking and scalability at the network layer (L2, not the internetworking layer-L3). The first was to separate a physical LAN into many manageable virtual LANs (VLANs) by frame tagging

(IEEE802.1Q [184]), and the second was to attach several manageable LANs by bridges and switches among them (IEEE802.1D [183]). Higher layers for Ethernet networking scalability are described later, mainly the IP and metro Ethernet (or carrier class Ethernet). It should be noted that these directions were not taken specifically for the Ethernet (IEEE802.3), but can be used for any LAN, and therefore are defined by IEEE802.1. However, since practically the only relevant LAN technology is the Ethernet, we refer from here on just to the Ethernet.

2.6.1.4.1 Virtual Local Area Networks

Ethernet virtual LANs (VLANs) were created by simply amending the Ethernet II frame and creating a new Ethertype, 0x8100, that identifies the frames as tagged frames to be used for VLANs (Figure 2.23) [184]. Each created VLAN is identified by its 12 bits VLAN ID (VID). There are two reserved VIDs (0, which means a frame that is actually not a VLAN frame and tagging is used just for assigning priority, and 0xFFF, which is reserved for implementation use). Thus, there are 4094 (i.e., 2^{12}-2) possible VLANs in a network. This is quite enough to include all the LANs in even a big enterprise, but it is far from enough for public use, or carrier class Ethernet. Solutions are described in the following.

Hosts are assigned to different VLANs either according to the port of the VLAN equipment they are connected to (*port-based*), the host's MAC address (*MAC-based*), the applications (*protocol-based*), or otherwise. Having many VLANs allows the physical separation of LANs, since each VLAN is a broadcasting domain, and hosts across the VLAN are normally inaccessible to one another. Figure 2.24 is an

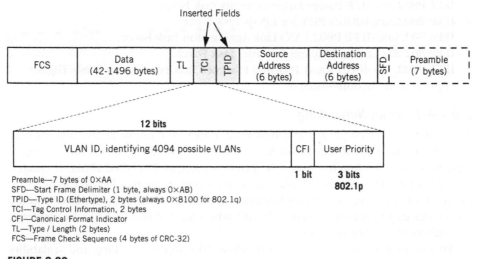

FIGURE 2.23

Ethernet II tagged VLAN frame (from right to left)

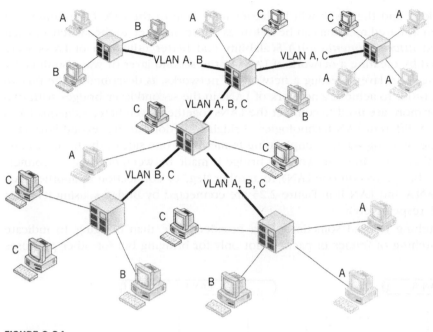

FIGURE 2.24

VLAN example

example of a VLAN network composed of three virtual LANS (A, B, and C) to which hosts are connected; each host belongs to one of the three LANs. These LANs can be functional, organizational, geographical, or simply match the assigned IP subnetworking. In some cases such as protocol-based VLAN, a host can be connected to several VLANs, as each application in the host can be assigned to a different VLAN according to the upper-layer protocol it uses. *Trunks* in VLANs are segments of the LAN that carry the tagged frames (the thick connections in Figure 2.24), and usually connect network nodes (VLAN equipment).

Access and *trunk* ports are two port modes that are defined for each port of the VLAN equipment. Hosts are connected through the access ports, and usually use untagged Ethernet frames. The trunks between the VLAN equipment are connected to the trunk ports of the equipment.

2.6.1.4.2 Bridging and Switching

Bridges and switches[27] were introduced for two purposes: (1) allowing scalability in terms of having as many hosts as required share the LAN, as was noted above, and (2) reducing the number of hosts on a shared medium so that collisions can

---

[27]Bridges are switching devices by function, so the term "switch" is widely used to describe what the IEE802.1 calls bridges (the term "switch" is not used in any of the IEE802.1 standards).

be avoided and the result will be better utilization of the LAN. (The number of hosts on a shared medium can be as low as one, connected to the switch, in what is called *micro-segmentation*.) Scalability and better utilization of LANs were achieved by creating a network of LANs in the second layer (it was also done by IP in the third layer, creating a network of networks, as described in the following). In order to achieve a network of LANs in the second layer, bridges with two ports or more are used to connect the LANs together (even heterogeneous LANs based on different LAN technologies). Bridging[28] is done at the second layer (L2) of the networking stack by duplicating frames from one side of the bridge to one of its other sides. In other words, a bridge is made between one "leg" (a connection the bridge has on one LAN), to another "leg," or connection to another LAN (e.g., LAN A and LAN B in Figure 2.25 are connected by Bridge A, using its ports 1 and 2, respectively).

Switching is used sometimes in a broader sense than bridging, to indicate the switching of frames or packets not only for bridging but for other purposes

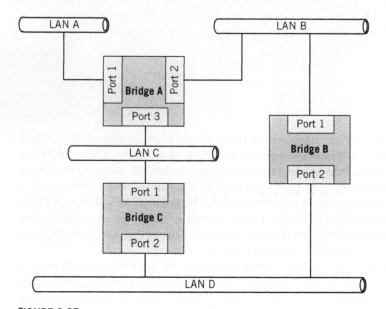

FIGURE 2.25

Bridged LAN

---

[28]There are basically two types of bridging—Source-Routing and Transparent Routing. This subsection describes only transparent bridging. Source-routing bridging is based on a specific route that is appended to each of the frames at and by the source, so that bridges know how and where to forward the frame. Transparent bridging relays the native frames, based on spanning tree algorithms and protocols that are described in this subsection. Source-Routing Transparent (SRT) bridging is a mixed operation of the two types [185].

as well; for example, optical switching described in the telecommunication equipment. Two basic switching mechanisms exist: store-and-forward, and cut-through. In store-and-forward switching, the entire frame or packet is received, buffered, analyzed and then switched, or forwarded. In cut-through switching, on the other hand, the frame or the packet is switched as soon as it enters the switch, delayed just long enough to allow the switch to analyze the head of the incoming frame and to identify where to forward it.

Going back to the basic bridging mechanism described previously, one can note that there might be a problem. In a network with many bridges or switches, some topologies might create multiple paths between a source node and its destination. Although multiple paths may be good for avoiding redundancy and for creating alternative paths between sources and destinations, they can also pose the problem of loops. Since frames are copied from one side of the bridge to its other side, frames may be duplicated and start looping in the network, multiplying themselves further on, until a *broadcast storm* may happen.

An example of such a *broadcast storm* can be seen in Figure 2.25, where frames received from LAN B by port 1 of Bridge B are copied into LAN D, then to LAN C by Bridge C, and again to LAB B by port 2 of Bridge A, and then they continue looping endlessly.

To prevent this, a spanning tree must be constructed over the network in order to eliminate and disable such "loops" (rings, or closed paths). A spanning tree should be configured and maintained throughout the entire life of the network, and all bridges should take part in it. The resulting tree, based on the network nodes and links, is called the *active network*. The Spanning Tree Protocol (STP) is used for establishing the active network, and has many variants. Since the spanning tree is an important algorithm and protocol for networks of LANs (actually, bridged networks), it deserves elaboration, and is described in the next subsection.

In principle, bridging contains three main processes: (a) establishing and maintaining the spanning tree, (b) learning MAC address locations, and (c) forwarding frames accordingly. A spanning tree is created with all bridges and LANs in the network by assigning (or configuring) a state to each port (*port state*) in every bridge in the network; the fundamental port states are *discarding*, *forwarding*, or *learning*,[29] where the forwarding state includes learning activity. Then each port of each of the bridges either forwards the frames from an incoming port to another port (by simply relaying the frames), or it discards the incoming frames. This bridging decision is based on filtering information that is kept in a filtering repository (called a *filtering database)* and is an outcome of the learning process and administrative configuration. The learning process is based on the source MAC address of

---

[29]Several other port states exist for the sake of the spanning tree mechanism to function, some of which were used by STP and are no longer used by Rapid STP (RSTP). They are also recognized by the IETF Bridge MIB description [95]. Such states include disabled, blocking, listening, and broken, which describe various reasons for the discarding state.

the incoming frames and the port they came through, so the bridge knows which hosts are reachable from this port; that is, where hosts with certain MAC addresses reside.[30] If the frame is not filtered, then the frame is forwarded to the port from which the destination host is reachable (if its MAC address is found in the filtering database; otherwise, it is broadcasted to all ports). The filtering information is either static (explicitly configured) or dynamic (collected by the learning process during normal operation of the bridge). In order to prevent data explosion in the filtering database, as well as to enable updated operation of the bridge, there is an aging process that erases old entries that exceed some age threshold. This filtering information enables the forwarding process to query about the bridge's outbound port for an incoming frame with a given destination MAC (or to discard the frame).

An extension of the filtering services enables registration of a group of MAC addresses by the Multiple Registration Protocol (MRP) [182] used by the MRP MAC Address Registration Protocol (MMRP) described in this chapter's Appendix A. MRP creates, updates, and removes groups of MAC addresses from the filtering database once allowed to do so by the bridge management services.

In a bridged LAN, broadcasting is still required and happens when frames with broadcasting addresses are used, or when frames with unknown (or aged) MAC addresses enter the bridged LAN. In a big network, when thousands of hosts are connected, big filtering databases are required for each of the bridges to allow unicast relaying of frames. In other words, a MAC destination address will be used in each of the bridges along the path to relay the frame to its destination, without the need to broadcast it. This becomes unmanageable, if not impractical, when a bridged LAN becomes large enough, or broadcasting of frames happens more frequently than required, which decreases LAN utilization.

Bridging can change, since there are many forces that can cause current bridging techniques to be insufficient. Frame forwarding has to consider priorities, class of service and drop eligibility,[31] as well as various types of higher IP layers' classes of service (e.g., Integrated Services, IntServ; Differentiated Services, DiffServ, as described in the following). The concept of "network span" is also modified, and VLANs separate the network to subnetworks that enclose traffic and allow customer separation not only in the enterprise, but also for service providers. Finally, learning from source addresses also changes from learning from every port and frame to independent learning in each VLAN, and eventually, for scalability purposes, learning only when forwarding decisions are affected (as described in the carrier class Ethernet).

---

[30]This assumes a symmetric traffic; that is, if a host with a MAC address X sends a frame, it is "registered" to some port in every bridge along the frame's path, and then, any other frame returning to this host will use the same path to access this host. In other words, frames from address A to B use the same path that frames from B to A use.

[31]These issues are detailed in Chapter 6.

2.6.1.4.3 Spanning Tree

The basic spanning tree algorithm assigns the following roles to bridges and bridges' ports:

- Identifying a unique *Root Bridge* (RB) for the <u>entire network</u>,
- Identifying a unique *Root Port* (RP) on <u>each bridge</u> (a specific port through which the bridge is connected to the RB), and
- Identifying a *Designated Bridge* (DB) and its *Designated Port* (DP) for <u>each LAN</u> (a specific port of a bridge that is connected to the LAN, through which the LAN prefers to be connected to the rest of the LANs).

Once these roles are established, a spanning tree is created with a unique path from each LAN to the RB (through the designated bridge and port and then through the root ports of every bridge along the path). This, then, obviously creates a unique path between any LAN to any other LAN, with no loops (closed paths). For example, the bridged LAN of Figure 2.25 would look like the network shown in Figure 2.26 after assigning the right roles to the bridges and to the ports, and deciding on the right states of the ports (forwarding and discarding).

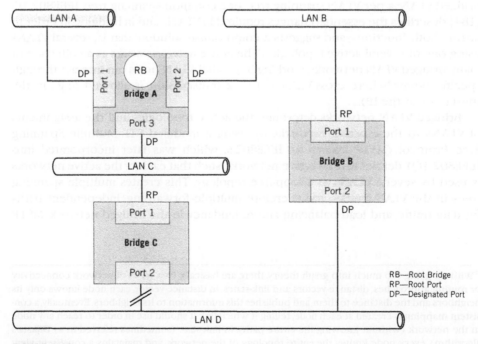

FIGURE 2.26

Spanning tree of the bridged network depicted in Figure 2.25

There are, however, many possibilities for establishing the above roles and the resulting spanning tree, and several protocols exist to carry the required information, synchronize all of the bridges, and respond to modifications, failures, and other events in the networks.

The first and most important protocol was the STP [183, 350]. This protocol is a distance-vector[32] type protocol and is pseudo-static; that is, it is based on a timer used to change states. It requires almost a minute to reconfigure. STP was replaced by the Rapid Spanning Tree Protocol (RSTP), which was specified previously in IEEE802.1w (which was later incorporated into IEE802.1D), and is principally quite similar to STP, although it has some substantial modifications. Since STP is only used in legacy systems, and RSTP is backwards compatible to STP, we focus on describing RSTP in Appendix B. The main differences are that RSTP uses timer-free reconfigurations in point-to-point bridge connections (which are very common in current LAN deployments), and communicates ports' roles and states, thus allowing reconfiguration in as little as a few milliseconds.

2.6.1.4.4 Bridging, Switching, and Virtual Local Area Networks

Bridging or switching and LAN separation into VLANs can coexist, and usually they do; in other words, bridging is done in most networks that are composed of many VLANs. In principle, there are two basic possibilities for spanning trees in the bridged VLAN: a per-VLAN spanning tree, or a common spanning tree. IEEE802.1Q [184] describes the essential changes required in VLANs and in bridging in order to enable both functions, and suggests a compromise solution; that is, several VLANs using one of several active topologies. The entire network works as a collection of many bridged VLAN networks, working in parallel. VLAN hosts can interact through specific points where cross-VLAN network transmission is allowed (e.g., in the third layer, or the IP).

Bridged VLAN networks determine the active topologies and the assignments of VLANs to these active networks by using a modified STP. Multiple Spanning Tree Protocol (MSTP, known as IEEE802.s, which was later incorporated into IEEE802.1Q), defines several active networks, such that each of the active networks is used by several VLANs as a loop-free topology. This creates multiple spanning trees in the VLAN trunks, and therefore multiple forwarding, independent paths for data traffic, and load balancing and redundancy in the bridged network. MSTP

---

[32]Without going too much into graph theory, there are basically two types of network connectivity or routing approaches: distance-vectors and link-states. In distance-vector, each node knows only its neighbors and the distance to them, and publishes this information to its neighbors. Eventually, a consistent mapping is created at each node, telling it which link it should use in order to reach any node in the network (without knowing the entire path). In link-state (sometimes referred to as Dijksta's algorithm), every node knows the entire topology of the network, and maintains a consistent view of the network topology by exchanging its distance information—that is, its link states—with all the other nodes.

actually extends the RSTP algorithm to handle multiple spanning trees, and uses all RSTP improvements.

In a Multiple Spanning Tree (MST) bridged VLAN network, the learning process of MAC addresses, and ports, described previously, is also influenced by the VLAN association of the port, protocol, or whatever defines the VLAN. The learning process can either be used for each VLAN independently (called Independent VLAN Learning, IVL) or shared between VLANs (called Shared VLAN Learning, SVL). Additionally, the filtering database includes Filtering Identifiers (FIDs). Each FID identifies a set of VLAN Identifiers (VIDs) that share the same learning process of MACs and ports and each is assigned to one of the spanning trees, called Multiple Spanning Tree Instance (MSTI).

An *MST region* is identified by all of the bridges that participate in the MST definitions (by sharing the MST and being configured within it), and each of the bridges in an MST region is identified by a MST Configuration Identifier (MCID) specific to this MST region. All MST regions and "Single-Spanning-Tree" (SST) networks that are connected together compose a single Common Spanning Tree (CST). In a CST, an MST region appears as a single, virtual bridge to its surrounding MST regions and SST networks. In order to cover all bridges, including those in the MST regions, and to enable forwarding paths throughout MST regions and SST bridges, a Common and Internal Spanning Tree (CIST) is established, which is composed of the spanning tree paths in the CST created by SST bridges (calculated by STP or RSTP) and internal paths in the MST regions.

The internal connectivity throughout the MST bridges of an MST region is called the Internal Spanning Tree (IST), and is defined by the MSTP as MSTI number 0. This is a sub-tree that is a logical continuation of the "external MST region" spanning tree that connects all bridges in the MST region. The end result, therefore, is that CIST is a collection of the CST, SST bridges, and all ISTs of the MST regions, and correlates to every IST inside an MST region and the CST outside the MST regions.

This entire configuration is demonstrated in Figures 2.27 and 2.28. Figure 2.27 shows a four-region case, in which two regions are MSTs and two are SSTs. The dashed lines indicate links that are not used for the spanning tree. In each MST the IST is shown, and for MST Region A, two more examples of MSTIs are shown (MSTI 1 and 2). Figure 2.28 shows the CST (on the left) and the detailed CIST (on the right) of the same MST shown in Figure 2.27.

Detailed explanations of MSTP are beyond the scope of this book, and interested readers can refer to the standard [184] or the vast literature describing MSTP. Further development of spanning trees and other means to scale up the bridged, switched network happens continuously, and some of it is revisited later in this chapter when we discuss the carrier-class Ethernet.

2.6.1.4.5 LANs Networking Evolution

There has been continuous effort by the IEEE802.3 standards committee to keep updating the Ethernet networking protocol suite. As of 2008, the following

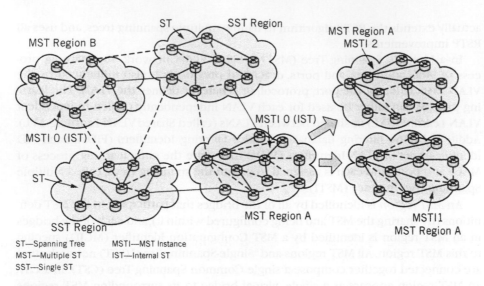

FIGURE 2.27

Multiple spaning tree example

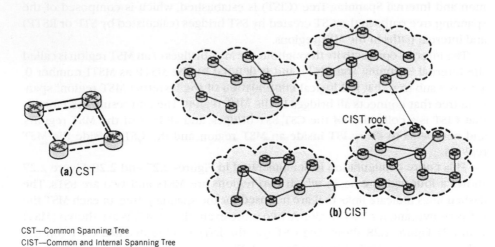

CST—Common Spanning Tree
CIST—Common and Internal Spanning Tree

FIGURE 2.28

The CST and CIST of the MST of Figure 2.27

standards were completed; some are integrated into the current IEEE802.1D-2004 and IEEE802.1Q-2005 standards:

802.1D (1998)—MAC bridges and Spanning Tree Protocol.
802.1D (2004)—MAC bridges and Rapid Spanning Tree Protocol.
802.1G—Remote MAC bridging.

802.1p—Traffic Class Expediting and Dynamic Multicast Filtering (Priority tags, published in 802.1D-1998).

802.1Q (2005)—Virtual LANs.

802.1s—Multiple Spanning Trees (published in 802.1Q-1998).

802.1v—VLAN Classification by Protocol and Port (published in 802.1Q-1998).

802.1w—Rapid Reconfiguration of Spanning Tree (fast spanning tree, published in 802.1D-2004).

802.1X—Port Based Network Access Control (authentication).

802.1AB—Station and Media Access Control Connectivity Discovery.

802.1ad—Provider Bridges (link aggregation control protocol).

802.1AE—MAC Security.

802.1af—MAC Key Security (Authenticated Key Agreement for MACSec, incorporated into 802.1X).

As of 2008, there are also still several study groups working on various issues and drafts that are in process, as follows:

- Interworking
  - 802.1AC—Media Access Control Service revision (VLAN tagging).
  - 802.1ag—Connectivity Fault Management.
  - 802.1ah—Provider Backbone Bridges.
  - 802.1aj—Two-port MAC Relay.
  - 802.1ak—Multiple Registration Protocol.
  - 802.1ap—VLAN Bridge MIBs.
  - 802.1aq—Shortest Path Bridging.
  - 802.1Qaw—Management of Data-Driven and Data-Dependent Connectivity Faults.
  - 802.1Qay—Provider Backbone Bridge Traffic Engineering.

- Security
  - 802.1AR—Secure Device Identity.

- Audio/Video Bridging
  - 802.1AS—Timing and Synchronization.
  - 802.1Qat—Stream Reservation Protocol.
  - 802.1Qav—Forwarding and Queuing Enhancements for Time-Sensitive Streams (correlates to 802.1Qaz).

- Data Center Bridging
  - 802.1Qau—Congestion Notification.
  - 802.1Qaz—Enhanced Transmission Selection for Bandwidth Sharing between Traffic Classes (correlates to 802.1Qav).

2.6.1.4.6 Layer 2 (Ethernet) Summary

In this subsection, we described the Ethernet, which provides layer 2 communication that enables two (or more) hosts to communicate in a local network by using the MAC addresses to reach each other. In order to do so, MAC addresses must be known to both of the participating hosts. This creates two problems: (a) it is a huge scalability issue, as it is simply impractical to maintain a database of all MACs of all network interfaces of all servers and all applications running on them as well as peer computers and their applications, and (b) it is a highly inflexible solution even in small-scale networking. For example, each time a server-application or a client-application changes its MAC address (e.g., replaces a hardware platform), a new MAC address must be associated with it and distributed to all peers that want to communicate with it.

To solve this issue, layer 3 offers IP addresses that can be configured easily to the hardware platform running the application, thus binding the required MAC addresses needed for layer 2 communications. It is important to note that L3—or, specifically, using IP—is efficient, even required, not only for internetworking, but also inside LANs (without L3 routing), as it deals with the isolation and independent use and addressing of upper layers applications. This is described in the next subsection.

2.6.2 Internet Protocol

Internet Protocol is the most common platform for data-communications applications (e.g., e-mail, file-transfer, WWW), and recently even for many telecommunications services (e.g., Voice over IP [VoIP] and TV over IP [IPTV]). The fourth generation of the cellular phone system (4G) is also supposed to be based on IP infrastructure. From being a data communication network-of-networks, as it was until 2000, IP has become a multiservice network that is used for web applications, triple-play platforms (3-play: Internet, telephony, and TV) or quadruple-play platforms (with additional enterprise-class data applications, such as storage access), Peer-to-Peer (P2P) services, and more.

Internet Protocol is also the dominant inter-networking standard, or layer 3 protocol, just as the Ethernet dominates layer 2 networking. IP emerged at the beginning of the 1980s as a result of internetworking research at the U.S. DoD. Although there is an "IP" standard [355], the IP network we usually refer to is based on thousands of standards, but only very few of them actually define most of the IP network operations. The Internet suite of standards, IP, or TCP/IP as it is sometimes called, is being standardized by the Internet Engineering Task Force (IETF), which is the protocol engineering and development arm of the Internet Society.[33] The IETF publishes Requests for Comments (RFCs) that arrive from Internet experts, companies, and research institutes, and some of them become standards.

---

[33]The Internet Society (ISOC) is a professional membership organization of Internet experts who comment on policies and practices and oversee other boards and task forces dealing with network policy issues. These include the Internet Assigned Numbers Authority (IANA), which is in charge of all "unique parameters" on the Internet including IP addresses, and the Internet Architecture Board (IAB), which is responsible for defining the overall architecture of the Internet, providing guidance and broad direction to the IETF.

The Internet is not one homogenous network, but rather an aggregation of many thousands of networks of all types—wireline and wireless, LANs and core networks—that are used for data communications, multimedia, and telecommunications applications and services.

2.6.2.1 *Internet Protocol Overview*

The idea of the Internet is to append a header in L3 to the frame of L2, which will allow routers in the IP network to route packets from network to network and from host to host. IP network uses a connectionless, packet-switched, hop-by-hop routing mechanism of datagrams. Layers above the IP are responsible for creating various connection types for many applications (e.g., HyperText Transfer Protocol [HTTP] for web surfing and more). The two connection types defined in the transport layer (L4) are Transmission Control Protocol (TCP) and User Datagram Protocol (UDP), as described in the following. TCP is a connection-oriented transport service that provides reliability and flow control. UDP is a connectionless (datagram) transport service.

The Internet Protocol uses 32-bit address fields in IP version 4 (IPv4) [355], or 128 bits in IP version 6 (IPv6) [98]. These addresses are easily configured, and each host can have one or more IP addresses. IPv4 addresses are known by their four numbered "dotted decimal" notation; for example, 132.71.121.67. The motivation for using a 128-bit address resulted from a lack of addresses for current needs; however, intermediate solutions for IPv4 address shortages caused IPv6 to be adopted slowly. IP addressing is discussed in more detail in Chapter 5.

The Internet Protocol binds IP addresses to L2 MAC addresses (for transmitting packets on the Ethernet L2) by using Address Resolution Protocol (ARP) [352], which runs directly on the Ethernet (EtherType is 0x806). The transmitting host broadcasts an ARP packet, asking for the L2 MAC address of the receiving host that has the destination IP address. Either the MAC address of a host on the local network is returned (and the L3 IP packet is directly transmitted to it), or the MAC address of a router is returned (and the L3 IP packet is transmitted there for further routing in the IP network).

2.6.2.2 *Internet Protocol Packet Routing*

When the L3 IP packet leaves the local network at the beginning of its journey in the Internet or in the IP network, it is forwarded by a router toward its destination host. It's crucial to understand the router function in a nutshell, as the router is the principal instrument that enables the Internet. Routers are used to connect islands of local networks, whether they be tiny or enormous, that broadcast and bridge L2 frames internally.

Each router maintains a routing table that contains as many entries as required to perform routings in the IP network. Each entry contains at least a destination network address and the router's egress interface that matches this destination (the egress interface is the actual router's port through which packets leave the router). In other words, a routing table is no more than a road sign that shows which road to take in order to get to a destination. There is also one default entry, which dictates which egress interface to take if there is no matched destination in the routing table.

Packet forwarding is quite simple: each routing node examines the IP header, and a routing decision is made according to the Destination IP address (DIP) and the routing table maintained in that node. Based on this decision, the packet is routed to the next hop router or to its host destination. However, the processing required for such a routing decision might be quite demanding, especially since it has to be done for each packet individually. In the simplest case of a "leaf" router—that is, a router that connects a network to just one other network—there is no problem in routing, and all packets are forwarded to the "default" next-hop. However, as routers become more central and have many destination addresses to handle and many egress interfaces, it can become a problem to choose (best match) among the interfaces for the appropriate destination address. IP addressing itself is not trivial, and the entire issue of IP addresses and the way they should be matched to the routing tables is discussed in more detail in Chapter 5.

Since the Internet is huge and growing, and includes hundreds of thousands of networks, a hierarchy approach is used for routing and networking. Autonomous Systems (ASs) are defined as aggregations of networks (and routers), and used to build an "overlaid" IP network of AS "nodes."[34] This overlaid network enables Internet scalability by dividing the Internet into partitions, with each handling the routing internally and independently (using routing protocols called Interior Gateway Protocols—IGPs), while using AS routing protocols (called Exterior Gateway Protocols—EGPs) between them.

Most IP operations (actual routing) are done in the data plane. In other words, IP forwarding using routing tables at routing nodes in the network is part of the data plane mechanism. As mentioned before, this might be a heavy computation burden on the equipment in the critical data plane, and several solutions exist to assist; for example, network processors in the routing equipment, and other forwarding mechanisms—usually switching-based mechanisms—that allow routing to be done faster and less painfully (such as Multi-Protocol Label Switching—MPLS, described later). Maintaining the forwarding databases (e.g., the IP forwarding tables) is a control plane process. Routing protocols for IP, both IGPs and EGPs, are used to build and maintain these databases, and are described in the following subsection.

2.6.2.3 Internet Protocol Routing Approaches and Protocols

There are many routing approaches and protocols that can answer the required packet routing described in the preceding subsection. These approaches and protocols can be classified according to various criteria. We describe here a few protocols and approaches that we find relevant to network processing. We begin with plain,

---

[34]Every Internet Service Provider (ISP) as well as major organizations maintain their own Autonomous System (AS) or several ASs. Each Autonomous System is identified by a unique number that, until 2007, was 16-bits long (i.e., approximately 64 K ASs were possible in the Internet). In 2007, the length was extended to 32-bits long (enabling 4 G unique ASs). This was done due to the growth of the Internet, the consumption of ASs (about 10 a day), and the expectation that the 16-bit AS range would be exhausted by 2010.

unicast routing protocols, continue to selective packet routing approaches (i.e., in which packets are not equally treated), and conclude with multicasting protocols. It is important not to confuse routing protocols (in the control plane) with packet routing (in the data plane), despite the fact that we describe them in this subsection together.

2.6.2.3.1 Shortest-Path Routing Protocols

Routing algorithms are based on shortest-path algorithms, either *vector distance* or *link state* algorithms. A vector distance algorithm is based on the distributed Bellman–Ford algorithm, according to which every node exchanges updates of its routing table with its adjacent neighbors. Based on the distance ("cost") of the link between the updating node and the receiving node, the receiving node can calculate the best routing table to all nodes, and transmit the updates further on to its adjacent neighbors. In link state algorithms, each node exchanges its distances ("costs") to its adjacent neighbors with all participating nodes (and not just its adjacent neighbors). This allows each node to calculate its routing table independently, according to shortest-path algorithms such as Dijkstra, for example.

The results of the routing protocols, the "best" route from each network to every other network, are kept in the routing tables in each of the routers connecting the networks. In the data plane, the routers forward each packet according to these routing tables; hence, packets coming from any network are routed to their destinations along the shortest path possible.

The common routing protocols for IGP are Routing Information Protocol (RIP)[35] [167, 293, 294] and Open Shortest Path First version 2 (OSPF2) [325], which are based on distance vector and link state algorithms, respectively. The common protocol for EGP is the Border Gateway Protocol 4 (BGP-4)[36] [365], which is a distance vector algorithm based protocol.

These routing protocols use various layers and transport means in the IP network: RIP uses port 520 of UDP, OSPF runs directly on the IP using protocol 89, and BGP uses port 179 of TCP (ports and protocols are described in subsections 2.6.2.5 and 2.6.2.4, respectively).

2.6.2.3.2 Selective Packet Routing

The IP routing protocols described before are insensitive to the packets' identity and type, and as a result, they forward all packets similarly. However, it is clear that not all packets should receive the same treatment by routers in the IP network, because different kinds of packets represent different applications (e.g., e-mail versus

---

[35]There are actually three versions of RIP, the earliest routing protocol used in the Internet and its preceding Arpanet. The first version, RIPv1, does not support classless addresses (described in Chapter 5), or subnets. RIPv2 includes the subnets issue and adds authentication to improve routing security, and RIPng extends the RIP to handle IPv6.

[36]BGP is an extremely important protocol for interconnecting all networks, and it makes the Internet a network of all networks. BGP principle is also widely used in protocols for other purposes than interconnecting the ASs. BGP is described in the Appendix of the next chapter.

real-time telephony). A comprehensive discussion of service types, requirements, and handling is given in Chapter 6. At this point, it is important to note that the need to differentiate between types of packets according to their "importance" or service requirements has resulted in several routing models and approaches, which can be categorized as: relative priority marking, service marking, label switching, Integrated Services, static per-hop classification, and differentiated services. These are described next. It is also important to note, however, that the actual routing protocols do not necessarily change in selective packet routing; it is the packet handling in each of the routers that may be modified. This is in contrast to non-selective treatment, where just comparing the destination address with the routing table is sufficient for routing.

Relative priority marking is used when an application simply selects the relative priority ("precedence") for the packet and marks the packet accordingly. Examples of this model include VLAN tagging in Ethernet with a user priority indicating relative precedence (IEEE802.1p, as can be seen in Figure 2.23), or the first three bits of the Type of Service field of the IP header in the IP protocol standard [15, 355].

Service requirement marking is another way to differentiate between packets. In this model, explicit service requirements are marked in each packet; for example, the four bits following the three precedence bits of the Type of Service field of the IP header in the IP protocol standard [15, 355]. These bits indicate "minimize delay," "maximize throughput," "maximize reliability," or "minimize cost" service requirements.

The label switching model enables different treatment for each stream of packets by using a forwarding tunnel or a path that represents a specific service requirement from the network. This model is implemented in the layer under the IP layer in the network that can carry IP traffic, such as Frame Relay, Asynchronous Transfer Mode (ATM) and Multi-Protocol Label Switching (MPLS).

An Integrated Services (IntServ) model [60] maintains the legacy routing mechanism, but adds a Resource Reservation Protocol (RSVP) [57, 432] that dictates to each router how to handle each packet according to its classification and the reserved resources. IntServ is complicated and does not scale well, since there should be an updated record in every router for each flow or stream of similar packets with regard to their profile and demands. The RSVP protocol that reserves resources (usually bandwidth), maintains a resource database and keeps it updated. RSVP requests resource reservation in one direction. Despite the fact that RSVP is not a routing protocol,[37] it is important, as it is used as a base for other protocols for call set-up and traffic engineering (e.g., RSVP-TE in MPLS, as described in the following).

---

[37]RSVP runs at the same layer as IP, although it is encapsulated in IP packets (the IP protocol field is 46 for RSVP).

The Static per-hop classification model is a variant of the Integrated Services model, where packet classification and forwarding are static, and are only updated by periodical administrative operations.

Differentiated Services (DiffServ) [54] is a refinement of the relative priority marking model. It is a simple model, according to which each packet is classified at the boundaries of the IP network and assigned to a "behavior" aggregate. Each behavior aggregate is identified by a code point that replaces the Type of Service field in the IP header. Routers follow a per-hop behavior (PHB) that is associated with the code points, and is used by the router to forward the packets according to their assigned behavior.

2.6.2.3.3 Multicast Routing

Multicasting is becoming more important as IPTV applications seem to become the next "killer applications" of IP networks. In the 32-bit long IP address space (detailed in Chapter 5), there is a range of group addresses, called class D IP addresses, that start with "1110" as the high-order four bits. In IP address standard, the "dotted decimal" notation, group addresses then range from 224.0.0.0 to 239.255.255.255. Several of the first addresses in this range are reserved; that is 224.0.0.0 to 224.0.0.255 is reserved for routing protocols, topology discovery, and so on where 224.0.0.1 is used for the permanent group of all IP hosts in the directly connected network, and 224.0.0.2 is used only for the multicasting routers.

In the lower layer, 2L, the Ethernet also has multicasting addresses that map to a group of Ethernet hosts, as mentioned before. One-to-one mapping can be done between an Ethernet multicast MAC address and an IP multicast address by using the same lower 23 address bits (and having the upper 3 bytes of the Ethernet MAC address be 01-5E-00).

Multicast routing in the data plane is similar to unicast routing; in other words, instead of forwarding a packet by a router to one destination (the best one, according to the routing table), packets may be forwarded to a number of destinations, depending on the multicasting group. In the control plane, however, multicasting is concerned first with defining the multicasting groups (assigning, joining, leaving, and releasing hosts to and from these groups), and then with routing protocols to create the appropriate routing tables for packets to be multicast later to the proper multicasting groups.

Multicasting groups are handled by the Internet Group Management Protocol (IGMP) [66, 97, 126], which runs at the IP layer, although it looks like an upper layer protocol, encapsulated in the IP packet (the IP protocol field is 2 for IGMP).

Multicasting protocols are mostly experimental, and are based on two approaches: Source-based trees and Group-shared trees. A source-based tree approach extends the unicast protocols of OSPF and the distance vector principle to support multicasting. Each router has a shortest path tree for each group. This approach is used by multicast extensions to OSPF (MOSPF) [324], Distance Vector Multicast

Routing Protocol (DVMRP) [424], and Protocol Independent Multicast[38]—Dense mode (PIM-DM) [7]. The group-shared tree approach assumes a core router that has a shortest path tree to all groups, and is the only one doing the multicasting. This approach is used in the Core Based Tree (CBT) protocol [42] and in Protocol Independent Multicast—Sparse mode (PIM-SM) [116].

2.6.2.3.4 Internet Control Message Protocol

A control mechanism for an IP network is implemented in the Internet Control Message Protocol (ICMP) [356]. IPv6 introduced a matched version of ICMP, called ICMPv6 [87]. ICMP is used mainly by routers to indicate network and routing errors. ICMP runs at the same level as IP in terms of networking layers, although in terms of packet structure (header and packet encapsulation) it looks like a higher layer protocol (the Protocol field of the IP header is 1 for ICMP).

2.6.2.4 Internet Protocol Headers

Internet Protocol packets have two versions (IPv4 and IPv6); the structure of an Internet IPv4 packet is shown in Figure 2.29. The fields of the IPv4 header are in the table that follows.

Version	4 bits	Indicate the header version (4)
Internet Header Length (IHL)	4 bits	Length of the header in 32 bits words, points to the IP payload
Type of Service	8 bits	Used for quality of service desired (described in Chapter 6)
Total Length	16 bits	Total length of the packet in bytes
Identification	16 bits	Sequence number of the packet, used to identify the packet or its fragments
Flags	3 bits	Fragmentation flags (described in Chapter 5)
Fragment Offset	13 bits	Indicates the position of this fragment in the original packet, if it is fragmented, in units of 8 bytes (described in Chapter 5)
Time to Live (TTL)	8 bits	The maximum hops the packet can stay in the network, decremented by one in each routing, and if zero, the packet is discarded. This helps avoid loops in bad routing situations.
Protocol	8 bits	Indicates the upper layer protocol,* for example, 6 for TCP or 17 for UDP (although some protocols may be at the same logical layer as IP, like ICMP, as described in subsection 2.6.2.3.4)
Header Checksum	16 bits	Checksum of the header (described in Chapter 5)

---

[38]PIM runs at the same layer as IP, although it is encapsulated in IP packets (the IP protocol field is 103 for PIM).

Source Address	32 bits	Source IP address (SIP)
Destination Address	32 bits	Destination IP address (DIP)
Options and Padding	Variable	Optional information and options in the packet, padded to the 32 bits boundary

*A complete list of all protocols is can be found at: *http://www.iana.org/assignments/protocol-numbers.*

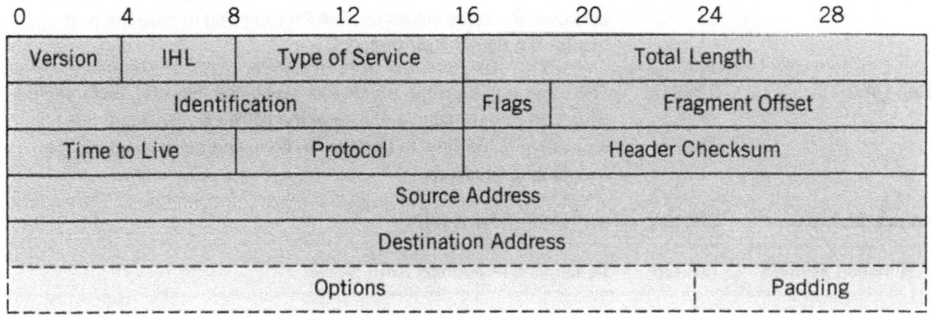

0	4	8	12	16	20	24	28
Version	IHL	Type of Service		Total Length			
Identification				Flags	Fragment Offset		
Time to Live		Protocol		Header Checksum			
Source Address							
Destination Address							
Options						Padding	

FIGURE 2.29

IPv4 header

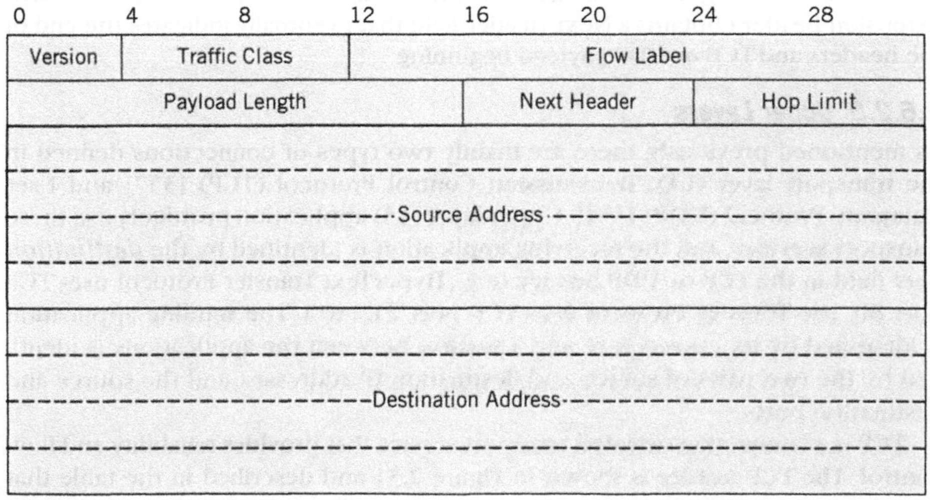

0	4	8	12	16	20	24	28
Version	Traffic Class			Flow Label			
Payload Length				Next Header		Hop Limit	
Source Address							
Destination Address							

FIGURE 2.30

IPv6 header

The structure of an Internet IPv6 packet is as shown in Figure 2.30. The fields of the IPv6 header are shown in the table that follows.

Version	4 bits	Indicate the header version (6)
Traffic Class	8 bits	Used for quality of service desired (described in Chapter 6)
Flow Label	20 bits	Flow label, used to indicate handling instructions of a stream of packets (e.g., real-time packets)
Payload Length	16 bits	Length of packet's payload in bytes (not including IPv6 header)
Next Header	8 bits	Identifies the type of header immediately following the IPv6 header, and uses the same values as IPv4 Protocol field in case the next header is a higher-layer protocol.
Hop Limit	8 bits	The maximum hops the packet can stay in the network, decremented by one in each routing, and if zero, the packet is discarded. It is equivalent to the Time to Live field in IPv4, and helps avoid loops in bad routing situations.
Source Address	128 bits	Source IP address (SIP)
Destination Address	128 bits	Destination IP address (DIP)

In IPv6, there are additional, optional headers as well—some of which replace fields of the IPv4 header and some of which enhance the IPv6 functionality. Such headers include fragmentation information, routing directives, and options. Each extension header contains a Next Header field that eventually indicates the end of the headers and TCP or UDP payload beginning.

2.6.2.5 Upper Layers

As mentioned previously, there are mainly two types of connections defined in the transport layer (L4): Transmission Control Protocol (TCP) [357] and User Datagram Protocol (UDP) [354]. Upper layer (L5) application protocols use these transport services, and the receiving application is identified by the *destination port* field in the TCP or UDP header. (e.g., HyperText Transfer Protocol uses TCP port 80; File Transfer Protocol uses TCP port 21, etc.). The sending application is identified by its *source port*, and a *session* between the applications is identified by the two pairs of source and destination IP addresses, and the source and destination ports.

TCP is a connection-oriented transport service that provides reliability and flow control. The TCP header is shown in Figure 2.31 and described in the table that follows it.

User Datagram Protocol is a connectionless (datagram) transport service, and has a simple header that does almost nothing. (Its main function is to indicate

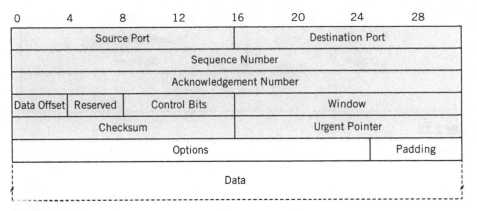

FIGURE 2.31

TCP header

Source Port	16 bits	The Source Port number (identifies the sending application)
Destination Port	16 bits	The Destination Port number (identifies the receiving application)
Sequence Number	32 bits	The sequence of the first data byte in this segment
Acknowledge Number	32 bits	The next sequence that the sender of the segment expects to receive; that is, acknowledging data received up to this value
Data Offset	4 bits	The number of 32 bit words in the header (indicates the beginning of the data)
Control Bits	8 bits	From left to right: CWR: Congestion Window Reduced ECE: ECN-Echo URG: Urgent Pointer field significant ACK: Acknowledgment field significant PSH: Push Function RST: Reset the connection SYN: Synchronize sequence numbers FIN: No more data from sender
Window	16 bits	Window size (number of bytes beginning with the one indicated in the acknowledgment field that the sender of this segment is willing to accept)
Checksum	16 bits	Checksum of the header and the pseudo header (detailed in Chapter 5)
Urgent Pointer	16 bits	Value of the urgent pointer to urgent data, a positive offset from the sequence number in this segment
Options and Padding	variable	Optional information and options in the packet, padded to the 32 bits boundary

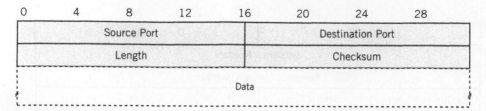

FIGURE 2.32

UDP header

Source Port	16 bits	The Source Port number (identifies the sending applications)
Destination Port	16 bits	The Destination Port number (identifies the receiving application)
Length	16 bits	Length of the header and data
Checksum	16 bits	Checksum of the header and the pseudo header (detailed in Chapter 5)

the destination port number, to provide the source port of the calling application, and to identify a session like in the case of TCP.) The UDP header is shown in Figure 2.32 and is described in the table that follows it.

2.6.2.6 *Internet Protocol Summary*

Internet Protocol is the main platform for data communications and telecommunications, used for computer communications, telephony, videoconferencing, entertainment, business applications, news, and more. IP therefore is used on many transport layers, not just the common Ethernet, to serve as the communication platform. IP over ATM (IPoA), IP over SONET (IPoS), and IP over MPLS (IP/MPLS) are just few examples.

As described in this subsection, the basic data packets that flow through the Internet have many headers, each of another layer; these packets are usually Ethernet packets that encapsulate IP packets that encapsulate TCP or UDP packets that carry data payload (application layer data). The general picture that the network processor sees when it receives a data packet of, for example, an HTTP type, is depicted in Figure 2.33.

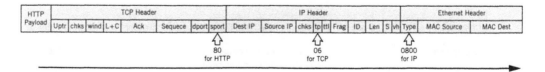

FIGURE 2.33

HTTP Packet

2.7 SUMMARY

In this chapter we discussed network fundamentals, and tried to generalize as much as possible in order to cover the essential networking concepts that are relevant to network processing. We described network models, the connection-oriented, streamed traffic-oriented telecom networks, and the connectionless-oriented, bursty traffic-oriented data networks, especially Ethernet and IP networks. In the next chapter we dive further into network technologies, and describe how data networks and telecom networks are being converged and used in the core and in the metro networks.

APPENDIX A
REGISTRATION PROTOCOLS

In order to distribute and manage information in an orderly manner among bridges, there are several registration protocols that are based on the Generic Attribute Registration Protocol (GARP), which was replaced by the Multiple Registration Protocol (MRP) [182] in 2007. MRP, known as IEEE802.1ak, specifies procedures and "managed objects" that allow participants to register various attributes with other participants of the MRP based applications.

GARP and MRP distribute (declare, register and withdraw) information (for example, about attributes and their values) among a group of participants within the GARP or MRP application. They are used for distributing bridging information between bridges of a network of LANs. One MRP application, for example, is MRP MAC Address Registration Protocol (MMRP), which distributes group MAC addresses and Group service requirements attributes, replacing the GARP Multicast Registration Protocol (GMRP). In GARP and MRP, attribute values are states, and GARP/MRP applications behave as state machines. Every change in the state of a GARP/MRP application triggers a GARP/MRP message for attribute distribution.

MMRP and GMRP addresses are universal addresses, 01-80-C2-00-00-20, and are used as a MAC destination address to carry MMRP/GMRP PDU between MRP/GARP entities. A MRP/GARP VLAN Registration Protocol (MVRP/GVRP) address is 01-80-C2-00-00-21 and it is used exactly like MMRP/GMRP, as described in the following. IEEE802.3 frames are used for carrying MRP/GARP frames; that is the LLC sublayer and header are used. The address of GARP in the LLC layer is 01000010, like the address of the STP, and the GARP PDU is encapsulated in the information field of the Unnumbered Information (UI) LLC type 1 command PDU. The GARP PDU contains a Protocol identifier (PID), equal to "1," to distinguish the GARP PDU from the Bridge PDU (BPDU) or other, unsupported protocols. MRP PDU is using specific Ethertypes for MMRP (0x88F6) and for MVRP (0x88F5).

In GARP, six operations, called events, dictate what should be done with the attributes; they are: leave all, join empty, join in, leave empty, leave in, and empty. The attributes distributed and managed by the GARP application are indexed and known to the GARP entities (there are 256 possibilities). A formal description of the GARP PDU structure is given in the following, using BNF [183]:

```
GARP PDU ::= Protocol ID, Message {, Message}, End Mark
Protocol ID SHORT ::= 1
Message ::= Attribute Type, Attribute List
Attribute Type BYTE ::= Defined by the specific GARP Application
Attribute List ::= Attribute {,Attribute}, End Mark
Attribute ::= Ordinary Attribute | LeaveAll Attribute
Ordinary Attribute ::= Attribute Length, Attribute Event, Attribute Value
LeaveAll Attribute ::= Attribute Length, LeaveAll Event
Attribute Length BYTE ::= 2-255
```

```
Attribute Event BYTE ::= JoinEmpty | JoinIn | LeaveEmpty | LeaveIn | Empty
LeaveAll Event BYTE ::= LeaveAll
Attribute Value ::= Defined by the specific GARP Application
End Mark ::= 0x00 | End of PDU
LeaveAll ::= 0
JoinEmpty ::= 1
JoinIn ::= 2
LeaveEmpty ::= 3
LeaveIn ::= 4
Empty ::= 5
```

A schematic description of the GARP PDU structure is given in Figure A2.1. The MRP PDU structure is similar but not back-compatible. The operations are different, the 2-bytes PID is replaced with 1-byte Protocol version, and the attribute structure is replaced with a vector attribute. Since MRP was finalized after writing of this appendix, GARP is detailed here, and the interested reader is referred to [182].

The exponential growth of Ethernet networks in size, spread, and connectivity requirements, especially in the Provider Bridged Networks (discussed in the next chapter), greatly increases the use of bridges. This, however, also dramatically increases the number of VLANs and Group MAC addresses used. The performance of GVRP and GMPR, in terms of time and resource consumption (e.g., bandwidth), has become a dominant factor in fault recovery time for these big networks. GARP also may cause disruption of traffic in a very large network, when a topology change happens. MRP, which replaced GARP, improved performance and reduced traffic disruption by localizing the topology change to a small portion of the network. MVRP also improves GVRP by rapid healing of network failures without interrupting services to unaffected VLANs.

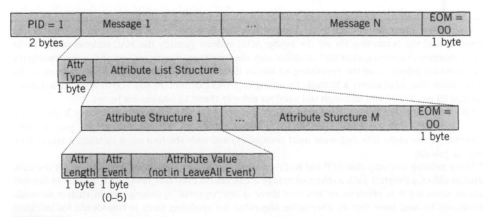

FIGURE A2.1

GARP PDU format

APPENDIX B
SPANNING TREE PROTOCOLS

Spanning Tree Protocol and Rapid STP (RSTP) are used to define the ports' states (e.g., forward or discard) and roles (e.g., root or designated), and to create the active topology of a spanning tree based on these states and roles. STP and RSTP also respond quickly by reconfiguring the active topology in case of changes or failures, and they maintain a connected spanning tree. These activities are done in a transparent operation, and are minimal in resource consumption (bandwidth, processing, and memories). RSTP was specified in IEEE802.1w, and superseded STP in IEEE802.1D-2004. RSTP is backwards-compatible with STP or, more precisely, it can work with legacy bridges that still run STP.

In this appendix we refer mainly to RSTP, as it is more relevant today; however, many of the features and operations described here are relevant to STP as well, and where differences exist, they are mentioned.

RSTP OPERATION

For RSTP to operate, every bridge should have a unique identifier,[39] all bridges should recognize and response to a unique Group MAC address, and a unique port identifier should be defined for each port in a bridge.[40] In order for RSTP to create an efficient[41] active topology, the relative priority of each bridge in the network should be preconfigured, as well as the relative importance of every port in every bridge, and the relative cost of using each port (Port Path Cost [PPC], which typically relates to the speed of the attached LAN, and can never be 0).

---

[39]The bridge identifier is eight bytes long, where the first two bytes are used to indicate bridge priority and the remaining six are the bridge MAC address (usually the MAC address of its lowest port number). However, as of 802.1D-2004, only the four most significant bits of the first two bytes are used for priority, and the remaining 12 bits of these two bytes are used as a locally assigned ID Extension. This field allows a Multiple Spanning Tree (MST, defined in 802.1Q) to generate a distinct bridge ID per VLAN, and although still used for priority, these 12 bits cannot be set.

[40]The port identifier is two bytes long, where the first byte (8 bits) serves as priority. It should be noted, though, that as of 802.1D-2004, the 12 least significant bits are used for the port number (instead of the eight bits that were used previously) and only the four most significant bits can be set for priority.

[41]Many assume wrongly that STP and RSTP produce a minimum spanning tree. This is not the case and, in effect, a shortest path to the root bridge is calculated. Even this, however, is subject to debates about whether it is efficient, or whether there is any "optimal" spanning tree in such a network. It should be said here that an alternative algorithm for spanning trees is considered for bridging, instead of—or side-by-side with—the RSTP, namely the Shortest Path Bridging (SPB), IEEE802.1aq. SPB calculates many active topologies; that is for each bridge, it calculates its shortest paths to destination bridges.

Root ID	Root Path Cost	Bridge ID	Port ID

Rapid STP uses a basic structure of information, called the *priority vector*, which contains four fields,[42] and can be ordered by priorities.

The first two components describe the Root Bridge (RB) identifier (or what the bridge believes to be the RB identity, and initially it uses its own identifier as the RB identifier) as well as the root path cost from that bridge to the root (initially, it is zero, of course). This cost is the sum of all costs of ports (PPC, as defined previously) along the shortest path from the sending bridge to the RB. The next two components identify the sending bridge and its sending port.

The priority vector can be ordered; the order is determined by comparing the fields one by one, starting from the RB identifier, and the "better" priority vector is the one that has a lower value in it (if two values in a field are the same, then the next field is compared). Rapid STP works as follows:

1. Initially, each bridge declares itself as a Root Bridge, and sends a Configuration Message (CM) through each of its ports in the Bridge Protocol Data Unit (BPDU, described as follows). The CM contains the priority vector and, as mentioned before, initially points to itself as the RB.

2. If a bridge receives a CM, and the received Root Bridge is "better" than the known Root Bridge (i.e., lower), then (i) the received RB is selected as the RB, (ii) the root path cost (RPC) is calculated (the new RPC is the sum of the received RPC and the port path cost, PPC), and (iii) appropriate CMs are sent to the bridges around it, if they had previously transmitted a "worse" priority vector.

3. For every CM that the bridge receives, it calculates the "best" cost to the RB (i.e., the lowest) along the shortest path (comparing its current RPC with the received RPC plus the PPC of each port). The port[43] that has the lowest cost to the RB, is assigned as the Root Port (RP) of the bridge.

4. For every LAN, one bridge that has the "best" route to the RB (i.e., the lowest RPC), is assigned as the Designated Bridge (DB).[44] The port that connects this LAN to the bridge is assigned as the Designated Port (DP).

5. Designated Ports periodically transmit CMs as "keep alive" messages, in order to detect failures.

---

[42]There are actually five fields, but the last field (the receiving port ID) is not always used and is not relevant to the explanation.

[43]If two ports have the same cost, then the port with the lower port identifier is selected.

[44]Since all bridges send CM BPDUs when they have a better path to the RB, all bridges connected to a LAN can know each other's RPC, so the bridge with the lowest RPC, through which other bridges assume a path to the RB, is considered as the designated bridge. If two bridges have the same cost, then the bridge with the lower bridge identifier is selected as the DB.

6. A port that is neither the RP nor the DP is assigned as a Backup Port (BP) if the bridge is the designated bridge (DB) to the LAN through this port, or as an Alternate Port (AP) otherwise. These ports are then used in cases of an RP or a DP failure, where an AP replaces an RP (connecting the bridge to the RB via a different path than the one the RP uses), and a BP replaces a DP (since it is connected to the same LAN or segment, it therefore backs up the DP).

All root ports and designated ports (DPs) are put in the forwarding states. The initialization and creation of the spanning tree in RSTP is like in STP, which means that the first four steps of the previous description apply to STP as well. The major change is that the process of the RSTP convergence is faster than the STP. When there is some failure of a link or a bridge, the spanning tree becomes invalid (usually disconnected, but loops might be created, especially in recovering attempts). The filtering database ages relatively slowly, and temporary loops or back-holes might be created. To avoid this, a topology change mechanism is incorporated into the STP and RSTP protocols, which enable faster aging of the filtering database, and faster recovery of the spanning tree.

It should be noted that in order to prevent temporarily loops, RSTP changes the ports' states after some delay, and even sets ports to discard states temporarily. RSTP speeds up the spanning tree reconfiguration by allowing RPs and DPs that are connected in point-to-point connections (dedicated links) to change to forwarding states quickly, without having to wait for protocol time-out events to happen. In addition, RSTP defines bridges' ports that are alone on their attached LANs as edge ports, and sets them as designated ports.

Rapid STP allows a DP that is attached to a point-to-point LAN to change to a forwarding state when it receives an explicit role agreement from the other bridge on that attached LAN. Using a port handshake to negotiate roles means faster configuration and reconfiguration, rather than waiting for the protocol time-outs to happen.

BRIDGE PROTOCOL DATA UNIT

A Bridge Group Address is a universal address, 01-80-C2-00-00-00, and is used as a MAC destination address to carry BPDUs between STP entities. IEEE802.3 frames are used for carrying BPDU frames (Figure B2.1), that is the LLC sublayer and header are used. The address of STP in the LLC layer is 01000010, and BPDU is encapsulated in the information field of the Unnumbered Information (UI) LLC command PDU. The BPDU contains a Protocol identifier, PID, to distinguish BPDU from GARP PDU or other, unsupported protocols.

The fields of the BPDU are divided into three parts: the first part identifies the protocol, the BPDU, and its roles. The second part is the priority vector described in the preceding subsection. The last part of the BPDU is a group of fields that are related to the STP and RSTP mechanism, as described in the following.

PID = 0	Protocol Version	BPDU Type	Flags	Root Identifier (8 bytes)	Root Path Cost (4 bytes)	Bridge Identifier (8 bytes)	Port Identifier	Message Age	Max Age	Hello Time	Forward Delay	Version 1 Length

Protocol Version ID = 0 (STP), 2 (RSTP)
BPDU Type = 0 (STP BPDU), 2 (RSTP BPDU)
Flags bit 8 = topology acknowledgment, bit 1 = topology change, bit 2 = proposal (in RSTP),
    bit 3–4 = port role (in RSTP, 0-unknown, 1-alternate or backup, 2-root, 3-designated),
    bit 5 = learning (in RSTP), bit 6 = forwarding (in RSTP), bit 7 = agreement (in RSTP)
Version 1 length (only in RSTP BPDU) = 0

FIGURE B2.1

STP and RSTP BPDU (from left to right)

Rapid STP communicates the port role and state in the flags field of the RSTP BPDU. These were not communicated in the STP, and they are helpful for faster spanning tree reconfiguration.

The BPDU contains several timer values that are required by the STP and RSTP protocol, such as the "Message Age," "Max Age," "Hello Time," and "Forward Delay." These fields are filled by default values (usually in seconds) and by bridges, according to the ongoing protocol status.

In STP, the RB sends CM BPDU every "Hello Time" with zeroed "Message Age," and triggers CM BPDU messages in the bridged LAN. Any bridge that does not receive a CM within the "Max Age" assumes that the RB is dead, declares itself RB, and start transmitting CMs around until the bridged LAN reaches stabilization again. In RSTP, on the other hand, every bridge sends RSTP BPDUs as "keep-alive" messages, and an adjacent neighbor bridge is considered disconnected after three "Hello Time" if no RSTP BPDU was received.

When a BPDU is relayed (informing other bridges through the designated ports about a better RB received by the bridge), the "Message Age" increases (usually by 1 s), and the BPDU is treated only if the "Message Age" is less than the "Max Age" of the BPDU. This ensures the decay of the BPDU in a nonstabilized spanning tree. Bridges not only relay received BPDUs from their RP to all their DPs, but they also reply to a received BPDU in any of their DPs, with their own BPDU, causing the other bridges connected to the DPs to potentially update their ports' roles and states. Any transition to forward state occurs only after the bridge waits for the "Forward Delay" to avoid too-rapid changes that may lead to intermediate loops.

In STP, Topology Change Notification (TCN) BPDUs (Figure B2.2) are sent by the bridges that sense a topology change, toward the root bridge (RB). The RB acknowledges the TCN BPDU, and informs all other bridges of the topology change. Following a topology change, all CM BPDUs sent from the RB will be flagged by a Topology Change flag in the CM BPDUs, for a duration of "Forward Delay" plus "Max Age."

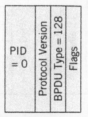

Protocol Version ID = 0

FIGURE B2.2

Topology change notification BPDU

In RSTP, every bridge that senses a topology change sends a RSTP BPDU and sets the Topology Change flag. Every bridge that receives a RSTP BPDU with the Topology Change flag set, forwards this RSTP BPDU.

Lastly, an RSTP port handshake (described previously) for deciding on ports' roles locally is done by using the RSTP BDPU flags bits of "proposal" and "agreement" (bits 2 and 7, respectively). In STP, when a port becomes a designated port (DP), it has to wait twice the "Forward Delay," to shift to a forwarding state (usually about 30 seconds). In RSTP, it simply sends a proposal RSTP BPDU, and the receiving bridge sets its port to a root port (RP) and sends an agreement RSTP BPDU.

Converged Networks

3

In the previous chapter we described basic networking, which includes everything from concepts to data networks (e.g., Ethernet and Internet Protocol networks) and telecommunications networks (e.g., PDH and SDH/ONET networks).

This chapter focuses on converged networks. In these networks, data and telecommunications applications interface and mix, despite their very different natures. Data networks are bursty, and oriented toward connectionless traffic, whereas telecommunications networks are streamed, and connection-oriented. We divide this chapter into two parts, each of which corresponds to a particular path of evolution: the first, from telecom networks to converged networks, and the second from data networks to converged networks.

We briefly describe a legacy networking technology called Asynchronous Transfer Mode (ATM), which was used for about a decade in the 1990s as a solution for mixed data, video, and telephony network. We deal with ATM since, first of all, ATM is still with us—and in several places even dominates networking implementations—but more importantly, because ATM has had a significant impact on networking algorithms, standards, and development. The most important technology that absorbed ATM ideas is the Multiprotocol Label Switching (MPLS), which is also described in this chapter.

Metro technologies and Next Generation SONET/SDH technologies are then described, as they seem to be the main transport backbone networks of the future (at least in the view of the telecom industry). Metro technologies as well as IP and Ethernet evolutions are then described, as they seem to be the main information backbone of the future (at least in the view of the datacom industry).

In the next chapter we describe access and home networks. These networks also use converged technologies, and connect customers and their devices to the long-haul networks.

3.1 INTRODUCTION

After describing network fundamentals, models, and pure telecom networks and data (enterprise) networks, we turn to converged networks; that is, data networks that are used for telecommunication applications and, conversely, data communication networks that are used for telecommunications applications.

Network technologies that are most likely to use high-end Network Processors applications in network nodes (devices) are Ethernet (enterprise and carrier class), MPLS, and IP. Therefore, most of the description in this chapter is focused on these networks, with some discussion of other network technologies required for an overview of how networks are combined, relate to each other, and evolve.

3.2 FROM TELECOM NETWORKS TO DATA NETWORKS

As noted before, one convergence approach between data networks and telecommunications networks is to adopt various techniques that enable transmission of data-patterned traffic over telecommunications networks. The idea of such convergence has also led a move toward more data-centric traffic in traditional telecommunications networks, equipment, and services.

As of today, SONET/SDH is the technology that has the greatest chance of dominating data and telecommunications networks, mainly due to the huge installed base and investments made in these networks. Nevertheless, the supporters of this technology are seriously threatened by the alternative Ethernet-based core networks, mainly due to their low-cost, simplicity, and natural extension from the residential and enterprise networks, as described in the next subsection.

We start with a discussion of the Asynchronous Transfer Mode technology despite the fact that it is hardly in use now, since it still carries several important services, and contributed important networking concepts. We continue with next generation SONET/SDH, which, as mentioned previously, is the main telecommunication platform, and end with Resilient Packet Ring (RPR), which some categorize as a data network platform. We present the general concept of Multiservice Platforms (MSPs) that are referenced in the telecom industry for coping with data services alongside telecommunications.

3.2.1 Asynchronous Transfer Mode

The Asynchronous Transfer Mode introduced a new concept of cell switching (or relaying) that defines an architecture, protocols, cell formats, services, and interfaces. The idea was to allow data and video services to be transmitted over circuit-switched networks together with traditional voice services, and to integrate the three into one network technology. A fixed cell size of 53 bytes was defined to carry 48 bytes of payload,[1] with a 5-byte header. The fixed, small-sized cells supported high-speed and low-cost switching equipment, but created difficulty in interfacing with the variable-length, long packets that are common in packet-switching networks.

---

[1] Some say the size was determined by choosing the average of $2^5 = 32$ and $2^6 = 64$ bytes, or by compromising between long packets for data requirements and short packets for voice requirements.

ATM was promoted by the telecom industry, and standardized by the ITU-T [221, 222] as a Broadband Integrated Services Data Network (B-ISDN) at the beginning of the 1990s. As such, ATM fit smoothly into the SONET/SDH framework and its frames' structures. Another important contribution to ATM standardization came from the ATM forum (now part of the MFA forum, which combines the ATM, Frame-relay, and MPLS forums). However, despite the huge investments, interest, research, and publications, it received relatively little attention from the Internet community, and although ATM was offered for data networks infrastructures, and even used for a while, it was rejected eventually by the datacom industry. As of today, ATM technology has almost vanished (except in Asymmetric Digital Subscriber Line [ADSL] infrastructure). Nevertheless, it is important to understand ATM concepts because they were used for similar approaches in data networks such as MPLS, which is described in the following.

ATM is a connection-oriented technology, like most telecom technologies, according to which a call set-up and the establishment of a connection must take place prior to data transfer. The connection-oriented approach and supporting protocols allow a Quality of Service (QoS, explained in Chapter 6) to be used for each of the ATM sessions, which enables voice, video, and real-time data to be supported. The connections are called Virtual Channels (VC), and they are bundled into Virtual Paths (VP), which are groups of VCs that share a common link with the same endpoints. These two layers of connections, which are called ATM layers, are places in a Transmission Path (e.g., fiber), which belong to the physical layer. Each cell's header uses a VC identifier (VCI) and a VP identifier (VPI) to indicate to which VC and VP it belongs. In other words, VCI identifies the particular VC link for a given Virtual Path Connection (VPC), whereas VPI identifies a group of VC links that share the same VPC (see Figure 3.1).

There are two main applications of VPCs; the first is the User-to-Network VPC and the second is Network-to-Network VPC. ATM headers are defined according to the interface used: User–Network Interface (UNI) or Network–Network Interface (NNI).

The Generic Flow Control, which appears only in the UNI header, is by default 0, and is used for protocol procedures in controlled equipment. The VPI and VCI fields are used for routing and switching, with some preassigned VCI values for

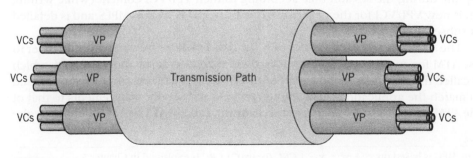

FIGURE 3.1

ATM connections

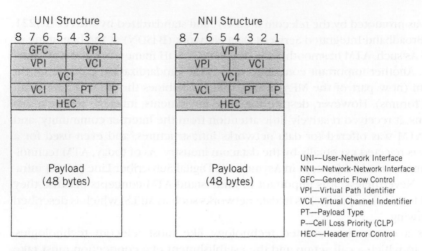

FIGURE 3.2

ATM cell structure; for all fields, the first bit sent is the MSB

specific uses (e.g., various signaling types). The Payload Type (PT) field specifies the type and some attributes of the cell (e.g., Operation and Maintenance [OAM], user cells with or without congestion experience). The Cell Loss Priority (CLP) field indicates how cells should be handled when network conditions request that cells should be dropped (e.g., in congestion); if CLP contains a '1' bit, then this cell should be dropped before cells with CLP = 0. The last field of the header is the Header Error Control, which contains the calculated Cyclic Redundancy Check (CRC) of the header fields, for validating the header by the receiving node (see Figure 3.2).[2]

The usage of VPI/VCI in the header tremendously simplifies the cell forwarding and switching in an ATM network. At set-up time, a path is constructed by a routing protocol,[3] such that each switching node in the routed path maintains a "routing table" that associates the incoming VPI/VCI with an outgoing port, and potentially with a new VPI/VCI path identifier that is known to the next switching node along the path. In this way, each switching node knows how to relay cells rapidly during the session time according to their VPI/VCI content (while writing their new VPI/VCI for the next hop). This principal is used in MPLS, and is detailed in subsection 3.3.1.1.

The ATM layer by itself defines only the data link layer functionalities. In order to use ATM for applications and services, there is another layer above the ATM, which is called the ATM Adaptation Layer (AAL). Various types of AAL mechanisms are used to match the applications' data frames/packets and service requirements to that of the ATM cells. There are such five mechanisms, called AAL1 to AAL5, defined by the

---

[2]The HEC is based on $x^8 + x^2 + x + 1$ CRC (or 0x07); CRC is explained in Chapter 5.

[3]The common routing protocol used for ATM signaling is called ATM Private Network–Network Interface (PNNI), which is defined by the ATM-Forum (today MFA forum) [30].

ITU-T. AAL1 is used for voice and other TDM E1/T1 circuit emulations, characterized by a Constant Bit Rate CBR and connection-oriented type of traffic. At the other end of the spectrum, AAL5 is used for IP and LAN data type traffic, characterized by a connectionless orientation and Variable Bit Rate (VBR) type of service. In between, AAL2 was used for real time VBR data traffic (usually compressed TDM data such as video and voice), while AAL3 and AAL4 target data services with VBR and asynchronous connection and connectionless-oriented types of traffic such as frame relay, X.25, and Switched Multimegabit Data Service. AAL3 and AAL4 were used only rarely, while AAL5 was the main platform of most data services and protocols for carrying IP and LAN over ATM (e.g., classic IP over ATM [273], LAN Emulation—LANE [28], and Multi-Protocol Over ATM—MPOA [29]). As mentioned before, ATM is not used for high-speed networks any more, but still it can be found in the Mbps range WANs, mainly in ADSL and wireless (cellular) infrastructure.

3.2.2 Next Generation-Synchronous Digital Hierarchy/ Synchronized Optical Networks

The data applications that drove most of the traffic in the core networks forced SDH/SONET architecture to be upgraded to better serve this type of traffic. The enhancements offered, collectively called next-generation SDH/SONET (NG-SDH/SONET) [247], include Generic Framing Procedure (GFP) [206], Link Capacity Adjustment Scheme (LCAS) [207], and Virtual Concatenation (VCAT) [208]. These were standardized by the ITU-T as G.7041, G.7042, and G.7043, respectively, and are integrated into recent versions of G.707 [195] and G.803 [197] (SDH standards), as well as in G.709 [196] (Optical Transport Networks—OTN Standard). These solutions are only implemented at the source and the destination equipment, leaving the core network equipment intact, and therefore offer advantages in terms of costs and implementation, but some disadvantages in terms of network utilization.

Returning to the discussion of SONET/SDH that we began in the previous chapter, contiguous concatenation of containers was used to carry higher bandwidths of data. For a TDM environment, this contiguous concatenation is quite efficient. However, continuous, fixed rates of data streams, mapped directly to SDH/SONET, might waste the space of SDH/SONET containers (tributaries). For example, putting a Gigabit Ethernet (GbE) in SDH containers requires the use of C-4-16c, which carries 2.4 Gbps, as the C-4-4c is not enough; therefore, 1.4 Gbps out of the 2.4 Gbps are wasted (58% of the bandwidth). If it is not continuous data stream, as is often the case with bursty Ethernet or IP traffic, the waste can be even higher.[4]

In order to solve the problem of wasted space, a flexible concatenation of SONET/SDH payload is used, called VCAT (in contrast to contiguous concatenation).[5]

---

[4]Further analysis of data usage efficiency over SDH, with and without VCAT, is provided in RFC 3255, "Extending PPP over SONET/SDH" [241].

[5]Concatenation can be regarded as "inverse multiplexing," since it uses many channels for one session between a pair of entities. This is the inverse operation of multiplexing, where one channel is used for many sessions between many pairs of entities.

First, the data streams are segmented into groups of virtual containers (VCs) that are smaller than the data stream rate, and then sent in "VC paths." For example, the GbE is mapped into seven C-4 containers (each of 150 Mbps). These groups of VCs, called Virtual Concatenated Groups (VCGs), are then transmitted in the SDH network, each VC in its own path. In other words, each member of the VCG (that is, each VC) may be transmitted over some STM-n/OC-n link, not necessarily contiguously in the same link, and maybe even in different links. The source node arranges these VCGs, and after their arrangement, only the destination node is aware of the VCGs and is responsible for building the original data streams. Intermediate nodes are not required to be aware of any of this. However, due to the potential use of multiple paths by these VCGs, the destination equipment may encounter unordered VCs, as well as differences in delays, and must be prepared to handle this.

VCGs are notated in the SDH terminology as VC-m-nv, where m is the VC container number, n is the number of VCs used in the VCG, and v signifies the use of VCAT rather then contiguous concatenation. In the case of GbE, for example, the VCG is called VC-4-7v (which means 7 VC-4's).

VCGs can also be arranged with PDH channels; for example, E1-nv, where n stands for the number of E1 channels (each of 1.98 Mbps) concatenated together (up to 16). VCAT can also be implemented over OTN, and Optical Payload Unit (OPU) frames can be concatenated; for example OPU3-nv, where n stands for the number of OPU3 channels (each of 43 Gbps) concatenated together (up to 256), yielding a 10 Tbps link.

The LCAS is a protocol based on VCAT. Its purpose is to add flexibility to the VCG-created links by dynamically changing the SONET/SDH pipes used for VCGs (i.e., increasing or decreasing VCG capacity) [51]. This is done without affecting the existing VCG services. The source and destination nodes exchange LCAS messages; for example, to cope with bandwidth-on-demand requirements (according to time of day, special events, service granularities, etc.) by requesting the addition of more VCs to the VCG in one direction (capacity control is unidirectional). It is possible to save unused bandwidth (e.g., resulting from a failure of some member of a VCG, or from "right sizing" bandwidth for some application) by using LCAS, which automatically reduces the required capacity. LCAS can be used also to enhance load sharing functionalities, as well as QoS differentiation, or traffic engineering (TE), described in Chapter 6.

The GFP is a traffic adaptation protocol [172] that provides a mechanism for mapping packets and circuit-switched data traffic onto SONET/SDH frames, while accommodating diverse data transmission requirements. Traffic mapping is done at the source node into general-purpose GFP frames, and then the GFP frames are mapped into VCs or VCGs and onto the SDH network. The intermediate modes are not aware of the GFP mapping, and only the destination node de-maps the original traffic. GFP uses two modes, referred to as Frame-Mapped GFP (GFP-F) and Transparent-Mapped GFP (GFP-T). GFP-F is optimized for a packet-switching environment and a variable-length packet, whereas GFP-T is intended for delay-sensitive traffic, bandwidth efficiency, and transparency to the line code data.

Typical applications of GFP-F are Point-to-Point Protocol (PPP), IP, Ethernet, MPLS, RPR, or Digital Video Broadcast (DVB) (according to ITU-T G.7041 [206]). The entire data frame is mapped into one GFP-F frame, and the resulting GFP-F frames are variable in size.

GFP-T is used for Storage Area Networks such as Fiber Channel, Enterprise Systems Connection, and Fibre Connection, or video applications such as Digital Visual Interface and DVB, or even GbE (according to ITU-T G.7041 [206]). The data may span over many fixed-length GFP frames. It is mapped into the GFP-T frames byte by byte and organized in N 536-bit superblocks, each of which is constructed from eight 65-bit blocks and 16-bit CRC.[6] For optimal usage of SDH link capacities (i.e., minimum GFP-T overhead), ITU-T recommends using a minimum number of superblocks in a GFP-T frame. For example, for a 3.4 Gbps Fiber Channel on a VC-4-24v VCG path size, at least 13 superblocks per GFP-T frame are required, or for GbE on VC-4-7v, at least 95 superblocks per GPF-T frame are required.

The structure of the GFP frame is shown in Figure 3.3. The two transport modes are different in the information field of the payload area. These two transport modes can coexist in a transmission channel. Generic Framing Procedure is a very useful adaptation layer, and one that is not necessarily restricted to SONET/SDH or OTN.

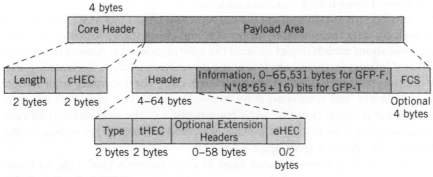

HEC—Header Error Check (CRC-16)
cHEC—Core HEC
tHEC—Type HEC
eHEC—Extension headers HEC
FCS—Frame Check Sequence (CRC-32)

FIGURE 3.3

GFP frame structure (from left to right)

---

[6]The organization of the superblock and its coding is beyond the scope of the book, and can be reviewed in ITU-T.7041 [206] and [172].

3.2.3 Resilient Packet Ring

Resilient Packet Ring (RPR) was standardized by the IEEE (IEEE 802.17) [190] for enabling packet-oriented data networks to function efficiently on optical rings.[7] RPR offers a new Medium Access Control (MAC) scheme that provides efficient bandwidth utilization, a fast protection mechanism, distributed fairness algorithms, differential QoS, and network survivability similar to SONET/SDH (on which RPR resides). RPR can work above SONET/SDH or Ethernet links, and can support up to 10 Gbps bandwidths, up to 255 station attachments on rings that can span up to 2000 km, and it is scalable and interoperable. Scalability is achieved by using bridges attached to the ring, as well as specific support of RPR in bridging (and remote stations).

RPR's MAC is based on spatial reuse; in other words, at every segment of the ring topology, data can be transmitted simultaneously with other data over other segments, and is stripped at its destination. Data is transmitted in a dual, counter-rotating ringlet; that is, two close unidirectional rings, each forming a path of an ordered set of stations and interconnecting links.

The physical layer underneath RPR's MAC layer, as well as the upper layers above RPR's MAC, are agnostic to RPR, since well-defined interfaces are used. RPR addressing is 48 bits wide (the usual IEEE addressing scheme), of which 24 bits make up an Organizationally Unique Identifier. RPR addressing supports unicasts, multicasts, and broadcasts. An additional mode of addressing is to flood the frames to a set of bridges that are possibly associated with an individual MAC address. This is like multicasting to all bridges, but where the destination address is a unique station's address rather than a multicast address.

Each station controls two ringlets (called west and east) in its MAC layer, which are both used for data transmission, protection, and congestion avoidance. The station's decision about which ringlet to send the information on is subject to the ring condition and known state, and obviously the destination address. Each station can add information to the ring (in either direction), copy the incoming information (for its upper layers, and retransmit it further on), strip it (i.e., take it off the ring; stop transmitting it further), or discard it (retransmit it to the next station). In case of a ring failure, the information is quickly wrapped or steered according to the failure location on the ring (see Figure 3.4).

Some traffic management modules exist in each station's MAC. One of these mechanisms is a fairness control that is aware of the ring usage and congestions (due to periodical broadcasted fairness control frames). These fairness modules control some of the stations' traffic by shaping it, such that they will inhibit the stations from disproportional usage of the ring, which would prevent "downstream"

---

[7]Although we categorize RPR as technology that goes from telecom to data networks on the road to convergence, many categorize RPR as data network technology that goes to the telecom for the same purpose (and remember its origin—the IEEE802). We position it here, since the telecom industry that traditionally manufactures, operates, or are used to optical rings, pushes RPR.

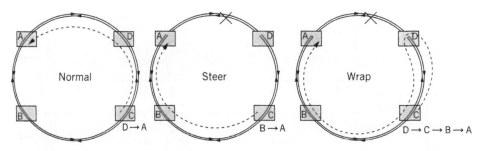

FIGURE 3.4

RPR failure protection

stations from using it. Each station maintains two such fairness modules, one for each ringlet. Each of these fairness modules sends its control frames to its upstream neighbor via the opposing ringlet. In other words, each RPR station processes the incoming fairness control frames, and adjusts the rate at which its active queue is drained onto the ring. It further monitors its sending queues, and when they start to be congested, the node notifies the "upstream" stations to slow down by using the other ringlet.

There are three service classes defined for RPR:[8] Class A is for real time applications; this class receives a guaranteed bandwidth, has a low jitter, and is not subject to the fairness control mechanism. Class B is for near real time applications, and has two subclasses: Committed Information Rate (CIR) and best effort transfer of the Excess Information Rate (EIR) beyond the CIR. The CIR subclass receives a guaranteed bandwidth, has a bounded jitter, and is not "fairness eligible" (subject to the fairness control), while the EIR subclass receives the same treatment as Class C. Class C is a best-effort class, which has no guaranteed bandwidth and is subject to fairness control.

There are five RPR frame formats, as depicted in Figure 3.5, which are:

- Data frame, which has two formats:
  - Basic data frame (used for local-source frames)
  - Data-extended frame (used for remote-source frames)
- Control frame (used for various control tasks)
- Fairness frame (used for broadcasting fairness information)
- Idle frame (used to adjust rate synchronization between neighbors)

As noted previously, RPR can reside on top of Ethernet, on SONET/SDH directly, or in the GFP adaptation layer. RPR architectural positioning, layers and interfaces are depicted in Figure 3.6.

To conclude, RPR has the potential of being an important and efficient transport mechanism, residing on top of the very common SDH rings or on future OTN rings.

---

[8]Service classes are described in more detail in Chapter 6.

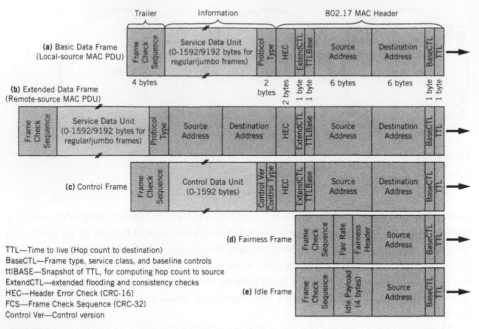

FIGURE 3.5

RPR frame formats (from right to left)

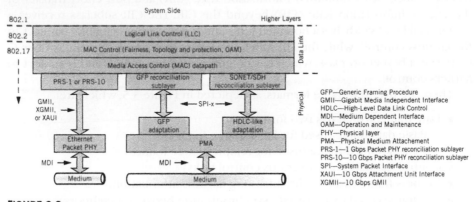

FIGURE 3.6

RPR architectual positioning

However, as of today, it is not widely adopted, and it might either end up like Fiber-Distributed-Data-Interface or be accepted extensively for packet data networks. NG-SONET/SDH technologies are certainly not helping RPR's acceptance, since they are not only more familiar to the carriers, but they can also potentially provide an answer to the data-centric traffic for their already existing SONET/SDH rings, with less investment.

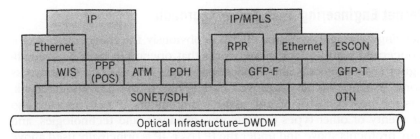

FIGURE 3.7

Networking technologies over transport networks

3.2.4 **Multiservice Platforms**

Although Multiservice Platforms (MSPs) do not present any specific technology, the term MSP is widely adopted by carriers and vendors, and is often mentioned in relation to data services in the telecommunications environments. MSPs are network elements (NEs)—Multiservice Provisioning Platforms (MSPP), Multiservice Transport Platforms (MSTP), or Multiservice Switching Platforms (MSSP)—which are actually composite, SONET/SDH based equipment, geared for packet-oriented services as well as TDM-oriented services. MSPP is associated with NG-SONET/SDH, and can be regarded as its implementation; MSTP is basically MSPP plus DWDM capabilities, while MSSP is MSPP that integrates cross-connect and other high-end switching capabilities. MSPs function as an important interim phase in the migration process from telecommunications infrastructures toward data (and TDM) service infrastructures.

In general, MSPs are NEs that essentially implement NG-SONET/SDH, support CWDM and DWDM, and offer diverse optical and electrical interfaces, including GbE, IP, OTN, DS-n, and E-n PDH signals.

The shift from telecommunications networks to data networks through NG-SONET/SDH technologies and MSPs can be summarized in Figure 3.7, where all relations between technologies are shown. The abbreviations used in this figure are explained throughout this chapter.

3.3 **FROM DATACOM TO TELECOM**

The second convergence approach between data networks and telecommunication networks is to use data networks (i.e., frame- or packet-oriented networks) as core global and national carrier networks, as metro carrier networks, and for metro access networks, carrying both TDM and data applications.

This subject is quite new and, as of this writing, is only a couple of years old. The field is still quite unstable, with many changes and new approaches and many standards being offered by many standardization bodies—not all of which will be adopted. This part tries to provide the most relevant and stable information.

3.3.1 Internet Engineering Task Force Approach

The Internet Engineering Task Force (IETF) is obviously Internet-centric, and it is mainly concerned with IP for data networks that operate in layer 3, strengthened by higher layer protocols and applications (e.g., TCP, HTTP). IP networks and backbone IP networks are deployed extensively today, providing remarkable answers for most data sharing and usage requirements. IP networks are also adapted for plenty of other types of networks, channels, and technologies, and support many protocols above and under the IP layer. IP is principally used above Ethernet networks by enterprises, as well as above many other channels, links, and transport networks (e.g., Frame Relay, ATM, leased lines, dial lines, SONET/SDH).

The industry (and IETF) have extended pure IP-based networks to other technologies in response to requirements for bridging multiple networking technologies and protocols, interoperability, and integrated support of multi-protocols, including "telecom" type of communications (i.e., streaming, as well as real-time, scalability, reliability, and manageability demands from these networks, and the associated QoS). One major result is the introduction of a shim layer between all kinds of layer 2 networks and other layer 2 and layer 3 networks (hence multi-protocol). This shim layer uses switching instead of routing (resulting speed and scalability) in preassigned "tunnels" in the network (hence, the ability to engineer the traffic and assure QoS). This shim network, the MPLS [370], became a fundamental concept, a network, and a tremendously successful IETF solution for carriers and service providers, mainly in their core and long haul networks.

Carriers and service providers enable virtual separation of their Packet Switch Networks (PSN)—their IP or MPLS networks—by providing seemingly private networks for their customers. These Provider Provisioned Virtual Private Networks (PPVPNs) are provided according to several techniques suggested by the industry and standardized by the IETF.

This part focuses on two main technologies: MPLS and PPVPN. Just before diving into these two technologies, however, it should be noted that current IETF-based networks are actually interoperable combinations of technologies and architectures: PSNs (such as IP or MPLS) and other networks (such as ITU-T's SONET/SDH, IEEE802's Provider Backbone Bridged Networks [PBBN] and Provider Backbone Bridges-Traffic Engineering [PBB-TE], both of which are described in the next part). IETF-based carrier networks mostly originated from a routing standpoint rather than bridging and switching,[9] as in the case of IEEE 802 driven networks.

3.3.1.1 Multiprotocol Label Switching

MPLS [370] is a forwarding paradigm for label switching, which came about as an effort to speed up IP datagrams in backbone networks. Instead of handling,

---

[9]Considering where IEEE802 and IETF came from (i.e., layer 2 LANs, bridges and switched Virtual LANs, and layer 3 Internet Protocol, respectively), their networking approaches are quite obvious.

routing, and forwarding each IP datagram separately at each node (router) along the datagram path to its destination throughout the network, MPLS suggests implementation of label-switched paths, or tunnels in the backbone networks, over the various existing link-level technologies, through which similar IP packets can flow. MPLS is a transmission mechanism that incorporates IP and ATM ideas into a new sublayer, under the third layer and above the second one. In other words, it is a 2.5-layer protocol, or a shim layer, which can carry IP, Ethernet, ATM, or TDM sessions (hence, the Multi-Protocol), and it can be implemented on top of L2 SONET/SDH, Ethernet, Frame-relay, ATM or Point to Point links. MPLS allows precedence handling (Class of Service, CoS), QoS assignments, TE, and Virtual Private Networking (VPN) services, as described in the following.

MPLS merges the connectionless type of traffic with traditional connection-oriented transmission, thus allowing streaming and multimedia applications to coexist with legacy IP applications.

3.3.1.1.1 Multi-Protocol Label Switching Overview

The basic idea in MPLS is that packets can be classified into Forwarding Equivalence Classes (FECs), where each FEC contains packets that can be forwarded in the same manner (for example, they have similar destinations, similar paths, or similar service requirements from the network). When packets enter the MPLS network, they are classified by the edge, ingress router (called Label Edge Router, LER) and assigned to an appropriate FEC, which is represented by a short, fixed-length label. This is referred to as a Label Push, in which a label is attached to the packet by either replacing a field that exists in the L2 frame (e.g., the VPI/VCI fields in ATM cell), or as an additional field between the L2 header and L3 headers (e.g., IP), as shown in Figure 3.10. The labeled packet is forwarded (switched) by the core routers (called Label Switching Routers, LSRs) according to the label; that is, the label is used to point both to the next hop and to a new label that replaces the incoming (original) label for the next hop (this is called Label Swapping). In other words, a label is a result of an agreement between any two consecutive routers on the path of the packet, and it represents the original FEC along the entire path, despite the label swapping in each hop. Each LSR maintains a Label Information Base (LIB) for each FEC that includes the next hop (to which the packet is forwarded) and the swapped label to be used (an example is given later). At the edge, egress router (the LER), the label is removed (Label Pop) and the packets continue to their destination outside the MPLS network. The route that was used in this procedure is called a Label Switched Path (LSP). The architecture of the MPLS network can be seen in Figure 3.8, which gives an example and is further described.

Figure 3.8 provides an example of two packets, A and B, that enter the MPLS network with destination addresses 132.72.134.2 and 143.34.2.121, respectively. The LSP starts at LER1, ends at LER4, and goes through LSR2 and LSR3. The packets are processed first by the ingress LER (LER1), which has to decide with which FEC each of the packets should be associated. This is a complex task that involves classifying each incoming packet and associating it with an allocated, appropriate

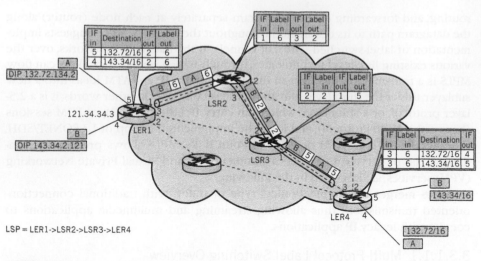

FIGURE 3.8

MPLS network (from right to left)

FEC (or potentially creating a new FEC if there is no adequate one). Let us assume that in this example there is such an FEC, to which a "binding" of label 6 is agreed by LER1 and LSR2, as indicated by the LIB that LER1 maintains. (This label value, 6, is arbitrarily chosen by LER1 and LSR2, and it is local; that is, it can be used in other segments by other LSRs to indicate this or another FEC.) In other words, packets A and B are sent from LER1 to LSR2 carrying label 6, which indicates that these packets belong to a specific FEC—one that carries packets toward IP networks 132.72/16 and 143.34/16[10] in this example. LSR2 is called the upstream router in this segment for these packets, and LER1 is called the downstream router. The downstream router (LER1 here) is responsible for setting the label. It can use a label associated with the FEC as it finds fit, or it can use a label provided by an upstream router, either after requesting it ("*downstream-on-demand*"), or as an initiative of the upstream server ("*unsolicited downstream*").

A packet can carry a "label stack" of several labels that are organized as last-in, first-out. The label at the bottom of the stack is considered a level 1 label, and the label at the top of the stack is level m label, where m is the stack's depth. In any MPLS network, the top level label is considered first, which allows several MPLS network hierarchies to coexist in a packet's path. An example of this is given in Figure 3.9, where we show MPLS subnetworking inside an MPLS network.

The result of MPLS's use of label hierarchy is the creation of MPLS tunnels at each level of the hierarchy (including the very first, level 1, label hierarchy). There are situations in which one router needs to send packets to another router, but the second router neither follows the first router on the path of the packet, nor is

---

[10]IP Network addresses in IP packets are further described in Chapter 5.

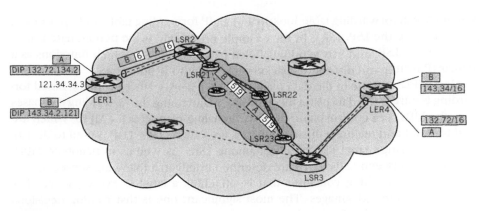

FIGURE 3.9

MPLS hierarchy

it the ultimate destination of the packet. Such a situation may occur, for example, when routers are using different networking mechanisms. The created path is a tunnel through which packets are sent. Every LSP, by this definition, is a tunnel. Tunnels, on the other hand, can be comprised of networks other than MPLS (such as LSPs), routers, or MPLS network segments. When tunnels in an MPLS network are made of several MPLS routers (i.e., a sub-MPLS network), then a hierarchy of MPLS networks is created, as in Figure 3.9. Instead of using an LSP that is composed of just one level (for example, LER1-LSR2-LSR3-LER4, which is a level 1 LSP), an additional sub-LSP, or level 2 LSP, is used between, say, LSR2 and LSR3, and is composed of LSR2-LSR21-LSR22-LSR23-LSR3, so that the entire path would be LER1-LSR2-LSR21-LSR22-LSR23-LSR3-LER4. The second level LSP would then use double labels, or a label stack of depth 2.

The Router LSR23, which is called the *penultimate router* in the level 2 LSP, pops the level 2 label from the label stack, handing the level 1 labeled packet to LSR 3 to continue the packet's journey in the level 1 LSP. Since the last router usually requires no label at all, it makes sense to have the penultimate router—the one before the last—pop the label rather than swap it. The penultimate router forwards the packets to their destinations based on non-MPLS network addresses, or else on higher label and LSP levels in MPLS networks. The label's purpose is to point the LSP up to the last router, and once the packet is sent by the penultimate router to the last one, there is no longer any need for the label, unless, in some cases, the last router does in fact use information from the label to make forwarding decisions (e.g., based on the FEC). Because of this, when setting up the LSP, the penultimate router receives clear indication from the egress LER, through the Label Distribution Protocol (LDP), as to whether it should do *penultimate hop popping*, or leave the label to the last router (LER). Penultimate hop popping improves forwarding efficiency by allowing the egress LER to avoid performing both an MPLS forwarding table lookup and an IP forwarding table lookup for each packet exiting the LSP. LSR3, in the example previously, is the penultimate router in the level 1 LSP, and may send unlabeled packets to LER4, which functions as a regular IP router and distributes packets according to IP addresses.

Most MPLS is done in the data plane. In other words, label-swapping-based forwarding is part of the data plane mechanisms. Maintaining the forwarding databases (e.g., the LIBs) is a control plane process. Signaling protocols for MPLS, which are used to build and maintain these databases, are called LDPs. LDP is used to distribute and exchange label/FEC bindings among LSRs, as well as to negotiate MPLS capabilities between LSRs. LDPs are described briefly in a following subsection.

The label-switching concept—that is, attaching a label to every packet—has several important advantages. The most significant one is that routing decisions are simplified and sped up at the core network, since the only thing required to forward a packet is to swap the label and to switch the packet according to the packet's label and the LSR's LIB. Second, label-switching allows the forwarding mechanism to further classify and prioritize packets by more than their IP headers; for example, by their originating ports. Third, the label can be used to indicate a desirable, specific path for the packets to traverse the network (sort of a "source routing"[11] mechanism), for various traffic considerations. Fourth, the label carries precedence information, or "CoS," that allows the NEs (e.g., routers, queues) to apply appropriate scheduling disciplines, overflow discarding thresholds, and so on. Finally, label-swapping allows scalability of the MPLS network since there is no need to "globalize" the labels throughout the entire MPLS network.

3.3.1.1.2 Multi-Protocol Label Switching Label

An MPLS label (Figure 3.10) [369] is 4 bytes long, with a 20-bit label field, which is the actual label value.[12] The Time-to-Live (TTL) field is similar to the TTL in the IP header,[13] and is used for loop-avoidance (in each swap operation, the TTL is decremented by 1, and when it reaches 0, the packet is discarded). The last field was originally left as an experimental-use field of 3 bits, but usually it is used as a

---

[11]In source routing, each packet carries with it a list of nodes that it should pass while traversing the network toward its destination, and each node along the path complies with this "instructions list" by forwarding the packet to its next hop and removing itself from node-list (like a pop operation).

[12]Values 0–15 are reserved for specific uses, as defined in *http://www.iana.org/assignments/mpls-label-values*: A value of 0 indicates that the label stack must be popped, and the packet should be forwarded based on the IPv4 header (called "IPv4 Explicit NULL Label"). This value is valid only at the bottom of the label stack (the last label). A value of 2 is the same for IPv6. A value of 3 is used only in the LDP and is not actually sent as a label in any frame; instead, it indicates to the LSR the use of a "null" label. In other words, it instructs the LSR to POP a label rather than to SWAP it when applied (hence, it is called "Implicit NULL label").

[13]Actually, the TTL in the label in the ingress LER is taken from the TTL field of the IPv4 or the hop limit field of IPv6, as the case may be. The TTL result in the egress LER then replaces the TTL or hop limit fields in the IPv4 or the IPv6.

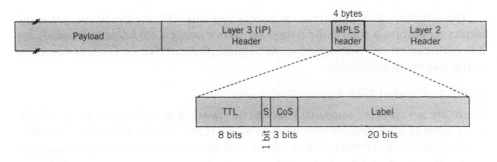

FIGURE 3.10

MPLS header (from right to left)

CoS field indicating the service requirements of this labeled packet. Field S indicates the bottom of the stack (the last label) when it is set to "1."

3.3.1.1.3 Label Information Base

MPLS uses information stored at each LSR and LER to determine which labels the LSR or LER should be using for the forwarded packets in their next-hop, and what to do with the labels. These tables are manipulated by the LDPs, as described before; in other words, they indicate whether the labels should be swapped, popped, or pushed. This information repository, called the LIB in the MPLS terminology, is not standardized, and can be designed and built in many ways. There are three defined terms in the MPLS standard that are necessary for constructing the LIB: Next Hop Label Forwarding Entry (NHLFE), Incoming Label Mapping (ILM), and FEC-to-NHLFE (FTN).

The NHLFE is an entry in the LIB. It contains the information required for the packet to be forwarded: the packet's next hop and the operation that the LRS router has to perform on the packet's label stack. The operations could be swapping the label, popping it, pushing a new one, or a combination of these.

The question of how the LSR assigns a packet with a specific label to a NHLFE is answered by the ILM, which actually assigns a set of possible NHLFEs, from which just one entry is chosen based on specific requirements. Being able to consider multiple entries is useful, for example, for load balancing, backup transmission, or TE applications.

An analogous mapping of packets without labels to NHLFE is done by the FTN mapping, which is usually done in the ingress LER. Again, FTN can map packets, classified and assigned to some FEC, to a set of suitable NHLFEs, from which just one entry can be used.

An example of an LIB can be seen in Figure 3.8. Each LSR in the LSP uses the ILM to find the right NHLFE and, according to it, decides on the next hop and operation (usually to swap with a new label). The LER has to analyze the packet first, determine its FEC, and then use the FTN to find the right NHLFE and execute the operation in the entry (usually to push a new label).

Once again—it is important to understand that LIB, along with its parts and implementations, is used in the data plane (while packets are being forwarded), but it is constructed and maintained in the control plane, usually by LDPs, as described in the following subsection.

3.3.1.1.4 Label Distribution Protocols

There are several LDP standards that can be classified according to various parameters, such as who decides which labels to use (LSP control for the required FECs) and how it is done (routing). More specifically, there is the ordered versus independent LSP control, and hop-by-hop versus explicit routing, as described as follows.

In ordered LSP control, the egress LER binds a label to an FEC, and every downstream LSR (a preceding LSR on the LSP) binds a label only after receiving the label binding for the specific FEC. Ordered LSP control can be initiated by any LER; that is, the ingress LER can ask the egress LER to start with the label binding process. In independent LSP control, each LSR (or LER) binds a label as it finds it necessary to do so for a specific FEC; thus segments of the LSP can be built while the packet is forwarded.

Hop-by-hop routing is similar to IP routing; that is, each node along the path (LSR in the case of MPLS) decides independently what will be the next hop. Explicit routing is similar to source routing.[14] The difference from traditional source routing is that packets remain as they are (apart from attaching a label for each hop), and are carried in an explicit LSP tunnel that was created by LER (usually by the ingress LER). The route is created by the LER by specifying the sequence of LSRs in the LSP and manipulating the LIBs in each LSR along the LSP such that labels along the LSP will forward the packets through the required explicit route.

Although we refer in this subsection to LDP as a generic name for a group of protocols, there is one specific LDP protocol [22] (its origin [21] was introduced together with the MPLS, and is briefly described in Appendix A of this chapter). A simple LDP ("Vanilla" LDP) is based on creating a "tree" of unidirectional paths that connects a destination to every possible source, by hop-by-hop routing. It uses the existing IP forwarding tables to route the control messages required to build the LSP ("Vanilla" LSP). Another LDP mechanism can build an Explicitly Routed LSP (ER-LSP) by following a path that the source initially chooses, on which the control messages required to build the ER-LSP are routed.

Other techniques and protocols are used for TE in MPLS, by extending topology databases to include constraints, and then running an LDP that does not violate the topology constraints. There are two common ways to do this: (1) by extending the LDP to Constraint-based Routed LSP using LDP (CR-LDP) [235], which is LDP with additional explicit routes and constraints, and (2) by using Resource Reservation Protocol—Traffic Engineering (RSVP-TE) [38], which adds label distribution to the RSVP.

---

[14]See footnote 11 for more about source routing.

3.3.1.1.5 Multi-Protocol Label Switching Layering

Up until now we described a "clean" layered network model, according to which each layer is used above or under another layer, in proper order. However, to make things more complicated, MPLS can be used not only above Layer 2 and underneath Layer 3. MPLS can also carry L2 frames above it, or be carried above L3 packets.[15]

Carrying MPLS over IP: It is possible to encapsulate MPLS packets (which are of layer 2.5) into IP packets in IP-based networks (which are layer 3), for example, this can be done to interconnect two MPLS networks by an IP network. Two IP-based encapsulations are offered [431]: MPLS-in-IP and MPLS-in-GRE. MPLS-in-IP is done by replacing the top level label with a "regular" IP header (with protocol number 137, which indicates an MPLS unicast payload), and the result is that an MPLS tunnel is replaced by an IP routing. Similarly, MPLS-in-GRE is done by using an IP header followed by a Generic Routing Encapsulation (GRE) [120] header.[16]

Carrying L2 over MPLS: A set of drafts, called the Martini-drafts, defined how to transport L2 frames; for example, Frame-Relay, ATM AAL5, ATM cells, Ethernet, Packets over SONET (POS), TDM, HDLC and PPP, across MPLS network. "Draft Martini" was known as Cisco's Any Transport over MPLS (AToM), or sometimes also as Any Protocol over MPLS (APoM). These drafts were standardized for historical and documentation purposes as RFC 4905 [306] and RFC 4906 [307], and were superseded by the Pseudo Wire Emulation Edge to Edge (PWE3) Working Group specifications [309] and related standards (described in the following).

3.3.1.2 *Using Packet Switched Networks*

Initially enterprises used point-to-point links between their sites, and between their sites and service providers. These links were based on dedicated telecom lines (low rate TDMs, like T1/E1 or T3). These point-to-point links were used for providing point-to-point services, and to create private networks for these enterprises. However, using dedicated telecom lines, especially for private network setup, was extremely expensive (and still is). The next phase, then, was to use the public network infrastructure of Service Providers (SPs) (i.e., the PSN), and to create a VPN as an overlay network to which the enterprises are connected. When the VPN service, or interface, is based on Layer 2 frames (e.g., ATM, Ethernet), the VPN is called L2VPN, and when the VPN service (i.e., interface) is based on Layer 3 packets (e.g., IP), it is called L3VPN.

---

[15]It should be said, however, that this is not just an MPLS capability; IP (Layer 3) packets can encapsulate and carry L2 frames, such as Ethernet, even though these IP packets are encapsulated in some L2 frames. In other words, it can happen that L2 frames carry L3 packets, which contain other L2 frames again for their end-point applications. Some examples of this, which we'll describe in the following, are Layer Two Tunneling Protocol (L2TP) [409], Generic Routing Encapsulation (GRE) [120], and more.
[16]Generic Routing Encapsulating is used to encapsulate an arbitrary network layer protocol over another arbitrary network layer protocol. In the case of GRE over IP, the IP header's protocol field is 47, indicating a GRE payload, and the GRE header's protocol field contains the Ethertype of unicast MPLS, 0x8847, or multicast MPLS, 0x8848, as the case may be.

The IETF approach to carrier-provisioned PSNs covers the two basic services of point-to-point and VPN (in addition to the PSN networks themselves). These two service models are the result of several working groups of the IETF, and the two relevant working groups are described next.

The first working group, Pseudo Wire Emulation Edge to Edge (PWE3), offers a set of standards for a point-to-point link emulation over a PSN (e.g., IP or MPLS). Using PWE3 technologies, layer 2 interconnections are done transparently between any two edge nodes of the PSN.

The second set of working groups, PPVPN, provides a set of standards for networking that relies on the three layers of PPVPN.[17] VPN is a subnetwork of either a private or public network whose group of users (customers or sites) are separated from other groups (and subnetworks). This group shares this network, and its users communicate among themselves in a subnetwork and maintain it. The method and extent of the separation between the groups' VPNs result from the technology used; that is, the isolation between traffic belonging to different groups can be achieved by using mechanisms such as Layer 2 connections (Frame relays, ATM, Ethernet, etc.), or Layer 3 tunnels (IP, MPLS, etc.).[18]

3.3.1.3 *Pseudo Wire Emulation Edge-to-Edge*

Pseudo Wire Emulation Edge to Edge[19] specifies the encapsulation, transport, control, management, internetworking and security of services that are emulated over IETF PSNs (mainly IP and MPLS networks). A pseudo-wire (PW) emulates a point-to-point link, which appears to be a private link, or a circuit of some service to its users. As its name implies, PWE3 is an edge-to-edge facility that has nothing to do with the underlying PSN control. Several technologies can be used by PWs; for example, Ethernet, Frame-Relay, PPP, HDLC, ATM, or SONET/SDH, which are defined as PW types.[20] PW setup, configuration, maintenance, and tear-down require specific control and management functions that are described in the PWE3 and are based on tunneling protocols; for example, LDP [309] or Layer two Tunneling Protocol version 3 (L2TPv3)[21] [272].

The reference model of PWE3 contains the main elements used in the IETF terminology for PSN: A piece of Customer Edge (CE) equipment (e.g., router, switch, or host) is connected to the network's Provider Edge (PE) device by the Attachment

---

[17]Instead of the PPVPN working group, there are Layer 2 VPN (L2VPN), Layer 3 VPN (L3VPN) and Layer 1 VPN (L1VPN) working groups, each dealing with a different layer than the one on which the VPN can be based.

[18]It should be noted though, that traditional IP (Layer 3) VPN services emerged from security needs; that is, VPN-based IPsec [249].

[19]PWE3 is the third evolution made by IETF, following Pseudo-wire over Transport (PWOT), which itself followed Circuit Emulation over Transport (CEOT).

[20]RFC 4446 [301] defines a 15-bit coding of PW types (e.g., 1 for Frame-Relay, 4 for Tagged Ethernet [VLAN], 5 for Ethernet, 7 for PPP).

[21]The L2TP extensions (L2TPext) workgroup defines L2TP specifics for L2TPv3-based PW.

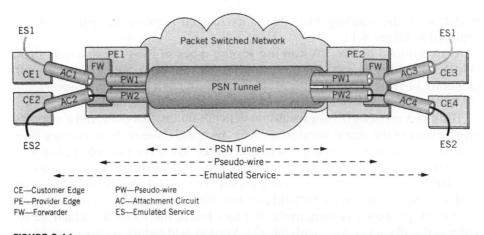

CE—Customer Edge   PW—Pseudo-wire
PE—Provider Edge   AC—Attachment Circuit
FW—Forwarder   ES—Emulated Service

FIGURE 3.11

Pseudo-wire reference model

Circuit (AC; e.g., Ethernet port, Frame-Relay or ATM VPI/VCI). Figure 3.11 shows this reference model.

The PEs create PSN tunnels among themselves, and establish PWs inside these tunnels to convey the emulated service PDUs through these PWs. The CE is unaware of the emulated service it receives from the PE, and sees it as a native service. The frame, or the bit stream, that originates with the CE, transverses the AC and is received by the PE. The PE uses an internal forwarder that matches a PW to the incoming frames either according to the AC they were received from, the frames' attributes (payload content, address, etc.), or statically/dynamically configured forwarding information. The PW protocol is based on three layers:

- Emulated service (e.g., TDM, ATM),
- Pseudo-wire (PW, i.e., Payload Encapsulation), and
- PSN Tunnel (PW demultiplexer, PSN Tunnel, and PSN and Physical layers).

The Emulated service is maintained between the two CEs, the pseudo-wire is maintained between the two PEs, and the PSN tunnel is maintained only in the PSN, from the PE to the other edge PE. The tunnel header takes the PDU across the PSN, from PE to PE. The PW demultiplexing field distinguishes between different circuits in a tunnel (e.g., MPLS label). The PW encapsulation layer is comprised of three sublayers, that encapsulates the service PDU payload, adds timing information (e.g., for real time or synchronized channels), and carries sequencing tasks (i.e., it ensures ordered and unduplicated frame delivery). The PW identification/demultiplexing layer enables the usage of one PSN tunnel for multiplexed PWs. The PSN convergence layer provides a consistent interface to the PW, making the PW independent of the PSN type. The protocol-layering model, with the resulting PDU that is exchanged between the pair of PEs, is depicted in Figure 3.12.

As mentioned before, PW can use several types of PSNs (or ride over them [328]): MPLS (LSP is used for tunneling the PW packets), MPLS/IP (MPLS-in-IP tunneling [431] is used, with the MPLS shim header as a PW demultiplexer), and L2TP-IP (where L2TPv3 [272] is used for tunneling the PW packets).

The exact service-PDU encapsulation depends on the PSN used, and a mapping between any of the above mentioned PWE3 layers and those of the underlying IP or MPLS tunnel layers is defined in PWE3 and related RFCs. For example, Figure 3.13 shows the Generic PW MPLS Control Word (PWMCW) that is required for PWE3 over MPLS PSN [62], and the resulting packet. The tunnel label is the MPLS outer label, and the PW label is the MPLS inner label. MPLS distinguishes between IP packets and PW packets according to the first four bits of the PWMCW (which are 0), whereas the IP packet starts with the 4-bit Version field (which is either 4 or 6).

It should be noted that establishing the tunnel in the PSN and establishing the PW are separate and independent tasks, even though, for example, they might both be done by using LDP.

Although PWE3 is used for PPVPN (described as follows), PWE3 is very important by itself, since it enables many heterogeneous technologies to utilize the many deployed PSNs, especially MPLS-based PSNs. For example, Ethernet

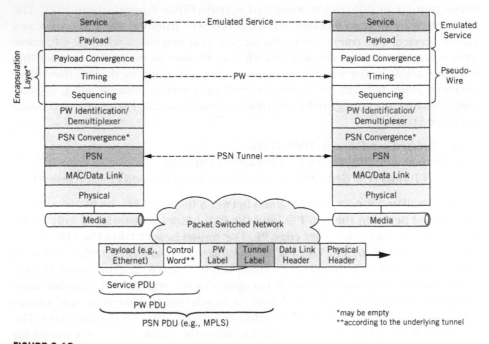

FIGURE 3.12

PWE3 protocol layering model with the resulting PDU

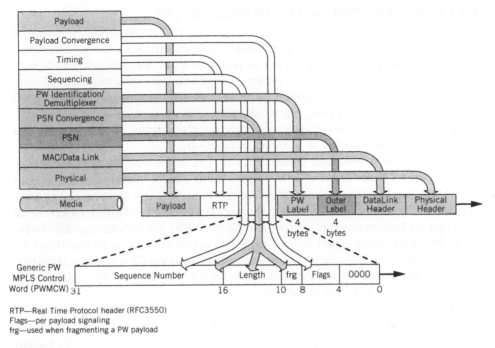

FIGURE 3.13

PWE3 over MPLS

services can be provided over these PSN networks using encapsulation of the Ethernet frames[22] [308]; PDH circuits (T1/E1 and T3/E3 TDM) can be emulated over PSNs [415]; PPP, HDLC and Frame-Relay can be transported over PSN [305, 310]; ATM services can use PSN infrastructure [304]; and even SONET/SDH services can be offered over PSN, using PWE3 [292].

By employing the PWE3 mechanism [302], Ethernet frames can use PSN networks as a backbone infrastructure, even for the IEEE802 solution for metro Ethernet described in the next subsection (i.e., PBBN, 802.1ah). In other words, PWE3 service can be used to encapsulate PBBN-tagged frames and carry them over MPLS networks. This enables providers to connect "islands" of PBBNs by MPLS links, and to offer "emulated" Ethernet services.

It should be noted that PWE3 is backward-compatible with the so-called "draft-martini" protocol [307], which describes the transport of Layer 2 frames over MPLS PSNs (discussed briefly in the MPLS section). "Draft-martini" is used in many deployed implementations, but it was superseded by the PWE3 specifications described previously [62, 292, 302, 304, 305, 308–310, 415].

---

[22]See also footnote 20. Two distinct types of Ethernet PW are defined: a raw mode (PW type 5) and an 802.1Q-tagged mode (PW type 4).

3.3.1.4 Provider Provisioned Virtual Private Network

From a service point of view, Provider Provisioned VPN (PPVPN) is the second generation evolution of enterprise networks. First, enterprises used to lease point-to-point links from service providers to build their private networks. VPN later allowed enterprises to exploit the public, shared networks infrastructure. The traditional VPN networks were based on layer 2 virtual circuits (e.g., Frame Relay, ATM), and offered lower cost, higher security, QoS support, and the use of well-known technologies. Their limitations were poor scalability of provisioning and management, as well as low IP network integration.

Virtual Private Networking is based on either network-wise, PE to PE tunnels, or on end-to-end, CE to CE tunnels (see Figure 3.14). PE-PE tunnels can use various IP/MPLS technologies (e.g., L2TP, IPSec, GRE, MPLS, etc.), and allow layer 2 or layer 3 connectivity and services to the customers by packet forwarding based on L2 or L3 header information, respectively. In CE–CE VPNs, the service provider (who owns the VPN) offers either layer 2 services (virtual circuits, as in traditional L2 VPNs), or layer 3 services (e.g., IP), by creating end-to-end, CE to CE link layer tunnels or IP/MPLS tunnels, respectively. These CE–CE tunnels may use independent PE–PE tunnels that the VPN maintains.

Provider Provisioned VPN [329] refers to a VPN provisioned and managed by a Service Provider (SP), or several SPs or operators, to which the customers are connected and from which they receive VPN services, as described in the following. According to IETF PPVPN, the entire networking model is based on tunneling (e.g., PWE3) in a Provider Provisioned PVN, as depicted in Figure 3.14. The entities in the reference model include:

- *Customer edge equipment*—May be any customer device; for example, a router or a host that has no VPN-specific functionality but to access the PPVPN services.
- *Provider edge equipment*—A device that supports one or more VPNs and interfaces with customers; for example, a router or a switch.

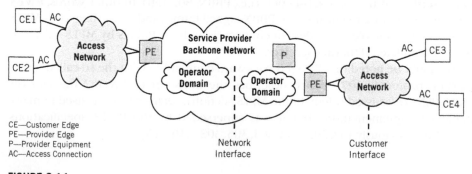

CE—Customer Edge
PE—Provider Edge
P—Provider Equipment
AC—Access Connection

FIGURE 3.14

IETF provider provisioned virtual private network reference model

■ *Provider equipment*—A provider device (usually a router) within a provider network that interconnects PE and other P equipment, and does not have any VPN functionality.

■ *Service provider network*—A network administered by a single service provider (operator), but which can use networking facilities of several SP networks.

■ *Access connection, or attachment circuit*—An isolated layer 2 link between a CE and a PE; for example, a dedicated physical circuit, a logical circuit (such as Frame-Relay, ATM, and LAN's MAC), or an IP tunnel (such as IPsec, or MPLS).

■ *Access network*—A network that provides the ACs. It may be a TDM network, a layer 2 network (e.g., Frame Relay, ATM, or Ethernet LAN), or even an IP network over which access is tunneled (e.g., MPLS).

The terminology of PPVPN is suggested in RFC4026 [19], which clarifies the various technologies and services used in PPVPNs. L2VPN services include Point-to-Point (P2P) Virtual Private Wire Service (VPWS), Point-to-MultiPoint (P2MP) Virtual Private LAN Service (VPLS), and P2MP IP-only Like Service (IPLS), while L3VPN [68, 71] offer services that interconnect equipment based on their IP addresses (e.g., IP-based VPN [148]). L1VPN services include connections set-up between CE devices [402], and are not treated in this context of carrier networks.

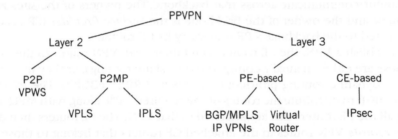

3.3.1.4.1 Layer 3 Provider Provisioned Virtual Private Network

L3VPN uses layer 3 information (e.g., headers and addresses), so the shared network infrastructure (VPN backbone) must be an IP or MPLS network. This implies that the SPs determine either at the PEs or at the CEs how to route the VPN traffic, and they control the QoS that the customer's traffic receives. It also means that the customers are restricted by the L3VPN to just one protocol (IP) that they can use.[23]

L3VPN can be as simple as a backbone L2 or L3 network that is owned by some SP, that interfaces with its customers by L3 (e.g., IP), and is used by

---

[23]As mentioned before, IP can be used to carry L2 frames, which relieves this limitation.

these customers as an "overlay network."[24] This solution is hardly sufficient, as scalability issues, conflicts in cross customers' address spaces, and massive administrative work in configuration and modifications prevent practical usage of this solution for carrier class infrastructure. A better solution is suggested by Border Gateway Protocol (BGP)/MPLS IP VPN defined in RFC4364[25] [368], using MPLS tunnels in the SP infrastructure network, which is isolated from the customers' VPNs that use it.

It should be also noted here that L3VPN, a successor of the original PPVPN concept, was tremendously successful in service providers' core networks, particularly since 2000. A very common infrastructure network is the BGP/MPLS IP VPN.

L3VPN technologies such as Virtual Router [259] (which did not catch on) and IPsec [249] (which is used extensively in secured scenarios) will not be discussed here. BGP/MPLS IP VPN [368], however, deserves description due to its massive deployment.

BGP/MPLS IP VPN—General Description   BGP/MPLS IP VPN (RFC4364) [368] networks are so named because CE routers[26] sends IP datagrams to the PE routers, and because BGP [46, 365] is used to distribute the VPN information, while MPLS is used to forward the VPN traffic across the provider's VPN backbone.

BGP/MPLS IP VPN terminology defines a VPN as a collection of *sites*, connected through a *backbone*, that have IP interconnectivity. Sites that have no VPN in common cannot communicate across that backbone. The owners of the sites are the *customers*, and the owner of the backbone is the *Service Provider* (SP). Each site is connected to the backbone's PE routers by its CE router.

To establish a VPN, the CE routers send their *local* VPN routes to the PE routers. These are either manually configured (i.e., static routing), or they use an IGP[27] or EGP[28] dynamic routing protocol (e.g., OSPF,[29] RIP,[30] eBGP[31]). The PE routers then use BGP to distribute the routes of the required VPN, along with MPLS labels, among all the PE routers that are attached to that VPN. The PE routers then distribute the *remote* VPN routes to the attached CE routers that belong to those VPNs (again, by using any dynamic routing protocol).

---

[24]An overlay network is built on top of an underlying network in such a way that the overlay-network's nodes represent one or more nodes of the underlying network, and these overlaying nodes are interconnected by virtual (logical) paths, which are composed of one or more physical links of the underlying network.

[25]RFC4364 [368] used to be known as RFC2547bis [11], which replaced the RFC2547 network [367].

[26]In BGP/MPLS IP VPN, any L2 CE (e.g., switch) is not regarded as part of the CE, rather it is considered as part of the Attachment Circuit (AC), between the SP VPN's PE routers and the customer's CE routers.

[27]Interior Gateway Protocols.

[28]Exterior Gateway Protocols.

[29]Open Shortest Path First.

[30]Routing Information Protocol.

[31]External Border Gateway Protocol (BGP connection between external peers).

The customer's data packets are encapsulated with the MPLS label that matches the route of the packet's destination in the customer's VPN. The resulting MPLS packet is again encapsulated for tunneling it across the service provider's backbone toward the egress (exit) PE router (e.g., by another level of MPLS or GRE encapsulation). By doing this, we have a two-layer label stack of MPLS, and the backbone routers are unaware of both the VPN they serve and its routes (as they are represented by the inner, bottom label).

The CE routers are peered only with the PE routers, not among themselves; that is, they do not exchange routing information with each other. This means that the CE does not manage the backbone or handle any inter-site routing. It further means that the created VPN is not an "overlay" network on top of the service provider's backbone network. Extended addressing mechanisms and unawareness of the core routers of the VPNs, allows participants of nonoverlapping VPNs to use whatever addressing scheme is required in each of the VPNs, even if the address spaces may overlap. This creates a very scalable and flexible solution for the customers and the service providers, as well as a very simple network for usage and configuration.

BGP/MPLS IP VPN—Internal Operation   There are two basic operational modes in BGP/MPLS IP VPN: the control plane flow of routing and forwarding information, and the data plane flow of traffic. Control plane operations can be further categorized into four main tasks, which are required to achieve BGP/MPLS IP VPN functionality:

- Separation and distribution of VPNs' forwarding tables.
- Extended addressing, for overlapping IP addresses in different VPNs.
- Distribution of routing information.
- Managing the MPLS forwarding mechanism.

Separation and Distribution   A PE router maintains several VPN Routing and Forwarding (VRF) tables, one of them being a "default forwarding table." ACs between the CEs and the PE (usually identified by the physical incoming PE ports) are mapped to the PE's VRFs. The destination IP of any packet coming through such an AC is compared with the associated VRF to determine its route to the egress (outlet) PE. Careful building and maintenance of the entries in these VRFs keeps the VPNs separated as required. Clear identification of the AC on which each packet enters the PE is mandatory for choosing the correct VRF, and sometimes it is even important to distinguish between "virtual ACs" (e.g., different incoming VLANs or classification by the source IP addresses), in order to choose the "right" VRF. A BGP/MPLS IP VPN example is shown in Figure 3.15.

PE_1 in Figure 3.15, for example, maintains two VRFs—one for each VPN it is attached to (VPN_A and VPN_B), each through a different AC. VRF_B contains all the IP addresses and their forwarding addresses (next-hop), as in any routing table, in order to allow routing in VPN_B. PE_1 distributes the routes of VPN_B by using BGP to all other PEs that are attached to VPN_B (just PE_3 in this case).

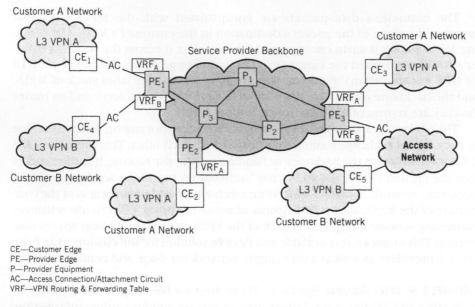

CE—Customer Edge
PE—Provider Edge
P—Provider Equipment
AC—Access Connection/Attachment Circuit
VRF—VPN Routing & Forwarding Table

FIGURE 3.15

BGP/MPLS IP VPN example

Extended Addressing The potential overlapping of IP address spaces (used by
different VPNs) might "confuse" BGP when distributing the VPN routes in the
SP backbone (since the same IP address can be used for different VPNs, and
the BGP might treat it as one address, neglecting the other VPNs). In order to
solve this, a separation between VPN address spaces is achieved by taking advantage of the BGP Multiprotocol Extensions [46], which extend IPv4 addressing
to multiple "address families." This enables generalized VPN-IPv4 addresses to be
advertised as a special address family by BGP Multiprotocol Extentions. Every
VPN-IPv4 address is 12-bytes long, composed of an 8-byte Route Distinguisher
(RD) (which is just a number), and ending with a 4-byte IPv4 address. The VPN-
IPv4 address is shown in Figure 3.16.

The Type field of the RD (which contains 0, 1, or 2) determines the length
and meaning of the two other fields of the RD; for example, Type 0 determines
that the administrator subfield is 2 bytes long, containing an Autonomous System Number (ASN), and the 4-byte Assigned Number subfield contains any
number assigned by the service provider to whom the ASN belongs. No two
VPNs have the same RD. BGP, however, disregards this structure, as it simply
compares address prefixes.

It should be emphasized again here that the VPN-IPv4 addresses are used only
in the control plane by BGP, only in the SP backbone, and only to disambiguate

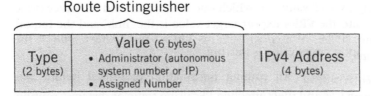

FIGURE 3.16

VPN-IPv4 address (from left to right)

IP addresses—not for constrained distribution of routing information (route filtering).

Distribution of Routing Information   IP routes from any given CE to its attached PE must be associated, in the PE configuration, with a particular RD. The IP routes that the PE "learns" from its attached CEs (as described before, by manual configuration or by any dynamic routing protocol) are inserted into the PE's VRFs. These routes are then translated to VPN-IPv4 routes (with the associated RDs) and routed by BGP to all other relevant PEs. At the receiving PEs, the VPN-IPv4 routes are converted back to IP routes, and inserted into the relevant VRF tables of the PEs (and possibly also sent to the attached CEs).

The question of how to decide which of the PEs are relevant is answered by using Route Target (RT) attributes, which are analogous to the Route Distinguisher (RD), and have a similar structure. These attributes define a set of sites, customers, or VPNs, and are assigned to VRFs and to distributed VPN-IPv4 routes. A VPN-IPv4 route, carrying an RT, is distributed to all PE routers by BGP, along with an MPLS label. BGP distribution inside the backbone can be fully meshed, or it can be done by the BGP reflectors concept, which saves BGP messages and assists in scalability (see this chapter's Appendix A).

At any rate, only those PE routers that possess VRFs associated with the transmitted RTs can use these routes to update their VRFs. The mechanism for updating is, on the one hand, to use a set of "*import route targets*" that the PE maintains for filtering the routes received from other PEs and potentially to "import" them into the VRF, and, on the other hand, to use a set of "*export route targets*" that the PE attaches to routes belonging to a site (a VPN). In some cases, customers are allowed to provide route targets to be used by the PEs, thus enabling the customers to have some control over the distribution of VPN routes.

For example, when a set of VPN sites and their corresponding VRFs are assigned with a unique RT value, which is both the import route and the export route and not used by any other VPN, a closed user group (i.e., one that no outside customer can communicate with), is created over the SP backbone. This becomes a fully meshed network (i.e., one in which each site communicates directly with any other). Another example might be a "hub and spoke" VPN network topology that is created by using two RT values, in which one is the "hub" and the other is the "spoke." At a hub site, the VRF's export target value is the "hub" and the import target value is the "spoke." At a spoke site, the VRF's export target value is the "spoke," and the import target value is the "hub."

Managing MPLS Forwarding   All routing information described before was exchanged between CEs and their attached PEs using static or dynamic routing, and among PEs across the SP's backbone using BGP.

When a PE advertises a VPN route (from its VRF, converted to a VPN-IPv4 address) using iBGP, it attaches an MPLS label and assigns its loopback address as the BGP next hop for the route. This attached MPLS label ("VPN route label" [368]) is used as the next-hop by this PE when it later receives data packets; for example, for forwarding the packets back to the correct CEs.[32]

Following this, LSPs are established by LDP between the PEs at the SP's VPN edges, across the backbone, and using another MPLS label ("tunnel label" [368]).[33] Several LSPs can be established between any two PEs for TE by specific LDP protocols, to enable various classes of QoS between those PEs. In non-MPLS SP backbones, the customer's packet carrying just the VPN route label is tunneled to the BGP next-hop using other tunneling techniques; for example, encapsulating MPLS in IP or GRE [431].

The result is that every customer's packet gets an inner (bottom) label (which is the VPN route label, for remote PE forwarding) and an outer (upper) label (which is the tunnel label, for the SP's VPN tunneling).

Data Plane Flow   Customers' packet forwarding is simple once the VPN routes are established, and the SP's backbone is only dealing with internal tunnels between PEs that connect the customers' VPN. In other words, each packet that arrives at the PE from some CE is compared (according to longest prefix, as described in Chapter 5) with the appropriate VRF (according to the AC the packet arrives through), and the matched entry is used to:

- Attach the VPN route label of the route that was advertised by the destination PE.
- Attach the tunnel label, according to the advertised "BGP next-hop" (the loopback address of the destination PE).
- Forward the packet on the interface port of the PE that is pointing to the right path.

---

[32]This is necessary, since in the data plane flow, packets do not carry VPN-IPv4 addresses. When the PE receives two packets that belong to different VPNs that have overlapping IP addresses, it will not be able to forward them to the correct CE unless there is some mechanism to distinguish between their VPN belonging. VPN-IPv4 addressing is required only for the control plane, for the BGP to advertise correctly the IP routes of different VPNs among all PEs.

[33]The reason for another, outer label is the requirement to separate the "VPN knowledge" from the backbone; that is, the PE–PE paths are maintained by the SP backbone, and are independent from any VPN modifications that are done in the CEs and PEs by the customers and the SP.

The MPLS labels and the "next hop" are first installed in the VRF when the PE receives the iBGP advertisements from the remote (destination) PE (and when it was advertised, it used VPN-IPv4 VPNs routes). The packet is then tunneled to the egress PE, using the outer, upper tunnel label. There, the packet is further forwarded to the appropriate CE, according to an inner, bottom VPN route label.

The SP's VPN P routers are simply used as Label Switching Routers (LSRs) in an MPLS network, completely unaware of any VPN. Only the PE routers (acting as LERs) are aware of VPNs, and then only those VPNs that are attached to them, thereby allowing scalability, flexibility, and ease of configuration and management.

3.3.1.4.2 Layer 2 Provider Provisioned Virtual Private Network

L2VPN uses Layer 2 information (e.g., headers and addresses) to separate between VPNs, and to interconnect users within VPNs. This means that any Layer 2 shared network infrastructure (e.g., ATM, Ethernet, or Frame-Relay) can be used as a VPN backbone, or can at least be interfaced. This grants the customers total control on their networks, including routing, QoS, prioritization, and bandwidth control. It further allows customers to use any protocols they want, provided they are encapsulated in Layer 2 frames that are forwarded by the PPVPN (according to the L2 addresses and information; for example, MAC, VPI/VCI). L2VPN uses the encapsulation mechanisms defined by the PWE3, and is targeted mainly toward defining three services, as noted previously: VPWS, VPLS, and IPLS. An example of an L2VPN network is shown in Figure 3.17.

Provider Edges are interconnected in the L2VPN by Pseudo-wires (PWs) that are carried in tunnels. These tunnels can be MPLS LSPs, Layer Two Tunneling Protocol (L2TP) [409], IPsec, MPLS-in-IP tunnels, and so on, and they all must be able to

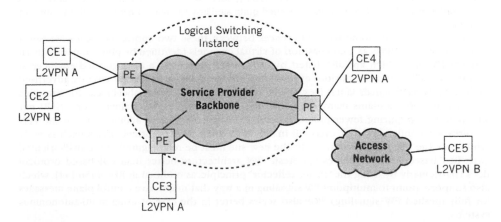

FIGURE 3.17

L2VPN network example

multiplex and demultiplex the PWs that use them. Specific auto-discovery[34] and signaling protocols (for set up and tear down) are used for the PWs. These Point-to-point PW auto-discovery and signaling protocols can be based on LDP [271, 309], L2TP, or BGP [264].[35] A short summary of both LDP and BGP protocols is given in Appendix A of this chapter.

It should be noted, however, that the VPN tunneling signaling protocols (that setup and maintain the tunnels in the VPN) are absolutely independent of the signaling protocols of the services offered over them (e.g., P2P PW setup for VPLS use, as described below, can use LDP, and the tunnels in the utilized MPLS VPN can also use LDP for setting up the tunnels).

Resiliency can be achieved in L2 PPVPN by assigning a provisioned protection PWs, and switching to these protection PWs when a fault is detected. (A more detailed discussion of Operation, Administration and Maintenance, OAM, in L2 PPVPN and in general, is provided in a following subsection and in [321]).

Virtual Private Wire Service   Virtual Private Wire Service [20] is a L2VPN service. It is also Metro Ethernet Forum's (MEF's) E-Line service type described in the following; that is, it is a point-to-point service, connecting two CEs. Frames received on an AC from a CE are forwarded by a PE's "VPSW forwarder" to a PW that connects this

---

[34]Provisioned configuration is a possibility; however, it might be an extensive management task, particularly in a fully meshed network with a large number of PEs. Protocol-based auto-configuration is based on auto-discovery protocols.

[35]Two L3 control plane protocols, BGP-based and LDP-based, are used by L2PVN. There is a dispute in the industry regarding these two approaches. In a nutshell, these approaches originated from either the PWE3 framework, which is based on draft-Martini and is LDP-oriented, or from the L3VPN framework—such as BGP/MPLS IP VPN (RFC4364 [368])—which is based on BGP and is referred to as draft-Kompella. Both approaches are used quite similarly to create a flat network by different routers.

There are, however, some basic differences between these two approaches. Luca Martini, a Cisco fellow, sees VPNs as a composition of virtual channels identified by point-to-point Virtual Circuits ID tags, which are distributed by LDP. Kireeti Kompella, a distinguished engineer at Juniper, sees VPNs as a composition of point-to-point and point-to-multipoint circuits, advertised by BGP. Since LDP signals labels for circuits, whereas BGP can signal label-blocks for sites, this means that LDP maintains many FECs (two per circuit), while BGP maintains one NLRI per site, therefore requiring fewer protocol states. It further means that adding a site to the VPN requires a reconfiguration of every PE in the VPN when the LDP-oriented approach is used, while only the PEs that are attached to the new site must be reconfigured in the BGP-oriented case. BGP-based protocol serves full mesh VPN architecture better than LDP-based protocol does (particularly by using the "router reflector" principle, as outlined in RFC 4456 [44], which also supports point-to-multipoint PW signaling in a way that might save control-plane messages for fully meshed PW signaling). BGP also scales better in the case of using multi-autonomous systems.

On the other hand, the BGP-oriented approach is more complex than the very simple protocol offered by the LDP-oriented approach. The LDP-oriented approach saves the advertised reachability information distributed via the control plane as BGP requires, and converges more quickly than BGP using the standard data plane bridge learning, and in a more secure way.

ingress PE to the egress PE in the VPN, and then the frames are forwarded to the AC that is connected to the other CE.

Virtual Private LAN Service   Virtual Private LAN Service[36] [20, 37, 264, 271] is a L2VPN service, and is a MEF's Ethernet LAN (E-LAN) service type (described in the following). This service emulates the full functionality of a LAN across the relevant PEs and CEs, and enables the operation of several remote LAN segments as if it they are one single LAN. A VPLS instance supports one emulated LAN; that is, a single LAN or VLAN, or multiple VLANs, and each VPLS must be assigned a globally unique identifier.

VPLS can thus be perceived as being used as a bridged LAN service over networks (some of which are MPLS/IP), as the LAN itself, the underlying MPLS/IP network to which other LANs are attached, or as LAN and VLAN emulation [321].

Frames received on an AC from a CE are first handled by a bridge module at the PE, and then are forwarded by a "VPLS forwarder" module to the required PEs through PWs, using a mapping between MAC addresses and PWs. Unicast frames are forwarded on a single PW to a single PE, while broadcast frames and frames with unknown destination MAC addresses are broadcasted to all PEs in the emulated LAN. At the egress PE, the VPLS forwarder learns the frames' source MAC address for the future forwarding decisions, and the frames are forwarded to the AC that is connected to the other CE that is part of the same LAN (see Figure 3.18).

The PEs' "VPLS forwarders" and the PWs among them compose the emulated LANs. An emulated LAN instance consists of a set of forwarder modules (one per PE per VPLS instance) interconnected by the PWs, that goes through PSN tunnels,

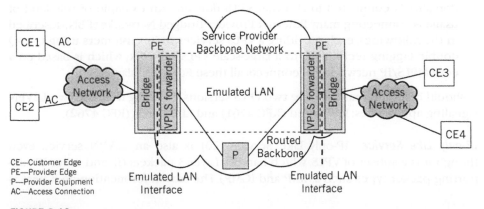

CE—Customer Edge
PE—Provider Edge
P—Provider Equipment
AC—Access Connection

FIGURE 3.18

VPLS and E-LAN reference model

---

[36]Other names of VPLS are Transparent LAN Service (TLS) and Private Switched Network Service.

over a routed backbone. This set of PWs, interconnecting the PEs, can form a fully meshed topology or a tree structure topology of an "overlay network" (or combinations of the two topologies).

A VPLS PE behaves like any bridge or switched bridge; that is, it learns and ages out MAC addresses and makes forwarding decisions accordingly.[37] This, plus the requirement of potentially fully meshed PWs connecting many PEs in the PPVPN, creates a scalability issue.

In order to solve the scalability issue resulting from the fully meshed PWs topology and the massive PW signaling involved, Hierarchical VPLS (H-VPLS) is used [271]. There are two basic usage scenarios that can use H-VPLS:

Hierarchical connectivity A two-tier H-VPLS, based on a hub and spoke model is used for cases of hierarchical connectivity. A hub (core) made of PW interconnects the central PEs in the VPN (creating a basic, full mesh VPLS). Access devices such as aggregation PEs or Multi-Tenant Units (MTUs) are connected by spoke PWs to the central PEs. The spoke PWs use encapsulation techniques (or Q-in-Q logical interfaces, described in the next part); therefore, each access device can create a single, tunneled PW to the single PE it is attached to in the core and not have to create many PW connections to each of the PEs in the VPN.

Multidomain VPLS H-VPLS (two tier or more) can be used either to divide a large, single domain, or to create a large scale VPLS domain that spans over many VPLS domains that have sparse interconnections between themselves. The principle here is to connect two fully meshed VPLS domains using a single tunnel between VPLS "border" devices. All domains are interconnected by a higher-level hierarchy of fully meshed tunnels; that is, each of the connected VPLS domains is connected to all other VPLS domains. An example of this kind of usage is connecting many Ethernet Provider Bridged Networks (PBN, described in the following), each providing a VPLS services to its customers using Q-in-Q double tagging technology, to a large-scale VPLS network, which is based on a core MPLS/IP network that connects all these Ethernet islands.

It should be noted that there are two VPLS versions, resulting from the two L2VPN signaling approaches: BGP-based (RFC 4761) and LDP-based (RFC 4762).

IP-only Like Service IP-only Like Service [20] is also an L2VPN service even though it is a subset of VPLS, which handles Layer 3 packets (IP and other IP supporting packet types such as ARP and ICMP). The reason for mentioning it here is

---

[37]Usually, a spanning tree topology is not required between the PEs, since their connections are manually configured and do not contain loops in the VPN. However, if "backdoor" connections are allowed—that is, CEs that are connected to two PEs (in a "multiple-homing" arrangement) and that run bridging functionalities among them—then STP must be done, as with any other bridges.

that IPLS may use other PE platforms that are simpler than the VPLS-enabled PEs, and when it does, it might become a full-fledged service on its own, including its own architecture and protocols.

3.3.2 Carrier Class Ethernet

Carrier Class Ethernet (CEN) refers mainly to standardized services that offer scalability, reliability (i.e., the resiliency to protect customers against network failures), CoS (i.e., assurance of a service level agreed on by the carrier and its customers), and management capabilities. Although CEN mainly addresses Metro Ethernet Networks (MENs), it is not confined just to metro networks, and can be used also in access and core networks. There are four standardization players in CEN that try to coordinate standards and services for the CEN: The MEF, the IEEE802 LAN/MAN Standards Committee, the International Telecommunication Union (ITU)—Telecommunication Standardization Sector (ITU-T), and the IETF.

CEN are not just Ethernet: some use of MPLS or IP routing might be included in various versions, approaches, or implementations of CENs. Moreover, there are many terminologies that overlap in meaning, resulting from the various approaches to CENs (e.g., [157]). The main theme of CEN is providing and supporting "services"; that is, applications at higher layers than the CENs, which relate to various technologies. As usual with networking, many of these services are "virtual" or use virtual facilities. We shall try to put some order in all this virtualization by transforming it into actual, concrete terms and services.

Ethernet, described in the previous chapter, and referred to as "Enterprise Ethernet," is excellent for enterprise LANs and WANs, but when it comes to "public," multi-enterprise and disjoined Ethernet services, Enterprise Ethernet is not sufficient. To begin with, maintaining bridging system for millions of users is impractical, if not impossible, and having 4094 VLANs is far from enough for carrier class service. Then, the Ethernet infrastructure should have the service assurance of carrier class services, such as 99.95 to 99.999% availability and sub-50 ms protection (i.e., in case of a connection fault, finding and using an alternative connection within 50 ms).

We start with the MEF approach, since it summarizes and describes the services and the architecture of the MEN, rather than its technology. As a matter of fact, MEF focuses on the usage perspective of the network, and does not provide details of implementation.

3.3.2.1 Metro Ethernet Forum Approach

Metro Ethernet Forum (MEF) models the network in three service layers (Transport, Ethernet, and Application) and three planes (data, control, and management). Although MEF refers to access networks, metro networks, and core networks, the standards of MEF relate to MENs.

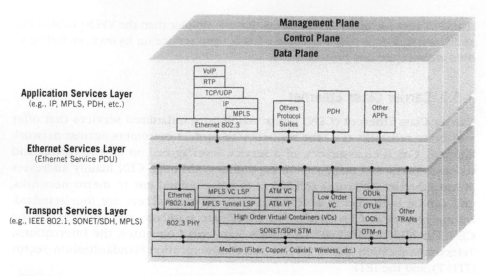

FIGURE 3.19

MEF network model with sample decomposition into layer networks and protocol stacks

The Ethernet layer (ETH layer) is the main focus of the MEF approach, and is responsible for the Ethernet-MAC oriented connectivity and the delivery of Ethernet frames throughout the network. The transport layer (TRAN layer) supports the connectivity among ETH layer elements. Various technologies may be used to support the transport requirements for the ETH layer. The application layer (APP layer) supports various applications that are carried on the ETH layer, across the network. The MEF network model, with sample decomposition into layer networks and protocol stacks, is depicted in Figure 3.19.

As we discussed in the previous chapter, the data plane defines the means of transporting information across the network. The control plane defines the means for the Customer (subscriber) and the Service Provider (the network) to communicate in order to make use of the data plane; that is, connection, signaling and control. The management plane defines the means to configure and monitor the operation of the data and control planes; that is, service provisioning, static service discovery, service load balancing, service protection and restoration, and Operations, Administration, and Maintenance (OAM) services. MEF assumes nothing about how to implement the network, or its details, and it can be either a single switch or multiple networks of many different technologies.

Each customer is attached to the network at the CE via User-to-Network Interface (UNI), which must be standard IEEE802.3 Ethernet PHY and MAC. Providers are connected to the network at the PE by PE equipment. NEs, autonomous networks (and service providers) are connected in the network by using Network-to-Network Interfaces (NNI). A schematic diagram of the MEF network is shown in Figure 3.20.

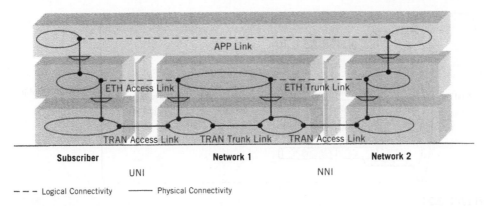

FIGURE 3.20

MEF network model; reproduced with permission of the Metro Ethernet Forum

The UNI is a fundamental concept through which MEF actually focuses on the user's perspective of the CEN. Three UNI types are defined, with various functionalities: in UNI type 1, the CE equipment is manually configured for network connections, where type 1.1 is used for nonmultiplexed services and type 1.2 is used for multiplexed services. In UNI type 2, the network may automatically provision, configure, and distribute a network connection's information to the CE, and in UNI type 3, the CE equipment may request and negotiate connections in the network.

The second fundamental concept is the *Ethernet-Virtual-Connection* (*EVC*) in the Ethernet layer. EVCs are established between two (point to point) UNIs or more (multipoint to multipoint, or rooted-multipoint).[38] The EVC actually defines the association between UNIs, and Ethernet frames are exchanged only among these UNIs. EVCs are used for end-to-end subscriber services across one or more service providers' networks, and are represented by instances of the Ethernet Services Layer of the MEF model. Two types of service attributes are used for the connections—those that apply to UNIs and those that apply to EVCs.

MEF maintains IEEE802.1 bridging concepts and IEEE802.3 Ethernet frames' format and functionality, including addressing and dedicated addresses,[39] and Virtual

---

[38]Rooted-Multipoint EVC was defined in MEF 10.1 standard [317], to allow point-to-multipoint connections; that is, one or more Root UNIs to or from many Leaf UNIs.

[39]By dedicated addresses we mean bridge and MRP/GARP blocks of addresses (described before, i.e., Bridge Group Address, 01-80-C2-00-00-00 through 01-80-C2-00-00-0F, and 01-80-C2-00-00-20 through 01-80-C2-00-00-2F for MRP/GARP applications), as well as the "All LANs Bridge Management Group Address" (01-80-C2-00-00-10).

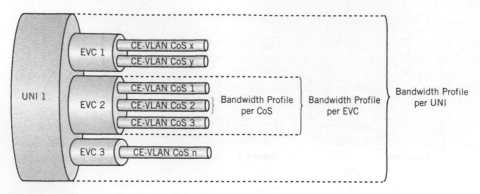

FIGURE 3.21

Ingress bandwidth profiles per UNI, EVC, or CoS identifiers

LANS (VLANs, where the IEEE802.1Q VLAN tag is termed by MEF as CE-VLAN Tag). VLANs are supported end-to-end by assigning VLANs to the EVCs. Several VLANs may be mapped into one EVC (termed *bundling*[40] by MEF), if they share the same path and service attributes. In addition, Layer 2 control frames (e.g., BPDUs frames) can be tunneled through the network by using a dedicated EVC, which can filter some of the control frames as the service provider determines. UNIs identifications (UNI-ID) and EVCs identifications (EVC-IDs), as well as the UNI-EVC-ID (concatenated UNI-ID and EVC-ID strings that identify an EVC at the UNI) are used solely for control and management purposes, and are not used anywhere in the frames for transport purposes.

Each UNI, its EVCs, and their assigned VLANs can have a predefined ingress bandwidth profile, as depicted in Figure 3.21, and a predefined egress bandwidth profile per UNI or EVC, as shown in Figure 3.22 for EVC.

MEF defines Ethernet Local Management Interface (E-LMI) protocol based on the ITU-T Q.933 standard, and uses it to communicate the EVC status (along all its assignments, service attributes, etc.) to the CE [318]. Two basic service types are defined by MEF: *Ethernet Line* (*E-Line*) service and *Ethernet LAN* (*E-LAN*) service.

- E-Line is point-to-point EVC, that can be used for Ethernet Private Line (EPL, for connecting *nonmultiplexed* UNI type 1.1, using the same EVC;[41] for example, for replacing a private TDM line), or Ethernet Virtual Private Line (EVPL, or virtual wire service, for connecting *multiplexed* UNI type 1.2,

---

[40]All to One Bundling means that all VLANs are mapped onto the same EVC. This is defined for all UNIs that use this EVC.
[41]EPL is expected to be transparent, and to have a high quality of service parameters; that is, low Frame Delay, low Frame Delay Variation, and low Frame Loss Ratio. EPL uses a dedicated UNI (physical interface), and because of its transparency, all CE-VLAN ID are mapped to a single EVC at the UNI.

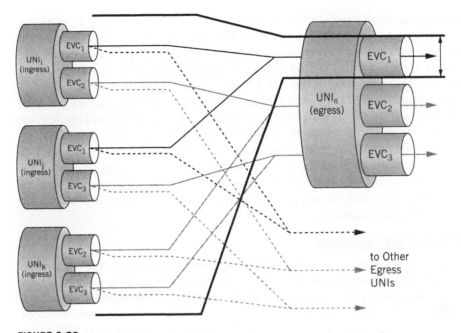

FIGURE 3.22

Egress bandwidth profile per EVC; reproduced with permission of the Metro Ethernet Forum

using several EVCs;[42] for example, two routers at the enterprise network). E-Line can also be used for Ethernet Internet access, or upper layer
point-to-point services (for example, Layer 3 Virtual Private Network; that
is, IP-VPN). EVPL is used for MP2P (multipoint-to-point), or more correctly,
multiple P2P.

- E-LAN is a multipoint-to-multipoint (MP2MP) EVC, creating in effect an
Ethernet LAN over the carrier Ethernet network. E-LAN can be used for
multipoint Layer 2 Virtual Private Network (VPN), Transparent LAN service
(TLS), or multicast networks (point-to-multipoint [P2MP], for video and IPTV,
for example).

Since the network should serve telecommunications applications, there are three
major service requirements that are critical for operator adoption: (a) supporting
TDM traffic (e.g., telephone services), (b) supporting protection, and (c) supporting
OAM. MEF's OAM model is described in a subsequent subsection that describes
carrier Ethernet OAM support.

---

[42]EVPL does not provide full transparency like EPL does. Layer 2 Control Protocol (L2CP) frames (e.g.,
BPDU containing STP/RSTP/MSTP packets, authentication, MRP/GARP, etc.) should be discarded in
EVPL, whereas most of theses control frames must (some should) be tunneled by EPL (with one
exception, the IEEE802.3x PAUSE frame, that should be discarded in both cases).

Circuit Emulation Service (CES) is used to "tunnel" TDM traffic (e.g., PDH and SONET/SDH signals, from DS0 to OC-12/STM-4). CES should be transparent to the TDM source and destination equipment, even though it uses a packet network. CES runs on E-Line service that should provide a certain (strict) level of service quality to ensure proper operation of the CES. CES uses the network as a "virtual wire."

Circuit Emulation Service over Ethernet (CESoETH) is used to provide a TDM Line (T-Line) service (e.g., least line replacement), or a TDM Access Line Service (TALS) (e.g., access to a remote Public Switched Telephone Network, PSTN).

MEF defines sub-50 ms protection in hop-by-hop and end-to-end schemes. Two protection types are defined: the $1 + 1$ protection type (replicating resources and traffic, and selecting one copy of the frames at the protection merge point for further forwarding), and the m:n protection type (using m protection resources to back up n working resources).

3.3.2.2 IEEE 802 Approach

As was described in our discussion of the Enterprise Ethernet, the LAN grew into WAN by combining several LANs in Layer 2 bridges and switches, to have a large enterprise network. However, using such a network in the "public" domain is impractical, particularly using it instead of and for telecommunications services. It is important to note, however, that the Ethernet-based network a la IEEE802 is based on several Ethernet frame mutations.

The Ethernet IEEE802.3 frame starts at the CE enterprise (or home) equipment, which is connected through the IEE802.3ah (now part of the IEEE802.3) access network (called Ethernet First Mile, EFM, or Ethernet Passive Optical Network, EPON). The frame then goes through the access interface (e.g., OLT) into the metro network, which is defined by IEEE802.1ad (now part of the IEEE802.1Q) technology, the PBN. From the metro network, the frame continues to the core network, defined by IEEE802.1ah technology (which is still under work), the PBBN.

The general topology of the CEN, described before, is depicted in Figure 3.23. In the next subsection, each of these network's technologies will be described briefly. (Access networks, both the xDSL with its DSLAM interface, and the IEEE802.3ah, are described in the next chapter).

3.3.2.2.1 Provider Bridged Network

The natural step in expanding Ethernet into the public domain was to allow service providers to use their own bridged network to interconnect their customers' Ethernets (enterprises as well as SOHO and residential customers). The major issue here is to allow each customer to manage its own networks, including its virtual LANs (VLANs), and to create an Ethernet cloud that is transparent to the internal MAC services and internal VLAN management of each of these customers' networks. Furthermore, it must segregate each customer's MAC and VLAN services and management from those of the other customers, and from those of the provider's Ethernet network. This, plus the fact that only 4094 VLANs are allowed in the entire Ethernet space (according to IEEE802.1Q), which is nowhere near

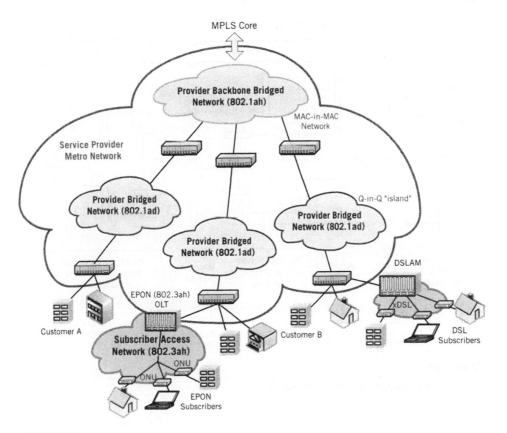

FIGURE 3.23

IEEE802 metro Ethernet

enough to support hundreds or more customers, suggested the usage of another hierarchy of virtual-LANs. The result was the IEEE802.1ad standard, in which a Service VLAN (S-VLAN) tag (S-Tag) is added to the customer VLAN (C-VLAN) tag (C-Tag) in the Ethernet frames, in a way termed Q-in-Q, double tagging, or VLAN stacking. Using this double tagging, a PBN is constructed, comprised of Provider Bridges (S-VLAN bridges and Provider Edge Bridges) and attached LANs, under the administrative control of a single service provider [179].

The PBN behaves internally as a "LAN" by itself; that is, maintains its spanning tree, its VLANs, and all MAC layer services. Frame forwarding is based on MAC addresses and the S-Tag, exactly in the same way as any 802.1Q virtual bridged network. The S-Tag is assigned by the provider at the ingress edge bridges (the interface equipment in Figure 3.24), and is attached to the customer frame before the C-tag (see Figure 3.25). Every incoming frame, from every customer, must go through a "customer" port of a Provider Bridge (PB), or through a Provider Edge Bridge, and must be pre-appended by the assigned S-Tag. Upon departing, the egress PB strips

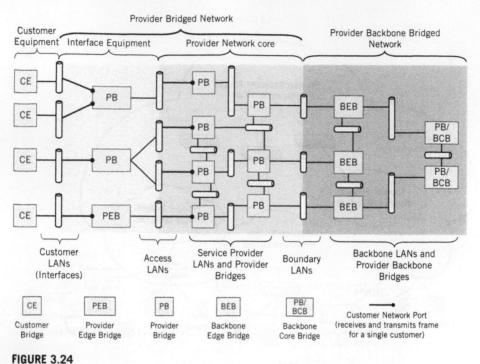

FIGURE 3.24

Basic provider network model and interfaces

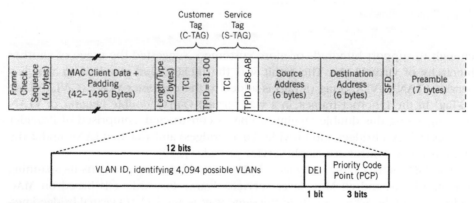

SFD—Start Frame Delimiter (1 byte)
S-Tag TPID—Service Tag Identifier, 2 bytes, always 0x88-A8
C-Tag TPID—Customer Tag Identifier, 2 bytes, always 0x81-00
TCI—Tag Control Information, 2 bytes
DEI—Drop Eligible Indicator

FIGURE 3.25

PBN (802.1ad, Q-in-Q) frame format (from right to left)

off the S-Tag, leaving the customer frame as it was in the PBN ingress point, and forwarding the frame to the customer VLAN network.

Roughly speaking, in the PBN, each VLAN represents a customer. This is important to understand, as in the next subsection we describe a higher level of hierarchy in which the VLANs have different meanings.

In addition, in order to ensure that the customers' VLANs are indeed separated from the PBN and cannot impact the PBN, any customer frame that arrives at a PBN ingress port and is addressed to any of the common bridge addresses (the Bridge Group Address or any MRP/GARP application address) must be conveyed transparently to the egress port of the PBN, to the destination LAN, and it cannot be used by any of the PBs. A new PB Group Address (01-80-C2-00-00-08) and a new PB MRP/GARP VLAN Registration Protocol (MVRP/GVRP) address (01-80-C2-00-00-0D) have been defined to allow independent PBN MAC service operations.

The Provider Bridged VLAN thus supports two tags, the Customer VLAN tag (C-Tag) for a customer's internal networks, and the S-Tag for the provider VLAN, indicating the customer ID. The S-Tag is used to identify the customer's traffic and to separate it from other customers and the internal provider bridged "LAN." It means, by the way, that up to 4094 customers can be supported by any PBN, hence the need for an additional hierarchy level, described in the following subsection. Moreover, although the PBN can reduce the requirement for learning end station MAC addresses, every switch in the PBN must still maintain a huge MAC address learning-table, which leads to another scalability issue.

3.3.2.2.2 Provider Backbone Bridged Network

Adopting the hierarchy principle and adding another, higher level, solves the limitations of the PBNs mentioned previously. However, the additional hierarchy is achieved according to a slightly different mechanism, which is more aggressive in terms of the bridging concept. The IEEE802.1ah proposed mechanism[43] [181] tunnels each service provider's traffic (frames) throughout the PBBN in backbone VLANs (B-VLAN). In addition, and more importantly, this mechanism assigns a specific Backbone MAC (B-MAC) address to each Point of Presence (POP) in the provider network (the interface point between the PBN and the PBBN). This allows PBBN to see each and every provider network (PBN) separately as one entity, to build a PBBN spanning tree, and to relate to the PBBN internally as a "regular" bridged LAN. To this end, the PBN's frame is encapsulated by a higher layer of Ethernet frame, or it can be regarded as if another Ethernet shim is added to the PBN's frame, making the customer and the service provider Ethernet frames just payloads in the PBBN's Ethernet.

The result is IEEE802.1ah, in which Backbone MAC (B-MAC) addresses are added (source and destination), as well as a Backbone VLAN tag (B-Tag) and a service instance tag (I-Tag). This procedure is called MAC-in-MAC, or tunneled MAC.

---

[43]As of the beginning of 2008, the standard is still under discussion; thus, we are still using principles, terms, and quantities from various earlier drafts.

To illustrate the difference in the hierarchy models, we can think of 802.1Q enterprise Virtual LANs and subsequently 802.1ad provider VLANs as mechanisms that were used just to partition the forwarding plane of the common, single Ethernet layer network. The 802.1ah divides the Ethernet layer into two very similar Ethernet (MAC) layers, each partitioned into virtual networks.

B-Tag is identical in its format and its tag identifier to S-Tag, since the Backbone Core Bridges (BCBs) of the PBBN are essentially the Providers Bridges (PBs) described previously; hence, B-Tag has a role similar to that of S-Tag in PBN networks. The service instance tag (I-Tag) includes a service instance identifier (I-SID), 24 bits in length, which is used to extend the limited VLAN space the PBNs have, which might be insufficient for the service providers in order to support all their customers. The resulting frame may be of various formats, as it is used by various components and interfaces in the PBBN bridges, sometimes without some fields or tags. The full format is shown in Figure 3.26.

As far as the PBBN is concerned, all the devices and networks supported by the PBN, including their spanning trees, MAC services, MAC addresses, VLANs, and so on can be ignored (or the relevant PDUs are transparently transported over the PBBN). This can save the requirement of learning all the MAC addresses of all the devices as well as saving the requirement of maintaining a huge MAC address table. As for the provider networks and customers, they can regard the PBBN as one internal, bridgeless LAN connecting their bridges; hence, the VLANs they use, MAC addresses, spanning trees, MAC services and everything these services do, is transparent to the PBBN and isolated from each other.

Another consequence is that instead of focusing on creating a PBBN Ethernet LAN that is topologically loop-free, a carrier-class-centric infrastructure, controlled by provisioning and control plane options, can be deployed. This is done by letting the

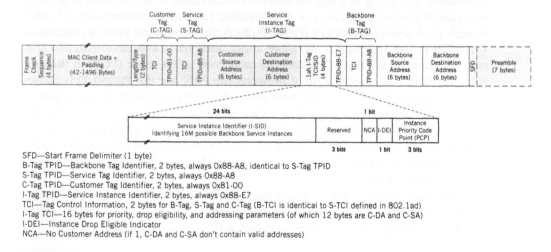

SFD—Start Frame Delimiter (1 byte)
B-Tag TPID—Backbone Tag Identifier, 2 bytes, always 0x88-A8, identical to S-Tag TPID
S-Tag TPID—Service Tag Identifier, 2 bytes, always 0x88-A8
C-Tag TPID—Customer Tag Identifier, 2 bytes, always 0x81-00
I-Tag TPID—Service Instance Identifier, 2 bytes, always 0x88-E7
TCI—Tag Control Information, 2 bytes for B-Tag, S-Tag and C-Tag (B-TCI is identical to S-TCI defined in 802.1ad)
I-Tag TCI—16 bytes for priority, drop eligibility, and addressing parameters (of which 12 bytes are C-DA and C-SA)
I-DEI—Instance Drop Eligible Indicator
NCA—No Customer Address (if 1, C-DA and C-SA don't contain valid addresses)

FIGURE 3.26

PBBN (802.1ah, MACinMAC) frame format (from right to left)

PBBN bridges independently learn MAC addresses coupled with the PBBN VLANs, using various control plane and management utilities, rather than the common flooding mechanism (the way spanning trees and MAC address learning are done). Then, forwarding frames using the created paths, or tunnels, can support resilience and traffic engineered infrastructure (this will be developed further in the next subsection).

Internally, PBBN works exactly like PBN; that is, its core bridges (BCB) are Provider Bridges. The main difference lies in the Backbone Edge Bridges (BEB), which are of two types (plus a hybrid one): The I-Type BEB encapsulates customer's frames and assigns each to a backbone service instance, with both a backbone destination address and a backbone source address. The B-Type BEB relays encapsulated frames to other BEBs. The PBBN network model is shown in Figure 3.27.

The service instance tag, I-Tag, is used for a new type of tagged frame introduced by PBBN which is not a VLAN. The service instance tagged frame is identified

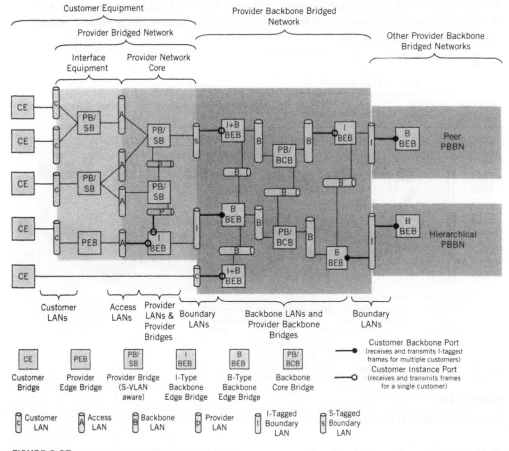

FIGURE 3.27

Basic provider backbone bridged network model and interfaces by the Service Instance Identifier (I-SID), which is contained in the I-Tag (along with the customer source and destination MAC addresses). This service instance is extended service, identifying the Provider Bridges' Service VLANs (S-VLANs)[44] and the customers' LANs and customers' VLANs (C-VLANs) that are interfacing the BEB. The S-VLANs and their Service VLAN Identifiers (S-VIDs) are local to the PBNs, while the service instance domains and their I-SIDs are local to the PBBN and are defined by the PBBN operator. In other words, in a given PBBN, the I-SID uniquely identifies a set of BEBs and their interconnections that support the attached service providers' S-VLANs and customers' LANs and C-VLANs. This creates a separate virtual medium for transporting the frames of the attached service provider S-VLANs and customers' LANs and C-VLANs. Several such virtual media (identified by I-SID) may be used by each BEB for each combination of B-MAC address and VLAN identifier. The BEB uses a provisioned service instance table that maps I-SIDs to B-VIDs, to select a B-VLAN that will carry the frames in the PBBN. A basic scheme of the relations among customer VLANs (C-VLANs), service VLANs (S-VLANs), Service Instances, and B-VLANs is given in Figure 3.28 for point-to-point connections of these VLANs.

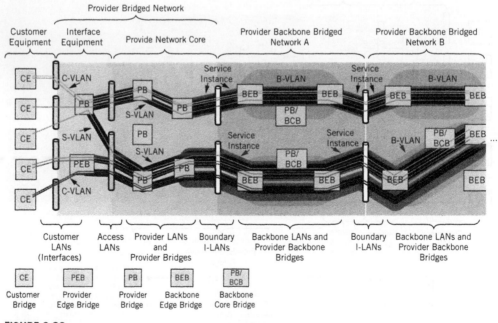

FIGURE 3.28

Relations between VLANs and service instances for P2P links

---

[44]It can be either a one-to-one mapping of Service VLAN Identifier (S-VID) to I-SID, or bundling all S-VIDs to I-SID.

To conclude, it is worth noting that although PBBN Ethernet frames are meant to transverse in Ethernet networks, the IETF defined the PWE3 mechanism, described in the previous subsection, according to which these frames can also use PSN networks as backbone infrastructure [302]. In other words, PWE3 service is used to encapsulate PBBN tagged frames and carry them over MPLS networks. This will enable providers to connect "islands" of PBBNs by MPLS links, and offer "emulated" Ethernet services.

However, some scalability issues still exist in the PBBN. BEBs have to learn the customer MAC (C-MAC) and the B-MAC addresses. C-MAC address learning is required, for example, for encapsulation and de-encapsulation of service frames. This encapsulation and de-encapsulation is done by BEBs by using the connection identifiers that are stored in the bridges' filtering database and correlating C-MACs to B-MACs addresses. The RSTP/MSTP also limits the size/span of PBBN (due to overhead required), and the use of ST reduces the available links to be used (reduce network utilization). PBB-TE emerged to address limitations related to scalability and reliability of PBBN. PBB-TE may be deployed in place of, or in parallel with, PBBN.

3.3.2.2.3 Provider Backbone Bridges—Traffic Engineering

As discussed in the preceding subsection, using PBBN (defined by IEEE802.1ah) for transport networks makes Ethernet not as simple as it is perceived to be. Moreover, PBBN does not provide answers to all the requirements of transport networks that are already solved by the alternative telecom networks (e.g., SONET/SDH) or MPLS.

One such example is the deficiency in link utilization provided by the Spanning Tree Protocol (STP) or the Multiple STP (MSTP) used in PBBN (i.e., STP can result in unused links when it is just targeted at preventing loops). In order to increase utilization of the backbone links that the regular STP provides, a better scheme of Shortest Path Bridging (SPB) is now being standardized by IEEE802. The principle (under work in IEEE802.1aq) assumes a rooted shortest path tree from each source toward every destination.

A somewhat alternative approach to Ethernet-based transport networks based on PBBN (as well as on PBN and VLAN networking, of course) is proposed by Provider Backbone Bridges—Traffic Engineering (PBB-TE,[45] or IEEE802.1Qay [375]). This approach mainly provides a simpler alternative for path creation.

In PBBN, customers' interconnections are made essentially in a two-layer Ethernet model, where C-MAC address learning and spanning trees are done in the first layer (802.1q and 802.1ad), and backbone paths are assigned in the second layer (802.1ah) by the operators of the PBBNs. PBBN further suggests using MSTP and MAC learning to forward the backboned frames between the BEBs, including flooding in cases of unknown destinations, correlating C-MAC

---

[45]PBB-TE used to be called Provider Backbone Transport (PBT).

addresses and B-MAC addresses, and mapping of S-VLANs and customers' LANs and C-VLANs into service instance domains.

PBB-TE describes an additional way for the provider to assign and deploy backbone paths in BPPN (i.e., to determine B-MAC addresses and VLANs that create the paths in the PBBN), and then to use these backbone paths to forward frames. PBB-TE essentially replaces the Multiple Spanning Tree Protocols (MSTP) control plane with a provisioned management plane or control plane that populates the filtering tables of the 802.1ad and 802.1ah bridges with static entries, creating point-to-point unidirectional Ethernet Switched Paths (ESPs).

PBB-TE identifies a method for splitting the B-VLANs between provisioned control and MSTP control, provides extensions to Connectivity Fault Management (CFM) that supports Continuous Check and Loopback protocols on the provisioned paths, and provides extensions to PBBN (802.1ah) for one-to-one protection switching [14, 55, 375]. To accomplish that, the active topology of the networks are not exclusively controlled by Spanning Tree Protocols (STPs) or Shortest Path Bridging (SPB) agents, but by some external agents that set up active topology using a specific Multiple Spanning Tree Instance ID (MSTID, which contains 0xFFE). Forwarding does not imply learning any more, when forwarding is controlled by the PBB-TE external agent, and frames with unknown destination addresses are discarded and not flooded. The forwarding decision is based only on preassigned forwarding tables, set up across PBBN bridges that are on the PBB-TE paths (ESPs), where these tables contain entries that are composed of backbone addresses and B-VLANs. This is further described in the following.

First, it should be noted that PBB-TE uses VLANs in a very different sense than regular bridged networks do; instead of a global meaning of VLANs, B-VLANs in PBB-TE have just a local meaning, assigned by PBB-TE.

PBB-TE creates trees that are rooted in the backbone destination MAC address (B-DA); these routed trees are not spanning trees, as they connect only the required backbone source MAC address, B-SA. ESP is a path on a tree identified by a selected B-VLAN (by its B-VID) to a destination B-DA. The number of trees that access any B-DA is the number of B-VIDs assigned to PBB-TE (up to 2^{12}). Since every tree is identified by the tuple B-DA and B-VID, and B-VID can be reused for each B-DA, the total number of trees that can exist and be managed by PBB-TE is the number of B-DA multiplied by the assigned number of B-VIDs. The result is that a unique tree can be identified by a concatenation of 46 bits of the B-MAC (allowing for the multicast and local reserved bits in the MAC space) and all 12 bits of the B-VID. In other words, each B-DA termination can sink 2^{12} different routing trees (which is far more than required, since it is not necessary to have so many alternate paths to a single destination), and potentially about 2^{58} ESPs can be defined. The primary application of the resulting alternate paths is to allow a protection path in case of failure in the working path. Several such alternate paths can be reserved for various scenarios of network failures. All the protection and alternate paths are pre-calculated, and the resulting forwarding tables' entries are configured into the PBB-TE bridges' forwarding tables that are relevant on these paths.

For each B-SA that has an ESP to B-DA via one of these trees, PBB-TE creates a reverse path that does not necessarily have the same B-VID, and uses it for CFM to monitor the ESP (described in subsection 3.3.2.3). Once faults are detected (e.g., loss of Continuity Check) and forwarded to the B-SA, this triggers a swap of the faulty B-VID with the already reserved alternate B-VID in the frames' MAC (leaving the B-MAC intact), and the subsequent frames use the alternate, protection path immediately (it can be in few milliseconds, less than the required 50 ms).

An example that clarifies the principle behind PBB-TE is provided in the following. Assume that two backbone-VLANs, B-VIDs 52 and 98, are allocated to PBB-TE in a PBBN that runs PBB-TE together with internal PBBN MSTP (on other B-VIDs). PBB-TE set up all B-VIDs 52 and 98 forwarding tables in the BEBs and BCBs that are part of the PBBN that runs PBB-TE, as shown in Figure 3.29. PBB-TE also causes these BEBs and BCBs not to flood frames anymore, and to use the static entries provided; that is, not to learn any new MACs for these VIDs. The rest of the B-VIDs are used for parallel, regular PBBN MSTP operation, even in the BTT-TE assigned bridges.

The tuples <B-DA, B-VID> that were set up by PBB-TE in the forwarding tables described before are used to access BEBs identified by the B-DA and to forward frames throughout the PBBN in the PBB-TE created ESPs, identified by the <B-DA, B-VID> pair. Figure 3.29 shows such three ESPs, terminated at two B-DAs,

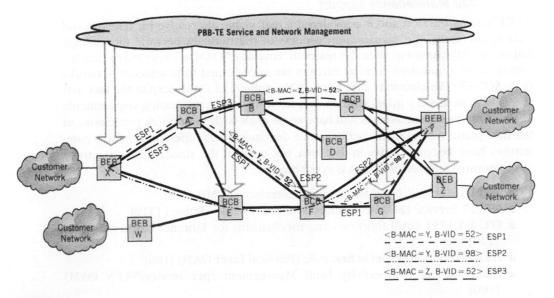

FIGURE 3.29

PBB-TE example

BEBs Y and Z. The two B-VIDs that are used (52 and 98) in creating these three ESPs, two for BEB Y and one for BEB Z, have no traditional VLAN meaning anymore, and generally in PBB-TE they can be regarded as a path number <B-VID> to a B-DA. It results that the BCB node F in Figure 3.29 can distinguish between two paths terminating at the same destination B-DA (BEB Y), even though the two paths cross it, and forward frames to two different BCB nodes, depending on the B-VID. Figure 3.29 also shows how two ESPs that are targeted to the same B-DA (BEB Y) diverge at BCB node F into two different paths, since they use different B-VLANs, and two other ESPs that use the same B-VID (52) also diverge at node A, since they are targeted to different B-DAs.

3.3.2.2.4 IEEE 802 Summary

IEEE 802 is currently working on metro Ethernet quite extensively, so we can expect more standards and technologies suggested for metro Ethernet that are based on Ethernet infrastructure, as well as some modifications and abandonment of the technologies described in this subsection.

The goal of IEEE 802 CENs is to provide an Ethernet-based bridged network infrastructure, starting from the personal devices and Personal Area Networks (PANs, 802.15), the desktops and servers, the LANs and Wi-Fi (802.11) or the Wi-MAX (802.16) connection of a nomadic terminal, down to a peer or to a content service provider.

3.3.2.3 Carrier Ethernet Operations, Administration, and Maintenance Support

As CEN gains traction and is considered a real, viable alternative for telecom and data applications, the management aspects of the network become increasingly important. Management, or in the telecom "language," OAM, is expected to provide functionality equivalent to what carriers are accustomed to in telecom networks (e.g., SONET). We elaborate a bit here on the subject of OAM, despite the fact that this field is in its very initial stage in data networks, since processing requirements of these management aspects will have to be dealt with in network processors, in the data plane, due to heavy performance demands. Many different standards committees have become active in this area, but most of the standardization activity (architectures and terminology) is synchronized:

- MEF16 Ethernet Local Management Interface (E-LMI) [318]
- MEF17 Service OAM Requirements & Framework—Phase I [319]
- ITU-T Y.1731 OAM functions and mechanisms for Ethernet based networks [229]
- IEEE 802.3ah Ethernet in first mile (Physical Layer OAM) [186]
- IEEE 802.1ag Connectivity Fault Management (per service/VLAN OAM) [180]
- IETF draft L2VPN OAM requirements and framework [321]

Starting from the access networks, we describe the OAM used in EPONs, and then we go to the aggregation networks, metro networks, and backbone, core networks.

Link level OAM in Ethernet First Mile (EFM), or EPON (IEEE802.3ah), is used for monitoring link operations, such as remote fault indication and remote loopback control. OAM control and information is conveyed in "*slow-protocol*" Ethernet frames, with a MAC destination address 01-80-c2-00-00-02, slow-protocol Ethertype (which is 0x88-09), and OAM subtype (which is 03), as shown in Figure 3.30(a).

The information and link events are carried in a Type, Length, and Value (TLV) type format, as shown in Figure 3.30(b), whereas variables requests and responses use the branch and leaf sequence numbers in the relevant Management Information Base (MIB) structure to handle these variables. We shall also see use of TLVs in other OAM PDUs, as described in the following. We turn now to the aggregation, metro and core networks, going from the link level OAM to a network level OAM.

| Frame Check Sequence (4 bytes) | OAM data + Padding (42-1496 Bytes) Composed of sequence of TLVs (for Information and events) and variables | Code (1 byte) | Flags (2 bytes) | OAM subtype (03) | Length/Type (2 bytes, 88-09) | Source Address (6 bytes) | Destination Address (6 bytes) 01-80-c2-00-00-02 |

(a) IEEE802.3 EFM (EPON) OAMPDU Format

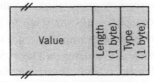

(b) Generic EPON OAMPDU TLV Format

Flags:
  bit 0—Link Fault was detected
  bit 1—Dying Gasp (unrecoverable local failure) occurred
  bit 2—Critical Event was detected
  bits 3&4—Local Evaluating and Stable status
  bits 5&6—Remote Evaluating and Stable status

Code:
  00—Information
  01—Event Notification
  02—Variable Request
  03—Variable Response
  04—Loopback Control

Type:
  00—End of TLV market
  01—Local Information
  02—Remote Information
  0xFE—Organization
        Specific Information

FIGURE 3.30

IEEE802.3ah (EPON) OAM frame format

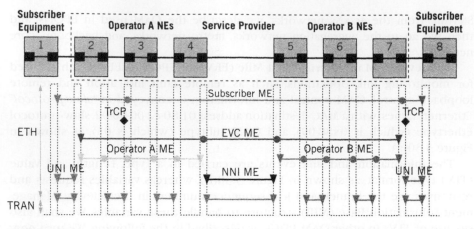

FIGURE 3.31

Point-to-point MEs at ETH layer (reproduced with permission of the Metro Ethernet Forum)

MEF describes OAM requirements and framework [319] and Ethernet Local Management Interface (E-LMI)[46] [318] without detailing the implementations of OAM; that is mechanisms and protocols. The IEEE802.1ag CFM project [180] and the ITU-T definition of OAM functions and mechanism [228, 229] detail implementations, protocols, frame formats, and so on. IETF provides an OAM framework and requirement for Layer 2 Virtual Private Networks (L2VPN), but this is surely just the beginning of many more OAM standards of IETF services.

The main functions of OAM that are covered by these standards are fault management, performance monitoring, and auto-discovery. Some standards only focus on fault management, while others also focus on performance monitoring.

Each OAM entity that requires management in one Maintenance Domain is defined as a Maintenance Entity (ME); this OAM entity (ME) is essentially an association between two maintenance endpoints that each requires management. MEs that correspond to the ETH layer are shown in Figure 3.31, at different levels: the subscriber, the service provider, and the network interfaces (user and network). If the service provider uses two different network operators, each of the operators manages its own OAM domain, using operator MEs.

Maintenance Entities (ME) are grouped into a ME Group (MEG) if they belong to the same service (in point-to-point, it is just one ME, but there are many MEs in multipoint connections). MEGs are also called Maintenance Associations (MA)

---

[46]MEF defined E-LMI messages that are transferred across the Ethernet User-Network-Interface (UNI), encapsulated in untagged Ethernet frames, using MAC destination address 01-80-C2-00-00-07, and Ethertype 0x88-EE.

(by IEEE802.ag CFM). OAM frames are imitated and terminated at the MEG End Point (MEP), shown as triangles in Figure 3.31 (two MEPs in point-to-point; many in multipoint connections). OAM frames can be treated and possibly reacted to at the MEG Intermediate Points (MIPs), represented by "circle" symbols in Figure 3.31. Figure 3.31 shows the MEGs organized in various levels to distinguish between OAM frames that belong to different nested MEs.

ITU-T Y.1731 and 802.1ag define three common fault management functions that are described in this subsection: Continuity Check (CC), LoopBack (LB), and LinkTrace (LT).[47] The CC function is a proactive fault management utility, aimed to detect loss of continuous link between any pairs of MEPs. It can also be used to detect mis-linkages and other configuration defects. Loopback function is an on-demand utility that is used to verify bidirectional connectivity between any MEP to another peer MEP or MIP. It can also be used to diagnose peer MEPs connections; for example, out-of-service, throughput, or bit-error-rate. Link Trace function is also an on-demand utility that is used to retrieve adjacency relationships between a MEP and a remote MEP or MIP, that include a sequence of MAC addresses of the MIPs and MEPs hops on the path. This can be used, for example, for fault localization; that is, in case of a fault, the retrieved sequence of MIPs and MEPs can provide information about what went wrong and where the error occurred.

The function of the CC is highly resource-demanding, as described in the following, and thus requires extremely fast processing. The other functions—loopback and link trace—are on demand, expecting replies within five seconds, and therefore demand fewer resources than the CC function.

Each function is executed by sending an OAM PDU message either periodically or on demand, and expecting an OAM PDU reply in the case of LB and LT functions. This results in three main types of OAM frames:

- Continuity Check Messages (CCM), exchanged between MEPs
- Loopback Messages and Replies (LBM/LBR), exchanged between MEPs and between MIPs
- Link Trace Messages and Replies (LTM/LTR) exchanged between MEPs and between MIPs

These OAM frames are encapsulated into standard Ethernet frames with a maximum message length of 128 bytes. A specific Ethertype (0x89-02) is defined to identify the OAM packets. The OAM message frames can have either unicast or multicast destination addresses, whereas replies can have just unicast addresses. The payload of the Ethernet frame—that is, the OAM frame—has the format as shown in Figure 3.32. The Maintenance Level (or Maintenance Association Level) is 3-bits long, which means that seven layers can be used to distinguish between nested layers of OAM entities (the relevant MEPs and the MIPs are configured as the

---

[47]There are other fault management and performance measurement functions that are beyond the scope of the book; for example, performance measurement functions include Frame Loss and Frame delay.

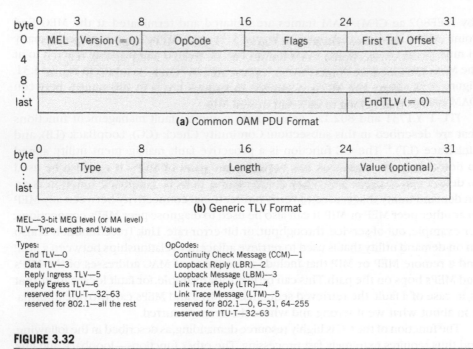

MEL—3-bit MEG level (or MA level)
TLV—Type, Length and Value

Types:
End TLV—0
Data TLV—3
Reply Ingress TLV—5
Reply Egress TLV—6
reserved for ITU-T—32–63
reserved for 802.1—all the rest

OpCodes:
Continuity Check Message (CCM)—1
Loopback Reply (LBR)—2
Loopback Message (LBM)—3
Link Trace Reply (LTR)—4
Link Trace Message (LTM)—5
reserved for 802.1—0, 6–31, 64–255
reserved for ITU-T—32–63

FIGURE 3.32

Common OAM PDU format (from left to right)

expected maintenance level, and intercept these OAM frames; lower levels OAM frames are dropped, and higher levels are forwarded).

The CC function deserves more description here, since it is expected to be executed by network processors, in data plane processing. CCM messages are generated periodically, per session, at a rate determined by the relevant application (either fault management, performance monitoring, or protection switching). There are seven transmission periods that are defined for CCM, from 3.33 ms (300 frames/s) to 10 min (6 frames/h). The recommended (and default) transmission period for protection switching is 3.33 ms, for performance management it is 100 ms, and for fault management it is 1 s. An MEP detects a loss of connection when it does not receive an expected CCM from a peer MEP within an interval of 3.5 times the transmission period (roughly the expected time to receive three consecutive CCMs). Given that the transmit period can be as low as 3.3 ms, this can result in a very high frequency, if we consider thousands of connections or more. Thus, supporting CCM transmission and monitoring on a per VLAN basis, for example, requires processing to be performed on a network processor, in its data plane processing.

Tunnels in PBB-TE are monitored through CCM messages, which enable protection switching to back-up tunnels (protection paths) within tens of milliseconds, much like in the SONET infrastructure. This increases the criticality of fast transmission and time-out detection times for CCMs, and hence the sensitivity for rapid OAM processing. Poor or slow handling of OAM frames will cause network topology outages.

It should be noted that ITU-T distinguishes between protection and restoration of connections [198, 212]. The differences could be summarized as:

- Protection makes use of preassigned capacity between nodes (also known as *make before break*).
- Restoration makes use of any capacity available between nodes (also known as *break before make*).

3.3.3 International Telecommunication Union-T Approach

Generally speaking, the Telecommunication Standardization Sector of the International Telecommunication Union (ITU-T) is well aware of the IETF and IEEE initiatives with regards to next generation telecommunications networks, mainly transport networks, and synchronizes its standards with these bodies.[48] The two study groups (SGs) of ITU-T that are relevant to our discussion here are SG13 and SG15; SG13 deals with next generation networking, while SG15 focuses on optical and other transport network infrastructures.

In most of the technologies described previously, we also mentioned ITU-T contributions with references to its standards; however, there is one specific ITU-T SG15 contribution that deserves a description of its own. This is MPLS over transport networks, called Transport MPLS (T-MPLS) [214–216].[49]

T-MPLS is a subset of IETF MPLS that is tailored for and aligned with ITU-T G.805 and G.809 definitions of transport-layered network. This reduced MPLS creates a connection-oriented subset of the MPLS packet-switched network, which significantly simplifies MPLS, and fits the connection-oriented world of the telecom industry. This means that L3 issues (e.g., IP) are not part of the T-MPLS, which means cheaper equipment and a further reduction of deployment costs to the carriers.

In a sense, T-MPLS is an evolution of Ethernet over SONET/SDH, in which packet-centric traffic is required to support a variety of services, including Ethernet, IP, and real time TDM. T-MPLS is supposed to support existing applications as well as the rapid mobile back-haul growth. It is well integrated into SONET/SDH, and its main "clients" are Ethernet and IP/MPLS services, as T-MPLS provides these services with

---

[48]More specifically, Y-Series standards deal with Global information infrastructure, Internet protocol aspects and next-generation networks (in which Y.1000–Y.1999 deal with Internet protocol aspects and Y.2000–Y.2999 deal with Next Generation Networks). G-Series deal with Transmission systems and media, digital systems and networks (in which G.600–G.699 are Transmission media and optical systems characteristics, G.700–G.799 deal with Digital terminal equipments, G.800–G899 deal with Digital networks, and G.8000–G.8999 are for Packet over Transport aspects, where the first G.80xx are for Ethernet over Transport aspects, and G.81xx are for MPLS over Transport aspects).

[49]SG13 contributed to T-MPLS and standardized the Operation and Maintenance aspects.

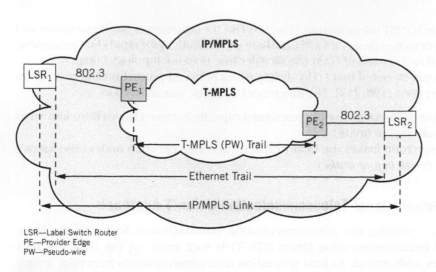

LSR—Label Switch Router
PE—Provider Edge
PW—Pseudo-wire

FIGURE 3.33

IP/MPLS via Ethernet over T-MPLS network [215]

an end-to-end "tunnel," much like PWE3[50] (see Figure 3.33). In other words, T-MPLS is intended to be a separate layer with respect to MPLS. This "tunneling" feature conflicts with other approaches taken by IEEE and IETF for provider backbone networks.[51] It means, among other issues, that plain MPLS cannot peer directly with T-MPLS, that is, an LSP initiated from one of these types of networks will have to be encapsulated when it transits the other type of network. It also means that the control plane of each of these types of networks will work independently of the other. This layering principle means also that T-MPLS can support any packet-oriented services, including higher hierarchies of T-MPLS itself.

T-MPLS retains the essential nature of MPLS, and works pretty much like MPLS, including DiffServ treatment, for example, which is essential for supporting CoS. However, it also lacks some of MPLS' capabilities (not just L3 and connectionless support), such as the Penultimate Hop Popping,[52] Equal Cost Multiple Path (ECMP),[53] and LSP merging.[54] Since transport connections can hold for very long times (in contrast

---

[50]For a complete discussion, the reader is referred to Appendix I of ITU-T Rec. G.8110.1/Y.1370.1 (11/2006) [215, page 22].

[51]And, as a matter of fact, both Cisco Systems and Juniper Networks expressed reservations concerning the work on T-MPLS and the approval of it [215].

[52]Penultimate Hop Popping provides the option of popping the label one hop before the last Label Switched Router (LSR), to ease the egress router processing (usually, it does not require this label anymore).

[53]This option makes it possible to split traffic that belongs to one LSP path into many routes of equal cost.

[54]This option makes it possible to merge several labels, belonging to different streams that travel at the same path, to one label.

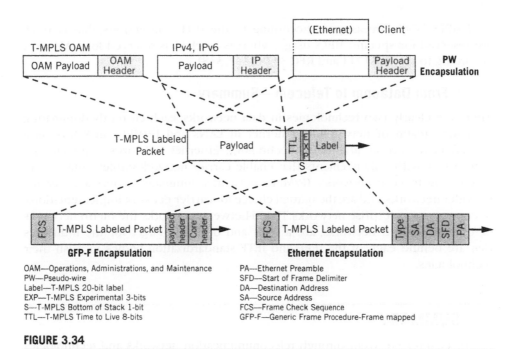

FIGURE 3.34

T-MPLS frame structure and layers (from right to left) [215]

to data communications), T-MPLS includes features that traditionally are associated with transport networks, such as protection switching and OAM functions.

It should be noted that as of this writing, the control plane of T-MPLS has not been defined yet, but it can use the same SONET/SDH provisioning mechanism, that is management plane systems, such as Generalized Multi-Protocol Label Switching (GMPLS) [234, 297, 408] or Automatically Switched Optical Network (ASON) [212, 213], which are both mentioned in the previous chapter.

T-MPLS is a transport network that can interface directly with upper L3 layers such as IPv4 and IPv6, or it can serve upper L2 layer clients through common inter-networking functions (convergence, timing, and sequencing) such as Ethernet, PDH,[55] SDH/SONET, ATM, Frame-Relay, HDLC and PPP. T-MPLS interfaces also with lower layers such as T-MPLS over PDH (MoP) or over SDH (MoS) through GFP or PPP in HDLC, T-MPLS over OTH[56] (MoO) through GFP, T-MPLS over Ethernet (MoE) or T-MPLS over RPR (MoR). The relation between these layers as it appears in the frame encapsulation structures is shown in Figure 3.34, which depicts several protocols at each layer.

---

[55]Plesiochronous Digital Hierarchy.
[56]Optical Transport Hierarchy.

T-MPLS labels are assigned according to the IETF conventions, that is, 0–15 are reserved for specific MPLS usage, whereas label 14 is reserved for OAM alert according to ITU-T Y.1711 and RFC 3429 [227, 338].

3.3.4 From Datacom to Telecom—Summary

There are clearly two technologies in data networks, each currently dominating a separate realm of networking: Ethernet in LANs and IP/MPLS in WANs. As a cost-effective, strong, and viable technology, Ethernet networks dominate the enterprises, with enhancements that enable carrier network implementations in connecting to core networks. IP/MPLS networks dominate carriers and service provider networks, and are the natural choice for carrier network implementations, in connecting enterprise networks. The in-between networks, the metro networks or the aggregation networks, are a clear and desirable target for the industries that are behind each of the IEEE and IETF standardization bodies to push their technologies.

3.4 SUMMARY

In this chapter we went through telecommunication networks and technologies that adopted packet traffic in order to fit the next generation networks, and we also described how the data communication networks were modified to become carrier class, for telecommunication applications. Both types of networks are continuously merging and converging into one global PSN, that serve applications that are using the IP or the Ethernet model as underlying transport mechanism.

Mobile, cellular wireless networks are excellent examples of how these two technologies and applications are converging. These networks, along with other access and home networks are described in the next chapter.

APPENDIX A
ROUTING INFORMATION DISTRIBUTION PROTOCOLS

There are two important protocols that are used for distribution of routing and forwarding information: BGP and LDP. BGP is more comprehensive than LDP and is used for routing in L3, whereas LDP is simpler, and is aimed to support MPLS and other tunneling protocols in L2 forwarding. These two protocols are described in this appendix since they both are used in various inter- and intra-networking mechanisms. Although they are very different in their goals, readers will find many similarities, particularly in the protocol formats.

BORDER GATEWAY PROTOCOL

BGP is an inter-Autonomous System (AS) routing protocol, which is used primarily to interconnect TCP/IP networks to the entire Internet. Its principles are used also for other networking purposes that either require inbound communications between edge devices (e.g., BGP/MPLS IP VPN), or outbound information exchange, mainly routing information.

BGP has four versions, starting with BGP-1 [288], followed by BGP-2 [289], BGP-3 [290], and finally BGP-4 [364, 365]. BGP-4 has gone though a number of improvements and extensions for wider and better operation, and it can be regarded as a "core protocol" (RFC 4271 [365]), enhanced by extensions,[57] and applied along with routing policy and inter-domain routing management.

BGP allows the Internet to decentralize and work with distributed architecture, by supporting the AS separation. Every AS in the Internet has a unique identifier, called AS-ID.[58] BGP routers exchange network reachability information with other BGP routers (called *BGP speakers*). The communicating BGP routers can be either in a different AS (where they are referred to as external peers), or in the same AS (when there is more than one boundary BGP router in the AS, and where they are referred to as internal peers). Internal and external BGP sessions (iBGP and eBGP) have different functions; iBGP synchronizes data among the BGP speakers themselves, whereas eBGP exchanges network reachability routing information. A BGP speaker advertises to its peers the routes that it uses (i.e., the most preferred routes that are used for forwarding). A route in BGP is defined as

---

[57]The two most important enhancements are Multi-protocol extensions and Route Reflection. Multi-protocol extensions (MP-BGP) mean the ability to interconnect not only IPv4 networks, but other kinds of networks and addressing schemes as well (e.g., IPv6), including overlapping IPv4 address spaces [45]. BGP Route Reflection (BGP-RR) [44] relieves the required full-mesh connectivity of BGP routers in an AS TCP/IP network. Other extensions include Route Flap Damping [421], AS Confederations [410], BGP Support for 4-byte AS Number Space [422], and Extended Communities Attribute [376].

[58]AS IDs are termed AS numbers, and are in accordance with RFC 1930. AS numbers are assigned by IANA (*http://www.iana.org/assignments/as-numbers*).

the information that pairs a set of destinations that have the same prefix of IP address, with a set of identical parameters (called attributes in BGP) of a path to those destinations. Routes that are exchanged between BGP routers using the BGP protocol are stored in the Routing Information Bases (RIBs) in the BGP routers.

BGP is a vector-based algorithm, which carries the complete AS path vector between the communicating BGP routers, that is, the list of the AS-IDs that creates the path. This allows BGP to detect loops (when an attempt is made to reenter an existing AS-ID into the AS path vector). This enables BGP routers to maintain a routing map of the network, assuming the Internet is a set of ASs, but it does not provide the exact path a packet travels within an AS. Figure A3.1 demonstrates several AS networks connected by AS paths, for accessing destinations whose IP address prefix is 132.72 (for a detailed description of IP addressing, please refer to Chapter 5.

BGP uses TCP (port 179) as its transport protocol, which offers reliable transmission of BGP transactions. It is important to note that BGP targets specific IP addresses when it is used, that is, messages are exchanged between specific BGP routers, using IP destination addresses, obeying some routing protocol that affects the created paths. BGP-4 supports the Classless Interdomain Routing (CIDR) addressing scheme [140, 363], and the aggregation of routes and AS paths.

Border Gateway Protocol originally supported AS identifications (AS-ID) that were 16-bits long (i.e., up to 64 K AS IDs were allowed), and this is clearly insufficient nowadays. Therefore, AS-IDs have been extended to 32-bits long, and BGP supports the extended AS numbers alongside the "old" 16-bits AS numbering [422].

The BGP protocol contains five message types, each with its own packet format. The protocol also defines about 20 different parameters of paths, called path attributes, each of which consists of encoded list in the <Type, Length, Value> (TLV) format.

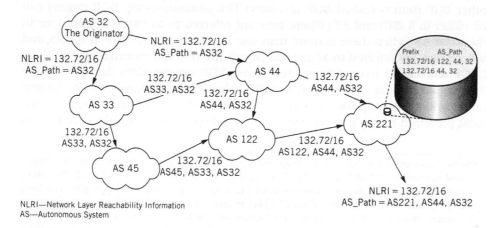

NLRI—Network Layer Reachability Information
AS—Autonomous System

FIGURE A3.1

BGP network example with AS networks and AS paths

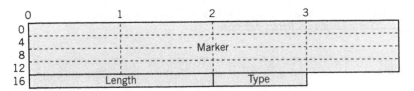

FIGURE A3.2

BGP header

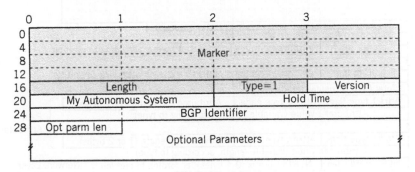

FIGURE A3.3

Notification message format

Border Gateway Protocol Message Formats

BGP message can be anywhere from 19 bytes (just the header) to a maximum of 4096 bytes, as shown in Figure A3.2.

The *Marker* is a 16-byte field, set to ones. The *Length* is a 2-byte field indicating the total length of the message in bytes. The *Type* field is 1-byte long, and indicates the purpose and format of the message, as follows:

- Open message (message type 1)—starts a BGP session.
- Update message (message type 2)—exchanges information about the routes.
- Notification message (message type 3)—is sent when an error is detected and the BGP connection should end.
- KeepAlive messages (message type 4)—periodic messages that verify the continuous connection of the BGP peer.[59]

The BGP formats for the various messages are described in the following.

Open

The Open message (see format in Figure A3.3) is the first message sent by both BGP routers (called *BGP speakers*) when a TCP connection is established.

---

[59]Additional message type 5, "Route Refresh," was proposed in RFC2918 [77], and is not discussed here.

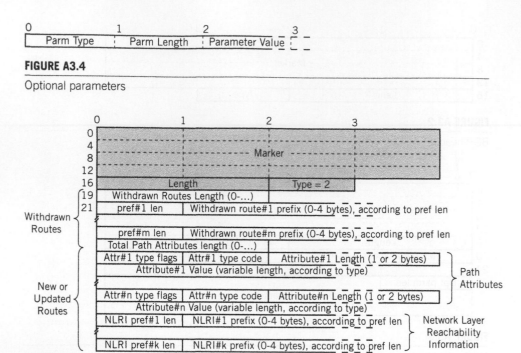

FIGURE A3.4

Optional parameters

FIGURE A3.5

Update message format

In addition to the header fields, the Open message contains the following fields: *Version* is a 1-byte field that indicates the version number of the message protocol (currently 4). *My Autonomous System* is a 2-byte field that indicates the AS sender number (AS-ID). *Hold Time* is a 2-byte field that indicates the maximum elapsed time between successive KeepAlive message and/or Update message receipt (in seconds). *BGP Identifier* identifies the sender by its IP address, and *Optional Parameter Length* is 1-byte long and indicates the length of the following *Optional Parameter* field in bytes. The *Optional Parameter* field consists of an encoded list of parameters in the <Type, Length, Value> format, as shown in Figure A3.4.

Update

The Update message advertises routes to a set of destinations that have the same prefix of IP address. The message is used by a BGP speaker to advertise a previously received route, to add or modify the path attributes before advertising it, or to "delete" (withdraw) an already advertised route.

The update message is shown in Figure A3.5, and includes a variable number of fields in two parts of the message; some of the fields are themselves of variable length. After the header there is a part that defines the withdrawn routes, which is followed by the second part that describes the new or updated routes. This latter part is also composed of two subparts, the first of which describes the common path attributes of these routes, while the second defines the IP address prefix that all target destinations have.

Withdrawn Routes Length is a two-byte field that indicates the total length of the Withdrawn Routes part of the message in bytes (if 0, no withdrawn routes exist in that message). Each of the Withdrawn Routes entries is a variable-length description of an IP address prefix to be withdrawn, consisting of a pair of <length, prefix>. Length is one byte that contains the prefix-length (in bits) of the IP address, and Prefix is the prefix of the IP addresses (which is 0–4 bytes long, depending on the prefix length).

Total Path Attributes Length is a two-byte field that indicates the total length of the Path Attributes field in bytes (if 0, neither Path Attributes nor Network Layer Reachability Information [NLRI] are distributed by this message). *Path Attributes* are the common path parameters that are distributed in the Update message, and they are represented in the <Type, Length, Value> format. Each is of a variable length according to its type. *Attribute Type* is two bytes long; the first byte contains the Attribute's Flags[60] and the second byte is the Attribute Type Code, described as follows.

Attribute Length is one or two bytes long (according to the Extended Length flag), indicating the length of the Value field in bytes, and the attribute's Value field is interpreted according to the Attribute Type Code. The three well-known and mandatory attributes are:

- ORIGIN (Type code 1)—defines the origin of the prefix (0 means from interior routing in the originating AS; 1 means for NLRI acquired from inter-AS routing, or 2 from somewhere else).
- AS_PATH (Type code 2)—describes the associated AS path vector, which is composed of a set or sequence of AS path segments through which routing information has passed. Each segment is described by the TLV format.[61]
- NEXT_HOP (Type code 3)—provides the IP address of the border router that should be used for the next hop to reach the destinations listed in the NLRI.

The NLRI field consists of a list of IP address prefixes. IP address prefixes are defined by one byte containing the length of the network mask in Classless

---

[60]The flags are 8 bits, where just the first four most significant bits are currently used. The first, the high-order bit, is the optional (1) or well-known (0) attribute, and in the case of an optional attribute, the second bit instructs the BGP receiver to accept the path and distribute it if it is transitive (1), and to set the next, third bit, the partial bit. This partial bit indicates to BGP routers that somewhere along the path there was at least one BGP router that wasn't familiar with the optional attribute. The last bit, the fourth high-order one, is the extended length flag that is used to indicate a long attribute, and the Length field is two bytes long (rather than just one) for Value lengths greater than 255 bytes.

[61]The AS_PATH attribute is composed of a set or sequence of AS path segments; each segment is represented by Type, Length, and Value parameters. Type is a 1-byte field, containing "1" (indicating an unordered set of AS's in the path) or "2" (indicating an ordered sequence of AS's in the path). Length is a 1-byte field, containing the number of AS's in the Value field, and the Value field is one or more AS's in a 2-bytes long field for each of the AS numbers.

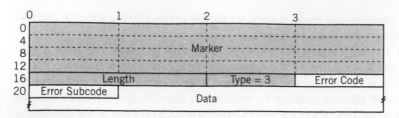

FIGURE A3.6

Notification message format

Interdomain Routing (CIDR) terminology, and up to four bytes that contain the address prefix itself (IP addressing, including CIDR, is described in Chapter 5. The NLRI has no explicit encoded length in the message, and should be calculated by subtracting 23,[62] the *Withdrawn Routed Length* and the *Total Path Attributes Length* from the Update Message *Length*.

Notification

The Notification message is sent before terminating the BGP connection, to notify that an error condition was detected. Three fields exist in the notification message, as shown in Figure A3.6; *Error Code* is a 1-byte field that indicates the primary error condition (containing 1 to indicate an error in the message header, 2 for an error in the Open message, 3 for an error in the Update message, 4 for expiration of the *Hold Time*, 5 for a BGP state machine error, or 6 for an unexplained session termination). *Error Subcode* is a 1-byte field that provides more details about the error. The *Data* field is of variable length, and it is used to diagnose the reason for the Notification message. No explicit, encoded length of the data exists in the message, and it must be calculated by subtracting 21 from the Notification message length.

KeepAlive

The KeepAlive message is used for keeping the Hold-Time from expiring, thereby maintaining a keep-alive mechanism that ensures peers' reachability. The KeepAlive message format is just the header, with type of message equals 4.

Broader Gateway Protocol Implementation Issues

As noted previously, BGP avoids AS looping by watching the AS_Path vector, and stopping when reaching an AS that is in the AS_Path. However, with iBGP sessions this method will not work (since all transmissions are within the same AS), and routes

---

[62]There are 19 bytes for the header, a mandatory two bytes for the *Withdrawn Routed Length* field, and a mandatory two bytes for the *Total Path Attributes Length* field, totaling 23 bytes.

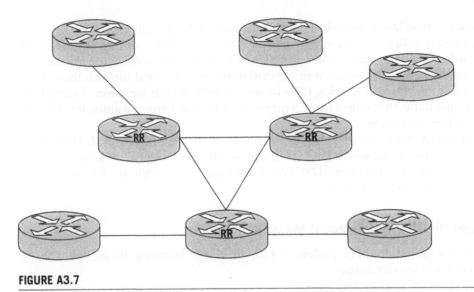

FIGURE A3.7

Route reflectors in a BGP network

received via iBGP must not be distributed to any other iBGP session, or a loop may be created. This creates a need for full mesh connections between all BGP routers, which in turn results in a huge number of iBGP sessions (for n routers, $n(n$-$1)/2$ peering sessions are required). This poses a serious scalability issue, let alone the load problem and the difficulty in adding a new BGP router and configuring all BGP routers as a result.

Route reflectors (RR) were introduced to solve the full-mesh iBGP connections and load problem. A route reflector is a standard BGP speaker at the core of the AS that is allowed to re-advertise the incoming iBGP sessions to its Client peers after summarizing the routes and picking only the best ones (reflected routes). Together, they form a cluster [44] (see Figure A3.7).

Alternatively, an AS can be subdivided into smaller sub-ASs, where each sub-AS can use a route reflector (or not use it). However, the federation of the sub-AS's works as one homogeneous AS to the other ASs.

Support in protocols other than just the IPv4 protocol was added in the Multiprotocol Extensions for BGP-4 (MP-BGP) [45]. Attribute types Next-Hop, Aggregator, and NLRI had to be defined differently in order to support other *Address Families* (of other protocols). Last, BGP had to support 32-bit AS numbers [422], and new attribute types were added, for example, *AS4_Path*.

LABEL DISTRIBUTION PROTOCOL

Label Distribution Protocol [22] is a protocol that originated and is mainly used in MPLS, as described in this chapter. LDP distributes label binding information between Label Switched Routers (LSRs). The protocol maintains incremental updates of the LIB, when only the differences are exchanged. Its principles, however, are used in other systems in which tunnels or end-to-end paths are required. LDP includes a set of procedures and messages that enable LSRs to exchange information about the labels that are used to forward traffic between and through these LSRs. In other words, LDP enables LSRs to set up LSPs, which are Layer 2 (data-link) switched paths that reflect Layer 3 (network) and upper layers' routing information and other constraints.

In MPLS, the labels that LDP distributes are associated by LDP with FECs [370], which are groups of similar packets, that is, they have the same destinations, or treatment, or priority. Therefore, LDP enables mapping of specific packets to specific LSPs, by their FECs, or labels.

Label Distribution Protocol Message Format

Label Switched Routers exchange LDP messages between them that can be categorized into four types:

1. *Discovery messages*—Announcing the presence of an LSR in the network by sending periodic UDP Hello messages.
2. *Session messages*—Establishing, maintaining, and terminating LDP sessions between LDP peers (including Initialization and KeepAlive messages).
3. *Advertisement messages*—Creating, changing, and deleting label mappings assigned to or associated with FECs.
4. *Notification messages*—Providing information (about the session or the status of a previously received message) and error signaling.

Following the discovery process, an LSR can establish an LDP session with a discovered LSR by using the LDP initialization procedure over TCP. The LSRs that maintain an LDP session are known as "LDP peers." LDP protocol is bidirectional, that is each LDP session allows both peers to exchange and learn label mappings from the other.

Label Distribution Protocol uses TCP (port 646) for the session, advertisement and notification messages for reliability demands and in-order delivery of messages. For the discovery mechanism, LDP used UDP (port 646) in order to distribute the LDP Hello messages without the requirement to establish a TCP session ahead.

Each LDP PDU[63] contains an LDP header followed by one or more LDP messages. LDP header is described in Figure A3.8. *Version* is a two-byte field that indicates the protocol version. *PDU Length* is also a two-byte field that indicates the total length of the PDU in bytes (excluding the first four bytes of the *Version and PDU Length* fields). *LDP Identifier* is a 6-byte field that is used for assignment and distribution of

---

[63]All message types, as well as TLV information types, are detailed and updated in *http://www.iana.org/assignments/ldp-namespaces.*

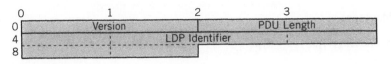

FIGURE A3.8

LDP header format

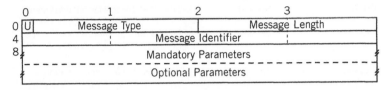

FIGURE A3.9

LDP message format

labels. The first four bytes of the *LDP Identifier* identify the LSR (e.g., its IP address), and the other 2-byte field (called *label space)*[64] can be used to identify an interface within the LSP platform.

An LDP message is described in Figure A3.9. *Message Type* indicates the type of message, and is a two-byte field, starting with a *U (Unknown message)* bit. If the received *Message Type* is unknown, and the *U* bit is set to "1," the message is silently ignored, otherwise, if the *U* bit is set to "0," a notification is sent to the LSR that originated the unknown message. *Message Types* and their classifications are grouped into the following categories, which are detailed in Table A3.1.

- General message types are 0x0001 to 0x00FF.
- Neighbor discovery message types are 0x0100 to 0x01FF.
- Initialization phase message types are 0x0200 to 0x02FF.
- Address message types are 0x0300 to 0x03FF.
- Label distribution message types are 0x0400 to 0x04FF.
- Connection-related messages have message types of 0x0500 to 0x05FF.

Message Length is a two-byte field specifying the length of the message in bytes, excluding the first four bytes of the *Message Type* and *Message Length*. The *Message ID* is a four-byte field that identifies the message in cases of notifications that refer to this message. *Mandatory Parameters* and *Optional Parameters* are variable length fields of required and optional sets of parameters. Parameters are specified in LDP by the TLV encoding scheme.

LDP's TLV has a typical structure as shown in Figure A3.10. An *U (Unknown TLV) bit* is used similarly to the *U* bit of the message; that is, the message is silently

---

[64]The *label space* is used to identify interfaces of the LSR (which is called per *interface label space*), or it contains zero for interfaces that can share the same labels (which is called *platform-wide label space*).

Table A3.1 LDP Messages

Message Name	Message Type	Type	Description
Notification	0x0001	Notification	Informs LDP peer of a significant event
Hello	0x0100	Discovery	Part of the LDP Discovery Mechanism
Initialization	0x0200	Session	Part of the LDP session establishment
KeepAlive	0x0201	Session	Part of a mechanism that monitors the LDP session transport connection
Address	0x0300	Advertisement	Advertises the LSR interface address
Address Withdraw	0x0301	Advertisement	Used by an LSR to withdraw previously advertised interface addresses
Label Mapping	0x0400	Advertisement	Advertises LSR FEC-label bindings to the LDP peer
Label Request	0x0401	Notification	LSR requests binding for a FEC from a LDP peer
Label Withdraw	0x0402	Notification	Used by LSR to notify a LDP peer that it cannot use previously advertised FEC label mapping
Label Release	0x0403	Notification	Used by a LSR to notify a LDP peer that a previously requested or advertised FEC label mapping is no longer needed
Label Abort Request	0x0404	Notification	Used to abort a Label Request message
Call Setup	0x0500	LDP extensions for ITU-T's Automatic Switched Optical Network [5]	
Call Release	0x0501		
Vendor-Private	0x3E00-0x3EFF		Used to transport vendor-private information between LSRs
Experimental	0x3F00-0x3FFF		

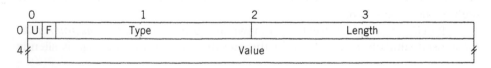

FIGURE A3.10

TLV encoding

ignored when the received *Type* is unknown and the *U* bit is set to "1"; otherwise, if the *U* bit is set to "0," a notification is sent to the message originator. An *F* (*Forward unknown TLV*) *bit* is applied only when the *U* bit is set and the *Type* is unknown, and it determines that the message is to be forwarded with the contained message (when the *F*-bit is set to "1"), or not forwarded (when the *F*-bit is "0").

The *Type* field is 14 bits defining the *Value*, and *Length* is a two-byte field specifying the length of the *Value* field in bytes. *Value* is a variable length field that contains information according to the *Type* field. The *Value* field may contain nested TLV information. The most commonly used TLVs that are defined in the LDP protocol are listed in Table A3.2, although there are additional TLVs

Table A3.2 TLV Types and Usage

TLV	Type	Contained in messages
Forwarding Equivalence Class (FEC)	0x0100	Label Mapping, Label Request, Label Abort Request, Label Withdraw, and Label Release
Address List	0x0101	Address, Address Withdraw
Hop Count	0x0103	Label Mapping, Label Request
Path Vector	0x0104	Label Mapping, Label Request
Generic Label	0x0200	
ATM Label	0x0201	Initialization
Frame Relay Label	0x0202	Initialization
Status	0x0300	Notification
Extended Status	0x0301	Notification
Returned PDU	0x0302	Notification
Returned Message	0x0303	Notification
Common Hello Parameters	0x0400	Hello
IPv4 Transport Address	0x0401	Hello
Configuration Sequence	0x0402	Hello
IPv6 Transport Address	0x0403	Hello
Common Session Parameters	0x0500	Initialization
ATM Session Parameters	0x0501	Initialization
Frame Relay Session Parameters	0x0502	Initialization
Label Request Message ID	0x0600	Label Abort Request

in other LDP protocol enhancements, as described in the following subsection. The most commonly used TLVs are FEC, Address List, Hop Count, Status, and Path Vector.

Label Distribution Protocol Implementation Issues

As noted previously, LDP defines a set of procedures, which include peer discovery, session initialization and management, label distribution and handling events (i.e., sending and receiving various messages), and notification of errors and other information. The LDP Discovery process, for example, enables an LSR to discover potential LDP peers. The process is such that LSR periodically sends Hello messages to "all routers on this subnet" group multicast address (in the Basic Discovery process), or to a specific address (in the Extended Discovery process). The Basic Discovery process discovers LSR neighbors that are directly connected at the link level. LSR uses the received Hello message to identify a potential LDP peer, either at the link level (if it was a Link Hello message in Basic Discovery) or at the network level (if it was a Targeted Hello message in Extended Discovery). Following the discovery process, an LSR can establish a session with another LSR learned via the Hello message, by using the LDP initialization procedure. During an LDP session, various label distribution procedures take place (such as sending labels for FECs to an LDP peer), as well as procedures for handling events (such as receiving label mappings). All these procedures are based on message exchanges.

In addition, Loop Detection in LDP is required in order to prevent Label Request messages from looping, and to avoid loops in LSPs. Path Vector and Hop Count TLVs that are carried in Label Request and Label Mapping messages are used for loop detection. Any LSR that transmits a message adds its LSR identifier to the Path Vector TLV if the message contains a Path Vector TLV, and increments the Hop Count TLV if the message contains a Hop Count TLV. An LSR that receives a message that contains its LSR identifier in its Path Vector detects a loop. If the maximum allowable Path Vector length is reached when the LSR tries to add its LSR identifier, or the Hop Count has reached a predefined threshold value, then the LSR assumes that a loop was detected.

Label Distribution Protocol Enhancements

Label Distribution Protocol uses the network topology in order to discover LDP peers, among whom it establishes LDP sessions for label exchanges that are used later for traffic forwarding by MPLS or other tunneling protocol. This, obviously, does not assume any specific traffic pattern or requirement or other constraints, and in fact ignores TE practices.

Label Distribution Protocol, however, can be enhanced such that it can create tunnels in a controlled way, using TE requirements, network constraints, and so on. This enhancement is known as Constraint-based LDP (CR-LDP) [235]. Other routing protocols were also enhanced to allow the creation and maintenance of tunnels by MPLS or other tunneling protocols such as Extensions to RSVP for LSP Tunnels (RSVP-TE) [38] and Open Shortest Path First Traffic Engineering (OSPF-TE) [248]. We briefly describe Constraint-based LDP (CR-LDP) in this subsection.

CR-LDP [235] is a signaling protocol that establishes and maintains explicitly routed LSPs. This protocol extends the LDP protocol with information used for setting paths by specific constraint-based routing, using additional TLV parameters and procedures. The additional parameters and procedures provide support for:

- Strict and Loose Explicit Routing
- Specification of Traffic Parameters[65]
- Route Pinning[66]
- CR-LSP Preemption[67]
- Handling Failures
- LSP Identifiers (LSPID)[68]
- Resource Class[69]

The protocol is usually used after a calculation of the route is done at the edges of the network (providing an explicit route by specifying the hops along the route). It can carry predefined traffic requirements (a set of edge traffic conditioning functions) that should be met and various other parameters that affect the required CR-LDP, as well as, the existing CR-LDP. CR-LDP has the following additional TLV parameters (types 0x800-0x8FF), as described in Table A3.3.

Since CR-LDP can use explicit routing, call set up can be a simple two-phase protocol: a Label Request and a Label Mapping. A Label Request message in CR-LDP may contain LSPID, Explicit Route, Traffic, Pinning, Resource Class, and Preemption TLVs (in addition to the FEC TLV used in LDP). The Label Map message in CR-LDP may include additional LSPID and Traffic TLVs.

There are additional extensions for LDP that are used for Pseudo-wire [303, 309], as discussed in the chapter. There are, however, many other LDP extensions and enhancements that are beyond the scope of this book that deal with a variety of network aspects or a wider operation of LDP, such as fault tolerance protection

---

[65]Traffic parameters include: Peak Data Rate (PDR), Peak Burst Size (PBS), Committed Data Rate (CDR), Committed Burst Size (CBS), and Excess Burst Size (EBS).

[66]Route pinning is used when it is undesirable to change the path used.

[67]If a route with sufficient resources that were signaled by CR-LDP cannot be found, existing routes are rerouted so that resources will be reallocated, potentially enabling the required route. Priorities are assigned to the existing and the required routes.

[68]LSPID is a unique identifier of a CR-LSP within an MPLS network, which is composed of the ingress LSR Router ID (or any of its own IPv4 addresses) and a locally unique identifier. This local identifier relates to the way this CR-LSP is recognized in the Ingress LSR, the one that originated this CR-LSP.

[69]Resource classes are also known as "colors" or "administrative groups."

Table A3.3 CR-LDP TLVs

TLV	Type
Explicit Route*	0x0800
IPv4 Prefix ER-Hop	0x0801
IPv6 Prefix ER-Hop	0x0802
Autonomous System Number ER-Hop	0x0803
LSP-ID ER-Hop	0x0804
Traffic Parameters	0x0810
Preemption	0x0820
LSPID	0x0821
Resource Class	0x0822
Route Pinning	0x0823

*The Explicit Route TLV specifies a path, defined by a sequence of nested Explicit-Route-Hop TLVs. Each may be one of the 0x0801 to 0x0804 TLV types.

[121], VPLSs [271], signaling unnumbered links [265], Maximum Transmission Unit (MTU) signaling [53], optical UNI signaling [360], Inter-Area LSP (multiple IGP-areas within an AS) operation [96], and many others. There are important CR-LDP extensions for control systems, such as for ASON [5] and for GMPLS [26].

Access and Home Networks

In the previous chapter, we described converged networks technologies that are used mainly in the core networks. In this chapter, we discuss access and home networks, those networks that are attached to the networks described in the previous chapter. Access and home networks are the simplest in term of technology and complexity, yet they are the most expensive part of the entire infrastructure. These networks also use converged technologies, and they make the connections between customers and their devices with the long-haul networks. We dedicate a chapter to these networks because equipment in these networks is increasingly based on network processors, as bandwidth and complexity reach a degree that justify it. This chapter is divided into two parts—access networks and home networks.

4.1 ACCESS NETWORKS

Access networks include many kinds of networks and various technologies that enable customer-to-network connection. These networks provide legacy telecommunications services (e.g., telephony) as well as data services (e.g., Internet). Historically, access networks were always the most expensive segment of the telecommunications facilities due to the huge physical spread and the potential difficulty of accessing the Customer Premises Equipment (CPE) or the users' terminals (e.g., mobile phones). However, since access networks used to be on the "low" bandwidth spectrum of networking, their networking process demands were relatively low, and they did not have such complicated processing technologies.

This is changing rapidly, as access bandwidth has risen sharply (from the Kbps range to the Gbps range) and some networking functionalities have been pushed downwards to the terminal ends of the networks (e.g., routing and tunneling, scheduling and Quality of Service [QoS], security and firewalls). To cope with the increasing processing demands of access aggregation equipment, access network processors are now used more frequently in various places and equipment in the access network.

Access networks use wireless (fixed and mobile), and wire-line technologies (fiber, coax, power-line,[1] and regular, two-wire copper telephone connections). In the following, we describe both wired and wireless access networks so the reader can understand the requirements for implementing access network processors.

4.1.1 Wired Access Networks

The very first access networks were deployed for the telephone services (on the Public Switched Telephone Network, PSTN), and then for the cable TV (CATV) services. Later on, these networks were utilized for additional and advanced digital services, along with dedicated fiber access networks—Passive Optical Networks (PONs). Wired access networks are schematically described in Figure 4.1, and briefly discussed in this chapter. Please note that terminology is not always consistent, despite the fact that these networks are many decades old. In the following subsections we describe the three media used for access networks (coax, twisted pairs, and fiber-optics), and we finish with data services.

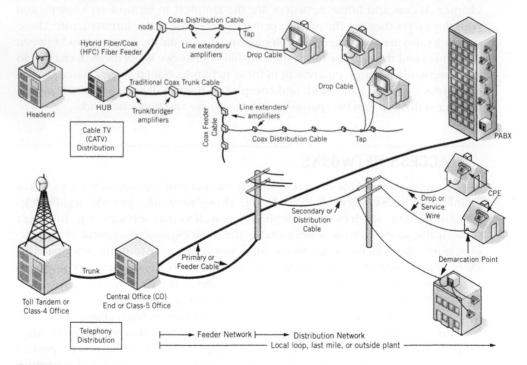

FIGURE 4.1

PSTN and CATV wired access network

---

[1]Power-Line Communications (PLC) will not be described here, since they are not common as of this writing.

We begin by discussing cable television (CATV) access networks (used mainly for CATV services) and PSTN Access network (used mainly for telephony services). Both of these networks are also used for data services, mainly Internet access to the residential customers and the Small Office Home Office (SOHO) customers.

Then, we discuss Fiber-in-the-Loop (FITL), which is used mainly for data access, but also for telephony services and TV services. It should be noted that both Voice over IP (VoIP) and TV services over the IP (IPTV) are technologies that are also used for telephony and TV services, but they are part of, and ride above, the Internet access, rather than being dedicated services such as digital telephony or digital CATV. We end the section on wired access networks with a discussion on data services, primarily Internet access.

4.1.1.1 *Cable Television Access Network*

Cable TV has its own cable plant, which is analogous to that of the PSTN cable plant. Traditionally it was based on all coax cables, but current networks are based on hybrid fiber and coax (HFC) cables. From the Headend (the television main distribution point), trunk cables drive the feeder cables, which then split to distribution cables, and are again split to drop coax cables. In a modern CATV cable plant, there is a transport network composed of primary hubs connected by a ring to the Headend. Each primary hub is connected to distribution hubs through secondary rings. The distribution network starts at these distribution hubs, and is composed of the feeders, distribution, and drop cables.

Cable TV access networks are also used for data communications, using Cable Television Laboratories[2] Data over Cable Service Interface Specifications (DOCSIS) standards [64, 65], which were also approved by the ITU-T [224–226]. DOCSIS version 3, released in 2006, allows IPv4/IPv6 Internet at high speed, both uplink and downlink, reaching an approximately usable speed of 150 Mbps and 110 Mbps, respectively, per channel (instead of a TV channel).[3] These bandwidths are shared by several customers that are connected to the same CATV cluster. Customer equipment (cable modems, CM) is connected to the coax demarcation point at one side of the access network (usually with an "F" type connector), and a Cable Modem Termination System (CMTS) is connected at the other side of the access network, working similarly to the Digital Subscriber Line Access Multiplexer (DSLAM) in the DSL systems used for twisted pair that is described as follows.

---

[2]Cable Television Laboratories, Inc. originally standardized Data over Cable Service Interface Specifications (DOCSIS).

[3]CATV uses the 5 to 860 MHz frequency spectrum of the cables. In North America, the NTSC TV format is broadcast over 6 MHz channels, whereas in Europe the PAL format is broadcast over 8 MHz channels. Therefore, the usable downlink speed in Europe reaches 200 Mbps due to the wider channel. The Upstream (Reversed) channels usually are allocated in the 5 to 42 MHz frequencies, whereas the Downstream (Forward) channels use the 54 to 860 MHz frequencies.

4.1.1.2 Public Switched Telephone Network Access Network

Access networks of the traditional Public Switched Telephone Network (PSTN) started at the carriers' network edge, that is, at the Central Office (CO) of an Incumbent Local Exchange Carrier[4] (ILEC) or at the Point-of-Presence (POP) of a Competitive Local Exchange Carrier (CLEC). These access networks were based on the twisted pair wire system,[5] in the local loop (also called the last mile) to the demarcation point. The demarcation point is where the telephone company's network ends and the wiring of the customer begins. The demarcation point is connected to the Customer Premises Equipment (CPE), which is usually a telephone or Private Branch eXchange (PBX), providing Plain Old Telephone Service (POTS).

The twisted pair access network is utilized not just for the POTS, but for digital services such as Integrated Services Digital Network (ISDN), using Digital Subscriber Loop/Line (DSL) technologies. ISDN was used mainly for telephony, but also for low-speed data communications. Other DSL technologies,[6] however, enable high-speed services, while utilizing the vast twisted pair outside the cable plant. DSL enables 24 Mbps or more, according to the specific DSL technology used and the distance between the demarcation point and the DSL termination point in the carrier's network. At the demarcation point, the customer equipment is connected to the DSL network through a DSL "modem,"[7] and at the other end of the access network (the central office), there is the DSL Access Multiplexer (DSLAM). The DSLAM is connected to the data networks, that is, the aggregation and core networks. Following the introduction of Digital Loop Carrier (DLC), optical fibers penetrated to the outside cable plant in several ways, as described in the next subsection.

4.1.1.3 Fiber-in-the-Loop

Optical fibers were introduced first into the interoffice trunks, and then into the outside cable plant. Fiber-in-the-Loop (FITL) can appear in many ways, starting from Fiber-to-the-Home/Premises (FTTH/FTTP), to Fiber-to-the-Building (FTTB), Fiber-to-the-Curb/Cabinet (FTTC), Fiber-to-the-Zone (FTTZ); these are more generally referred to as FTTx. All FTTx technologies usually use PONs, which are

---

[4] Local telephony companies in the United States are referred to as ILECs, or Local Exchange Companies (LECs) as a result of the breakup of AT&T into seven Regional Bell Operating Companies (RBOCs), known as "baby bells," according to the Telecommunication Act in the 1980s.

[5] The 2-wire cable system starts with feeder cables that include hundreds and thousands of 2-wires, split into distribution cables to which the drop wires connect the demarcation points of the customers.

[6] There are several DSL technologies, collectively called xDSL, that are continuously evolving and improving. The important ones include: Asymmetric DSL (ADSL), which uses the available bandwidth in an uneven way (more for downlinks than for uplinks), High bit rate DSL (HDSL) or Symmetric DSL (SDSL), and Very-high-speed DSL (VDSL). These technologies were standardized by the ITU-T; HDSL or SDSL is G.991 [203], ADSL is G.992 [204], and VDSL is G.993 [205]. VDSL defines several profiles; for example, one can handle 100 Mbps downstream and upstream on a 30 MHz bandwidth used for very short lines, and another is 50 Mbps downstream and 12 Mbps upstream for up to 2 km lines.

[7] The DSL modem is connected to the same twisted pairs used for POTS, through a splitter that filters and separates the frequencies used for telephony (below 4 KHz) from those used by the DSL system (100 KHz to 1 MHz).

point-to-multipoint and support up to 32 customers, or sometimes even more. For FTTZ, FTTC, or FTTB, complementary DSL access is required from the fiber end to the demarcation point, and electro-optic conversion is done in an active Remote Digital Terminal (RDL). Optical fibers, especially FTTP, allow high data bandwidths, reaching Gbps.

At the central office (CO) point, the PON is connected by an Optical Line Termination (OLT), while at the customer side (or near it) there are several Optical Network Units (ONUs) or Optical Network Terminations (ONTs), as shown in Figure 4.2. From the ONUs, there are various access channels (e.g., xDSL over copper) that reach the customers and are terminated by a Network Terminator (NT). The Optical Distribution Network (ODN) contains the fibers that connect the OLT to the ONUs and the ONTs, and it may use WDM to carry both transmission directions on the same fiber, in two wavelengths. There are several underlying technologies used in PON, and several generations of PONs. The common practice is to use Time Division Multiple Access (TDMA) in the upstream direction (from the customers to the CO), where the ONU is time synchronized to use these time slots properly. Other possibilities are WDM Access (WDMA), Subcarrier Division Multiple Access (SDMA), and Code Division Multiple Access (CDMA) [274]. WDMA seems to be the most likely future technology for PONs.

The first generation of PON was based on the Asynchronous Transfer Mode and thus called APON, which was standardized by ITU-T (ITU-T G.983). This standard evolved and was later called Broadband PON (BPON), which offered 155 Mbps, 622 Mbps, or 1.2 Gbps downstream and 155 Mbps or 622 Mbps upstream (but usually was used for 155 Mbps/622 Mbps upstream/downstream). Later, Gigabit PON (GPON) was standardized (ITU-T.984), and offered 1.2 Gbps or 2.4 Gbps downstream, and 155 Mbps to 2.4 Mbps upstream, encapsulating not just ATM as APON and BPON did, but also Ethernet and TDM (PDH). GPON also supports up to 128 splits in the ODN, but practically, its deployment is less in the reach and split ratio, depending on the optical link quality (called *link budget*) [111]. Therefore, up to

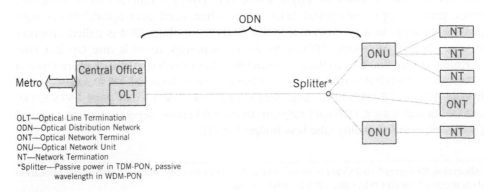

OLT—Optical Line Termination
ODN—Optical Distribution Network
ONT—Optical Network Terminal
ONU—Optical Network Unit
NT—Network Termination
*Splitter—Passive power in TDM-PON, passive
      wavelength in WDM-PON

FIGURE 4.2

Passive optical network

128 logical ONUs can be supported by the GPON transmission convergence layer, as defined in ITU-T G.984.3 [202]. The span of the GPON can reach up to 60 km (with a 20 km differential reach between ONUs).

The GPON frame is standardized, and is generalized to cope with multiservice needs (i.e., telephony, Internet, etc.). Every frame lasts 125 μs and can contain 19,440 bytes in a 1.24 Gbps system or 38,880 bytes in a 2.48 Gbps system.

Gigabit PON Transmission Convergence (GTC) is a sublayer in GPON, which defines a basic control unit called Transmission Container (T-CONT) that is identified by Allocation Identifier (Alloc-ID).[8] T-CONT is multiplexed into the ONUs,[9] which are multiplexed into the PON. T-CONT includes TDM blocks and data fragments that are encapsulated according to the GPON Encapsulation Method (GEM),[10] as well as ATM cells. To be more specific, the TDM and data traffic (e.g., Ethernet) are multiplexed into Ports (identified by Port-ID), which are in turn multiplexed into the T-CONTs. Similarly, ATM's Virtual Circuits (VC) are multiplexed into Virtual Paths (VP), which are in turn multiplexed into the T-CONTs.

Gigabit PON's downstream frames (those that flow from the OLT to the ONU or ONT) contain multiplexed ATM, TDM or Ethernet data for the ONUs/ONTs, as each of the Physical Control Block downstream (PCBd) headers of the frames defines in its physical length downstream (Plend) field (see Figure 4.3).

The GPON upstream frames contain multiplexed data from the ONUs for the OLT, where the allocation of each piece of the ONU's data is determined by the Upstream Bandwidth Map field in the PCBd of the downstream GPON frame, as shown in Figure 4.4. Each of the allocated data slots for an ONU contains ATM data and GEM frames; each encapsulates Ethernet or TDM data.

In parallel to GPON, Ethernet PON (EPON) was standardized by the IEEE as IEEE802.3ah [186], and is considered to be the "first mile" Ethernet. EPON offers dynamic bandwidth allocation at 1 and 10 Gbps. Some view it as the ultimate TDM-PON, while others see GPON as the ultimate TDM-PON (at least if the anticipated 100 Gbps symmetric links are made) [274].

Ethernet for subscriber access networks, also called Ethernet in the First Mile (EFM) [186], defines the Physical and MAC layers of Ethernet to be used over voice grade copper or optical fiber cable. When used over optical fibers with passive splitters in a point-to-multipoint (P2MP) topology, EFM is called Ethernet Passive Optical Network (EPON). In such a topology, there is one Optical Line Terminal (OLT; attached to the access network) connected to one or more Optical Network Units (ONUs; near the subscribers). This is depicted, for example, in the hybrid (P2P and P2MP) EPON topology shown in Figure 4.5 (see page 157). EPON can contain at least a 1:16 split ratio, up to even 64 splits, depending on the optical path components' quality (the link budget) [111].

---

[8]Allocation ID values 0 to 253 are used to address the ONUs directly, and if more than one allocation ID is required for an ONU, values 255 to 4095 are used.

[9]Identified by ONU-Identifier (ONU-ID).

[10]GEM's concepts and format is similar to the Generic Framing Procedure (GFP) [172], described in the previous chapter.

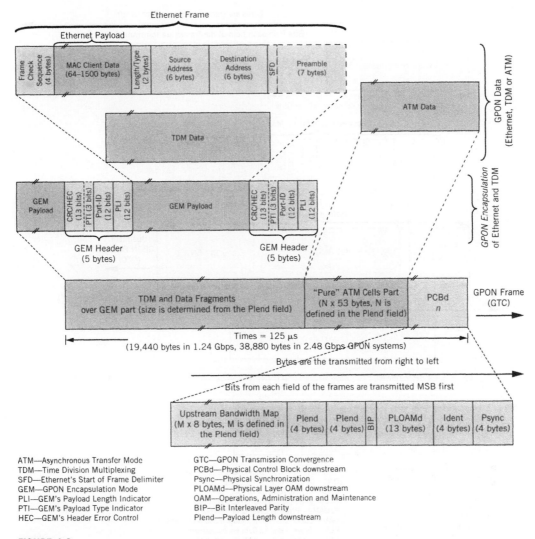

FIGURE 4.3

GPON transmission convergence—downstream frame (from right to left)

A Multipoint MAC Control Protocol (MPCP) is used to control the multipoint access between any ONU and the P2MP OLT. An MPCP frame format is shown in Figure 4.6, where the Op-Code,[11] TimeStamp, and data fields contain the required messages, timers, and statuses. Point-to-Point (P2P) emulation is achieved by the MPCP, to enable higher protocol layers to see the P2MP network as a collection of point-to-point links. This is done by replacing three bytes of the preamble with

---

[11]For instance, Op-Code 00–02 is used for the "GATE" command, which the OLT uses to allow ONUs to transmit frames at a specific time and for a specific duration.

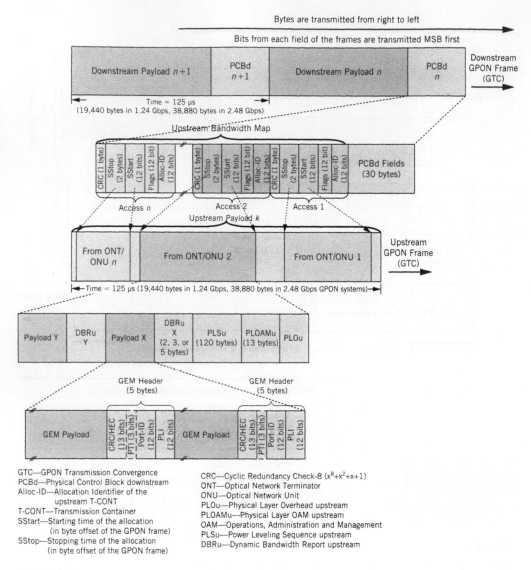

Bytes are transmitted from right to left

Bits from each field of the frames are transmitted MSB first

GTC—GPON Transmission Convergence
PCBd—Physical Control Block downstream
Alloc-ID—Allocation Identifier of the
    upstream T-CONT
T-CONT—Transmission Container
SStart—Starting time of the allocation
    (in byte offset of the GPON frame)
SStop—Stopping time of the allocation
    (in byte offset of the GPON frame)

CRC—Cyclic Redundancy Check-8 (x^8+x^2+x+1)
ONT—Optical Network Terminator
ONU—Optical Network Unit
PLOu—Physical Layer Overhead upstream
PLOAMu—Physical Layer OAM upstream
OAM—Operations, Administration and Management
PLSu—Power Leveling Sequence upstream
DBRu—Dynamic Bandwidth Report upstream

FIGURE 4.4

GPON transmission convergence—upstream frame (from right to left)

a two-byte Logical Link Identification (LLID) field, and a one-byte CRC-8[12] field (Figure 4.7). The frame's preamble is restored when the frame is used outside of the link between the OLT and the P2MP ONU.

---

[12]The CRC is computed from the 3rd byte of the preamble through the LLID field, the 7th byte of the preamble, and its generating function is $x^8 + x^2 + x + 1$. For a description of Cyclic Redundancy Check (CRC), see Chapter 5.

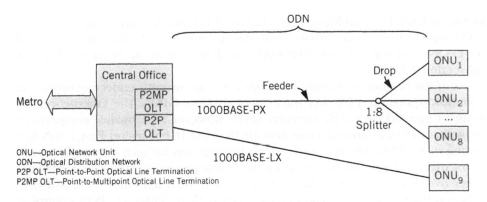

FIGURE 4.5

Hybrid EPON example

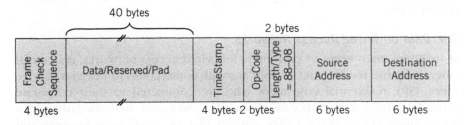

FIGURE 4.6

Generic MPCP frame (MAC control frame, from right to left)

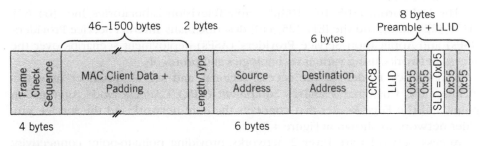

FIGURE 4.7

EPON frame format (from right to left)

Downstream frames are broadcast by the OLT, and each ONU discards the frames that are not addressed to it (using the LLID field). The upstream frames are transmitted by the ONUs, using the time slots assigned to them by the OLT (negotiating it with the MPCP).

An optional Operations, Administration, and Maintenance (OAM) sublayer may also be included in the EFM, to support network operations and troubleshooting in the link-level (in addition, IEEE802.1ag [180] standardized a network level OAM, which is the Connectivity Fault Management [CFM]). The link level OAM (of EPON) and the network level OAM are described in the previous chapter's subsection 3.3.2.3.

As of the beginning of 2008, a 10 Gbps EPON is currently underway, proposing a fourfold speed increase over the current 2.5 Gbps GPON. The Ethernet-centric usage of networks naturally supports EPON; however, performance comparisons show that GPON is more efficient than EPON in the 1 Gbps region. First, EPON line coding of 8B/10B (which requires 10-bit transmission for each 8-bit byte) causes a 25% overhead. Second, the upstream burst multiplexing is difficult in EPON, and larger overhead is required than in GPON (more than ten times the guard time between frames, preamble, and delimiter times) [330]. Third, any PON control in EPON requires an MPCP Ethernet frame, which is not the case in GPON. However, as can be easily seen, EPON is simpler than GPON, and in an all Ethernet network, EPON eliminates the need for the convergence sublayer and the required mappings and conversions.

4.1.1.4 Data on Wired Access Networks

There are three main types of customers for wired access networks: residential customers who use the Internet (usually through connections to Internet Service Providers, ISP); residential customers who are connected to their corporate networks (intranet); and corporations, usually Small and Medium Enterprises (SMEs) or Small Office Home Office entities that are using both Internet and intranet. Large corporations usually bypass the access networks and connect directly to the carriers' networks, that is, to their regional networks, beyond the CO or the POPs.

The DSL forum [105, 107, 108],[13] Cable Television Laboratories, Inc., [64, 65], ITU-T [224–226], and the IETF [25, 340] describe many Internet Service Providers' (ISPs) and Application Service Providers' (ASPs) deployment scenarios over the access networks, using various technologies and protocols.

A reference model for the access, regional, and service provider networks includes the Customer Premises Equipment (CPE),[14] Access Node, Aggregation point, and the Edge Router that connects the core network and the service provider network, as shown in Figure 4.8.

Access networks are Layer 2 networks, providing point-to-point connectivity, from the CPE to some aggregation point. Layer 2 networking in the access networks must deal with service requirements such as security, QoS, routing, address assignments, and more. In order to respond to these requirements, many implementation

---

[13]The DSL forums' TR-025 is the basic, traditional DSL network architecture. TR-059 is basically about BRAS/DSLAM collocation, and it specifies the next generation DSL network architectures in terms of multiservices, support of QoS, and IP services. TR-101 suggests the next evolution by upgrading the access network to support Ethernet transport and switching capabilities.

[14]CPE usually includes hosts and Customer Premises Networks (CPN), attached to a network interface device (e.g., modem, Ethernet switch or ONU) that is attached to the access network.

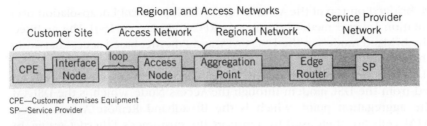

CPE—Customer Premises Equipment
SP—Service Provider

FIGURE 4.8

Access network reference model

scenarios are used, some more common than others, and they vary between operators and countries. All these scenarios can be grouped into three basic models[15] for data transport over the various types of Layer 2 access networks:

■ A bridged network that transfers plain Ethernet frames from the customer across the network to an Aggregation Point or the Edge Router. As is the case usually with Ethernet, a bridged network simply bridges the Ethernet frames.

■ A Point-to-Point Protocol (PPP) Terminated Aggregation (PTA) model, in which the old, common, and mature PPP remote-access method is used to connect the customer to an Aggregation Point. IP traffic completes the link and connects the Aggregation Point and the Edge Router.

■ A Layer 2 Tunneling Protocol (L2TP) Access Aggregation (LAA) model, in which a PPP tunnel is created between the customer and the Edge Router. An L2TP tunnel is established between the Aggregation Point and the Edge Router.

Bridged networks are very common in CATV and Ethernet deployments, and the other two models are used in DSL deployments, although they are also used for CATV and Ethernet deployments, due to the advantages that tunneling offers (e.g., greater security).

4.1.1.4.1 Bridged Model

This is the simplest way to transport Ethernet frames from the customer to the service provider, as bridging and plain routing are the only network functions required. In Ethernet access networks, where the entire access network is Ethernet-based, it is clear that bridging should be used. CATV access networks use cable MAC, which can easily carry the Ethernet frames, so it is also quite simple. However, DSL access networks impose the most difficult bridging mechanism, since they usually use the ATM on top of the physical DSL links. For these networks, Ethernet

---

[15]DSL Forum's TR-025 [105], which defines the network architecture, groups access networks into four categories: Transparent ATM Core Networks, Layer 2 Tunneling Protocol (L2TP) Access Aggregation (LAA), Point-to-Point Protocol Terminated Aggregation (PTA), and Virtual Path Tunneling Architecture (VPTA). We describe two of these models in this chapter, as the other two are end-to-end ATM-Based, and are either not scalable or not commonly used.

frames are bridged on top of the ATM, using the Multiprotocol Encapsulation over ATM Adaptation Layer 5 method (known as RFC 2684 layer) [154]. The architectures of the bridged models in the DSL, CATV, and Ethernet-based access networks are shown in Figure 4.9.

In DSL access networks, Permanent Virtual Circuits (PVCs) for ATM cells are established from the DSL modem through the Access Node, which is the DSLAM, and to the aggregation point, which is the Broadband Remote Access Server (BRAS). ATM cells are then used to transport the customer's Ethernet traffic, by using the Multiprotocol Encapsulation over ATM Adaptation Layer 5 (RFC 2684) [154] method. Finally, the BRAS aggregates the customer's traffic, and transports the customers' L2 or L3 traffic (IP) to the Edge Router, to the core network, and to the service provider.

In the case of CATV, Ethernet frames are carried in an easier way, as noted before, with minimal conversions and encapsulations. CATV uses DOCSIS' MAC to carry the Ethernet frames from the Cable Modem (CM) to the Cable Modem Termination System (CMTS), which aggregates the customers' traffic and uses IP packets to access the Edge Router, the core network, and the service provider.

Recently, Ethernet-to-the-Home/Business (ETTx) has become more and more popular, mainly in dense areas where Ethernet links are possible, such as multitenant buildings, large office buildings, or hotels. ETTx enables high bandwidth L3 (IP) and VPN services, and is highly efficient. Current ETTx deployments are based on either Ethernet in the First-Mile (EFM), FITL, or DSL-like Ethernet (*DSL generation* 3), such as Cisco's proprietary Long Reach Ethernet (LRE). Ethernet frames are aggregated in several levels of Ethernet aggregation switches, and are sent to the Broadband Network Gateway (BNG).[16] BNG can be used instead of the Edge Router, and it includes some of the BRAS functionalities. The difference between BRAS and BNG is that BRAS aggregates all ATM PVCs from the CPE through the DSLAM, while the BNG aggregates Ethernet frames through many Ethernet aggregation switches. Another difference results from service segregation, that is, allowing several, specific BNG for various services, such as Video-BNG, and so on [108].

There are, however, some permutations in these configurations, for example, in "*DSL generation 2*" (defined by TR-101 [108]), where the DSLAM terminates the ATM PVCs, and transmits Ethernet frames to a BNG, or more than one BNG.

4.1.1.4.2 Point-to-Point Protocol Terminated Aggregation Model

This model is the result of the vast use and experience gained with Point-to-Point Protocol (PPP)[17] [387] for remote access applications, which has been very common since the days of dial-up connections. PPP offers a mature and rich control-plane

---

[16]A BNG is an IP-edge router where bandwidth and QoS policies are applied; the functions performed by a BRAS are a superset of those performed by a BNG [108, 296].

[17]PPP [387] is based on the ISO 3309 HDLC [193] protocol and frame format [388]. HDLC is a well-known data-link layer protocol that includes a sliding window mechanism for Automatic Repeat Request (ARQ) handshake. PPP followed many protocols that were used to enable dial-in access and other applications, such as Serial Line Internet Protocol (SLIP).

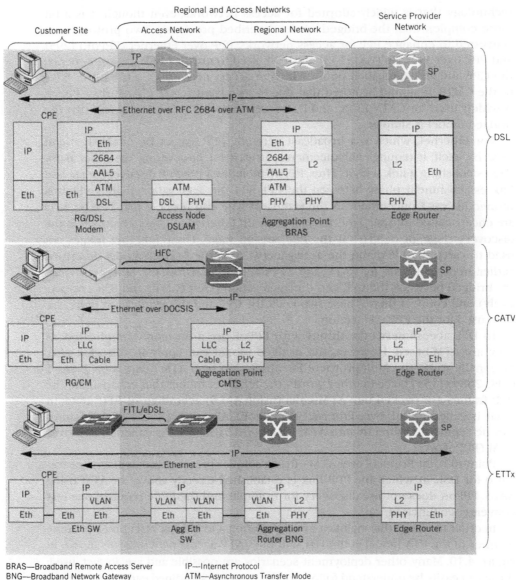

BRAS—Broadband Remote Access Server
BNG—Broadband Network Gateway
CPE—Customer Premises Equipment
SP—Service Provider
CATV—Cable TV
HFC—Hybrid Fiber Coax
CM—Cable Modem
CMTS—Cable Model Termination System
ETTx—Ethernet to the home/business/x
FITL—Fiber in the Loop
VLAN—Virtual Local Area Network (LAN)
Agg Eth SW—Aggregation Ethernet Switch

IP—Internet Protocol
ATM—Asynchronous Transfer Mode
AAL5—ATM Adaptation Layer 5
L2—Layer 2
Eth—Ethernet
LLC—Logical Link Control
Phy—Physical layer
DSL—Digital Subscriber Loop/Line
DSLAM—DSL Access Multiplexer
TP—Twisted Pairs telephone wires
eDSL—Ethernet DSL
RG—Residential Gateway

FIGURE 4.9

Bridged data access network architectures

technology that is widely adopted for access networks, even though it is a bit more complex than the bridged model described previously. Two protocols are suggested to enhance PPP for access networks: PPP over Ethernet (PPPoE) [295] and PPP over ATM (PPPoA) [153]. Both are used to create a PPP session between the CPE and an aggregation point, which in access networks are the BRAS, BNG or the CMTS. It should be noted that due to its advantages (e.g., security), PPPoE is widely used for CATV, ETTx, and DSL access networks, despite the fact that the bridged model is simpler.

For Ethernet, which is a broadcast network, PPP cannot be established and used by itself; it requires enhancements, at least for discovering the other party and establishing a link with it. Thus, PPPoE includes a pre-PPP discovery stage and link establishment phase between the CPE and the aggregation point (which is termed Access Concentrator in the standard). The end result is that PPP packets are encapsulated into the Ethernet frames, with Ethertype equal to 0x88-63 at the discovery stage, or 0x88-64 at the PPP session stage. These Ethernet frames are used to carry the PPPoE on the access network in the bridged model described earlier. In other words, for CATV and ETTx access networks, the Ethernet frames are bridged between the CPE and the aggregation point (CMTS and BNG, respectively), and connect them. These two ends, the CPE and the aggregation point, are then used for the PPP connection.

In DSL access networks, things with PPPoE are a bit more complicated, as they are ATM-based. However, the bridged model can apply in this case also, optionally using the Multiprotocol Encapsulation over ATM Adaptation Layer 5 (RFC 2684) [154] layer. In other words, the Ethernet frames that encapsulate the PPP frames are packed into ATM cells (using AAL5) according to RFC 2684, in what is called PPPoE over Ethernet (PPPoEoE). Alternatively, the PPPoE frames can be segmented directly into ATM (using AAL5), in what is called PPPoE over ATM (PPPoEoA).

Point-to-Point Protocol over ATM offers a simpler protocol and deployment scenario for DSL, in which the PPP link is established directly above the ATM using AAL5. PPPoA does not use discovery, for example, and therefore requires one less convergence function.

In order to clarify this multilayer encapsulation and links, several possible architectures of the PTA model in the DSL and the CATV access networks are shown in Figure 4.10. Many other deployment scenarios are possible and are, in fact, used; these can easily be understood following the principles outlined earlier and shown in Figure 4.10. It should be noted, though, that in the PTA model used for ETTx, PPP connections may begin in the CPE equipment and can be terminated either at the BNG, at the BRAS, or at the Edge Router, depending on the specific deployment scenario used for the ETTx.

As noted above, PTA models are often preferred over the bridged model despite their relative complexity, since they offer a better and simpler security scheme, authentication services, IP Quality of Service (QoS) control, accounting support, and possible dynamic address assignment, among many other advantages.

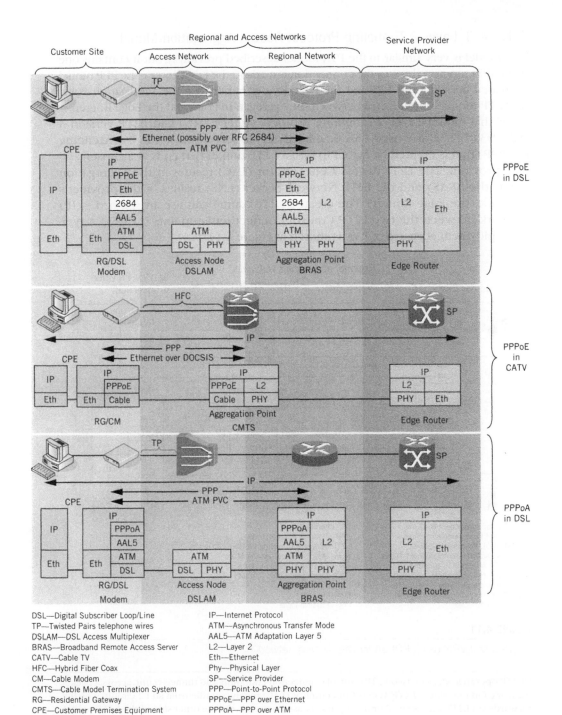

FIGURE 4.10

PTA model deployment examples

4.1.1.4.3 Layer 2 Tunneling Protocol Access Aggregation Model

This model is very similar to the PTA model described previously, but it contains one major difference in the way the connection is made to the Edge Router: the PPP tunnel is created between the CPE and the Edge Router rather than the Aggregation Point. An L2TP tunnel is established between the Aggregation Point and the Edge Router.

In this model, Layer 2 Tunneling Protocol (L2TP) [409] and L2TP version 3 (L2TPv3)[18] [272] are used. LAA enables dynamic service selection by extending the PPP (that is created by the CPE) into a L2TP tunnel that ends at a chosen service provider. The L2TP Access Concentrator (LAC) resides in the Aggregation Point (the BRAS), and the L2TP Network Server (LNS) resides in the provider's network (the Edge Router). The created L2TP tunnel can use any L2 or L3 (IP) network between the two L2TP Control Connection Endpoints (LCCEs). A typical deployment scenario of LAA is shown in Figure 4.11.

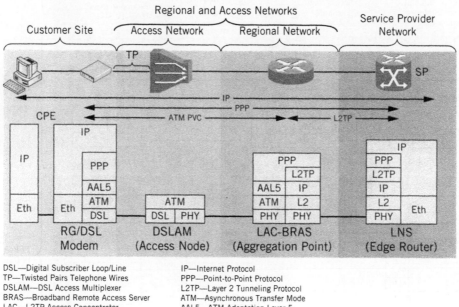

DSL—Digital Subscriber Loop/Line
TP—Twisted Pairs Telephone Wires
DSLAM—DSL Access Multiplexer
BRAS—Broadband Remote Access Server
LAC—L2TP Access Concentrator
LNS—L2TP Network Server
RG—Residential Gateway
CPE—Customer Premises Equipment
SP—Service Provider

IP—Internet Protocol
PPP—Point-to-Point Protocol
L2TP—Layer 2 Tunneling Protocol
ATM—Asynchronous Transfer Mode
AAL5—ATM Adaptation Layer 5
PVC—Permanent Virtual Circuit
L2—Layer 2
Eth—Ethernet
Phy—Physical Layer

FIGURE 4.11

L2TP-based VPN using PPPoA for DSL access network

---

[18]L2TP operates between two L2TP Control Connection Endpoints and tunnels traffic across a packet network. On one side, a L2TP Access Concentrator (LAC) receives traffic from an L2 circuit, which it forwards via L2TP across an IP or other packet-based network. On the other side of the L2TP tunnel, a L2TP Network Server (LNS) logically terminates the L2 circuit locally and routes network traffic to the ISP network.

4.1.1.4.4 Additional Access Networks Implementations

There are many other access network implementations, mainly for tunneling the data from the access and regional networks to the service provider networks. This can be done either through core networks that require tunneling, or for enabling dynamic service selection (e.g., between several ISPs, ASPs, or corporate services), or both. An example of such implementation is the Point-to-Point-Tunneling Protocol (PPTP) [166] (used mainly in Europe), in which a tunnel is created between the CPE and a PPTP server. Another example is a VPN usage, for example, MPLS, as demonstrated in Figure 4.12.

4.1.1.4.5 Ethernet Demarcation

Ethernet access networks seem to be the ultimate solution for connecting the Ethernets of enterprises, campuses, and multitalented buildings and houses with that of the providers' Ethernet. After all, the vast majority of data traffic originates and terminates in Ethernets, and all-Ethernet systems can simplify a lot networking, decrease costs, and increase functionality. In such an End-to-End Ethernet scenario, the access networks merge with the metro and the core networks, as can be seen in Figure 4.13.

As mentioned earlier, there are several technologies that, as of this writing, support this Ethernet-to-the-Home/Building/Office (ETTx) trend. In such ETTx deployments, the demarcation point becomes *Ethernet Demarcation*. Ethernet demarcation, however, is more than just the plain physical interface box used in the traditional demarcation point. That is, Ethernet demarcation is still the point

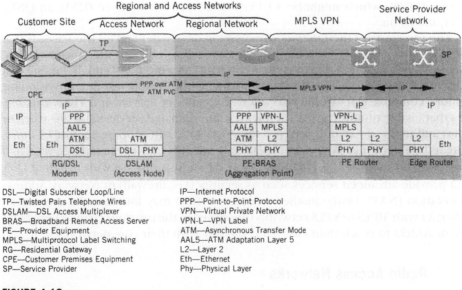

DSL—Digital Subscriber Loop/Line
TP—Twisted Pairs Telephone Wires
DSLAM—DSL Access Multiplexer
BRAS—Broadband Remote Access Server
PE—Provider Equipment
MPLS—Multiprotocol Label Switching
RG—Residential Gateway
CPE—Customer Premises Equipment
SP—Service Provider

IP—Internet Protocol
PPP—Point-to-Point Protocol
VPN—Virtual Private Network
VPN-L—VPN Label
ATM—Asynchronous Transfer Mode
AAL5—ATM Adaptation Layer 5
L2—Layer 2
Eth—Ethernet
Phy—Physical Layer

FIGURE 4.12

MPLS-based VPN using PPPoA for DSL access network

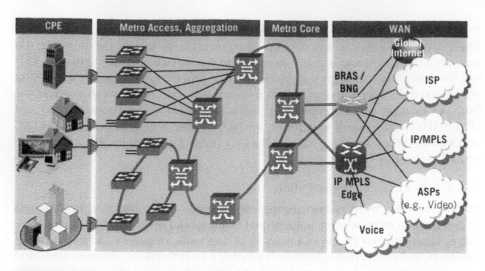

FIGURE 4.13

End-to-end Ethernet

where the provider's Ethernet ends and the customer's Ethernet begin. However, separating these two networks requires management of the operational and administrative aspects of these networks as well as the interface between them. Ethernet demarcation functions may even be attributed to an Ethernet Demarcation Device (EDD), which might be a CPE, a User-Network-Interface (UNI), an ONU/ONT, and so on.

The functions of the Ethernet demarcation device depend on the services that the customer requires from the provider, as well as on both the customer's and the provider's Ethernet deployment architecture. A simple EDD might be one that offers a clear point-to-point Ethernet channel with possibly just rate provisioning. A more complex EDD would also offer L2 Quality of Service and VLAN capabilities that enable multiple ports, each with a differentiated services (i.e., for voice or data) and traffic separation between the ports. EDD can support not just L2 differentiated services (i.e., Port and VLAN-based), but also L3- to L7-based differentiated services. This enables the EDD to support different protocols with matched QoS, and provide advanced services such as security (i.e., firewall) or Network Address Translation (NAT). Lastly, another category of EDD may interface the customer's Ethernet with MPLS or VPLS networks (described in the previous chapter), enabling the providers to reach their customers directly with their core networks.

4.1.2 Radio Access Networks

In mobile communications, Radio Access Networks (RANs) usually refer to the part of the cellular networks that interacts directly with the mobile equipment (and is sometimes also called the Base-station subsystem, BSS). The other part of these networks is called the core network (or the network subsystem, NSS), and is the part that is directly attached to the telecommunication network, for example, the Public Switched Telephone Network (PSTN), or the Internet.

Mobile communications began purely for voice applications, then data applications were added, and then these two converged to the point where voice has become just another data application of the converged network. Mobile communications has undergone several generations, and is still evolving; these generations are known as 1G, 2G, 2.5G, 3G, and 4G. The *first generation*, 1G, introduced analog cellular systems that only enabled voice telephony,[19] and used circuit switching. The *second generation*, 2G, revolutionized these networks by using digital communications.[20] The next generation was called 2.5G; it added packet switched based data transfer using the 2G infrastructure.[21] The *third generation,* 3G, significantly improved data services and offered higher spectral efficiency.[22] 3G is still evolving (as of this writing),[23] and offers interoperability with IP Multimedia Services (IMS) [69, 353] and with unlicensed mobile radio network (Wi-Fi, which is based on IEEE 802.11 standards, and is described in the next subsection—home networks).

The forthcoming *fourth generation*, 4G, is expected to offer higher data speeds that support real-time TV and video streaming, and is based on an all-packet network (voice and data).[24] 4G is also expected to support IMS and roaming to noncellular systems (Satellite, Wi-Fi, etc.) according to the Generic Access Network (GAN)

---

[19]The standards typically used for the first generation were the Nordic Mobile Telephone (NMT), C450, radiocom 2000, RTMI, Total Access Communications System (TACS) (used in the United Kingdom), Japan Total Access Communications System (JTACS), and Advanced Mobile Phone System (AMPS) along with its derived Narrowband Analog Mobile Phone Service (NAMPS) (both used in the United States).

[20]Known standards that were used for the second generation were Time Division Multiple Access (TDMA), Global System for Mobile (GSM), Digital AMPS, which were all based on time and frequency multiplexing, as well as Code Division Multiple Access (CDMA), known as Interim Standard 95 (IS-95), which was based on code multiplexing.

[21]Standards used for this intermediate generation were the General Packet Radio System (GPRS), Enhanced Data rates for GSM Evolution (EDGE), enhance GPRS, and 1xRTT, which is the CDMA data service.

[22]The standards implemented in the 3G originated mainly from two sources: the European-headed third Generation Partnership Project (3GPP) and the American-headed 3GPP2. The 3GPP's standards are Universal Mobile Telecommunications System (UMTS) or Wideband CDMA (WCDMA) with High-Speed Packet Access (HSPA), while 3GPP2 standards are CDMA2000 with Evolution-Data Optimized (EV-DO).

[23]The 3GPP, for example, published several releases of standards for 3G; for example, UMTS rel. 99, rel. 4, 5, 6, and 7 (the numbers refer to the years they were published; for example, rel. 4 was published in 2004). UMTS Rel. 8, termed Long Term Evolution (LTE) is 3GPP's first 4G standard to come, which is estimated to be finished by 2008.

[24]Standards developed for the fourth generation are UMTS Rev. 8 (Long Term Evolution, LTE) by the 3GPP, Ultra Mobile Broadband (UMB) by the 3GPP2, and WiMAX, which is based on IEEE 802 standards (802.16 working group).

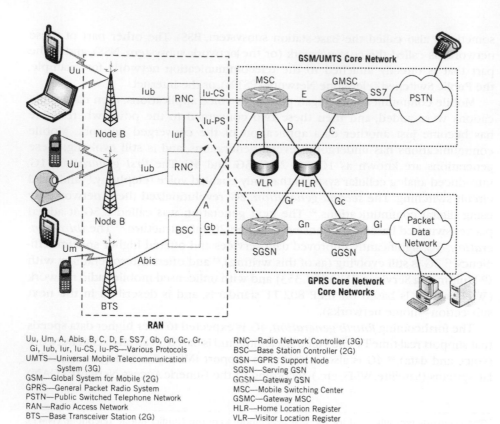

FIGURE 4.14

GSM/GPRS/UMTS network architecture

architecture (also known as Unlicensed Mobile Access UMA), which evolved under 3G. ITU-T organized and standardized the various 3G and 4G standards in its ITU Mobile Telecommunication (IMT-2000).

4.1.2.1 2G, 2.5G, 3G

Figure 4.14 depicts a generic architecture of RAN and the core network for 2G, 2.5G, and 3G (Global System for Mobile [GSM],[25] General Packet Radio System

---

[25]GSM originally stands for *Groupe Spécial Mobile* (in French), a body which was created in 1982 by the European Conference of Postal and Telecommunications Administration (CEPT). Later, in 1988, CEPT created the European Telecommunications Standards Institute (ETSI), and GSM, along with other standards' committees, was transferred to ETSI. GSM was changed to Global System for Mobile by ETSI in 1991, when the first GSM system became operational. In 1998, ETSI established the third Generation Partnership Project (3GPP) with other standardization bodies, to which GSM and its related standards were moved.

[GPRS], Enhanced Data rates for GSM Evolution [EDGE], and Universal Mobile Telecommunications System [UMTS]). The equivalent U.S. networks for 2G, 2.5G, and 3G, that is, the Code Division Multiple Access (CDMA) based Interim Standard 95 (IS-95) and CDMA2000, are principally similar, and we focus on the 3GPP networks since they are used more than their U.S. counterparts (about 85% of cellular networks worldwide are 3GPP based).

The network is based on GSM/EDGE Radio Access Network (GERAN), or on Universal Terrestrial Radio Access Network (UTRAN) in the case of UMTS, and a core network. A Mobile Station (MS), sometimes called Mobile Terminal (MT) or User Equipment (UE), is connected to base stations of the cellular network via a radio link, which is the only wireless channel in the cellular network.[26] A base station is termed Base Transceiver System (BTS) in GSM and Node B in UMTS; its controller is called Base Station Controller (BSC) in GSM, and Radio Network Controller (RNC) in UMTS. All these network elements can be interoperable in one cellular network, connected to the core network. Circuit Switching (CS) is carried by the Mobile Switching Center (MSC), which interfaces the PSTN through a Gateway MSC (GMSC). Packet Switching (for data) is handled by a Serving GPRS Support Node (SGSN), which interfaces the public Packet Data Network (PDN), for example, the Internet, through the Gateway GPRS Support Node (GGSN). The cellular system uses repositories that contain all customer information for authentication, authorization, billing, location, and other tasks. There is a Home Location Register (HLR) for keeping all local (home) customer information, and a Visitor Location Register (VLR) that serves roaming customers.

There are many more network elements that are omitted from the schematic description for clarity. The network elements run many protocols that are used for lots of control-plane tasks such as call set-up, messaging, circuit switching, radio-channels management, localization, billing, authentication, and OAM. Discussion of these tasks goes well beyond the scope of this book; however, as network processors reach the cellular networks for data applications, a description of the packet forwarding in the user plane is necessary.

For the sake of completeness, it should be noted that GSM evolved to UMTS release 99, which was updated to Release 4 (to include ATM or IP-based connections between network elements, and to separate the MSC into the MSC server—which does the control functions—and the Media Gateway, MGW—which provides media switching functions). Release 5 interfaces IP Multimedia Services (IMS), offers High Speed Packet Access (HSPA), and enables multimedia sessions, while Release 6 incorporates Generic Access Network (GAN) and uses Multiple Input and Multiple Output (MIMO) antenna arrays for doubling HSPA data rates.

---

[26]Additional microwave channels in the cellular network can be found for connecting various network elements of the cellular network, such as among the base station or between them and their controllers. These links are part of the cellular back-haul in the RAN. Other microwave links are possible between the base station controllers and the Mobile Switching Center (MSC).

Figure 4.15 demonstrates some of the protocol stacks that are used in the main network nodes that carry the User Equipment's (UE) packets to the Public Data Network (PDN), for example, the Service Provider (SP). It should be noted that since the systems are compatible, cross protocols and equipment can be used; for example, Abis protocol can be served on an RNC, thus allowing a UMTS system to work with GSM base stations. The edge protocol, that is, Gi, connects the Gateway GPRS Support Node (GGSN) to the Internet, ISP or to an IP Multimedia System (IMS) core for VoIP or other streaming applications.

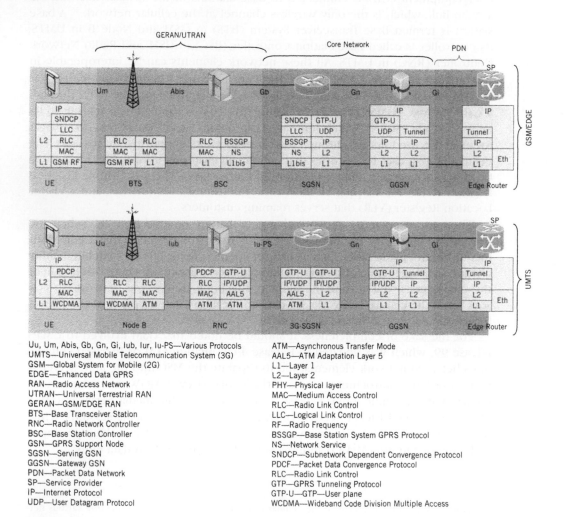

Uu, Um, Abis, Gb, Gn, Gi, Iub, Iur, Iu-PS—Various Protocols
UMTS—Universal Mobile Telecommunication System (3G)
GSM—Global System for Mobile (2G)
EDGE—Enhanced Data GPRS
RAN—Radio Access Network
UTRAN—Universal Terrestrial RAN
GERAN—GSM/EDGE RAN
BTS—Base Transceiver Station
RNC—Radio Network Controller
BSC—Base Station Controller
GSN—GPRS Support Node
SGSN—Serving GSN
GGSN—Gateway GSN
PDN—Packet Data Network
SP—Service Provider
IP—Internet Protocol
UDP—User Datagram Protocol

ATM—Asynchronous Transfer Mode
AAL5—ATM Adaptation Layer 5
L1—Layer 1
L2—Layer 2
PHY—Physical layer
MAC—Medium Access Control
RLC—Radio Link Control
LLC—Logical Link Control
RF—Radio Frequency
BSSGP—Base Station System GPRS Protocol
NS—Network Service
SNDCP—Subnetwork Dependent Convergence Protocol
PDCF—Packet Data Convergence Protocol
RLC—Radio Link Control
GTP—GPRS Tunneling Protocol
GTP-U—GTP—User plane
WCDMA—Wideband Code Division Multiple Access

FIGURE 4.15

GSM/EDGE/UMTS user plane data protocols

4.1.2.2 4G-Long Term Evolution

A generic architecture of the forthcoming 4G Long Term Evolution (LTE), is defined as shown in Figure 4.16. In this architecture [2, 3], the evolved Node B's (eNBs) of the evolved UTRAN (E-UTRAN) provide the user plane and control plane protocol termination on the side of the User Equipment (UE). The eNBs are connected to the Evolved Packet Core (EPC) (by the S1 protocol [4]) and are also interconnected among themselves (by the X2 protocol). The User Equipment (UE) communicates with the eNBs by Wideband Code Division Multiple Access (W-CDMA) radio links.

The Serving Gateway (S-GW) and the Packet Data Network (PDN) Gateway (P-GW) forward and route the user plane data. The S-GW is the mobility anchoring point, and the P-GW allocates UE IP addresses and enables service-level charging, gating, and rate enforcement. Therefore, network processors might become a

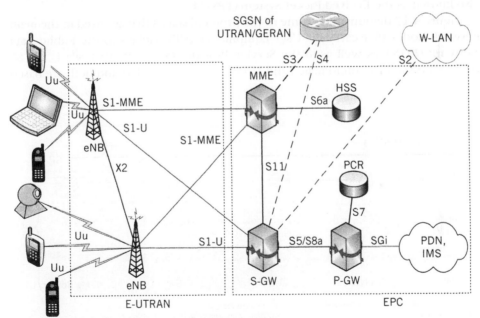

Uu, Um, X2, S1, S2, S3, S4, S5, S6a,
S7, S8a, SGi—Various Protocols
GPRS—General Packet Radio System
GSN—GPRS Support Node
SGSN—Serving GSN
RAN—Radio Access Network
UTRAN—Universal Terrestrial RAN
E-UTRAN—Evolved UTRAN
GERAN—GSM/EDGE RAN
eNB—Evolved Node B

MME—Mobility Management Entity
S-GW—Serving Gateway
P-GW—PDN Gateway
HSS—Home Subscriber Server
PCRF—Policy and Charging Rules Function
IMS—IP Multimedia System
EPC—Evolved Packet Core
W-LAN—Wireless Local Area Network (Wi-Fi)

FIGURE 4.16

4G (LTE) network architecture

necessity in P-GW devices, particularly when wideband and complicated IP tasks are involved. The S-GW and the P-GW can be on the same physical server, collocated, or on different and even distant servers. The Mobility Management Entity (MME) is responsible for signaling, selecting S-GW and P-GW, roaming, handover, and authentication. The function of the HLR of the GSM/GPRS networks is done in LTE by a Home Subscriber Server, which maintains all the customer information. Legacy networks (e.g., GSM) and non-3GPP networks (e.g., Wi-Fi) are interoperable with the LTE by means of protocols that tie them into the LTE network.

For the sake of clarity, many network elements and protocols are omitted from this schematic description, as in the GSM/GPRS scheme above. Protocols for packet forwarding in the user plane are described in Figure 4.17 in the same way as they were described above for the GSM/GPRS networks. In the case of LTE, however, the entire traffic is packet-based, rather than just the GPRS part (the voice traffic of GSM/UMTS networks is circuit-switched through the MSC and the GMSC). The packet network of LTE is called Evolved 3GPP Packet-Switched Domain, which is also known as the Evolved Packet System (EPS) [1].

Figure 4.17 demonstrates some of the protocol stacks that are used in the main network nodes that carry the User Equipment's (UE) packets to the Public Data Network (PDN), as well as to the Service Provider (SP). Note that S8a is the S5 protocol variant for inter-Public Land Mobile Network (PLMN), and both are based

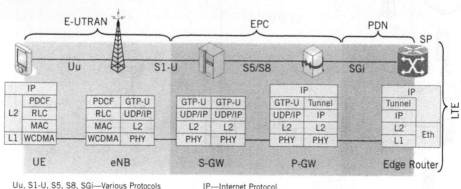

Uu, S1-U, S5, S8, SGi—Various Protocols
RAN—Radio Access Network
E-UTRAN—Evolved Universal Terrestrial RAN
EPC—Evolved Packet Core
PDN—Packet Data Network
SP—Service Provider
UE—User Equipment
eNB—Evolved Node B
S-GW—Serving GW
P-GW—PDN GW
L1—Layer 1
L2—Layer 2

IP—Internet Protocol
WCDMA—Wideband Code Division Multiple Access
PDCF—Packet Data Convergence Protocol
RLC—Radio Link Control
MAC—Medium Access Control
GPRS—General Packet Radio System
GTP—GPRS Tunneling Protocol
GTP-U—GTP—User Plane
UDP—User Datagram Protocol
PHY—Physical Layer
Eth—Ethernet
LTE—Long Term Evolution

FIGURE 4.17

LTE user plane data protocols

on the GPRS Tunneling Protocol (GTP). Since S4 is based on Gn protocol (which is also GTP), S4 can be used to attach legacy 2G and 3G networks to the LTE's S-GW instead of to the SSGN in the legacy networks (see Figure 4.16 for the protocol references).

4.1.2.3 4G-WiMAX

WiMAX was used to stand for Worldwide Interoperability for Microwave Access and is defined as a "standards-based technology enabling the delivery of last mile wireless broadband access as an alternative to wired broadband like cable and DSL."[27] Actually, WiMAX is perceived today more as an alternative to cellular systems, particularly 4G networks, and not so much as wireless fixed access. The WiMAX Forum, like other industry-based organizations promoting standards-based technology (such as Wi-Fi), defines a set of specifications that are based on standards developed by the IEEE 802.16 Working Group and ETSI's HiperMAN. The main purpose of the WiMAX Forum is to define and conduct conformance and interoperability testing to ensure WiMAX-certified products for the market. In addition, since most of the WiMAX Forum members are also members of the 802.16 working group, the WiMAX Forum implicitly affects the actual standardization process of its baseline standard.

The real ramp up of the WiMAX activity began with the IEEE 802.16-2004 standard [187] that promoted the Orthogonal Frequency Division Multiplexing (OFDM) for fixed technology, and then the Orthogonal Frequency Division Multiple Access (OFDMA) for mobile technology, which is based on IEEE 802.16e-2005 [188]. Both implementations concentrated at below 11 GHz bands. OFDM is a multiplexing technique that subdivides the bandwidth into multiple frequency subcarriers. The input data stream is divided into several parallel substreams of reduced data rate (thus increased symbol duration) and each substream is modulated and transmitted on a separate orthogonal subcarrier, that is, the subcarriers partially overlap but do not interfere with one another. OFDM modulation can be realized with efficient Inverse Fast Fourier Transform (I-FFT), which enables a large number of subcarriers (up to 2048 in IEEE 802.16e-2005) with low complexity. In an OFDMA system, resources are available in the time domain by means of OFDM symbols and in the frequency domain by means of subcarriers. The time and frequency resources can be organized into slots that represent a combination of time symbols and groups of subcarriers called subchannels, for allocation to individual users. Hence, OFDMA is a multiple-access/multiplexing scheme that provides a multiplexing operation of different streams to and from multiple users over time and frequency access units. OFDM is used also for ADSL, Wi-Fi, and Digital Video Broadcasting (DVB), for example, OFDMA is used also in DVB-RCT (return channel terrestrial) and recently in LTE.

---

[27]As stated in the WiMAX Forum site at *http://www.wimaxforum.org/technology/*. WiMAX is a registered trademark of the WiMAX Forum.

The baseline standards for WiMAX, as mentioned above, come from the IEEE 802.16 working group, specifically its published standards IEEE 802.16-2004 and IEEE 802.16e-2005. As of early 2008, the IEEE 802.16 working group is working on a revision project (Revision 2), which is supposed to be finished by the end of 2008 and that includes corrections (corrigendum) and some performance optimization. In addition, in October 2007, the ITU approved the inclusion of WiMAX technology in the IMT-2000 set of standards. The specific implementation, known as "IMT-2000 OFDMA TDD WMAN," is the version of IEEE 802.16 supported in a profile developed for certification purposes by the WiMAX Forum. Other future projects of the IEEE 802.16 include the relay task group (IEEE 802.16j), which focuses on developing the extension of IEEE 802.16 that supports multihop relay topology. Another project is handled by Task Group m (IEEE 802.16m or TGm [189]), which is charted to develop Advanced Air Interface specification. The TGm group mainly focuses on the approval of the IEEE 802.16 technology as part of the IMT-Advanced standards. Hence, its main charter is to provide performance improvements to the IEEE 802.16 standard that are necessary to supporting future advanced services and applications such as those described by the ITU in Report ITU-R M.2072. In practice, the requirements of the TGm group are very similar to the 3GPP's 4G Long Term Evolution (LTE) standard.

The IEEE 802.16-2004 standard was approved in July 2004. IEEE 802.16-2004 includes previous versions of the IEEE 802.16 standards (802.16-2001, 802.16c in 2002, and 802.16a in 2003) and covers both Line of Sight (LOS) and non-LOS (NLOS) applications in the 2-66 GHz frequencies. The IEEE 802.16-2004 was mainly focused on fixed and nomadic applications in the 2-11 GHz frequencies and supported two multi-carrier modulation techniques: OFDM with 256 carriers, and OFDMA with 2048 carriers.

The IEEE 802.16e-2005 amendment provided extensions and improvements to the IEEE 802.16-2004 to support combined fixed and mobile operations. This amendment extended the OFDMA physical layer by supporting additional FFT sizes (also called scalable OFDMA). It provided improved support for Multiple Input Multiple Output (MIMO) and Adaptive Antenna Systems (AAS). In addition, the Medium Access Control protocol was extended among other features, with power-saving capabilities for mobile devices, better security features, support of idle mode, support of hard and soft hand offs and specific network entry optimization procedures for optimized mobility support.

WiMAX architecture is based on a packet-switched framework, and apart from the above mentioned IEEE 802.16, it is based on IETF's IP and IEEE Ethernet standards [430]. The WiMAX Network specification targets an end-to-end all-IP architecture, optimized for a broad range of IP services and, as such, adopts IP packets for all its interfaces and protocols. This allows WiMAX to be highly modular and "open," such that WiMAX components, network elements, and mobile stations can interact and interface with many other systems and various players that support WiMAX's IP interfaces [239]. It means, for example, that WiMAX's AAA (authentication, authorization, and accounting) is based on IETF standards, such as RADIUS [366] and Diameter [67], and that the WiMAX control plane is based on the standardized Type-Length-Value (TLV) message primitives that are carried on UDP/IP port 2231 (the WiMAX port).

The WiMAX Forum extends the traditional Layer 1 and Layer 2 (PHY and MAC layers) definition of IEEE 802 standards with network architecture issues for mobile networks. The focus of the first network architecture specification (Release 1.0) is on delivering wireless Internet service with mobility.

Release 1.5 will add support for telecom-grade mobile services, supporting full IMS interworking, carrier-grade VoIP, broadcast applications such as mobile TV, and over-the-air dynamic provisioning.

The network reference model of WiMAX defines a logical representation of the WiMAX architecture and includes its entities and their reference points. The main logical entities are the Mobile Stations (MS), the Access Service Network (ASN), and the Connectivity Service Network (CSN). Figure 4.18 depicts the general network reference model of WiMAX. In addition, two main business entities are used in the context of WiMAX network architecture: Network Access Provider (NAP) and Network Service Provider (NSP). NAP provides WiMAX radio access infrastructure (using one or more ASNs) to one or more WiMAX NSPs. NSP provides IP connectivity and WiMAX services to WiMAX subscribers by establishing contractual agreements with one or more NAPs, or roaming agreements with other NSPs.

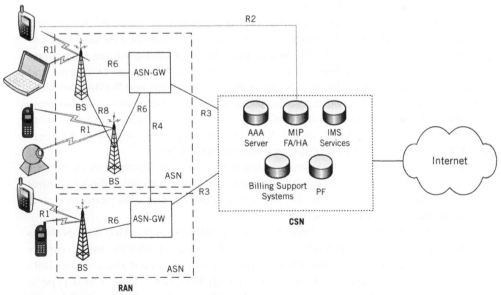

R1, R2, R3, R4, R6, R8—Various Protocols
RAN—Radio Access Network
ASN—Access Service Network
ASN-GW—ASN Gateway
BS—Base Station
CSN—Connectivity Service Network

AAA—Authentication, Authorization, and Accounting
MIP—Mobile IP
FA—Foreign Agent
HA—Home Agent
PF—Policy Function
IMS—IP Multimedia Subsystem

FIGURE 4.18

WiMAX network reference model

The MS is the end-user equipment, which, using the 802.16 protocol, communicates with the Base Station (BS) and, specifically, its OFDMA PHY layer (in the case of mobility).

The Access Service Network (ASN) is defined as a complete set of network functions needed to provide radio access to a WiMAX subscriber. The ASN is a logical boundary that represents the functional entities and the reference points necessary for access services. The main functions that must be implemented by the ASN are: WiMAX Layer-2 (L2) connectivity with WiMAX MS, transfer of AAA messages to WiMAX subscriber's Home Network Service Provider (H-NSP) for authentication, authorization, and session accounting for subscriber sessions, network discovery and selection of the WiMAX subscriber's preferred NSP, and relay functionality for establishing Layer-3 (L3) connectivity with a WiMAX MS and radio Resource Management. In addition, among other functionalities, the ASN supports the ASN anchored mobility, which refers to the set of procedures associated with the movement (handover) of an MS between two Base Stations (referred to in the IEEE 802.16 as Serving and Target BS), without changing the traffic anchor point for the MS in the serving (anchor) ASN. The Target BS may belong to the same ASN or to a different one. The ASN comprises network elements such as one or more Base Station(s), and one or more ASN Gateways. An ASN may be shared by more than one Connectivity Service Network (CSN).

The CSN is defined as a set of network functions that provide IP connectivity services to the WiMAX subscriber(s). A CSN may be comprised of network elements such as routers, an MS IP address and endpoint allocation, an AAA proxy or server, policy and admission control servers, WiMAX subscriber billing, and interoperator settlement, Inter-CSN tunneling for roaming and WiMAX services such as location-based services, connectivity for peer-to-peer services, provisioning, authorization and/or connectivity to IP multimedia services and facilities to support lawful intercept services. A CSN may be deployed as part of a Greenfield[28] WiMAX NSP or as part of an incumbent WiMAX NSP.

An ASN Gateway is a logical entity that represents an aggregation of control plane functional entities (paired with corresponding functions in its ASN or another ASN, or the resident function in the CSN). The ASN-GW may also perform bearer plane routing or a bridging function.

The bearer plane between WiMAX network entities such as two Base Stations or ASN Gateways is established by a Data Path. Each Data Path function is responsible for instantiating and managing the data bearer between it and another Data Path function. There are two types of Data Path Functions: Type 1 and Type 2. The Type 1 data path carries user payloads, either in IP packets or in Ethernet frames, tunneled or tagged (e.g., IP-in-IP, MPLS, GRE, or 802.1Q), between peers within the ASN or between ASNs. Like the Type 1 path, the Type 2 data path is also a generic Layer 3 tunnel (e.g., IP-in-IP or GRE) that carries a Layer-2 data packet, which is defined as an 802.16e MAC Service Data Unit or part of it.

---

[28]Greenfield means a new infrastructure deployment for service provisioning.

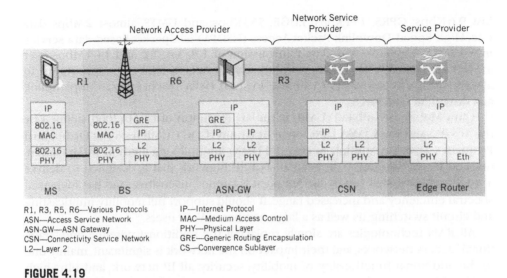

R1, R3, R5, R6—Various Protocols
ASN—Access Service Network
ASN-GW—ASN Gateway
CSN—Connectivity Service Network
L2—Layer 2

IP—Internet Protocol
MAC—Medium Access Control
PHY—Physical Layer
GRE—Generic Routing Encapsulation
CS—Convergence Sublayer

FIGURE 4.19

WiMAX IP protocol layer architecture

WiMAX is an all-IP system, and as such, adopts IP packets for all its interfaces and protocols. The Mobile IP (MIP, RFC 3344, and related RFCs for IPv4 and IPv6) is adopted as the mobility management protocol for all applicable usage/deployment scenarios requiring seamless inter-subnet/inter-prefix layer-3 handovers. For CSN Anchored Mobility Management, Client MIP and Proxy MIP protocols are supported.

The data plane in a WiMAX network consists of the transport encapsulation of the user payload within the mobile WiMAX network. Release 1.0.0 of the mobile WiMAX network specification assumes a routed transport infrastructure for all of the exposed network reference points. Therefore, user payload packets are encapsulated within IP packets when they are carried over the reference points R3, R4, and R6. User payload packets are encapsulated in 802.16 MAC frames when carried over R1. Data plane packets over R3 are encapsulated using IP-in-IP (in accordance to RFC 2003) or, optionally, using GRE encapsulation (in accordance to RFC 2784 and the extensions of RFC 2890). In addition, the GRE encapsulation is used as tunneling protocol for data plane packets over reference points R4 and R6. Figure 4.19 depicts the generic protocol layering of the data plane when applying the IP convergence sublayer in the 802.16 data transport reference.

4.1.2.4 Radio Access Network Summary

The bandwidth of each of the radio access technologies, that is, the radio capacities and the CE capabilities in terms of data transfer rates, is usually defined in "classes" of terminals. The peak data rates for LTE are 50/100 Mbps (uplink/downlink for class A) and 2 Mbps (for class C). Just for comparison, GSM was capable of transferring just 9.6 Kbps; GPRS, 171 Kbps; EDGE, 553 Kbps; and UMTS, almost 2 Mbps data rate. High Speed Downlink Packet Access (HSDPA) is a packet-based data service in W-CDMA downlink with data transmission of 1.8, 3.6, 7.2, and 14.4 Mbps over a 5 MHz bandwidth. High Speed Uplink Packet Access (HSUPA) supports up to almost 6 Mbps in the W-CDMA uplink. Evolved HSPA reaches 11/28 Mbps uplink and downlink data rates.

Ultra Mobile Broadband (UMB) is the 4G evolution of CDMA2000, defined by the 3GGP2 and the CDMA Development Group (CDG). UMB uses the Orthogonal Frequency Division Multiple Access (OFDMA) multiplexing method, MIMO antennas, and beam forming. UMB is an all-TCP/IP network that is designed to deliver at an ultra-high-speed (up to 288 Mbps), it is latency insensitive, and has increased spectral efficiency and increased range. It should support full mobility, multicasting and circuit switching, as well as a large number of VoIP users.

All RAN technologies are slowly replacing the traditional wireline and "traditional" access networks, and their impact on the network is significant, mainly due to the additional functionality of mobility, security, all IP network, and ultra-high bandwidth at their edge.

4.2 HOME AND BUILDING NETWORKS

Home networks are currently beyond the scope of the network processors, but as these networks are attached to access networks and equipment that might contain network processors, we shall describe them very briefly. Home networking originated with the need to attach printers to PCs, or to provide Internet access to home computers. It evolved to combine game and entertainment platforms that carry multimedia content such as video, TV, or music from a variety of sources, for example, cameras and consumer electronics, media and file servers, set-top boxes, or Internet and content providers. Home networking is defined by ITU-T as "the collection of elements that process, manage, transport, and store information, thus enabling the connection and integration of multiple computing, control, monitoring, communication, and entertainment devices in the home."

The DSL forum [106] outlined rough estimates of downstream bandwidth required by various home-applications. TV-focused services include MPEG2 based broadcast TV, Pay-per-view (PPV), Personal Video Records (PVR), and Video on Demand (VOD) (which requires 2–6 Mbps), High Definition TV (HDTV) (which requires 12–19 Mbps), and Interactive TV (which requires up to 3 Mbps). PC-focused services include live TV on PC and other video applications (such as VOD, games, remote education, and video conferencing), which require up to 750 Kbps, while all-purpose high speed Internet access requires up to 3 Mbps downstream bandwidth.

Like access networks, we can classify home networks as either wireless-based and wireline-based. Wireless Personal Area Networks (WPANs), which are defined by the IEEE 802.15 working group,[29] are not part of home networks, although some may consider them relevant.

Wireless networks include Wi-Fi and home-cellular networks, as well as several other proprietary wireless networks. Wi-Fi is an alliance that promotes Wireless Local Area Networks (WLANs) that are based on IEEE 802.11 standards. IEEE 802.11 standards define wireless technologies for data capacities ranging from 1 to 258 Mbps links, using various modulation techniques over 2.4 or 5 GHz carrier, security schemes, and Quality of Service standards. The range of a Wi-Fi access point is tens of meters in buildings (in an "indoor" environment), and up to a few hundred meters outside (an "outdoor" environment).

There are two basic configurations for Wi-Fi: infrastructure and Ad-Hoc. In an infrastructure configuration, mobile clients (such as laptops or IP phones) are connected to Access Points (APs), and all their communications are done through these APs (which function as Base Stations). In an ad-hoc configuration, there are no APs, and the clients communicate among themselves, with some relaying information to other clients.

Wi-Fi may be used in public areas such as restaurants and airports (in infrastructure configuration), to allow Internet access to bypassers, in a setup called "hot-spots." Recently, there have been attempts to create Wi-Fi coverage for larger areas (achieved by multiple access points), which are used as a sort of RAN that is attached to the 3G or 4G infrastructure.

Femtocells are tiny, residential "base-stations" of the cellular network, which resemble the access points of the Wi-Fi. Femtocells are extensions of the RANs and are attached to the cellular networks through the Internet links, with either DSL or the CATV infrastructure.

Wireline networks include LAN technologies (mainly Ethernet) in buildings and a variety of cable technologies in residential homes. These include Home Phone-line Networking Alliance (HPNA)[30] and Multimedia over Coax Alliance (MoCA) standards for data over the telephone's copper twisted pairs or TV's coaxial cables, and HomePlug standards for data over the electric wires (power lines). Other technologies include FireWire cabling for multimedia applications. Recently, the ITU-T initiated a Joint Coordination Activity (JCA) on Home Networking, called

---

[29]IEEE 802.15 includes several task groups that define wireless personal networks (WPANs) of various types: 802.15.1 defines Bluetooth, and 802.15.3 defines high-rate WPAN, which is referred to as Ultra Wideband (UWB). UWB includes various standards for Wireless Universal Serial Bus (USB) and wireless FireWire (IEEE 1394 serial interface), and is basically based on either the WiMedia alliance's Multi-Band Orthogonal Frequency Division Multiplexing (MB-OFDM) UWB or the UWB forum's Direct Sequence-UWB (DS-UWB). 802.15.3c, for example, defines a millimeter wave (57–64 GHz) based UWB, that is supported by the WiHD consortium, for delivering High-Definition (high bandwidth video) over WPAN. 802.15.4 defines low-rate WPAN, such as ZigBee. 802.15.6 defines even smaller range networks, called Body Area Networks (BANs).
[30]HPNA specifications were approved by the ITU-T [217–220].

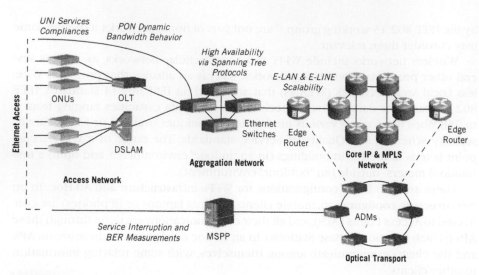

FIGURE 4.20

The Ethernet services delivery chain; reproduced with permission of the Metro Ethernet Forum

JCA-HN[31] that targets a unified standard for home networking over coax cables, twisted pairs, or power lines.

4.3 SUMMARY

This chapter describes access and home networks and completes our brief description of networks. These networks use converged technologies in terms of data and telecom applications, and serve as the last link between the core networks and the customers. Access and home networks are becoming faster and more complicated, as they integrate with aggregation networks (metro networks) and multiservice platforms, and require network processors for efficient functioning and implementation.

The Metro Ethernet Forum describes network integration and service convergence very clearly in its depiction of networks and architecture, as shown in Figure 4.20. By now the reader should be familiar with the terminology, technologies, and elements, as well as the way they interact and are interoperable.

---

[31]JCA-HN drafts several standards, such as G.hn—"Next generation home networking transceivers," G.hnta—"Generic Home Network Transport Architecture," H.ghna—"A generic Home Network architecture with support for multimedia services," J.290—"Next generation set-top-box core architecture," and X.1111 (formerly X.homesec-1), which describes a security framework for identifying threats and the necessary security functions in the home network security model.

• Frame and Packet Processing
• Frame Header Handling

and processing, with a concrete example of a network processor.

PART

Processing

2

In the previous part of this book, we described networks. Here, we turn to processors and processing and discuss the theory behind network processors. No specific network processors are described or mentioned, except when required as an example to clarify the theory.

This part begins with frame and packet processing, the algorithms used, data structures, and the relevant networking schemes required for packet processing. These include, for example, network addressing, classification, and look-up schemes. It then describes Class of Service (CoS) and Quality of Service (QoS) schemes, and the way that the network, the equipment, and the chips handle packet traffic accordingly (e.g., buffering and scheduling packets along network paths).

After dealing with packet processing and handling algorithms, this part turns to processors themselves. It begins with hardware (architecture), moves on to software (programming models, languages, and development platforms), and concludes with network processors' peripherals—devices that are usually adjunct to network processors.

This part contains the following chapters:

- Chapter 5—Packet Processing
- Chapter 6—Packet Flow Handling
- Chapter 7—Architecture
- Chapter 8—Programming Models
- Chapter 9—Network Processors' Peripherals

After the first two parts, which examined the two terms of the title separately, in the third part we turn to integrating the two subjects, networks and processing, with a concrete example of a network processor.

Packet Processing

The first part of the book provided a general description of networks, including specific protocols that are important to the understanding of network processing. From a networking and applications perspective, the requirements, challenges, and motivation of network processing should now be clear.

In this part of the book, we approach the network processor first by attempting to understand how a packet is processed and delivered, and then by looking at which architectures can feasibly carry out the tasks according to required speed and functionality.

This chapter discusses packet structure and all the processing functions that the packet must go through in order for the network tasks (i.e., framing, parsing, classifying, searching, modifying, etc.) to be performed.

The two main functions of network processing are searching (lookups) and classification. These two subjects are treated in this chapter in a detailed manner, and the knowledgeable reader can skip the introductory section.

5.1 INTRODUCTION AND DEFINITIONS

In this chapter, we shall concentrate on Ethernet packets carrying VLAN, IP, or MPLS[1] information, although the discussion here can be applied very easily to any packet processing.

As we discussed in the previous chapters, processing packets at wire speed imposes impractical constraints on general purpose CPUs, even with several multi-core CPUs combined. Just for the sake of demonstration, consider the case of a 1 Gbps Ethernet, running about 1 million packets per second. For each packet, classification based on complex parsing must be executed; for instance, getting the destination IP address, destination port, and in some cases, for some destination ports—also getting some field in layer 7 Protocol Data Unit (PDU), at an offset

---

[1]Virtual LAN (VLAN), Internet Protocol (IP) and Multi-Protocol Label Switching (MPLS) were described in Part 1 of the book, in the Networks chapter.

depending on the destination port. Then a search (or two) must be executed to retrieve a destination IP address and port; the search must be conducted among hundreds of thousands of possible addresses, and a longest (or best) prefix match is desired. All these parsing and searching activities must take no more than 1 μs, if we ignore packet modification and forwarding processing time. How much work, in terms of CPU instructions and memory accesses, can be done in 1 μs? Is one microsecond enough to carry all the work required by the tasks mentioned above for that packet processing? Now, what about 10 Gbps links? How much work can be done in, say, 50 ns?

Packet processing can be described in various ways, and often the same terminology is used to mean different things. Moreover, the trade press, the industry, and academic researchers not only use the terms in contradictory ways, but also categorize functions in different ways. In this chapter we examine three aspects of packet processing, focusing specifically on: tasks, path, and direction.

Packet processing tasks (or functions) include:

- Framing.
- Parsing and classification.
- Search, lookup, and forwarding.
- Modification.
- Compression and encryption.
- Queueing and traffic management (measurement, policing and shaping).

Packet processing can follow one of two paths:

- Data path (fast path).
- Control path (slow path).

Packet processing can be discussed according to direction:

- Ingress (entering the equipment or the network processor, from the network).
- Egress (exiting the equipment or the network processor, to the network).
- Combinations of Ingress and Egress.

As mentioned before, processing functions are separate tasks, each following the other. The process starts with the packet entering the network processor and immediately goes through framing, whose function is to make sure that the packet arrived correctly. (In the other direction, framing is the last task, and is targeted to ensure valid packet output.) The second phase is to parse and classify the packet, which simply means that the network processor must understand what the packet is, what type it is, and then must classify it according to the application requirements. Usually for this classification function, searching is required. Searching might also be required for other functions that the application dictates. The last function that the network processor carries is the required modification of the packet, which includes dropping the packet if required, multiplying it, or altering its content as required. Finally, transmitting the packet usually involves an extra function of queuing, prioritization, and traffic management of the packet to make sure that the receiver can receive the transmitted packet at traffic patterns that it expects. Queuing and traffic management sometimes happens inside the network processors and sometimes happens outside the network processors. Optionally, compression and encryption tasks are utilities that packets sometimes undergo and usually they are done outside of the network processor, although there are some network processors that contain an embedded security functional unit.

The main processing functions are classification of the packet (at real time or at wire speed), and searching for various values (e.g., next hop address) that correspond with some fields in the packets (e.g., IP address). These two functions have received extensive treatment in the industry, to the extent of special-purpose search engine coprocessors, and the development of parsing and classification languages. Due to their importance, these two functions receive more attention in this chapter than the other packet processing functions, and are outlined below.

A general framework of the three primary aspects of packet processing is depicted in Figure 5.1 [278]. The packets enter from left, in the ingress direction, and take either the slow path (through some kind of upper level processing, for example, updating routing tables of the network processor), or the fast path (going through the network processor functions of searching, modification, etc.). The packets are then forwarded either to a switch fabric or to the network (line interface) again, in the egress direction. Ingress and egress directions will be described in more detail in the next section, as they are a bit more complicated than described here.

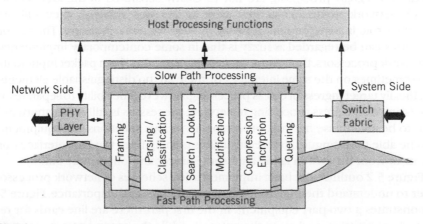

FIGURE 5.1

A general framework of packet processing

Although we are discussing the processing of "packets" here, it is important to note the formal definitions and differences between datagrams, frames, and packets—terms that are sometimes used interchangeably. We already discussed it briefly in Section 2.3 of Chapter 2 and described in Figure 2.8. RFC 1661 [387] provides good definitions:

Frame	The unit of transmission at the data link layer. A frame may include a header and/or a trailer, along with some number of units of data.
Packet	The basic unit of encapsulation, which is passed across the interface between the network layer and the data link layer. A packet is usually mapped to a frame; the exceptions are when data link layer fragmentation is being performed, or when multiple packets are incorporated into a single frame.
Datagram	The unit of transmission in the network layer (such as IP). A datagram may be encapsulated in one or more packets passed to the data link layer.

In essence, "packet" is a generic term for data that travels independently in the network, and is limited in size. "Datagram" is the data unit that applications use, packed in packets. "Frames" are sometimes defined as packets understood by hardware. In network processing we actually talk about and process frames, as we start working at layer 2 (the data link layer), although what we mean here is packets (and eventually datagrams).

5.2 INGRESS AND EGRESS

Ingress and egress processing are not as clearly separated in the architectures of today's network processors as they were in the past, although they still play an important role in some current network processors and equipment. The reason the categories can be regarded as fuzzy is that in some contemporary implementations of network processors, there is one processing direction, from packet input to its output (sometimes on the same interface), or there are no distinguishable elements that specifically target ingress or egress processing. However, the ability to separate ingress processing from egress processing in network processors is still very important if we want to be able to use network processors in various situations and equipment, that is, to be able to distinguish between packets coming from the line interfaces, on the way in, or from the switch fabric after some processing, on the way out.

Figure 5.2 outlines the basic implementation schemes of network processors, in order to understand the ingress and egress functions and importance. Figure 5.2(a) demonstrates a two-part equipment: in the first part there are line cards for receiving and transmitting packets to the network, while the second part is composed of switch fabric, service cards, and other forwarding and processing mechanisms that

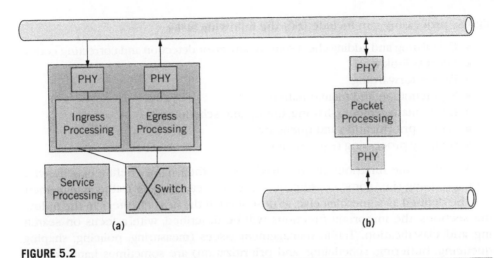

FIGURE 5.2

Network processors implementations

packets undergo internally. Figure 5.2(b) shows a plain in-line packet processing equipment that has one stage, and one direction of processing.[2]

Half-duplex processing, such as shown in Figure 5.2(a), can be done by two network processors, each dedicated to one of the directions, or it can be done by one network processor that works in both directions (such as shown in Figure 5.2(b), which is full-duplex processing). In half-duplex processing, we can distinguish between functions that a network processor executes on packets when they are in the ingress path, or those it executes on the packets when they are on their way out to the line again (egress). In some network processors, there are separated paths and functions in the architectures for ingress and egress, and in others they are combined.

Now, going back to packet processing in terms of functions, we can distinguish between typical ingress tasks and typical egress tasks. Usually, ingress processing can include the following tasks [86]:

- Error checking.
- Security checking and decoding.
- Classification (or demultiplexing).
- Traffic management (measurement and policing).
- Searching (usually address lookup).
- Header manipulations.
- Packet reassembly.
- Packet prioritization and queueing.
- Packet forwarding.

---

[2]Usually this kind of equipment comes in a "pizza box"-like packaging (form factor).

Egress processing can include [86] the following tasks:

- Calculating and adding checksum (or any error detection and correcting codes).
- Address lookup.
- Packet forwarding.
- Segmentation and fragmentation.
- Traffic management (shaping, timing, and scheduling).
- Packet prioritization and queueing.
- Security processing (e.g., encoding).

Not all of the functions are required, nor do their names define exactly what they are doing (i.e., some classifications include searching, or traffic management may be defined as something else, as described in the next chapter). In the following sections, the important functions will be described, with a focus on searching and classification. Traffic management issues (measuring, policing, shaping, queueing, buffering, scheduling, and prioritization) are sometimes handled by a unique processing element (a functional unit of the network processor or a dedicated coprocessor for traffic management). These functions are described in the next chapter due to their importance and complexity.

5.3 FRAMING

Received or transmitted frames should receive "framing" treatment, in order to assure that the correct and full packets or datagrams can be extracted from these frames. This means that incoming frames should undergo correctness tests (to make sure that the entire frames' bits are received without error), correcting attempts if required (i.e., using redundant information to fix incorrect bits, if there is enough information to do it), and integrity checks (to make sure that all packets' content arrived). Outgoing packets should be fragmented or segmented as required and "framed" correctly, that is, adequate headers should be attached or altered, proper terminators (trailers) should be appended or modified, and error detection and correction information should be added, when applicable, to enable later correctness tests and correcting attempts.

In case packets have to be transmitted in fragments, these packets should be fragmented, segmented, and reassembled again. There is a need to segment and reassemble proper headers to each of the layers, so that the end result is valid datagrams and packets.

There are situations where incoming and outgoing packets have to go through mapping procedures, as well as segmentation and reassembly, to meet specific standard interfaces, and these functions are done as part of the framing phase. Examples include IP to ATM, Packets Over SONET/SDH[3] (POS), or

---

[3]Synchronous Optical NETwork (SONET) is the North American and Japanese standard for synchronous data transmission over fiber-optic networks, while Synchronous Digital Hierarchy (SDH) is the ITU-T (usually European) standard.

packet over Plesiochronous Digital Hierarchy (PDH).[4] Some of these "mappers" (or "framers") are described in Chapter 7, where we discuss external interfaces (subsection 7.6.3).

There are many implementations of "framers" in hardware circuitry that are attached to the network processors (or to any other communication block), which perform a variety of framing functions, for example, error-detection and correction, or segmentation and reassembly (SAR).

The following subsections describe in more detail several kinds of algorithms for framing with regards to error detection and correction, as well as to segmentation, fragmentation, and reassembly.

5.3.1 Error Detection and Correction

Error detection is based on several algorithms, from complex Cyclic Redundancy Code (CRC) algorithms (briefly described below) to simple checksums. In the case of Ethernet, for example, the Frame Check Sequence (FCS) field of the packet is based on CRC, but it is usually done in Medium Access Control (MAC) circuitry and in dedicated chips external to the network processors. In the case of IP, on the other hand, the error detection is based on simple checksum, and must be done inside the network processors as part of the packet processing tasks.

5.3.1.1 Cyclic Redundancy Code

Error detecting codes are based on polynomial coding, known as CRC, and modulo-2 arithmetic (similar to XOR operations). These codes assume that the packet can be represented by a polynomial with coefficients of 0 and 1, as the packet's bit string. A k-bit packet is thus represented by a polynomial of degree $k-1$, having k terms at most, where its high order bit (the leftmost) is the coefficient of x^{k-1}. The next bit (the second after the leftmost) is the coefficient of x^{k-2}, and so on. The packet 101101, for example, represents a six-term polynomial that looks like $x^5 + x^3 + x^2 + x^0$.

The CRC idea is based on computing the checksum as the remainder of a modulo-2 division of the packet appended with r zero bits at the low-order end of the packet by some agreed-upon generator polynomial (of degree r, smaller than the original packet). This remainder, or checksum, is appended to the packet by the sender. The receiver, on the other side of the channel, then divides the received packet (appended by this checksum) with the same agreed-upon generator polynomial, and if the reminder is 0, can assume that the packet arrived with no errors. Otherwise, it is wrong.

---

[4]Plesiochronous Digital Hierarchy refers to networks that are almost synchronous (e.g., same clock rate but different clock source), like T1/J1 lines in North America/Japan, E1 lines in Europe, as described in Chapter 2.

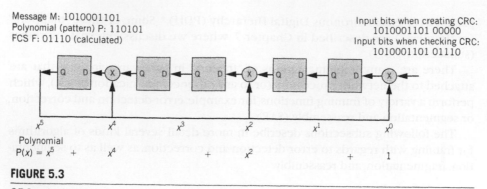

FIGURE 5.3

CRC calculation

It can be quite complicated to do this computation for each packet (either for checksum generation or for checking the packet), as well as demanding of processor power and memory resources. Because of this, special hardware assists have been used for decades now [351], based on shift registers like the one in Figure 5.3, which offloads CRC calculations from the processors.

For Ethernet packets, for example, the last field of the packet, the Frame Check (FCS), is CRC based. IEEE802.3 determines that the generator polynomial (or what is usually called CRC-32) is $x^{32} + x^{26} + x^{23} + x^{22} + x^{16} + x^{12} + x^{11} + x^{10} + x^{8} + x^{7} + x^{5} + x^{4} + x^{2} + x + 1$ (or 0x04C11DB7).

5.3.1.2 IP Checksum

Checksum, contrary to the CRC algorithm, is simpler to implement in software, although it also burdens the processing unit [59]. The main idea is to sum all the 1-byte, 2-byte, or 4-byte words in the packet or in the header into a byte, 2-byte, or 4-byte checksum, correspondingly. The calculated checksum is sent with the packet, and at the other end the receiver calculates the checksum again and compares his result with the checksum received from the sender. If the two are equal, the receiver can assume that the packet arrived with no errors. Otherwise, it is wrong.

IPv4 header [355], for example, uses a 16-bit checksum of the header (one's complement sum of all the 16 bits words in the header). For the purpose of calculating the IPv4 header checksum, the value of the checksum field in the IPv4 header is considered to be zero.

The TCP checksum field in the TCP header [357] is also computed by simple 16-bit one's complement sum of all 16-bit words in the header and the payload. (If the number of bytes to be summed is odd, which means that the last byte cannot be added by 16-bit sum, this byte is padded, just for the checksum calculation, by one more byte of zero to its right.) As with the case of the IPv4 header checksum, the checksum field itself is considered to be zero for the purpose

Source Address		
Destination Address		
Zero	Protocol	Length

FIGURE 5.4

TCP/UDP pseudo header for checksum

of calculating the checksum. Just to complete the description, for TCP checksum calculation purposes, a "pseudo" header is conceptually prefixed to the TCP header (but it is not transmitted). This pseudo header (see Figure 5.4) is 12 bytes long, and includes the 4 bytes of the IP source address, the 4 bytes of the IP destination address, one byte of zeros, one byte of the IP protocol field (for TCP the value is 6), and 2 bytes of the TCP length (header and payload, without the pseudo header).

One last word on checksums—that of the UDP [354]; it is optional, and in most cases no one uses checksum in UDP (its value is zero, indicating no use). However, if used, it functions exactly like the TCP checksum, where the pseudo header is included.

5.3.1.3 Error Detection Summary

It is important to note here that IP checksum as well as TCP or UDP checksum are traditionally done by software when packets are handled by software, or by network processors when packets are handled by processors. Lower level (layer 2) mechanisms are done externally, usually by specific hardware circuitry that performs the detection or the correction (or creates the codes for enabling detection and correction later).

5.3.2 Segmentation, Fragmentation, and Assembly

Fragmentation and segmentation are actually the same thing. The term "fragmentation" has traditionally been used in the IP world, whereas "segmentation" is the term used in the ATM world. In both cases, when we have longer datagrams than the network can transfer, then the datagram has to be sent in parts. In the case of IP, it happens when the datagram exceeds the Maximum Transmission Unit (MTU). In the case of ATM, it is (almost always) when the packet is larger than the cell size (which is 48 bytes of data allowed in the call payload). At any rate, once the original datagram has been fragmented or segmented, it must be reassembled at the other end of the transmission path.

In the fragmentation (or segmentation) process, a header containing all necessary information must be added to each fragment (or segment) to enable it to travel independently through the network, that is, making each fragment (or segment) a valid packet. It must also contain information that enables the other end of the path to reassemble the fragments (or segments) back into one complete datagram. In IP, it means that the entire IP header of the original datagram is copied almost completely into each of the datagrams' fragments (apart from IP datagram length, checksum, and fragmentation information, as described in the following).

5.3.2.1 IP Fragmentation and Reassembly

The IP protocol supports such fragmentation by including information and instructions in the IP header, for allowing or disallowing fragmentation down the packet's path, along with instructions as how to reassemble the packet at the other end of the transmission path (see Figure 5.5). In the FLAGS field of the IP header (3-bit field at offset of 6 bytes), there are two meaningful bits: DF (Don't Fragment, the second bit), which directs the equipment along the path whether this packet may (0) or may not (1) be fragmented, and MF (More Fragments, the third bit), which indicates whether this fragment is the last fragment (0) or there are more (1). The next IP header field that contains information is the FRAGMENT OFFSET field (following the FLAGS field in the IP header), which indicates the

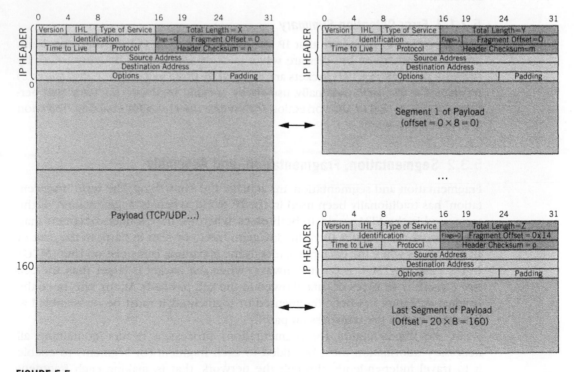

FIGURE 5.5

IP fragmentation and reassembly

position of this fragment in the original datagram, measured in units of 8 bytes. In other words, if, for example, there is "10" in the FRAGMENT OFFSET field, it means that this fragment begins at offset 80 bytes in the original datagram. Except for the last fragment, all IP fragments must therefore be multiples of 8 bytes. When a datagram is not fragmented, then both the FLAGS and the FRAGMENT OFFSET fields are zeroed.

To complicate things, there might be a situation where fragments have to be fragmented themselves. This can happen when a fragment crosses a path in which its MTU is smaller than the fragment's size (for example, if not all the path's MTUs were known when the datagram was originally fragmented). In this case, all sub-fragments have a More Fragments (MF) bit set to 1 (including the last subfragment, unless this fragment is the last one, in which case all subfragments contain 1 in their MF bit, but the last subfragment that contain 0). Additionally, all FRAG-MENT OFFSET fields of the subfragments are not related to the fragment they are fragmenting, but to the original datagram (i.e., the first subfragment FRAGMENT OFFSET field contains the same FRAGMENT OFFSET content of the fragment it is fragmenting).

When it comes to the reassembly of fragments back into one datagram, the process is quite simple if we ignore implementation issues, that is, if fragments arrive correctly, in order, and simply identified. In real-life situations however, fragments might appear duplicated (due to networking traffic issues, sometimes not even equal duplicates, for example, some fragments are further subfragmented), fragments might be lost, and above all—fragments might appear out of their original order. This requires the receiving side to maintain sufficient space in their buffers to receive and process all incoming fragments either until the datagrams can be reassembled from the out-of-order, duplicated fragments, or fragments are cleared once there are missing fragments or lack of room in the buffers. One additional issue is to identify those fragments that belong to the original datagram, and this is done by examining the identification field of the IP header (2 bytes field, at offset of 4 bytes, just before the FLAGS field), which, as mentioned above, is copied to all fragments headers, and identifying each fragment as belonging to a specific datagram. Because of the rare cases in which two IP sources use the same identification number for their original packets, fragment identification should also use the source address field of the IP header (4 bytes at offset of 12 bytes), with conjunction to the identification field.

5.3.2.2 ATM Segmentation and Reassembly

Since ATM is not detailed in this book, segmentation mechanisms related to ATM are covered very briefly, and just one common use of IP over ATM is described, which involves ATM mapping and segmentation/reassembly. The ATM framework includes an ATM Adaptation Layer (AAL), which isolates the ATM layer (which includes the protocol and functions) from the upper, or higher layer protocols and applications (like IP datagrams). These two layers (the ATM and the AAL) are together equivalent to layer 2 (data link layer). In some cases, the AAL layer is itself divided into two sub-layers: the

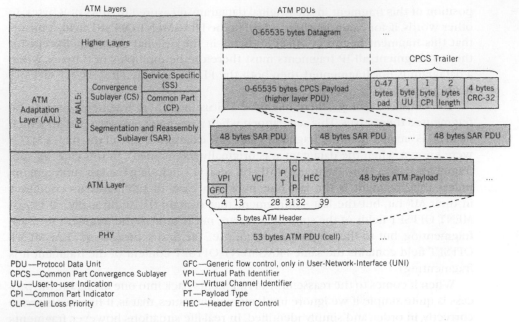

FIGURE 5.6

ATM layers and AAL5 SAR functionality (from left to right)

convergence sublayer (CS), which performs datagram identification, padding, time/clock recovery, and so on, and the SAR sublayer, which segments or reassembles the higher layer datagrams received from or transmitted to the CS sublayer. The ATM layers and how they relate to SAR of higher layer datagams are depicted in Figure 5.6.

AAL type 5 (AAL5) is specifically designed to handle and transmit IP packets through an ATM network, and is the most popular AAL, sometimes called Simple and Easy Adaptation Layer (SEAL). Its convergence sublayer simply adds an 8 bytes trailer field to the incoming datagram, which contains CRC-32 and the datagram length,[5] and pads this datagram to be 48 bytes boundary. The resulting Common Part Convergence Sublayer PDU is then segmented by the SAR sublayer into 48 bytes-long PDUs, which are sent to the ATM layer for transmission (after appending the ATM headers). The ATM layer uses the third bit of the Payload-Type (PT) field of the ATM 5-byte header to notify whether this cell is the last cell of a segmented datagram (1) or not (0). When the cells are received at the other end, they are reassembled back into the IP datagram in exactly the same steps, except in reverse.

---

[5]Although this 16-bit length indicator field allows using up to 64 K bytes IP datagrams, the standard limit for an IP datagram is 9180 bytes. If more than 9180 bytes have to be transmitted, IP must fragment the datagram before passing it to AAL5.

ATM is not commonly processed by network processors anymore, certainly not at the cell level. There are still, however, some traffic managers that include SARs that handle ATM networks and ATM traffic.

5.4 PARSING AND CLASSIFICATION

After a complete, valid packet is received and verified, the next step of packet processing is to look at the incoming packets in order to classify them for various treatments, as the processing requirements dictate. This step involves two combined subtasks: parsing and classification. Sometimes, searching and look-ups are combined with the parsing and classification task and are considered as one step, as classification requires some search operations; other times, searches and look-ups are considered separate issues, due to their complexity, implementation, and uses.[6] Here, these subjects are separated, although we leave algorithmic and implementation issues of classification to be described later, after we cover the searching and look-ups algorithms.

Packet processing has two basic architectures, or design philosophies, that are crucial for parsing and classification implementation; these architectures are *store-and-forward* and *cut-through*. These two architectures are fundamental, and require a short definition here.

In store-and-forward architecture, packets are first received in their entirety, stored temporarily, examined, analyzed, processed, and then, after a decision is made regarding them, transmitted, and the memory is cleared. Store-and-forward allows a complete, even complex treatment of the packet, before the packet is injected back into the network. This comes, however, at the expense of having to deal with buffering issues (hence, higher implementation costs), and delaying the packet, that is, adding end-to-end delay to the packet flowing from the ingress of the processing unit to its egress, due to the increased latency incurred by the buffering time that is required before processing starts.

The shortcomings of the store-and-forward are remedied by the cut-through architecture, where processing begins as packets flow into the processing units, and continues as bits continue to come in. The packet is transmitted after the required analysis, and the decision and processing is done "on-the-fly." This can be accomplished by examining specific bit patterns in various fields of the incoming packet, that is, parsing and classifying, as well as other decision and processing tasks that must be carried out in real-time. Although cut-through saves buffering and latencies, it allows only simple analysis, decision and processing tasks, and might even cause network overhead (e.g., transmitting a packet that eventually

---

[6]In [278] classification refers to two separate issues: (a) search (lookup) and forwarding, and (b) classification (sometimes referred to as deep packet classification, or deep packet inspection). Here we leave search (lookup) and IP forwarding to a separate section, and discuss classification in general, which includes generic forwarding.

turned out to be invalid, bad CRC, or, because a premature decision was made based on the first bits of the packet, the output-channels were loaded in vain, forcing packet retransmits).

Examples of these two architectures are, on the one hand, in low-end routers that use store-and-forward architecture (no "wire-speed" constraints), and on the other, VLAN or MPLS switches that use cut-through architecture, based on a tag in the beginning of the frame that allows forwarding decisions to be made instantly, while the packet is still inflowing.

5.4.1 Parsing

Each incoming packet must go through some sort of parsing to examine and understand what it is as well as its requirements, and then it must be classified, or handled according to its type and its required processing. Parsing therefore is the first analysis and action done on the packet content. Parsing can be very simple, trivial, and unnoticed during packet processing, or it can be a real and complex task that sometimes requires a unique language to describe the process. The task of parsing is sometimes even carried by a unique, dedicated processing element.

Parsing is basically identifying the relevant fields in the incoming packets, according to their place and type, and picking the field's values for continuing the parsing process, or using these values for classification. Therefore, parsing and classification, as described in the following, are tied together, and sometimes are not separated.

A simple parsing example in an IPv4 packet would be to detect its destination IP address, which is easy, since it is a fixed length field in the IP header, always at the same offset of the packet. A more complicated parsing example would be to detect the URL string of characters in the packet, which is a variable length field that exists only in some IPv4 packets, that is, several TCP packets running HyperText Transfer Protocol (HTTP). A complex parsing example would be to detect some Management Information Base (MIB) variable in some of the IPv4 packets, containing UDP packets running Simple Network Management Protocol (SNMP).

5.4.2 Classification

Classification means categorizing packets into "flows," in which they are processed in a similar way by the network entities. These flows are defined by rules that the packets obey, and the collections of these rules are called classifiers [161]. The rule database contains many entries, each of which is composed of a pair of a specific rule description and its appropriate action. These specific rules are matched with the incoming packets, and the best match determines the appropriate action to be taken on the incoming packet. Very often, the action is to mark the incoming packet with a notation, so that a subsequent process will take the appropriate action, based on this notation.

A specific case of packet classification is packet forwarding,[7] which deals with searching and packet lookups, and is described in a following subsection. In packet forwarding the rules are represented by the packet destination address fields, and the action is simply to forward the packet to its appropriate destination. In traffic management literature (as described in the next chapter), classification is sometimes considered part of the traffic management process, when it classifies packets solely for traffic management purposes (e.g., classifying incoming packets with regards to some threshold rate of arriving packets).

Classification is sometimes used interchangeably with demultiplexing, which is a different process, although both distinguish packets from one common incoming stream of packets. In multiplexing several flows of packets from different sources (applications or originators), are multiplexed, by identifying each packet in the resulting stream of packets by its source. This way, at the other end, packets can be demultiplexed to their appropriate targets. Such a multiplexing process, for example, happens very frequently when we use the port field of the UDP (or TCP) to define what application generated this packet at the origin, and to what application this packet should be sent at the destination.

In classification, the destination classifies the received packet based on a set of rules that it decides on, which may use several packet fields, the packet content, as well as one or more other parameters, regardless of the source's intentions (although it can consider source information supplied in the received packet). Thus, classification is much more abstract than demultiplexing. Another difference is that no "multiplexing" entity exists in the classification process on the packet originating side. More important, the classification is dynamic, operates on multiple layers of the packet (many fields, which might be of variable size and be based on other field's content), and can depend on the status of the system, or some status determined by a sequence of packets. None of these characteristics holds for the demultiplexing process.

Classification can be based on simple, quick indicators, such as the 3 bits of the Type of Service (TOS) field of the incoming IP packet, or the VLAN field in the Ethernet frame (e.g., the packet should be classified according to the VLAN or the TOS fields in the packet, which are compared with some values, or range of values, either constants or variables, which are the classification parameters). Classification can also be a result of calculation, for example, of the incoming rate of packets. Or, it can be the outcome of a complex set of rules applied to some pattern, which can be composed from several fields and subfields of the incoming packet. An example

---

[7]Sometimes the term packet classification is used interchangeably and ambiguously with packet forwarding, which is different.

of the latter case is given in Figure 5.7, which shows a classification of a packet based on its Ethernet Type field taken from the MAC layer, the IP Type field, IP Source and Destination addresses fields taken from the IP layer, and the TCP Source and Destination ports fields taken from the TCP field. The resulting classification is calculated, or searched, and put into the TOS field of the IP layer in the packet.

Complex classifications like the one just described are often called *deep packet inspection* (or *deep packet classification*). Deep packet inspection requires a store-and-forward architecture of packet handling, while simple classifications (like using just the TOS field, for example) can also be used in cut-through architecture.

Classification can be thought of as working at the single packet level, that is, deciding whether a packet belongs to this or that category, which is essentially a *selection* process rather than a classification one, or asking whether a packet holds some property, or what are its attributes, which are examples of *analysis* processes. Classification can also be applied to a stream of packets (i.e., marking every packet in the stream according to its destination, required priority, or session type). To clarify this difference, let's take an example of a classification in the processing of a single packet (a selection): if a packet is an IP packet, and is a TCP type, and is of a Telnet session, then mark this packet as one that matches the classification (selected). Otherwise, mark it as not matched. The process of classifying a stream of packets would look like this: if the incoming packet is not IP type, then mark it as type 1 (and later forward it accordingly, or discard it), otherwise, check if it is of TCP type and of a Telnet session; if it is, mark it as type 2 (and forward it), and if not, mark it as type 3. Thus, in the later example, we'll end up with three streams of classified packets, those that are not of IP type, those that are of some Telnet session, and those that are of IP type but don't belong to a Telnet session.

The two types of classifications are important: some network processors, for example, analyze (classify) each packet as it enters the processor, usually through a general classification, and assign some attributes to it. Later on, during processing, the stream of packets goes through a specific classification, according to the desired application. The mechanism for classification in both cases, though, can be very similar.

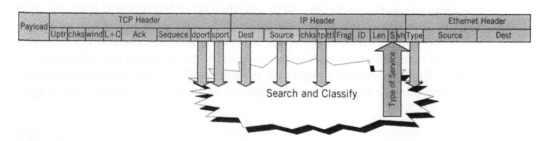

FIGURE 5.7

Searching and classifying

Searching in the rule database for a possible match between the specific incoming packet and the relevant rules (composing the classification), involves some knowledge of data structures and search algorithms. Applying rules might even require multiple searches in the packet and in the database rules, to figure out how to classify the packet. Since searching is described in the next section, we'll postpone the discussion of complex packet classification (i.e., based on searching) to the end of next section.

5.4.2.1 Classification Issues

There are several types of classification, based on the required packet processing. This subsection outlines some of the issues that are important in designing, implementing, and using classifiers and classifications.

5.4.2.1.1 Stateless and Stateful Classification

Stateless classification is any classification that is determined solely by the content of the packet it classifies, that is, it is independent of any previous packet flow or packet instances. It means that the stateless classification process is memoryless, or does not rely on states that should be maintained and stored in an orderly manner. Stateful classification is just the opposite, and requires keeping a state that is calculated or taken from previous packets, from the system's state or from other relevant state information (e.g., neighbors' congestion). One should be aware that stateful classification might result in different classifications when previous packets are reordered, and, of course, depends on the time of packet arrival, compared to other events that might influence the classifier state.

5.4.2.1.2 Different Kinds of Classification

The classification process can be subject to some pre-classification process, and different classification rules should be applied to a packet once it is found to be of a certain type. Clearly, IP packets should be classified in a different way than non-IP packets, although both types of packets might arrive in the same stream to the classifier. A WEB stream might also be classified differently than a TELNET stream, for instance, and rules that require examination of some field in one case, can be replaced by other rules that examine other fields in another case.

In some cases, the various classifications can be chained into one complex, multistage classification process, while in other cases that would be too complex or impractical; sometimes it calls for recursive classifications once some criterion is met. For example, classifying a packet as belonging to a specific flow might result in a decision that the incoming packet should eventually be classified as another type (belonging to another flow), or as a default packet (and may be reclassified). Another example might be a process that classifies a packet, determines its priority, and continues with the forwarding stage that, as mentioned before, is a classification process in and of itself; the forwarding might be based on the priority and the destination in order to map a proper output queue.

5.4.2.1.3 Variable Field Lengths and Offsets

In headers with all fields at predefined positions and of predetermined length, the classification process is straightforward with regards to accessing the various fields required from the header. When the examined fields are not at predetermined, fixed offsets, or they are of variable size, it creates a burden on the classifier that is supposed to get the required information rapidly and finish classification at wire speed.

In cases of variable field offsets or length, the classifier has to calculate the proper position it takes information from, and its length. For example, accessing the URL[8] in an HTTP[9] packet for a classification might be quite complicated. This is because part of the URL is in the IP destination address, and part of it is in the payload of the HTTP packet after a "GET" command, which comes after the IP header (which might have an option field that can determine its size), and the TCP header (which also might have an option field with variable size). And once the URL offset is determined (calculated), there is another difficulty in that the URL field itself is a variable length field, which is terminated by a CR character and an LF character (0x0D0A).

5.4.2.1.4 Static and Dynamic Classification

Static classification means that all classification criteria are predefined, and all rules are fixed. This allows easier implementation of classifiers, as some of the rules can be indexed and cached, and be used later quickly. Forwarding, for example, can simply use a flow classification index to access a port destination, priority, and next-hop address, all cached. Static classification is used for very clear distinguished flows that never change, for example, voice and video, different TOS filed contents, or different applications. However, in many cases, static classification is not good enough, as varying conditions may force change in the classification rules and actions. For example, imagine a source that is allowed to use the network up to some threshold, and once it exceeds this threshold, a different set of classification rules should be applied. Or, at different times, or under different loads, the classification rules and actions should again be modified. Another category is when the rules or the actions are computed, and are thereby subject to conditions that are dynamic by nature, or are the results of incoming packets, systems' state, and so on. A simple and common example of this might be a routing table that is continuously updated and, as a result, causes new types of flows to be recognized and classified accordingly.

Frequent and dynamic changes in classifiers are, of course, easy to implement by software means. Dynamic classifiers are hard to implement by hardware, even though the parameterization of hardware classifiers is possible, as shown in the following subsection.

---

[8]Uniform Resource Locator, that is, the compact string representation of a resource available via the Internet.
[9]HyperText Transfer Protocol.

5.4.2.2 *Classification Mechanisms*

Classification can be done in various ways and by various means. It can be as simple as plain or complex software processes, hardware circuitry, or a combination of all means. However, coping with the various classification issues detailed above, as well as with the basic requirement of classifying packets at wire-speed, at rates above millions of classifications per second, requires special mechanisms.

5.4.2.2.1 Simple Software Classification

Simple software classification is actually very similar to the description of the example given previously about the packet and stream classification (see page 198). Using the same example, we can write a pseudo-code describing the required classification:

```
if (Ethernet.type == IP && IP.protocol == TCP && TCP.dest_port == Telnet)
        then classified OK

else
        not OK
```

This example compares the packet's fields (Ethernet.type, IP.protocol, and TCP.dest_port) to constants that represents the required types (IP = 0x0800, TCP = 0x06, and Telnet = 23).

Clearly, when executing such a comparison with these instructions, it is a good practice to start by examining the most uncommon condition, so as to cease the comparison as soon as possible, and get to the next packet. In the preceding example, it is obvious that Telnet is part of TCP traffic, which is in turn a part of the IP traffic, so it is better to start comparing the TCP.dest_port with Telet value. Telnet traffic is a small fraction of IP traffic, so in most cases, comparison will execute just one instruction.

Another common way of dealing with such classification is to create one variable that is an aggregation, side by side, of all the fields, and to compare this variable once by one comparison statement like the one shown in the example above, that is,

```
if (composed.variable==composed.value) then
```

This variable can also be compared by using some search operation in a dedicated process, as described in the following subsection.

A pseudo-code describing the required classification for the preceding second example—stream classification—looks like this:

```
if (Ethernet.type != IP) then
                mark this frame 1
        else
                if (IP.protocol == TCP && TCP.dest_port == Telnet) then
                        mark this frame 2
                else mark this frame 3
```

Simple software classification is very flexible in setting and modifying classification criteria as the application changes, or some conditions on the application are changed. In other words, simple software classification enables dynamic classification criteria, thresholds, and values of classification parameters. The problem, though, is that simple software classification is not practical for high-speed packet stream, certainly not classification at wire speed when 1Gbps is used. For that, other mechanisms are required.

5.4.2.2.2 Complex Software Classification

Complex software classification should be used for deep packet inspection generally, or when there are multifield criteria for classification. Imagine there is a need to classify a priority of each packet, as described above, based on source or destination IP, application (port), and time of day. In software, it will look something like this:

```
if (Ethernet.type == IP && IP.protocol == TCP && TCP.dest_port ==
Telnet && (IP.SIP == 192.168.0.12 || IP.SIP == 192.168.0.22 ||
IP.SIP == IP_list) && (IP.DIP != 192.168.1.45 && IP.DIP != 192.34.4.5)
&& (time > 8am && time < 2pm) ... ) then
     if we did not receive a "time-out" (most unlikely), set this
     packets' priority to "1"
else
     ...
```

This is, of course, ridiculous.

A more appropriate way of executing such classification would be to arrange the rules' database we mentioned at the beginning of this subsection in a structure that will speed up the classification process. Some structures will even enable hardware classifiers to work on them.

The simplest structure is a linked list of all the rules, or a table of them, ordered (or sorted) by their priority (i.e., what rule, if matched, should apply first). Then, a linear search is done by comparing the incoming packet fields with each of the rules sequentially, until a match is achieved. This, of course, is not so different from the example of the software code given above, which means that linked list structure is not practical, nor scalable, despite its simple and efficient memory requirements.

More adequate data structures can be used to support classification at wire speed. The later Sections—5.5.3 and 5.5.4—describe such advanced data structures, so we postpone the discussion of deep packet inspection (or classification) until after these two subsections.

In the meantime, some insight into these types of classifications and the complexity of the problem can be obtained by using a simple two-dimensional geometric interpretation of a two-field packet classification [161].

A one-dimensional classifier can be represented by a number line, divided into several continuous intervals (each of range of values); each interval represents a distinguished classification rule and an assigned value (action). A packet's header content that matches an interval's rule (i.e., is in the range of that interval)

Table 5.1 A Two-field Example Classifier

Rule	Field 1	Field 2	Action
1	0–7	6–7	A
2	4–7	0–7	B
3	0–7	4–7	C
4	0–1	0–7	D

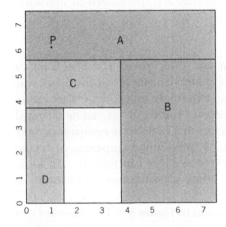

FIGURE 5.8

A geometric interpretation of the prioritized classifier of Table 5.1

is a point on this line (at the interval) and is classified by the interval's assigned value. This representation can be adopted for the multidimensional case. A two-dimensional rule can be represented by a rectangle, and a *d*-dimensional rule can be represented by a *d*-dimensional hyper-rectangle. A classifier is the collection of all the hyper-rectangles, ordered by their priority; classification is done by matching the packet's headers to the values of all *d* dimensions, thus creating a point in the *d* dimension that represent this packet. The classification value (action) is thus the attribute of the hyper-rectangle in which the packet (point representation) resides (the representation of the point).

An example can clarify the two-dimensional case: consider a classifier that has four rules, based on two fields of the packet, as described in Table 5.1.[10] Figure 5.8 represents, geometrically, this classifier, in which higher priority rules cover are actually covering lower-priority rules. A packet that contains, for example, values of 1 and 6 in its fields, is a point P(1,6) in Figure 5.8, which obey rules A, C, and D, but since rule A is the highest priority, this packet would be classified as A type.

---

[10]In this example, the rules are ordered by their priority, from the highest to the lowest.

Now, for estimating this problem's complexity, let us use the point location geometry problem, which finds the enclosing region of a point [161]. The best bounds, given nonoverlapping N rectangular regions and $d > 3$ dimensions are $O(\log N)$ time and $O(N^d)$ space, or $O((\log N)^{d-1})$ time and $O(N)$ space [344]. Since rectangles can overlap in packet classification, the complexity is at least as hard as point location, making the classification problem either huge in size or too slow to execute.

It is worth noting here that this complexity resulted in the offering of many special purpose processors (and coprocessors) that are specialized in classification. Coprocessors are described in Chapter 9.

5.4.2.2.3 Hardware Classification

If software classification mechanisms are at the one end of the spectrum of classification speed, then hardware classification mechanisms are at the other end: they are extremely fast, but, at the same time, extremely inflexible. Classification criteria can at most be parameterized (i.e., values of some criteria can be changed during operation). Hardware classifiers are circuits that constantly compare certain, predefined fields of the packets, to some values (predefined constants, or reconfigurable, or programmable parameters). These classifiers can just decode some predefined fields of the packets, to generate some classification information, or attributes describing the packet.

An example of this type of classifier is depicted in Figure 5.9. The classifier can analyze the packet while it is in the incoming packet memory, where the classifier actually copies the headers (or the required fields) into a packet-register (by a wide bus, so that it will not stall the packet memory). It can also, in some cases, even analyze the packet in a "cut-through" manner, that is, as the packet flies into the packet memory, it goes through a shift register in the hardware classifier, and get analyzed "on the fly." Some network processors or classification coprocessors (as defined in Chapter 9), have this kind of circuitry for pre-analysis (classification) of the incoming packets.

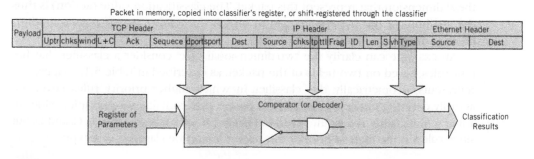

FIGURE 5.9

Classification by hardware circuitry

Complex hardware circuitries can pre-analyze or classify many fields of variable sizes and offsets, depending on the incoming packet instances.

5.4.2.2.4 Hybrid Classification Systems

In order to gain the benefits of hardware classification (speed) and software (flexibility), some hybrid solutions are used, whereas the classification process is divided into phases where the hardware classifier is doing some preliminary analysis, followed by more detailed, tailored-made software classification.

5.4.3 Parsing and Classification Languages

Defining exactly what rules should apply and how, to each of the arriving packets, as well as defining the rules themselves and the various parts of packets to be inspected, requires a detailed description, procedure, some sort of a database definition (for the rules), and access to this database, or a combination of these elements. Some classifying engines use procedural programming language that is the same as the programming language used by all of the NP processing units, while others use special-purpose classifying or parsing language that is specific to the classification task. Sometimes, a combination of a special purpose high-level classifying language that is pre-compiled into the general purpose NP language is used.

Two examples of unique parsing and classification languages are Intel's Network Classification Language and Agere's (now LSI's) Functional Programming Language. These two languages are described briefly in Appendix A of Chapter 8 for the sake of giving the reader some insight to parsing and classification complexity, requirements, definitions, and how these languages cope with it, as well as a closer look for those who need to implement parsing and classification techniques.

5.5 SEARCH, LOOKUP, AND FORWARDING

Search is one of the most important, complex, and common operations done in packet processing. Search is not a phase by itself in packet processing, like the other operations mentioned in this chapter (e.g., framing, parsing, classifying, forwarding, modifying, encrypting, and queueing); rather, it is an atomic operation. Search, or lookup, can be called in the classification phase or during forwarding, or even multiple times during any of the operations that are used for packet processing.

5.5.1 Introduction

Before dealing with search methodologies, this introduction describes the main requirements for IP lookups (which are IP address lookups) as well as describing what we are usually looking for. In order to understand IP address lookup, which is one of the main search targets, a detailed description of IP addressing is given.

5.5.1.1 IP Lookup—Background

The importance of rapid IP lookup cannot be exaggerated in the context of packet processing. IP packet forwarding, for example, is executed in any router or switch, where forwarding decisions have to be made in order to find the address of the next hop router and the egress port to be used to send the packet through. Although the most common IP lookups are done for forwarding, IP lookups are required also for classifications or other applications, for example, billing, access lists, and so on. In most cases, IP forwarding is based on IP addresses (but not always, e.g., some Quality of Service parameters). IP address lookups are obviously based on the IP address, that is 32-bit or 128-bit keys (IPv4 or IPv6, respectively), whereas other lookups can have much wider keys that can reach hundreds of bits, composed from multiple fields in the IP (or layer 2) packet.

Almost every packet processing activity starts with an IP-lookup, and therefore, in high-speed networks, the speed of IP-lookups is critical. Huge efforts are made to accelerate this specific task in any network processor or, as a matter of fact, in any networking device. In a 10 Gbps network, for example, the required time to forward a packet of minimal size is <50 ns; therefore, this is the maximum allowed time for looking for a match (or a longest match) of a specific IP address in a table of hundreds of thousands of entries (as well as doing some other things in this time frame). Other line-speeds (or the resulting range of packets per second, from minimal size to maximal size) and number of searches per packet, require search speeds as depicted in Figure 5.10.

Number of searches per packet can vary from one search for plain packet forwarding to numerous searches for additional IP lookups that are required for access control list checkups in firewalls, priority assignment for QoS demands (see next chapter), and network management tasks that are functions of the IP address, and so on. Number of packets per second depends on the technology used and the average packet size; for OC-768, for instance, the line speed is 40 Gbps, so the smallest 64-byte packets to the largest 1518-byte packets yield a rate of approximately 3 to 78 million packets per second, if we ignore the framing overhead, inter-packet gaps and packets preamble, and if we assume a continuous packet stream.

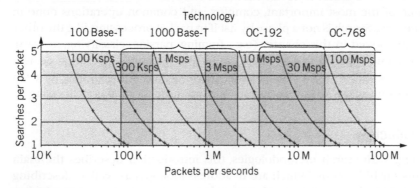

FIGURE 5.10

Searches per second

5.5.1.2 IP Addressing, Routing, and Forwarding

In most cases of IP lookups, address searches are conducted. Sometimes, IP-lookup refers specifically to IP routing and forwarding. In order to understand IP addressing, routing, and forwarding, a brief rehearsal is suggested here.

IPv4 [355] addressing is based on a 32-bit address field. The 32 bits of the IPv4 address are usually represented by the decimal value of each of the bytes in this 32-bit address, separate by a dot, that is, 131.44.2.1 means: 100000110010110000 00001000000001, or 832C0201 (hex). The IP address is further broken down to *network address* (or *netID*) part, *subnetwork* (or *subnet*) part, and *host address* (or *host ID*, which is the physical machine connected to the network); routing and forwarding takes place among the networks and the subnetworks.

The principal of routing and forwarding is quite simple: A routing table maintains all the addresses of the next hop and interfaces (ports) that should be used to forward an incoming packet. The chosen interface and next-hop address are functions of the incoming IP packet (its destination network address, as it appears in the IP destination address field of the header of that packet).

The network address part in the IP header went through some changes over the last decades of Internet use. Up until 1993, in what was later called *classful* (or *class-based*) networks, the IPv4 address was categorized in four address classes (A to D), with one reserved class (E). Originally,[11] the network address part of the IP address was defined by the IP class. A Class A address always starts with a "0" bit in its MSB, that is, addresses 0.0.0.0[12] to 127.255.255.255,[13] and the network ID is defined by the remaining 7 bits of the first byte of the IP address (i.e., a total of 128 networks). Class B addresses start with a "10" bit pattern in the MSB bits, and the remaining 14 bits of the first two bytes are the network ID, that is, class B contains addresses 128.0.0.0 to 191.255.255.255, and about 16 thousands networks. Class C address starts with a "110" bit pattern, and the remaining 21 bits of the first three bytes of the address define the network ID. Class C addresses thus are in the range of 192.0.0.0 to 223.255.255.255, organized in roughly 2 millions networks. Class D is used for multicast addresses, and their IP addresses always start with an "1110" bit pattern, followed by the remaining 28 bits for the multicast address. This class therefore is in the range of 224.0.0.0 to 239.255.255.255. The last class, E, is reserved, and its IP addresses start with an "11110" bit pattern. Some IP addresses (networks) are nonroutable, and are reserved for specific use (as outlined later in Table 5.3).

The rapid use and growth of the Internet and the consumption of IP addresses resulted in a severe shortage of IP networks and addresses with unique IP network addresses. Two million networks, most of them containing about 250 hosts, were simply not enough. To cope with this unbearable situation, many solutions were

---

[11] At the very beginning of the Internet, in the 1980s, the network address was simply the first byte of the IP addresses; that is, there were just 256 networks in the Internet …

[12] An IP address that ends with a "0" as its host ID means a subnet and not a specific host.

[13] An IP address that ends with all bits "1" in its host ID means a broadcasting address of all hosts in the subnet and not a specific host.

suggested, starting from many intermediate solutions to extending the IP address to 128 bits (IPv6 [98]). These intermediate solutions turned out to be so efficient that IPv6 was no longer required so desperately. However, the use of these various mechanisms to expand the original IP address space makes things a bit more complex. These mechanisms may include Border Gateway Protocol (BGP) [364], Classless Inter Domain Routing (CIDR) [363], and Network Address Translation (NAT) [395]. The most influential mechanism in solving the shortage of IP networks is NAT, but it is not impacting addressing mechanism, routing or forwarding. The most influential change in addressing, routing and forwarding is due to CIDR, which is described in the following.

Classless Inter-Domain Routing added hierarchy to the network addresses by defining subnets with a *prefix length* noted /n (where n defines the number of initial bits in the IP address that should be considered as the network ID and subnetwork), or using a *network mask*.[14] This hierarchy also allows aggregation (also called *summarization* or *supernetting*) or defines a range of networks by a simple notation and thereby shrinks the routing tables, and reduces routing advertises throughout the Internet. It also created the potential of overlapping address ranges, or more importantly, inclusions of address ranges within broader address ranges, a situation we call exceptions. This makes it possible to specify a broad range of addresses in a routing table in one entry, and to define networks that are exceptions within this range by additional entries in the routing table (that are providing specific routing rules, which are different from those defined by the entire range).

An example will clarify this concept: IP address 192.168.16.0/20 (or 192.168.16/20 for short, or 192.168.16.0 with a network mask 255.255.240.0) means all IP addresses from 192.168.16.0 to 192.168.31.255 (the first 20 bits of the address are masked, which leave the last 12 bits unspecified):

Obviously, 192.168.16/20 also contains 192.168.24/21,[15] 192.168.16/21[16] (which are two subnetworks having together the same span of 192.168.16/20), and 192.168.27/24.[17] Actually, any IP address with network mask bigger than

	IP Address	Network Mask
Regular notation	192.168.16.0/20	255.255.240.0
Hexadecimal presentation	C0 A8 1X XX	FF FF F0 00

---

[14]For the sake of historical correctness, network masks were used before CIDR, in 1985 [320], to specify subnetworks within the classful categories (i.e., the remaining bits of the masked address, after considering the network address, left the subnet address). Later on, in 1987, Variable Length Subnet Mask (VLSM) was used [58], to allow variable sized subnetting.

[15]192.168.24/21 includes 192.168.24.0 to 192.168.31.255

[16]192.168.16/21 includes 192.168.16.0 to 192.168.23.255

[17]192.168.27/24 includes 192.168.27.0 to 192.168.27.255

Table 5.2 Example of Routing Table

Row	IP Address	Next hop	Interface
1	192.168.16/20	10.10.1.1	2
2	192.168.27/24	10.16.54.2	3
3	192.168.0/17	10.16.1.6	1
4	0.0.0.0/0	10.1.1.0	7

255.255.240.0, or prefix length bigger than 20, of the same initial IP address C0 A8 1x xx, will be part of this network range. This means that in a routing table, a specification of a route (next-hop and interface) can be defined to a small range of IP addresses within the range of another group, having another specification route (next-hop and interface).

Assume, for example, a small routing table as shown in Table 5.2. IP destination address 192.168.156.5, for example, matches only row 4, by definition, the default routing. IP destination address 192.168.1.3 matches only row 3 (and row 4, by definition). IP destination address 192.168.18.1 matches rows 1 and 3 (and row 4), and IP destination address 192.168.27.56 matches all rows. This overlapping, or network exception, raises the question of how to route packets of IP destination 192.168.27.56.

Here comes the longest prefix match (LPM), or actually, the best matched prefix search. The longer the match is (with regards to the prefix length), the better the routing that will be chosen for this packet. In other words, a router needs to search more than just the right (matched) prefix; the router has to find the most specific match, which is the longest matching prefix in our case. The algorithm used to search for the best match is a bitwise comparison of the addresses, or more specifically:

```
    Assume the routing table like the one given in Table 5.2 (always, the
last row [entry] should be the default route, that is entry 0.0.0.0/0), and
an IP destination address, for which a next hop and interface is searched.
Make sure the routing table is sorted by the network mask (the prefix
length), from the most specific (the longest prefix) to the least specific
network (the shortest prefix).

For each row (entry) in the routing table do
{
    if (IP destination address AND entry's network mask == entry's IP address)
            {
            Use the entry's next-hop and interface
            Exit
            }
}
```

Please note that the above algorithm will always stop at the last row of 0.0.0.0/0. One of the biggest issues in maintaining such routing tables is the need to keep these tables sorted by the prefix lengths. In a large routing table, adding an entry of /31 for example (close to the top of the table), might require a shift for large portions of the table.

IPv6 addressing is based on an address field of 128 bits long, allowing $3.4 \cdot 10^{38}$ IP addresses (compared to the $4.3 \cdot 10^9$ IP addresses of IPv4). An IPv6 address is presented by 8 groups of 4 hexadecimal digits (16 bits) each, in a notation like 0123:4567:89AB:CDEF:0123:4567:89AB:CDEF.[18] The first 64 bits usually refer to the network (or subnet) address, whereas the last 64 bits are the host ID (usually drawn from the MAC address of the machine). IPv6 addressing uses the CIDR concept, that is, a range of addresses can be written by an address with a prefix length (/n, where n is the length of the network prefix). IPv4 addresses are recognized by an IPv6 addressing mechanism, for backwards compatibility, by attaching the IPv4 address to the ::/96 IPv6 prefix (all zeros), for example, ::192.168.0.1.[19]

Table 5.3 summarizes all unique and special addresses that usually are not routable, in IPv4 and IPv6.

5.5.2 Search Engines

In order to cope with demanding search operations—the many search operations per packet in complex applications carried out in high speed communication trunks that can end up in billions of operations per second, and the necessity to search at wire speed—search engines appeared for our salvation. Search engines here are by no means Internet search engines, like Google or the like. Search engines are software processes, hardware circuitry, or a combination of the two, sometimes designed as a functional unit on the network processor, and sometimes packed in a designated chip, search processor or coprocessor (as detailed in Chapters 7 and 9). The search engine functionality is simple: it returns a value (a *result*) when presented with a value (a *key*).

The simplest search engine is plain memory, where it returns a value when presented with the address of that value. As a matter of fact, memory returns a value contained in an address when presented with that address. The opposite operation happens with a special type of memory, called associative memory, or Content Addressable Memory (CAM), which, when presented with a value, returns

---

[18]The following rules apply to IPv6 address notation:

   (a) A sequence of one or more "0000:" can be replaced with a "::", but just once.

   (b) The leading 0 can be omitted in any of the groups.

   (c) The last four bytes can be represented by the decimal dotted notation, for example, ::DEF:192.168.0.1 is 0000:0000:0000:0000:0000:0DEF:C0A8:0001

[19]Another way is to map IPv4 addresses into IPv6 address by ::FFFF:192.168.0.1, so that IPv6 applications can handle IPv4 addresses, mapped into the IPv6 address-space.

Table 5.3 IPv4 and IPv6 Nonroutable Addresses

Addresses	CIDR	Purpose
0.0.0.0–0.255.255.255	0.0.0.0/8	Zero addresses
10.0.0.0–10.255.255.255	10.0.0.0/8	Private IP addresses
127.0.0.0–127.255.255.255	127.0.0.0/8	Localhost/Loopback Address
169.254.0.0–169.254.255.255	169.254.0.0/16	Zeroconf/APIPA
172.16.0.0–172.31.255.255	172.16.0.0/12	Private IP addresses
192.0.2.0–192.0.2.255	192.0.2.0/24	Documentation and examples
192.88.99.0–192.88.99.255	192.88.99.0/24	IPv6 to IPv4 relay Anycast
192.168.0.0–192.168.255.255	192.168.0.0/16	Private IP addresses
198.18.0.0–198.19.255.255	198.18.0.0/15	Network Device Benchmark
224.0.0.0–239.255.255.255	224.0.0.0/4	Multicast
240.0.0.0–255.255.255.255	240.0.0.0/4	Reserved
0000:0000:0000:0000:0000:0000:0000:0000	::/128	Any address
0000:0000:0000:0000:0000:0000:0000:0001	::1/128	Localhost/Loopback Address
	FC00::/7	Unique local unicast address
	FE80::/10	Local address (like zeroconf)
	FF00:/8	Multicast addresses

a value that is the address of the presented value. CAM, and the ways of using it, are described later in this chapter.

More complex search engines, as mentioned above, are special hardware assists. These search engines are internally based either on CAM memories, or on search algorithms materialized by hardware circuitry[20] that use regular memories inside these search engines, or are attached to them.

These external search engines (sometimes called network search engines [NSE], or search accelerators) can support, as of 2008, billions of searches per second, with search keys of hundreds of bits wide among millions of entries. These search engines can interface with the network processor like a regular memory, and work transparently with the network processor, or by using the standard LA-1 interface (described in Chapter 7).

---

[20]ASIC, FPGA or embedded processors, as described in Chapter 7.

Table 5.4 Classification of Matching Techniques

		Data entries in the table	
		Full	Partial
Search item	Full	Exact match	Classification
	Partial	Pattern match	Best match

As we have made clear, the most important factor in searching is finding an entry among hundreds of thousands of entries at maximum speed. However, the question of maintaining and modifying the data, mainly by inserting or deleting an entry, is of significant importance as well, as is the size of the data structure used to store all the entries. So when designing or using search engines, all of this must be carefully considered and despite the fact that it is not emphasized in this book, it plays a very significant role in network processing performance.

For our purposes, searching is done in network processors either in memory (internal or external) or by some hardware assist (internal or external). The simplest way, of course, is to run a search on data stored in the internal memory of the network processor or interfaced to the network processor in a transparent way. There are several types of memories that can be used in a network processor; they are described in Chapter 7.

Various kinds of searches are required in different situations. Table 5.4 [425] provides a summary. It is important to distinguish between different kinds of searches, as they require different data structures and algorithms. Full match means that the entire searched key must match the required data, either in full (exact match) or in part (classification, e.g., some of the fields of the required data must match the entire searched key). Partial search means that part of the searched key should match the required data, matching either all of the required data (i.e., pattern search), or part of it (best match).

To understand the way data is organized and stored, as well as how operations on that data are performed (e.g., search, insert, delete, or modify), some preliminary knowledge of data structure and algorithms is essential. In the following subsection, such an introduction is given; the knowledgeable reader can skip the next subsection.

5.5.3 Data Structures and Algorithms: Theory, Definitions, and Rehearsal

Data structure has significant importance for search time operations (searching, updating, etc.); thus, a great deal of research, patents, and literature cover this topic. This subsection describes the very basics of data structure, to remind and enable the reader to evaluate its importance, to match it to the data types and operations required, and to intelligently use the various options available in using or designing search engines, memory spaces, and procedures. The interested reader, or the one who is required to develop his or her own search mechanisms, can use the classical references on algorithms (e.g., [89, 260]) as well as the vast literature in this area.

Given a *set* of *elements* (values), each is identified by a *key*, organized in a *data structure*. The fundamental operations for handling the data are as follows:

- *Search*—the most important operation, in terms of time: Query the data structure and return the value associated with the given key. There are several ways of searching—exact (or full) match, longest prefix match, or pattern matching search.
- *Insert*—adding an element (value) to the set of the already maintained elements.
- *Delete*—extracting an element (value) from the data structure.
- *Modify*—changing the value of an element in the data structure.

Data structures in "plain" memories that can be used for lookups are based on lists, direct address tables, sorted tables, hash tables, skip lists, or various kinds of trees, as described below.

5.5.3.1 Metrics

In order to evaluate the data structure, three main metrics are used: (a) time complexity[21] to perform an operation (search or update); (b) size efficiency; and (c) scalability. Time complexity can be expressed by a function of the size of the data (or some constant if it is independent of the size of the data). Size efficiency is simply the ratio between the required space to maintain the data structure and the size of the data itself. Usually, time and space (size) complexities are trade-offs, that is, efficiency in one parameter causes deficiency in the other. Scalability refers to the ability to increase and scale up the data structure for varying requirements of data problems.

For network processing purposes, search time is of utmost importance, and size is of secondary importance. Since memory is expensive, especially fast memory, and since in most cases, memory size is restricted physically in a network processor,

---

[21]Complexities are often noted by a notation of $O(g)$, which means that the magnitude of the analyzed function has an asymptotic upper bound in terms of the function g, usually a simpler function. This means that for any n bigger than some n_0, the analyzed function is bounded by (i.e., less than) some constant c times the function $g(n)$. Formally, $f \in O(g) \Rightarrow (\exists c > 0, \exists n_0, \forall n > n_0) \{f(n) \leq cg(n)\}$. $O(n)$, for example, means that $g(n) = n$, and the complexity is linear with n, or, in other words, the analyzed function f is not growing faster than n. $\Theta(n)$ means that for any n bigger than some n_0, the complexity is bounded by c_1 times n and c_2 times n, (i.e., between them) where c_1 and c_2 are some constants. $\Omega(n)$ means that for any n bigger than some n_0, the complexity is bounded (i.e., greater than) by some constant c times n.

it is important to use highly efficient data structures without compromising the search time.

5.5.3.2 List Structures

A data structure that is extremely inefficient in carrying out most search operations in network processors is the *linked list* structure. Elements are ordered linearly with pointers to their next elements. Although the insert operation is extremely efficient (takes $O(1)$ time), all other operations (search, delete or modify) are very slow, and in the worse case it takes $\Theta(n)$ time for full match or any other search.

Skip lists are more efficient than linked lists for network processing purposes, but they are also not widely used since so many other lookup data structures are better. The principle of skip list is maintaining ordered linked lists, but with additional, randomly assigned, forwarding pointers that enable a search operation to skip quickly over parts of the linked lists. Several linked lists are constructed in parallel layers, starting with the lowest, fully sorted linked list; the probability for an element in linked list layer i to exist also in layer $i + 1$ is p. Searching is done by going from the upper layer linked list to the lowest one, scanning each layer for the element that is less than or equal to the target. The performance of any operation using this data structure, on the average, is $O(\log n)$, and in the worst case, it is linear, $O(n)$.

5.5.3.3 Table Structures

Direct addressable tables are simply arrays of the elements, addressed by their keys. These structures are the most efficient for full match (exact) searches, given the fact that all operations are of $O(1)$ time complexity (the value is available for any operation with one access attempt). The problem in this type of structure is that it requires a space (memory size) of the size of the potential key range, for example for possible key values of 2^{32} bits, like an IP address, a table of size 2^{32} entries must be maintained, regardless of how many entries actually exist. This type of table is efficient only for exact matches where a small range of keys are maintained, and for most keys there are associated values to be manipulated. Direct addressing lookups, in packet processing, can be applied in limited cases where the key range is relatively small, or the table size is small as a result. Such cases may include protocol classification, port assignments, VLAN determination, and so on.

To reduce the memory size (space), the elements can be placed in a table, in a sequence, sorted by their keys (resulting in a *sorted table*). Accessing the sorted elements is then done according to the required key, either through a simple *linear search*, which is inefficient,[22] through a *binary search*,[23] or through other methods (e.g., *Fibonaccian search* or *interpolation search*). In a binary

---

[22]Linear search in a sorted table will take on the average $\Theta(n/2)$ execution time, or $\Theta(n)$ in the worst case.

[23]Binary search has the best (most minimal) search time complexity in the worst case scenario, $O(\log n)$.

search, the required key is compared to the key of the element positioned in the middle of the table, and continues by ignoring the half whose values are irrelevant, repeating the search in the remaining relevant half in the same way until the element is found, if it is in the table. The other methods determine the relevant part of the table for the subsequent searches by other mechanisms, but the results are pretty much the same. All operations take longer for execution than in direct addressing, of course, and are of $O(\log n)$ time complexity on the average,[24] where n is the number of maintained elements. This data structure is very efficient in terms of memory size (space), and is very compact. However, it is not efficient enough in terms of operation time (search or update), and there are better ways to manipulate the elements.

5.5.3.3.1 Hash Tables

Hash tables are a compromise between direct addressable tables and sorted tables. Hash tables are very efficient in operations; the most advanced techniques yield operations of $O(1)$ time, on the average, while keeping the table size (space) fairly compact.

The principle behind hash tables is to create a direct addressable table in which the keys of the elements are expressed in such a way that there is an almost 1:1 relation between any required key and the address of its element in the table. The function that computes this transformation is called the *hashing function*. The problem in this idea is that a 1:1 relation is very complex to achieve, and some "collisions" may occur, that is, two keys might produce the same hash value. To overcome this problem, either a better hashing function can be utilized, or a solution to the expected collision must be provided. In order to improve the hashing function and to make it simple uniform (i.e., spreading the results uniformly over the space of the table), one has to know the key distribution, which is not always available. At any rate, apart from perfect hashing (described in the following), collisions are inevitable. For example, following the birthday paradox, there is a 99% chance of a collision when a simple uniform hash key is used in a 1-million-entry-sized table that is filled with just 3000 entries.

There are two basic methods for achieving hash functions: division and multiplication. The more common practice is the division method, where the key is divided by the table size m (or preferably by some prime close to it, definitely not a number close to 2^n for any n), and the remainder is used as the hash key k (i.e., the hash function $h(k) = k \bmod m$).

In the multiplication method, the key k is multiplied by some fraction A, and the fraction part is multiplied again by some m. The resulting integer is the hash

---

[24]Interpolation search, called sometimes also extrapolation search, takes on the average $O(\log \log n)$ execution time, assuming the keys are uniformly distributed. Worst case for any distribution in this case is $O(n)$.

number $h(k) = \lfloor m(kA \bmod 1) \rfloor$.[25] A affects the hash function, and one very good choice is a special case called the Fibonacci hashing, where A is the inverse of the "golden ratio,"[26] equal to $(\sqrt{5} - 1)/2 \cong Fib_{i-1}/Fib_i$, where Fib_i is the i-th Fibonacci number.

When there is a collision, there are two methods of resolving it. The simplest collision resolution algorithm in hash tables is to maintain a linked list (as described before) for any of the table entries where a collision occurs. In this way, the insert operation has a worse case time of $O(1)$, but, depending on the number of collisions on any single entry in the hash table, the worst case search, delete or modify can be $\Theta(n)$. In the case of a simple, uniform hashing function that distributes all keys evenly among all table entries, the average time to search, delete or modify will be $O(1)$.[27] In order to decrease the chance of having the worst case of chaining, binary-trees, dynamic arrays, and other techniques can be used.

The other collision resolution technique is open addressing. In open addressing, all elements are kept in the hash table (not as a linked list), and an efficient way to resolve collisions is by probing for another, available place in the table (for insert operation), or for the required key (when searching). In linear probing the other place is simply the next entry in the table; in a cyclical way, that is modulus the table size, or $h(k, i) = (h'(k) + i) \bmod m$ for any $i < m$. In quadratic probing, the next place examined is increased in a quadratic way—that is, $h(k,i) = (h'(k) + c_1 i + c_2 i^2) \bmod m$ for any $i < m$. Another way, double hashing, probes the next place in a distance that is a function of the key, like the hash itself—that is, $h(k, i) = (h_1(k) + i\, h_2(k)) \bmod m$. Linear probing performs well in terms of locality (cache performance) but is sensitive to clustering. On the other hand, double hashing has no clustering effects, poor locality, and requires a bit more computational effort. Quadratic probing is a good answer to these trade-offs. The performance of the open addressing collision resolution techniques is $O(1)$ on the average.[28]

Perfect hashing is possible when the entire set of keys that will be used is known ahead of time, during the design of the hashing function. In perfect hashing no collisions will occur, since every key is mapped to a different place in the table by the perfect hash function. Minimal perfect hashing guarantees that

---

[25]R mod 1 means the fraction part of R, *that is,* 5.762 mod 1 = 0.762

[26]The golden ratio for two integers x and y is φ, that is, $\varphi = x/y$, when $x/y = (x + y)/x$. Putting it in another way: $\varphi = (\varphi + 1)/\varphi => \varphi^2 - \varphi - 1 = 0$. The irrational value of the golden ratio is $\varphi = 1.618$ … It is interesting to note that this "magical" ratio appears in many natural and artistic phenomena, and has many mathematical characteristics. Among them, for example, is that the reciprocal of the golden ratio is the golden ratio minus 1, that is, $1/\varphi = \varphi - 1$ (equal to 0.618 …) and that the square of the golden ratio is the golden ratio plus 1, that is, $\varphi^2 = \varphi + 1$ (equal to 2.618 …), as it results directly from the definition of φ.

[27]More precisely, the time complexity is $\Theta(1 + n/m)$, where the hash table has m entries and occupies n elements (n/m is often called the load factor), since $m = O(n)$ in most cases.

[28]More precisely, the average number of probes in a collision is $1/(1-n/m)$, where the hash table has m entries and occupies n elements, and therefore, given constant n/m, the time it takes to resolve a collision is $O(1)$.

the hash table is minimal in size, that is every location of the table will be used. Perfect hashing also guarantees a constant time in the worst case for a lookup or key insertion.

5.5.3.3.2 Associative Arrays

The last type of data structure in this category is an abstract data structure called an *associative array*. Associative arrays are essentially definitions of lookup tables that are built by hash tables, self balancing search trees (described in the following section), skip lists, and so on, and their purpose is to find a value that is associated with a key. Association is a mapping, a function, or a binding, and there is a 1:1 relation between any key and any value in the associative table. As described in the following, special purpose hardware is available to offer associative table functionality, according to various mechanisms.

5.5.3.4 Tree Structures

Another category of data structures is the tree-family, although representation of various trees can also be done in tables. Some basic definitions are necessary to begin: A set of elements is organized in a *graph* G, where each element is a *node* in the graph, and the relations (or *binary relations*) between the elements are the *edges* in that graph. A *free tree* is a *connected* graph (i.e., a path between any two nodes in the graph) where no loops exists in any of the graph's paths (i.e., there is exactly one path between any two nodes).

A *rooted tree* is a free tree where one node is identified and is called the *root*. In rooted trees, there are relations between the nodes that result from their position in the tree: *Father* and *child* are two connected nodes where the father is closer to the root on the path from the child to the root. A node without any children is called a *leaf* (or *external* node). Nonleaf nodes are called *internal* nodes. Any node on the path from a node to the root is called an *ancestor*, and any node from a node away from the root is called a *descendant*. The *height* of the rooted tree is the longest path (in number of hops) from the root to the leaves. The *depth* of a node is its length of path (in number of hops) from the root. A *level* contains all nodes of the same depth. The *degree* of a node in a rooted tree is the number of children the node has (and not, as in a free tree, the number of connected nodes). An *ordered tree* is a rooted tree where the order of the children is important.

A *binary tree* is an ordered tree, where each node has a degree of maximum 2, and the right and left children are identified (and each child has a clear identification of being right or left even if it is the only child). A *full binary tree* is a binary tree that has no nodes of degree 1. A *perfect binary tree* (or a *complete binary tree*) is a full binary tree where all the leaves are at the same depth, and it always has *n-1* internal nodes and *n* leaves.

Now we reach the first data structure that maintains a binary tree: The *binary heap*, which is often used for *priority queues*. This structure is an array that represents a complete binary tree (or almost); the last (lowest) level can be empty in its rightmost leaves. The array contains the values of the tree's nodes, level after level, in the order of the nodes of the tree. The heap property is maintained, which means that a node can have a value that equals, at most, its father's value. Some results of this data structure are as follows: The place in the array of the parent of a node x is $\lfloor x/2 \rfloor$, its left child is $2x$ and its right child is $2x + 1$.[29] Inserting a value to the binary heap has a time complexity of $O(\log n)$, where n is the number of elements in the binary heap, but searching is not efficient in this structure at all.

A much better way to handle searches in terms of efficiency is with the *binary search tree* (BST). In BST representation, every node in the binary tree is kept in a structure, along with pointers to its parent node and the left and the right children nodes. In a BST, the keys always maintain the property that if x is a node and y is a node in the left sub-tree of x, then $key(y) < key(x)$, and the other way around; that is, if y is a node in the right sub-tree of x, then $key(x) \leq key(y)$.

Searching for a key in BSTs starts in the root and at every level a comparison is done on the node's key and the search may continue in one of the branches to a lower level. Thus, the search has a time complexity of $O(h)$, where h is the height of BST. If the binary tree is full, then the time complexity of conducting a search in the worst case will be of $\Theta(\log n)$, where n is the number of nodes in the tree, since h in this case is $O(\log n)$. However, if the BST is not representing a full tree, the time complexity is worse than of $\Theta(\log n)$, and in the case of a linear "chained" tree, it can reach the complexity of $\Theta(n)$. In case of a randomly built BST, that is every added node has a randomly chosen key, then the height of the tree h is again $O(\log n)$, and the search complexity is of $\Theta(\log n)$.

In order to better cope with the tree height problem (i.e., to minimize it as much as possible), a category of data structures called *balanced BSTs* (or *self-adjusted balanced BSTs*) is used. This type of tree contains more indications (bits, "colors," etc.) in each of the elements that are used to build the tree in a balanced way, that is making sure that the heights of the two child sub-trees of any node can differ by at most one. In a complete binary tree, there are exactly 2^h nodes in the h-th level, and exactly h levels. Balanced BSTs are not necessarily precisely balanced, since it involves repeated and expensive computation to keep the tree at minimum height at all times. Balanced BSTs keep their heights within a constant factor from the optimal height (it was shown that the height will never exceed 45%). Therefore, in this category of trees, the search time, as well as other operations on the trees, takes $O(\log n)$ execution time.

Examples of balanced BSTs include *AVL trees* (where the AVL stands for Adelson-Velsky and Landis, who, in 1962, suggested the use of two more bits per node [8]), *red–black trees*, and many others. They all possess the same attribute of search and other operations on the structure in an execution time of $O(\log n)$.

---

[29]Provided, however, that the root's index (position in the array) is 1.

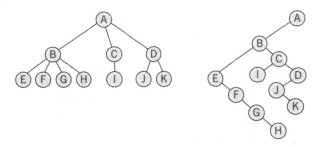

FIGURE 5.11

Binary representation of m-ary tree

A generalization of the binary tree is *multiway search tree* (*m-way tree* or *m-ary tree*), where each node can have *m* children, rather than just two, as in case of the BST. The most commonly used balanced tree among these multiway search trees is the B-tree, usually used in file systems or disk and other external storage systems. Most of these trees are suitable for various needs, but not really for packet processing; therefore, only the suitable category will be described in the following. It is worth noting that any m-ary tree (or even a forest) can be treated with a binary tree representation (where the leftmost child of a node in the m-ary tree becomes the left side child of the binary tree, and the right side child of the binary tree represents the next sibling node of the m-ary tree, as can be seen in Figure 5.11).

5.5.3.5 Tries

One category of multiway trees that is heavily used in networking applications is the *tries* [136], or *prefix trees*. Trie (which sometimes is pronounced like "tree" and sometimes like "try")[30] is an ordered tree that is used to represent strings, where the tree's edges are labeled by the characters of the strings, and the strings are represented by the leaf nodes (see Figure 5.12, for example). The internal nodes are used to "spell out" the strings following the path from the root to the nodes (up to the leaves, to define the represented strings).

An example of a trie holding a list of strings containing IP addresses given in Table 5.5, can be seen in Figure 5.12; please note that for the sake of explanation, the addresses are strings, spelled out character by character (and not as the usual IP 4-bytes representation). An exact match search is done by comparing one character at a time, level by level, between the searched string to the edges of the trie. This goes on, until the searched string is exhausted, the trie ends during the search, or there is no suitable edge to follow. If the search ends in a leaf simultaneously with the last character of the searched string—there is a match. For LPM purposes, the

---

[30] The term trie was derived from the ward "retrieve."

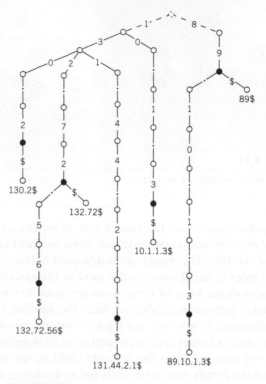

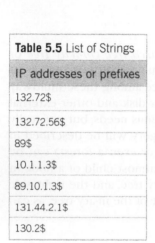

Table 5.5 List of Strings
IP addresses or prefixes
132.72$
132.72.56$
89$
10.1.1.3$
89.10.1.3$
131.44.2.1$
130.2$

FIGURE 5.12

IP addresses trie

terminator at the end of any string (at the edge leading to a leaf) indicates that the string, up to the node from which a terminator edge exits (dark node in Figure 5.12) is used as a prefix. This can be demonstrated, for example, in Figure 5.12: The string "89.10" will match "89$," since the last terminated node was "89." The string "132.7" will not match any of the strings in the trie, because no terminator was crossed.

Updating a trie is quite simple: Inserting starts with a search, and when a node with no correct edge to follow appears, then a node is added with the remaining string on the edge to this node. If, for example, "89.10.1.5$" is to be added, then the trie is searched until "89.10.1" node arrives, and an extra node with an edge equal to "5" is added. Deleting also starts with a search, until a match happens, and the matched node is either deleted (if a leaf, along with all its preceding nodes that remain without a terminating node) or unmarked as prefix node.

Trie is essentially a lexicographic order of strings, that is one can look at it as if it is a dictionary. The search complexity, in the worst case, is $O(m)$, where m is the key length. In the example of Figure 5.12, it is evident that this trie is not very efficient in terms of its search operation, due to its long strings. A better representation is a *compressed trie* or *compact trie*. A compressed, or compact[31] representation of a trie is one that merges all chains of edges that have no branches (the nodes between these edges are of degree one) to one edge, labeled with the string of characters of the merged edges, or labeling the resulting path. In the specific case of a compact binary trie, the total number of nodes is $2n-1$, like in a full binary tree, where there are n strings that the trie represents. An example of a compact trie is given in Figure 5.13, which represents the (nonbinary) trie shown in Figure 5.12 (again, for the sake of explanation, addresses are spelled out character by character and not as the usual IP 4-bytes representation). Searching for an exact match is similar to the search in a trie, as detailed above. For LPM purposes, the terminator at the end of any string indicates that the string, from the root to the node from which a terminator edge exits, can be used as a prefix. A prefix can also be the string created from the root to the leaf that has an incoming edge with a string terminated by the terminator (e.g., the dark node in Figure 5.13).

For networking applications, tries have the advantage of being able to help with LPM, that is searching for a value having the longest possible prefix similar to a given key. Tries can also be used for determining the association of a value with a group of keys that share a common prefix. Although tries are used mainly for searching (which is what we discuss here), they can also be used for encoding and compression.

Specific types of tries include the *bucket trie* (a trie in which leaf nodes are buckets—usually of fixed size—that hold several strings), and the *suffix trie* (a trie corresponding to the suffixes of given strings). *Compact suffix tries* (sometimes called *Pat-tries*) are suffix tries where all nodes with one child are merged with their parents). Suffix trees and compact suffix trees are very often used in fast, full text searches. Internet search engines use these techniques frequently,

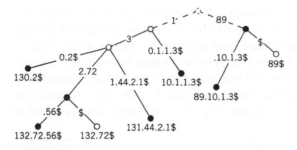

FIGURE 5.13

Compressed trie

---

[31]Sometimes compressed and compact representations are distinguished by the way edges are labeled and treated. The technique of compressing the one-child nodes is also known as *path compression*.

as opposed to network processor search engines, which use hashing techniques and prefix tries.

Tries occupy $O(n)$ space, where n is the number of objects represented, and operations on the tries take $O(km)$ time, where m is the string length and k is the "alphabet" size, that is the maximal degree of a node. Although it seems that $O(km)$ is efficient, in most cases a binary search is equivalent or better, since $O(km) \geq O(\log_k n)$ (for binary "alphabet," where n, the number of objects, is less than 2^m, the potential number of objects of length m). Tries, however, are simpler to handle than binary trees, especially balanced BSTs. Tries can even be considered a good alternative to hash tables, since usually they work faster, and simpler since they have no collision treatments, hash functions, and so on.

There are many advanced data structures based on tries that employ partitioning, hashing, heaps, links, and pointers, and other techniques in the tries, to make them faster and smaller (e.g., van Emde Boas tree [418, 419],[32] Willard's X-fast, Y-fast,[33] P-fast and Q-fast tries [428, 429]).

When keys in the trie are strings made of bits, as is usually the case, the trie becomes a *binary trie*, and searching becomes simple (so called *digital tree search* [84, 260]): If no match is established at some level, then, depending on the next bit, the search branches right to the next level if the next bit is 0, or branches left if it is 1. This is in contrast to the binary tree search, where comparisons have to be executed at each level. In the context of networking, where LPMs are desired, this data representation and searching is very useful. Binary tries were studied quite extensively, and the performance characteristics of these tries are known. The expected average depth of an element in the trie is $\Theta(\log n)$ if there are n elements in the trie and the trie was built randomly. This means that operations on binary tries perform similarly to those of the BSTs.

5.5.3.5.1 Patricia Tries

Another way to handle bit strings is through a *Patricia tree*[34] [163, 323], which is a binary compact trie,[35] where any node that has only one child is merged with its child. Patricia tries are therefore full binary trees. Patricia stands for "Practical Algorithm To Retrieve Information Coded In Alphanumeric." A Patricia trie, in its simplest form, does not store keys in its nodes; rather, the idea is to maintain in the nodes the number of bits that should be skipped over (in the case of an unmatched situation), to get to the bit according to which the next decision must be made, as shown and explained in

---

[32]van Emde Boas structure (known also as *vEB tree* or *vEB priority queue*) assumes a fixed known range of possible elements {0,1,2, … ,u-1}, sometimes referred to as the *universe* size, and achieves time complexity for search and update operations of $O(\log \log u)$ and size complexity of $O(u)$.

[33]Y-fast is a randomized algorithm that improves van Emde Boas size complexity to $O(n)$, and has similar time complexity for a search or update operation, that is, $O(\log \log u)$.

[34]Patricia Trie is often called Patricia tree, radix tree, or crit-bit tree (which means critical-bit, as can be understood from the way the tree works).

[35]There are definitions and algorithms that use m-ary Patricia tries in the literature, but we'll use the binary Patricia trie as the Patricia trie.

Table 5.6 IP Addresses

Pointer	String (IP address)	1234	5678	111 9012	1111 3456	1112 7890	2222 1234	2222 5678	2333 9012	3333 3456	3334 7890	4444 1234
						Bit position of the IP address in some bit representation						
P1:	132.72$	0001	0011	0010	1011	0111	0010	1100				
P2:	132.72.56$	0001	0011	0010	1011	0111	0010	1011	0101	0110	1100	
P3:	89$	1000	1001	1100								
P4:	10.1.1.3$	0001	0000	1011	0001	1011	0001	1011	0011	1100		
P5:	89.10.1.3$	1000	1001	1011	0001	0000	1011	0001	1011	0011	1100	
P6:	131.44.2.1$	0001	0011	0001	1011	0100	0100	1011	0010	1011	0001	1100
P7:	130.2$	0001	0011	0000	1011	0010	1100					

the following example. There are many ways to implement Patricia trees and related algorithms, depending on the need, but they all share the same concept. The concept of Patricia tries is also used for routing table lookup in the BSD kernel UNIX.[36]

An example will clarify the Patricia trees concept. Suppose we maintain a list of strings as described in Table 5.6. (These are the same IP addresses that we used in Table 5.5 and Figure 5.12; the addresses are the bit value of the IP strings, spelled out character by character, just for the sake of explanation, and not according to the usual 4 bytes IP representation.)

The Patricia trie that describes these entries is illustrated in Figure 5.14, where at every internal node there is the bit number that should be examined in order to make the next decision (turning to the right or the left child, 1 or 0, respectively). A more detailed description of the Patricia trie data structure and algorithm is given in Figure 5.15 and explained afterwards.

The data structure can be a linked-list or an array, where every node contains a pointer to the string list, right and left pointers to other nodes, and the bit position to be examined in this node. We begin the search from the header node (in this case, the one that points to P6), by going to the first node (in our case, the one pointing to P3). Every node we reach from now on tells us what bit position to check in the required value string. If this bit is 1 we go to the right node, and if this bit is 0 we go to the left node. Once we reach a node where the bit position to be examined is less than or equal to the bit position in the node we came

---

[36]The original Patricia [323] did not support longest prefix matching, and the scheme adopted for UNIX [389] modified Patricia to handle LPM and noncontiguous masks. Current UNIX-BSD implementations (called *BSD Tries*) are extensions of the Patricia to handle LPM [374]

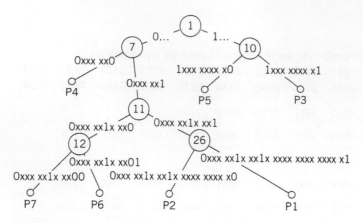

FIGURE 5.14

Patricia trie of the IP addresses

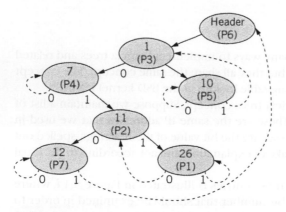

FIGURE 5.15

An example of patricia trie of the IP addresses

from, we stop, and compare the value we are searching for with the value that the node is pointing to. If it is a match, then we found the value, and if it is not, then the value is not in the trie. So, if, for example, we look for "132.72$" (0001 0011 0010 1011 0111 0010 1100, pointed to by P1), then we start at the node after the header, check bit 1 (which is 0) and go to the left. Next, we examine the 7th bit, which is 1, so we go to the next right node. There we have to examine bit 11, which is again 1, so, we again go right to the next node, and there we check bit 26, which is 1. Moving to the right, we are brought back to the same node, asking us to examine bit 26 again. This tells us to stop and compare the value we are searching for ("132.72$") to the value pointed to by this node, that is P1, which is what we are looking for.

It is interesting to note, that by using a presentation such as Figure 5.15, one can look for any or all keys starting with a particular value. Suppose, for example, we want to find all keys starting with "13" (0001 0011). We start as before, take the first node after the header, examine bit 1, take the left path since the bit was 0, check bit 7 and turn right (the bit was 1). Now, at the node pointing to P2, we are requested to check bit 11, which we don't have. So, if the value pointed to by this node matches the string we are looking for, then this node and all of its descendants (up to the header) are pointing to strings that start with "13" (and indeed, P2, P7, P1, and P6 all start with "13").

The behavior of Patricia tries for uniform-distribution has been studied extensively, and the average depth of the Patricia trie is $\Omega(\log n)$ [18].

To handle LPM with Patricia trees, prefixes are stored in the internal nodes and the leaves of the Patricia tries, in contrast to what was explained above. Since prefixes can be located on several one-edged nodes of some path of the trie before being compressed into Patricia, then some nodes in the Patricia presentation can contain a list of prefixes. Searching (and inserting new nodes or deleting nodes following the search) can be executed in one of the following two ways: (a) For each node that stores a prefix or a list of prefixes that is crossed as the trie is traversed, a comparison is done between the searched string and the prefix or prefixes, and is remembered if matched; this continues until termination (reaching a leaf or an unmatched node), and the last remembered prefix (if there was one) is the LPM. (b) The trie is traversed as much as possible, and then backtracked to find the longest matched prefix among the crossed prefixes. Again, an example will clarify this. A list of string prefixes is given in Table 5.7 (this time in normal, standard bit

Table 5.7 A List of Prefixes

Pointer	String (prefix)	Bit position of the prefix (IP address presentation)							
		1234	5678	111 9012	1111 3456	1112 7890	2222 1234	2222 5678	2333 9012
P1:	132.72*	1000	0100	0100	1000				
P2:	132.72.56*	1000	0100	0100	1000	0011	1000		
P3:	89*	0101	1001						
P4:	10.1.1.3*	0000	1010	0000	0001	0000	0001	0000	0011
P5:	89.10.1.3*	0101	1001	0000	1010	0000	0001	0000	0011
P6:	131.44.2.1*	1000	0011	0010	1100	0000	0010	0000	0001
P7:	130.2*	1000	0010	0000	0010				
P8:	132.72.57*	1000	0100	0100	1000	0011	1001		

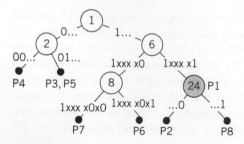

FIGURE 5.16

A Patricia prefix path compressed trie

presentation of IP addresses), and the Patricia trie that describes these prefixes is illustrated in Figure 5.16. Every internal node is the skip bit count, and nodes that contain a prefix or list of prefixes are darkened.

5.5.3.5.2 Multibit Tries, LC-tries, and LPC-tries

Other structures that are often used in networking applications are *multibit-tries* and *LC-tries* [18, 334–336], due to what is, on the average, their superior search performance over the Patricia trie. Extending the principle of the binary tries, Multibit and LC tries inspect not just one bit but several bits simultaneously at each step (at each node down the trie). Having said that, it is worth noting that due to their strict and complex data structure, most of these structures and algorithms increase the difficulty of frequently updating the data maintained. Nevertheless, they are frequently used in networking for IP lookup applications.

In *multibit tries* (or *multi-digit tries*), there are fewer branching decisions than in a binary trie because each branching decision is based on a multibit digit called a *stride*. In practice, there will also be fewer memory accesses, since the nodes' data is read from memory in access width, which is usually in multiple bytes. So, for example, a three level multibit-trie can be used for an entire IPv4 address (32 bits) by using a 16-8-8 stride pattern; that is 16 bits stride at the first level access, then 8 bits stride at the second level, and again 8 bits stride at the third level. The trie can be either *fixed stride*, if it has the same stride size at all nodes of the same level of the trie, or it can be *variable stride*. Any internal node in a multibit trie is of degree 2^i, $i > 1$ (i is the stride), and is a sub-multibit trie containing all the i-suffixes elements that start with an i-prefix digit. Multibit tries constructs, however, can yield empty-leaves if they are not built carefully, which means less efficient searches and bigger, unutilized storage.

Generally speaking, fixed stride tries are less efficient in memory utilization than variable stride tries. If the number of leaves of the multibit trie is the same as the number of leaves of a corresponding binary-trie, it is called a *dense trie*. Dense tries enjoy the desired properties and benefits of

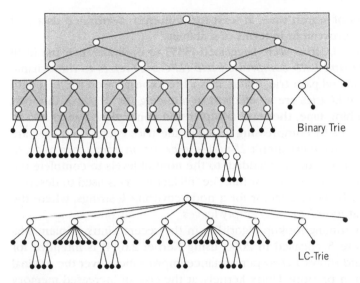

FIGURE 5.17

Binary trie and its equivalent LC-trie

binary tries with respect to space efficiency and proximity operations, while improving the cost of search operations when the branching factors are significantly high.

Nilsson's *Level-compressed trie* (*LC-trie*) is a dense trie, in which the degree of the root is 2^i, where i is the smallest number such that at least one of the children is a leaf. This is true for every child down to the leaves, so eventually every child and the entire tree become a level-compressed trie. An LC-trie is interchangeable with a binary trie, when the highest i complete levels of the binary trie (where in the next level there is at least one leaf) are reduced to a single 2^i-degree node in the corresponding LC-trie, and this reduction is executed top-down. In other words, LC-tries are recursive transformations of binary tries into multibit tries. An example is given in Figure 5.17, where the compressed levels are indicated by rectangles.

The expected depth of an element in the LC-trie is $O(\log^* n)$ where $\log^* n$ is the iterated logarithm function,[37] or $O(\log \log n)$ [18, 334] for an independent random sample taken from the uniform distribution. The behavior of LC-tries is better than

---

[37]The iterated logarithm function is used in computer science for algorithm analysis, and it is the number of iterations the logarithm function is applied before the result is less than or equal to 1.

It is defined as follows:
$$\begin{cases} \log^* n = 1 + \log^* (\log n) & \text{if } n > 1 \\ \log^* n = 0 & \text{otherwise} \end{cases}$$

The iterated logarithm function increases extremely slowly, and for all practical purposes can be considered as a constant.

Patricia tries in terms of search time, at least for uniformly distributed data, and shows a significant improvement in terms of scalability.

Several variations of LC-tries were suggested [335] to optimize their use in IP address lookup. *Level and Path compressed trie* (*LPC-trie*) refers to the dynamically compressed level and path trie, that is, it is a level compressed trie of a path compressed trie. Since LC-tries are so efficient in reducing the depth of the tries, and, hence, the searching time, the level compression concept was used also to create nondense LC-tries by compressing levels in the tries even when they are not complete, that is, when there aren't 2^k nodes to create an k-branches node. In other words, some empty nodes are "added" to the nonfull levels to complete the levels and to create a trie that is *almost* full. The "fill-factor" (x) is used to describe the highest degree of branching factor for a node covering k strings, where the branching produces at most k $(1 - x)$ empty leaves.

LC-trie is used for routing lookup algorithm in the recent Linux implementations (e.g., Fedora Core 5, released in 2006, using Kernel 2.6). It is intended for large routing tables and shows a clear performance improvement over the original hash implementation in previous Linux kernels, at the cost of increased memory consumption and complexity.

Updating LC-tries is more complex than the other structures described earlier. There are additional compression techniques for tries that are also complex to update [374] and are not described here; for example, the full expansion/compression scheme [90], the Lulea University algorithm [99], and others. There are more techniques for handling IP address lookups that based on the structures described above, and that are specifically aimed for prefix search or LPMs; these are described in [374].

5.5.3.6 Conclusions

Tries and self-balancing BSTs are generally slower in operations than tables—since their lookup time is $O(\log n)$, compared to the $O(1)$ that can be reached with hash tables—and are rather more complex to implement than hash tables. Nevertheless, various kinds of trees are used for specific search purposes, mainly tries, where string or nonexact matching (usually longest prefix) is required.

5.5.4 Hardware Search

In the previous subsection, we discussed data-structures and algorithms that are used for "software" search engines. In this subsection, we now turn to hardware search engines. As we have mentioned, there are many NSEs that look like hardware search engines to the network processor. These can be categorized into algorithmic NSE and those that indeed have hardware-based schemes that carry the searches within a few clock cycles. Some of these hardware-based schemes might also be used inside the network processor. *Content Addressable Memory* (*CAM*) is by far the most common way to execute searches by hardware; therefore, this subsection is focused around CAMs.

Before discussing CAMs, however, another word about search engine metrics: the hardware search engines (CAM included) have the same performance metrics that we used before (that is size of the data store and the number of searches per second that can be achieved). Nevertheless, hardware-based search engines should also be examined by their power consumption and the additional chip count required for their implementation. In CAM's case, for instance, power consumption is a real challenge.

CAM is a hardware search component that enables searching in a single clock cycle [346]. It is a memory composed of conventional semiconductor memory (usually like the static random access memory). CAM simultaneously compares a key (the searched data) against all of the table's entries stored in the CAM, and is thus capable of returning the address of the entry that matches the key in one single clock cycle. Although CAMs are fast, efficient, and flexible in terms of lookup functionality, they are also big energy consumers due to the large amount of parallel active circuitry. They are also big physically due to the large silicon area required for the memory cells and logic; they are small in terms of capacity compared to other types of memories; and they are very expensive.

A simplified scheme of a CAM having m words, each consists of n bits, is depicted in Figure 5.18. The CAM core cells are arranged into the horizontal words, where each bit is a CAM core cell that contains storage (the bit value) and the comparison circuitry. Typical implementations of CAMs have 36 to 144 bits per

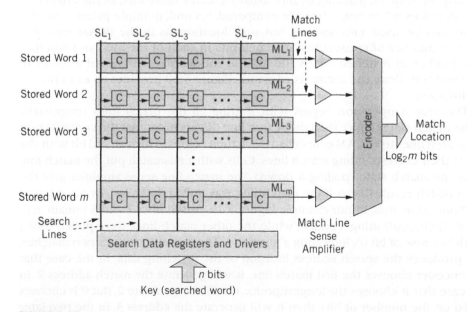

FIGURE 5.18

Simplified CAM scheme

word, and as many as from 128K rows (words) to 512K rows in recent CAMs. A key (a searched word) is broadcasted vertically on the CAM core cells, through the *search lines*. Each stored word has a horizontal *match line* that indicates whether the stored word is identical to the searched word (activated match line) or not (a *mismatch* status). This indication is the result of a logic "and" between all comparisons of the searched word bits with the CAM core cells of a row (stored word); any mismatch in any of the core cell of a row causes the match line of that row to be pulled down to indicate mismatch. All match lines are sensed and amplified separately, and fed into an encoder that produces the location (address) of the matched key (the searched word) in the table.

There are two kinds of CAMs, that is regular (or binary) and ternary. Binary CAMs (BCAMs) store and compare binary bits, that is zero or one. Ternary CAMs (TCAMs) support an additional *don't care* bit, which causes the match line to remain unaffected by the do not care CAM core cell, regardless of the searched bit. This ability of TCAM can be used for masked searches, and is perfectly useful for IP lookup applications [439], or complex string lookups, as described in the following.

It is possible in a BCAM to find more than one match between the search word and the stored words. In such a case, instead of using a simple encoder, as depicted in Figure 5.18, a priority encoder is used, which selects the highest priority location, usually defined by its address in the table (a lower position in the table grants higher priority). TCAM, on the other hand, works a bit differently. First of all, masking of bits usually creates more hits, as the chance of hits increases when fewer bits are compared. Second, multiple priority mechanisms can be used, emphasizing not only location, as in the BCAM case, but also the number of consecutive matched bits in each of the hits (and whether this number of matched bits is the LPM of the word), or the total number of matched bits along the entire word width. Figure 5.19 provides an example for clarification.

The search operation begins with putting all the match lines temporarily in the match state (high). Then the search line drivers broadcast the key 01011 on the search lines to all CAM core cells, which then compare their stored bit with the bit on their corresponding search lines. Cells with a mismatch put the match line in the mismatch state (pulling it down). The match line sense amplifier gets the word match result, that is if the match line was pulled down by any bit, it has at least one mismatch. In our example Figure 5.19, match lines 2 and 3 remain activated (high), indicating a match, while the other match lines were pulled down (both because of bit 0), indicating a mismatch. The encoder receives two matches, and produces the search address location of the matching data. In the case that the encoder chooses the first match line, it will generate the match address 2. In the case that it chooses the longest prefix, it will also generate 2. But if it chooses based on the number of hits, then it will generate the address 3. In the two later cases, the encoder requires more circuitry to enable counting the hits and the places of the hits.

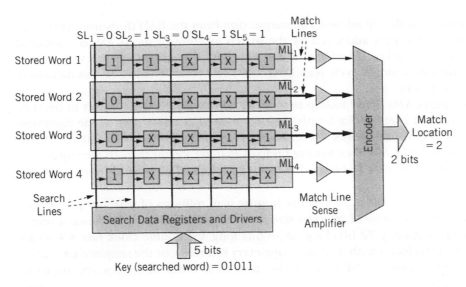

FIGURE 5.19

TCAM priority mechanism

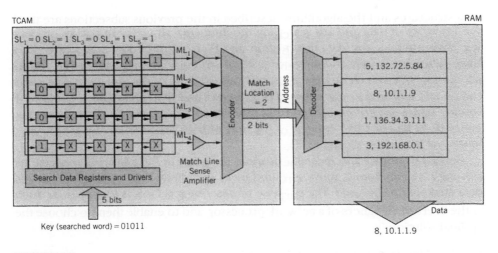

FIGURE 5.20

CAM/RAM system for IP address lookup

In most implementations, the output of the CAM is used to drive a regular RAM for receiving a value attached to the key that was searched in the TCAM. A common example, shown in Figure 5.20 is IP address lookup, using an IP address as an input key to the TCAM, while the resulted TCAM address location of that IP address is used to look in a RAM for a forwarding port and next hop IP address, which match that IP address. The output data from the RAM (forwarding port and next hop address), addressed by the match location of the TCAM, is associated with the TCAM key (IP address in our example). The combination is very powerful, enabling a single cycle searching, and two cycles updating of this dictionary lookup system.

Using TCAMs for IP lookup is done by storing all routing prefixes in decreasing order of their prefix length, padded with "don't care" bits in the rightmost side of the stored words. For example, a routing prefix 132.72.0.0/16 will be stored in the TCAM after routing prefixes that are longer than 16 are stored, and it will store the first 16 bits of this routing prefix and another 16 "don't care" bits to their right.

As of 2008, CAMs can execute hundreds of millions of searches per second (Msps) and are tens of Mbits in size. The key size, according to which a search is done, is usually 72 bits long or 36 bits long, for double clock rate. CAMs are usually interfaced with a 72-bit proprietary interface, or the standard LA-1 interface, defined by the NP forum. (This, as well as other interfaces, are discussed in Chapter 7.)

5.5.5 Summary of Using Search

Data structures and the algorithms described in the previous subsections are used for any types of data and any searching requirements. However, in packet processing environments, the data distribution and their required key searches make a huge difference when choosing the optimal structure and algorithms (i.e., direct addressing, BST, and so on). The most obvious example of the impact of data distribution and required key searches is IP addresses and IP address lookups, where IP routing tables are not composed from random entries evenly distributed over the entire IP 32 bits or 128 bits range of the IPv4 or IPv6, respectively, but are grouped in several IP address clusters.

Several examples and recommendations are given in the following subsections, although, as mentioned above, circumstances will dictate the optimal search and data structure to be used. The purpose of describing it here is to give some tools to the users or designers of a network processor, and to enable them to choose the optimal solution.

5.5.5.1 Simple Switching and Forwarding

When a simple decision is required for switching, forwarding, or queuing a packet, based on few numeric keys of a small range, then a small table residing in memory (RAM) is sufficient to make such decisions. This table can be direct addressable if the number of entries is about the same as the key-range, or hash table if the number of entries is small compared to the range of the keys. An example for using a direct addressable table for look-up is VLAN (Virtual Local Area Network) forwarding, for example, assigning an output port based on the VLAN. An example for using a hash-table for look-up is MAC addresses based-decisions, for example, assigning MAC addresses to specific VLANs, bridging devices operating in layer 2, and so on.

5.5.5.2 URL and Packet Content (L7) Lookup

URL lookup is usually done in layer 7 applications, like load balancers, and is usually based on nonexact matches, longest-prefix matches, or some pattern search (i.e., masked searches or keys that include "don't care" characters in the URL). There are many other applications that require content lookups at layer 7, such as the identification and targeting of specific phrases in packets, XML applications, access control applications, virus protection, intrusion detection systems, intrusion prevention systems, and the like.

In most of these cases, trees and tries are used for the searches and comparisons, although hashing tables can be used if the searches are carried on fixed-length, nonmasked keys. Since many of these applications generally use wildcards and prefixes, require a huge amount of data, and conduct multidimensional searches, then trees and tries residing in DRAM can be the only viable solution.

5.5.5.3 IP Address Lookup

This is the most common usage of searching in network processing. Generally speaking, there are several reasons for IP address searches. Based on the purpose of the search, a suitable data structure and algorithm should be used. If, for example, IP address search is conducted to match a specific decision to some specific IP address, and there are not many such IP addresses, then a hash-table can be used (e.g., for access control). If, on the other hand, a decision has to be made based on a range of IP addresses (e.g., for IP forwarding or routing algorithms) or partial IP addresses (i.e, prefix or pattern match of some IP addresses), then an exact match will not be efficient, and a routing table has to be constructed by some tree (or CAM), and will need to use LPM.

Searching in the routing table in the classful IP network environment was straightforward: In order to receive the next hop and the output interface, an exact match search of the fixed length network ID (according to the address class) should be performed. These tables were not very dynamic, that is they were not changed very often. Hash tables were very efficient in performing this task. The impact of CIDR on IP address lookup, however, was enormous; from fixed sized, hash table lookups, search took place in variable length prefixes of aggregated entries. The focus shifted to tree, tries, and LC-tries data structures, algorithms and their variants that best matched the IP address lookup in terms of data distribution and design, simultaneously with using TCAMs and other hardware search engines (coprocessors).

TCAM and other external search engines are targeted in IP lookup searches, and usually can provide an answer within one or two clock cycles, that is several nanoseconds.

For memory searches, M-tries is the best access method, since, as was mentioned above, bit access in memory for binary tries representations is very inefficient, since usually a byte is the natural, minimal access width anyway. In most cases, then, common M-trie-based searches require three memory accesses for a full IPv4 32-bit address search (using a three-level 16-8-8 stride pattern, that is 16 bits search, then 8 bits and lastly 8 bits of the stride pattern used for routing). These memory accesses determine the lookup latency and the number of lookups per second. Usually, however, IPv4 addressing is based on 24 bits or less, so just two memory accesses are required. Implementing a routing table in a memory—for example, for a 10 Gbps Ethernet running 75 bytes per packet—requires 10 ns access time memory, assuming two searches per packet.

IPv6's forwarding lookups are not different from today's IPv4 with CIDR, that is big routing tables, longest (the most specific) prefix match, and all the complexity involved. Here again, implementation of M-trie-based searches requires a 16 bits stride pattern for searches, which might be conducted up to eight times (using a eight level 16-16-16-16-16-16-16-16 stride pattern), and which might end up with eight memory accesses. Implementing a routing table in a memory for a 10 Gbps Ethernet running 98 bytes per packet, for example, requires about 5 ns access time memory, assuming again two searches per packet.

In both cases of IPv4 and IPv6 look-ups, the challenges are to cope with the update requirements of the routing tables, and to perform IP lookup for classifications (or other, nonforwarding tasks, that is those that are based not only on the IP address). These challenges require huge amounts of data being moved in the memory (for updates) and the use of wide keys for lookups. For example, just the average 100 route changes per second in the backbone routers, as reported in 1999 [267], yield enormous update burden and massive data moves in the routing tables of these routers.

The required bandwidth between the network processor and the memory in which the tables reside is another related issue that influences the design of network processors. Internal memory (on the network processor silicon) can work at aggregated bandwidth of terra bytes per second range, whereas external memory requires hundreds of I/O interface pins to achieve this bandwidth. This influences the design and cost of network processors, as silicon space becomes more and more efficient (in terms of logic, storage, and channels), while the I/O interface pins become the main factor for die size and its price.

A useful taxonomy for IP address lookup [425, 427] is based on a trie representation of IP addresses, and distinguishes between prefix length (the trie depth) and prefix value (the trie's leaves span) as possible search space, and between linear and binary searches in the search space. Searches are further distinguished by parallelism (that is either they are fully paralleled, or they are pipelined, which means that several different stages of the search are conducted in parallel, or they are serialized, which means that they are not parallel at all) and memory optimization (which includes caching, or using the most frequently required data in a faster memory, compression, so that the data requires less memory space and compacting, by using more efficient data structures). Similar taxonomy [374] for IP address lookup algorithms distinguishes between search-on-values approaches, and search-on-length approaches. This taxonomy is useful in classifying trie-based schemes, as described in the following.

IP address lookup schemes can be arranged in the following way [75, 374].

1. Cache-based schemes.
2. Trie-based schemes.
   - Simple, standard trie structures.
   - Path-compressed tries (Patricia) [323, 389].
   - Multibit tries.
   - Level-compressed tries (LC-tries and LPC-tries).
   - Other compact presentations of tries.
     - Lulea Algorithm [99].
     - Full expansion/compression [90].
     - Full tree bit map [109, 110]
   - Binary search on trie levels [426].
   - Prefix range search.
     - Binary range search.
     - Multiway range search [270].
     - Multiway range trees [400].
   - Two trie structure [24].
3. Hardware-based schemes.
   - DIR-24-8-Basic scheme [162].
   - SRAM based Pipeline Scheme [175].
   - CAM and TCAM.

Additional categories [412]—modifications to exact matching schemes, protocol-based solutions, and so on—were suggested.

5.5.5.4 Packet Classification

When choosing the right packet classifying scheme, it is not sufficient to consider only performance metrics attributes and those attributes that are necessary for a lookup. Additional metrics to be considered include the ability to handle large real-life classifiers as well as flexibility in specification [161].

Packet classification can be based on several fields or subfields of the packets, as described above. The keys, generated from the packet's fields, are used to search for rules in a database of packet classification rules. The resulting rule (indexed, numbered, or otherwise identified) is used for later decisions and actions. This rule-database is generally made from some sort of look-up table that allows exact match of a fixed length and format key, or partial match (best, longest, wildcard) of any key created from the packet's fields. Simple classification that is based, for example, on packet type or protocol, can be based on a direct addressable table. When the classification is based on several fields, or on a long key, then it is usually based on a hash-table. When the classification is somewhat fuzzy, or requires complex algorithms for classification, or is based on parts of the IP addresses, then nonexact matches and various schemes and algorithms based on trees and tries, or m-dimensional trees and tries (where m is the number of classification parameters), are often used to classify the packet.

A taxonomy of classification schemes [75, 161] categorizes the various algorithms and data structures as follows:

1. Basic data structures
   - Linear search.
   - Hierarchical trie.
   - Set-pruning trie [412].

2. Geometric algorithms
   - Grid of tries [394].
   - Cross-producting scheme [394].
   - 2-D Classification scheme [269].
   - Area-based QuadTtree (AQT) [63].
   - Fat Inverted Segment tree (FIS-tree) [122].

3. Heuristics
   - Recursive Flow Classification (RFC) [159].
   - Hierarchical Intelligent Cutting (HiCuts) [160].
   - Tuple space search [393].

4. Hardware-based algorithms
   - Bitmap intersection [269].
   - TCAM.

5.6 MODIFICATION

In all applications other than packet forwarding, modification of the packets is eventually the purpose of packet processing. Sometimes, some packet modification is required even for forwarding applications (which make up a substantial part of the applications), for example, changing the IP header may be required in the time-to-live (TTL) field, as well in the IP addresses, hence recalculation of the checksum header is also required, and so on.

Packet modification can include all of these operations:

- *Modification*—Changing the contents of a packet (usually changes in its header, or changes both in the header and in the payload, as a result of some processing, e.g., compression or encryption described in the next section)
- *Deletion*—Some of the packet contents or headers are deleted (e.g., de-encapsulation of packets)
- *Adding*—Additional information is added to the packet (e.g., encapsulation, authentication information)
- *Canceling the entire packet*—Simply removing the packet from the system (e.g., exceeded traffic, wrong addressing)
- *Duplicating the packet*—Copying the entire packet (e.g., multicasting, port-copy operations)

Since checksum is often changed in packet processing, some network processors contain an internal checksum functional unit to assist with maintaining the right checksums.

In addition, traffic analysis, some management tasks (e.g., accounting) and other applications can be performed in the modification phase, although most of these operations are expected to be performed in the control path. Statistics collection, however, must be done in the fast path, at the modification phase, to accommodate the packet rate.

5.7 COMPRESSION AND ENCRYPTION

Compression and encryption processing are optional phases in packet processing that are more typical to access network processors than to high-end network processors. The reason is that packets do not undergo any compression or encryption in the main trunks of the core and the metro networks, whereas these processes might happen at the edges of the network, at the access points. Usually, compression and encryption is executed by specific coprocessor, or in the mid-range and access network processor that contains an encryption functional unit.

Compression is used mainly in places where bandwidth is critical (mainly wireless applications), and most of the compression efforts focus on TCP/IP header (many protocols run very small packets, while most of these packets are just predictable headers) [326]. Other implementations where payload compression might be helpful are WEB services and HTTP applications, as well as the evolving XML-based protocols. When compression is used, it is usually based on the Lempel-Ziv algorithm (LZ) [443, 444] or derivatives of it, which take advantage of the recurring patterns of strings. The concept of LZ is to maintain a dictionary of the most used string patterns in the stream, and replace these string patterns by the dictionary index, when they occur again. In other words, as the data stream flows, character by character, LZ keeps track of repeating patterns in the stream, places them in a dictionary (it keeps the most frequent patterns), and replaces these patterns when they occur again with pointers to the place where they previously were transmitted in the stream. Upon arrival, these pointers are substituted with the already known string patterns, and the original stream is accurately reconstructed. For massive pattern matching, most implementations typically use CAM-based approaches.

Encryption is used for privacy, data integrity, and authentication to confirm the communicating parties' identities. There are many standards for security in the IP world; the main standard is Internet Protocol Security (IPsec) [249] framework, operating at layer 3, and SSL/TSL [102] operating above this layer (thus enjoying the TCP/UDP services in the connection, but leaving them unsecured). IPsec is used mainly in Virtual Private Networks (VPN), whereas SSL/TSL is used typically in client server applications (mail, web browsing, telnet, and so on).

The security protocols primarily use encryption for confidentiality, where encryption simply transforms unsecured information ("plaintext") to coded information ("ciphertext"), using some key and a transformation algorithm. The other components of the security suit are to make both sides know what encryption algorithm to use, and what key to use, and some specific protocols are doing just that. For our purposes, network processors that are involved in secured transmission have to take part in the various protocols, and the most demanding issue is that they have to execute the transformation from plaintext to ciphertext and vice versa at wire speed. Since SSL/TSL works above the network layer, network processors usually are not involved (unless deep packet processing or L7 processing is required). As for IPsec, there is one mode of IPsec, transport mode, in which only the payload (the data portion) is encrypted, therefore, most of the packet processing functions can be done without the need to encrypt any packet along the path, that is those packets that require only header manipulations like forwarding. Again, deep packet inspection or L7 applications will force the network processor to be involved in the encryption/decryption. The more secured mode of IPsec, the tunnel mode, encrypts the entire packet and encapsulates the encrypted packet in a tunneling packet, so in some network applications the network processor can avoid the involvement, while in others, obviously for deep packet inspection and L7 applications, the encryption/decryption is unavoidable [138].

5.8 QUEUEING AND TRAFFIC MANAGEMENT

Finally, the processed packet ends its quick journey in the network processor, and is about to leave. The last small but very complicated task is to decide how to get rid of this packet, or, more precisely, how to pass it on to the receiving party (equipment or communication link) on its way to its final destination.

As previous stages in the network processor (classifiers, modifiers, and so on) determined the output path (port), priority, and some handling parameters, the traffic management process forward this packet to an appropriate queue, and schedule it for transmission according to the conditions of the lines, the receivers, and the parameters that this packet has (e.g., priority). This process also meters the packets, and shapes the transmission pattern to a desired rate and burstiness. The traffic management process is complicated enough to be implemented in many network processors externally, by a dedicated traffic manager coprocessor. There are, however, some network processors that integrate the traffic management internally, thereby saving chips, power, and cost at implementation time, but more importantly, allowing faster and more integrated service to the packets.

In many cases, incoming packets also have to go through traffic management (metering, queuing, prioritization, and so on, which are determined by incoming port or communication lines or other parameters) in order to be processed in the network processor according to predefined scheduling scheme.

Since traffic management is a kind of flow handling and it involves definitions of traffic patterns and service quality that are beyond the processing of the single packet, this entire subject is discussed in the next chapter.

5.9 **SUMMARY**

This chapter discusses the theory behind network processors. Packet processing is the heart of network processing, and the various phases and operations that packets undergo in most of the network elements, are the subject of this chapter.

The next chapter is still theory, but it focuses on packet flow handling, which was only mentioned very briefly in this chapter.

CHAPTER

6

Packet Flow Handling

In the previous chapter, we dealt with the need to process packet content. Of similar importance, we have to manage the packet flow in the network. Moreover, very often we have to consider packet flows in designing equipment, in order to make the equipment function.

In this chapter, we address various aspects of packet flows, traffic management, buffers queueing, and other issues that impact equipment design and component functioning, as well as networking issues and functioning. The chapter is divided into two main parts: (1) Quality of Service (QoS) and related definitions, and (2) QoS control mechanisms, algorithms, and methods.

Handling packet flows is a key issue in providing services over the network. Not only is it important in network functioning, it is also crucial in the design and architecture of networking equipment, as well as, further on, in the chips and processors that handle packets.

Before diving into the issue of handling packet flows, we have to be aware of two basic kinds of traffic requirements—traffic that is sensitive to packet delays and traffic that is sensitive to packet losses. For example, packet delays can have a strong impact on voice quality, but if a few packets are lost, voice quality will not be seriously harmed. On the other hand, e-mail cannot tolerate any packet loss, but packet delays are more acceptable. These two very different traffic requirements are answered by various mechanisms, algorithms, processes, and methods that are defined to handle proper packet flow throughout the network, the equipment, or the chip.

It is not the intent of this chapter to provide an exhaustive review of all possible scheduling, buffer management, forwarding, or QoS schemes. Instead, this chapter provides theoretical background on packet flow, which will be used later in describing the internal architecture of network processors, and for functions of the network processors with regard to network requirements. Many of the algorithms and schemes discussed here rely on research papers and books that are cited in the reference pages at the end of the book.

6.1 DEFINITIONS

There are many terms in use in relation to packet and traffic flows, and they are sometimes contradictory. There are also multiple terms for the same thing, or the same term is used to define different things, both specific and broad. In the following, in the interest of clarity, I adopt terminology that I feel is appropriate, although some might argue for the use of alternative terminology.

Traffic management (sometimes called *traffic conditioning*, or *traffic access control*) usually refers to measuring, policing, and shaping the traffic, mainly to avoid congestion in networks. Transmit priority, bandwidth allocation, Call Admission Control (CAC), congestion avoidance, and selective packet loss can be associated with traffic management as well.

Traffic policing usually refers to handling the incoming traffic and allowing it into the network, or otherwise tagging it and handling it separately (discarding the traffic in certain circumstances, or allowing it in others). Traffic policing sometimes includes traffic measuring as part of handling it, as well as traffic admission control.

Traffic shaping means regulating the volume of packets released into the network by using buffering techniques, so that the transmitted traffic will fit a defined traffic behavior pattern. Traffic shaping is sometimes called traffic metering or traffic smoothing.

Flow control, or *pacing*, is a technique that makes sure that no receiving entity will overflow. It is sometimes also called *congestion control*. Usually flow control refers to handling the packet stream in the transport layer.

Traffic engineering refers to the ability to control traffic paths in a network, while avoiding and reducing congestion, and best utilizing the network resources. It usually refers to choosing nondefault routes for the traffic, for example, optimized routes in a congested network.

Network flows is a term used in graph theory to represent various algorithms for checking flows in a network; for example, measuring the capacity of possible flows, maximizing the flows in a given network, or calculating the cut capacities in a network.

Quality of Service (QoS) indicates the performance of a network (or equipment, or chip), and signifies its transmission quality. This definition will be elaborated more fully in the following subsection.

Traffic Conditioning Agreement (TCA) expresses the traffic management parameters and rules that are applied to the traffic streams generated by the user or the application, and are acceptable to the network (or the network providers). The agreement describes the various networking mechanisms required in order to handle packets according to the required QoS. TCA should be subject to the

Service Level Agreement (SLA), as defined in the following, and to the service level the network can provide [54].

Service Level Agreement (SLA) refers to the level of service the user is guaranteed to receive from the network, expressed in terms of QoS parameters over time, allowed exceptions, durations, and so on. SLA is usually a contract between a service provider and its customers, specifying the forwarding service that the customer should receive. SLA can be static (predetermined QoS parameters that can be modified periodically) or dynamic (negotiating Quality of Service parameters by some signaling protocol). Traffic conditioning rules and parameters that constitute part of the entire TCA may be part of the SLA [54]. SLA can be extended to describe and define more than just QoS parameters, but this is beyond the scope of this discussion (for example, service type and nature, availability, responsiveness, monitoring, accounting, etc.).

Class of Service (CoS), or *Type of Service (ToS)*, refers to the requirements of the application (or service) layer from the network layers underneath it. More generally, it defines how an upper-layer protocol requires a lower-layer protocol to handle its messages. This will be discussed more fully in the following.

6.2 QUALITY OF SERVICE

Quality of Service is an aggregate of measures that varies depending on the CoS requirements. It indicates how well packets behave when transmitted through the network, looking at issues such as packet loss, packet delay, and throughput. QoS determines how well we can run applications across the network; for example, the quality of voice transmission over IP networks (VoIP, or Internet telephony), the quality of TV service over IP (IPTV), how long it will take to retrieve a large file, or how much "real time" is indeed real time.

The International Telecommunication Union (ITU-T) defines QoS as "a set of quality requirements on the collective behavior of one or more objects." The ATM[1] forum defines QoS as "a term which refers to a set of ATM performance parameters that characterize the traffic over a given virtual connection." The Internet Engineering Task Force (IETF) defines QoS as capabilities of network infrastructure services and source applications to request, setup and enforce deterministic delivery of data. The European Commission (RACE D510 project) defines QoS as "the measure of how good a service is, as presented to the user. It is expressed in user understandable language and manifests itself in a number of parameters, all of which have either subjective or objective values."

---

[1]ATM (Asynchronous Transfer Mode) was described in Chapter 2, and refers to the entire concept of ATM networking.

Quality of Service became important and received systematic and comprehensive treatment since multimedia appeared on the network. Since then, QoS has developed and is integrated into any network. QoS is dependent on the statistical behavior of the network.

There are many parameters for QoS. The most commonly used are performance parameters, which include:

- *Bandwidth and throughput*—how many packets per second can flow through the network;
- *Packet loss ratio*—how many packets are lost due to discards, errors, and so on;
- *Delay*—what is the delay of a packet in the network, in various measurements (link, end-to-end, start-up [set-up], maximum round trip, etc.);
- *Delay variations* (*jitter*)—how stable the delay is;
- *Bit Error Rate* (*BER*)—how good the links are.

Quality of Service can be marked and treated in layer 2 and layer 3 by different mechanisms, as described in the following.

6.3 CLASS OF SERVICE

Class of service describes the characteristics of the network, or the nature of the traffic patterns required to carry some services (applications). There are various definitions of CoS, coming from the various network technologies. The main difference between QoS and CoS is that QoS specifies quantitative parameters, and any device that is compliant to QoS can act based on these parameters—even to the extent of discarding packets. CoS usually refers to the qualitative nature of the traffic, and can serve as guidelines on how to treat packets. In the following subsections, various types of CoS are described, separated according to the network technology with which they are associated:

6.3.1 ATM Services

Asynchronous Transfer Mode was described in Chapter 3, Section 3.2.1. ATM distinguishes between several classes of services, as described in the following subsections.

6.3.1.1 Constant Bit Rate, or Deterministic Bit Rate

Constant Bit Rate (CBR) (the ATM forum terminology), or Deterministic Bit Rate (the ITU-T terminology [223]), means a fixed bit rate, where data is sent in a steady stream. This CoS guarantees low packet loss ratio and defined delays. This CoS is expensive in terms of network resources, since it is a real time service, and it grants bandwidth to the application regardless of whether it is actually used or not. It is typically used for circuit emulation by applications that require constant delay and throughput, such as uncompressed audio or video information.

6.3.1.2 Variable Bit Rate

Variable Bit Rate (the ATM forum terminology) means an average bit rate and throughput capacity over time, where data is not sent at a constant rate. Like CBR, this CoS also requires low packet loss ratio, but delays can be variable and larger than in CBR. Real-time VBR (rt-VBR) is used for isochronous applications, like compressed audio and video; nonreal time VBR (nrt-VBR) is used for all other applications. The nonreal time VBR is termed SBR (Statistical Bit Rate by the ITU-T [223]). The difference between the two lies in their delay characteristics.

6.3.1.3 Available Bit Rate

Available Bit Rate (ABR) refers to the minimum network capacity available. When the network is free, higher capacity is used. The application that uses the network and uses ABR is required to control its rate, based on signals received from the network in cases of congestion. It is desirable to have minimal delay and packet loss in ABR, although it is not guaranteed. Although similar to nrt-VBR, ABR usually provides better throughput than VBR, and is less expensive than the CBR. ABR is the only service category (CoS) that uses flow control by the application and the network.

6.3.1.4 Unspecified Bit Rate

Unspecified Bit Rate (UBR) is a best-effort service. Nothing is guaranteed by the network, and higher layer services are required by the applications, such as flow and congestion control. UBR can be used by applications like file transfers, e-mails, and the like, which can manage with the leftover capacity of the network.

6.3.1.5 Guaranteed Frame Rate

Guaranteed Frame Rate is a combination of ABR and UBR (and has also been called UBR+). It acts like ABR without flow control, and like UBR with a minimal guaranteed transfer rate. The frame-based traffic is shaped by the network, and frames that are above the guaranteed traffic rate are treated like UBR (best effort). Packet loss is low when the traffic rate is below the guaranteed rate.

6.3.2 Integrated Services

Integrated Services (IntServ) is the best way to provide QoS in IP networks, although complicated and resource-demanding. It works on packet flows, as detailed in Chapter 2, Section 2.6.2.3.2. IntServ defines three CoS, as follows.

6.3.2.1 Guaranteed Service

This service [384] guarantees an upper bound on end-to-end packet delay. Traffic using Guaranteed Service (GS) is typically bound by packet peak rate and maximum packet size. This service guarantees both delay and bandwidth.

6.3.2.2 Controlled-Load Service

Controlled-Load Service, also known as predictive service, provides a service level similar to the one that will be experienced by the same packet flow crossing an unloaded network. This service [433], however, uses capacity (admission) control to assure that packets will receive this level of service even in an overloaded network.

6.3.2.3 Best Effort

Best effort—as the name implies—is a "service" where no guarantees whatsoever are provided, not even an admission control to examine the network status and to allow or disallow the service.

6.3.3 Differentiated Services

Differentiated Services (DiffServ) is the simplest way to provide QoS in IP networks, based on different treatment of packets of different classes, as was described earlier in Section 2.6.2.3.2. DiffServ defines three basic service levels, or CoS, as follows.

6.3.3.1 Expedited Forwarding

Expedited Forwarding (EF) is used for premium service, and provides a low loss, low latency, low jitter, and assured bandwidth end-to-end service. The expedited traffic is not affected by any other traffic streams in the network [232]. It appears as a "virtual leased line," point-to-point connection to the applications at the network boundaries.

6.3.3.2 Assured Forwarding

Assured Forwarding (AF) is the second-level service in DiffServ, between EF and default forwarding. AF offers four independent subclasses of packet delivery services. In each of these subclasses, packets are prioritized into three levels, which have different levels of drop precedence. The service level of a packet is determined by the AF subclass the packet receives, the available resources in the network to this subclass, and to the drop precedence of the packet in case of network congestion [170].

6.3.3.3 Default Forwarding

Default Forwarding (DF) means neither EF nor AF. In short, it is a best-effort service level, dependent on the leftover capacity of the network.

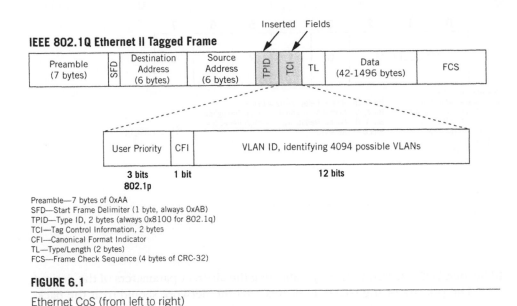

IEEE 802.1Q Ethernet II Tagged Frame

Preamble—7 bytes of 0xAA
SFD—Start Frame Delimiter (1 byte, always 0xAB)
TPID—Type ID, 2 bytes (always 0x8100 for 802.1q)
TCI—Tag Control Information, 2 bytes
CFI—Canonical Format Indicator
TL—Type/Length (2 bytes)
FCS—Frame Check Sequence (4 bytes of CRC-32)

FIGURE 6.1

Ethernet CoS (from left to right)

6.3.4 Specifying CoS

The following subsections describe the specification of CoS as it appears in the packets, organized by network layers:

6.3.4.1 Layer 2 CoS

Ethernet protocol, IEEE 802.3, which is the data link layer protocol, was amended to include 802.1Q VLAN (virtual LAN), which contains the 802.1p priority (or CoS) bits. As described in Section 2.6.1.4.1, four bytes were added between the source address field and the Type/Length field (Figure 6.1). The first two bytes (TPI) are type and always contain the value of 0x81-00h. The TPI is then followed by the 802.1Q tag bytes, which are headed by the 802.1p three bits of CoS.

The 802.1p CoS three bits are used by the network internally; these three bits simply indicate CoS superiority (000 is the lowest priority, 111 is the highest). Assigning these bits is the responsibility of the internal LAN, and usually reflects upper layer CoS (like the DTR bits of the TOS field of the IP header, as described in the following subsection).

6.3.4.2 Layer 3 CoS

The earliest implementation of CoS in layer 3, that is, in the IP layer, was suggested in the first IPv4 specification (RFC 791, 1981 [355]), in the TOS 1-byte field of the

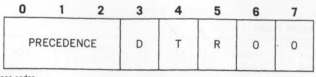

Precedence codes:
111—Network Control
110—Internetwork Control
101—CRITIC/ECP
100—Flash Override
011—Flash
010—Immediate
001—Priority
000—Routine

Bit D: 0—Normal Delay, 1—Low Delay
Bit T: 0—Normal Throughput, 1—High Throughput
Bit R: 0—Normal Reliability, 1—High Reliability
Bits 6–7: Reserved

FIGURE 6.2

IP TOS field

IP header. TOS referred to CoS by indicating the abstract parameters of the desired QoS. The priority of the packet indicates how the network should treat the packets in case of high load, according to its precedence. Precedence in TOS is defined in three bits, in increasing importance, as shown in Figure 6.2. In addition to the precedence, TOS defines a three-way trade-off between low delay, high throughput, and high reliability (DTR bits).

It was modified slightly by RFC 1349 in 1992 [15], so the TOS field of the IP header (Figure 6.2) is defined as a four-bit field, as follows (bits 3–6):

```
1000   -   minimize delay
0100   -   maximize throughput
0010   -   maximize reliability
0001   -   minimize monetary cost
0000   -   normal service
```

Later, IPv6 (RFC 2460, 1998 [98]) used the Traffic Class byte for CoS, similar to the usage of the TOS field in IPv4.

Then, at about the same time as IPv6, DiffServ [333] redefined the TOS byte of the IPv4 header and the Traffic Class byte of the IPv6 header to become the Differentiated Service (DS) byte. It superseded the "old" definitions, and defined the first six bits of the DS field as the Differentiated Services Code-Point (DSCP) to be used in the network. The DS field structure is presented below:

```
  0  1  2  3  4  5  6  7
+---+---+---+---+---+---+---+---+
|         DSCP      | CU  |
+---+---+---+---+---+---+---+---+

DSCP: differentiated services codepoint
CU:   currently unused
```

The Codepoint values of the AF CoS are presented in the following table:

	Class 1	Class 2	Class 3	Class 4
Low Drop Precedence	001010	010010	011010	100010
Medium Drop Precedence	001100	010100	011100	100100
High Drop Precedence	001110	010110	011110	100110

Codepoint 101110 is used for EF CoS.

6.4 QoS MECHANISMS

In this section, a more detailed description of QoS mechanisms is provided. The service of different kinds of QoS requirements is accomplished along two axes: control path and data path [155]. It is important to distinguish between NP definitions of these two terms[2] and their meaning here, although the analogy is very obvious. The packet flow handling functions introduced in this section are data path mechanisms, on which network QoS is built. Control path mechanisms deal with service definitions, the identification of entitled users to these services, network resource allocation to meet these requirements, and they also sometimes include signaling in cases of on-demand services rather than provisioning services.

Quality of Service mechanisms include many functions, such as traffic classification, flow control, traffic control, call admission, regulation, policing and shaping (or reshaping), signaling, routing, and more. Provisioning of a CoS, controlled by QoS mechanisms, requires coordinated use of all QoS mechanisms, specifically *admission control*, *traffic policing and shaping*, *packet scheduling*, and *buffer management*, coupled with *flow and congestion control* and *routing* [76]. The QoS functions that are described in this section include the main data path QoS functions (see Figure 6.3).

- *Admission control* (allowing connections into the network)
- *Traffic management* (measuring and regulating data flows by policing and shaping)
- *Packet queueing* (dropping, buffering and servicing packets along the network's path buffers). Packet queueing contains:
  - *Buffer management* functions (queueing or dropping packets along the network paths), and
  - *Packet scheduling* functions (service disciplines of the packets from the queues into the network segments).

---

[2]In network processors, data path, or the fast path, refers to packet analysis, modification and forwarding, while control path, or the slow path, refers to processing the packet content for administrative purposes (e.g., updating routing tables).

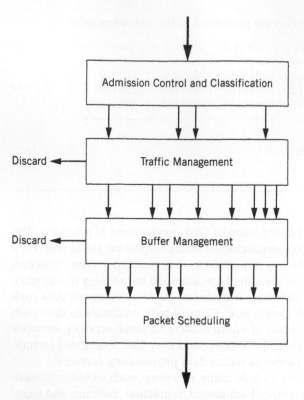

FIGURE 6.3

QoS functions (based on [278])

Packet classification is considered by some researchers to be also a function of packet handling, but we covered this topic while discussing packet processing. All other functions (e.g., routing, signaling, resource allocation, etc.) are beyond the scope of this book.

The above-mentioned functions are the building blocks of QoS treatment, and are based on many algorithms that are used for a variety of cases, circumstances and goals. These building blocks can be joined together, or they can be used in any order to result in the desired behavior of packets and traffic.

Packet flow handling is done at the edge and at the core of the network. Mechanisms of admission control and traffic access control (policing and shaping) are more prevalent at the edge of the network, while mechanisms of scheduling and buffer management are more dominant at the core. It should be noted, however, that shaping and regulating can also be found in the core, and buffer management can also be found in the edge (see Figure 6.4).

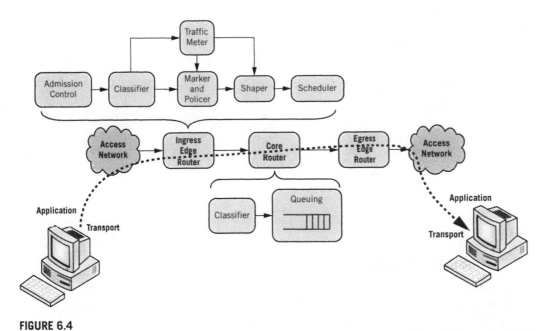

FIGURE 6.4

End-to-end QoS mechanisms

It must also be noted again that this section is not an exhaustive review of all possible shaping, policing, scheduling, and buffer management schemes. The purpose here is to enable the reader to understand the basic NP terminology, as well as the various classes of QoS schemes that are described in the research literature, in order to adopt them wisely.

6.4.1 Admission Control

Admission control restricts applications from requesting new or additional services (connections, or flows) from the network, at a time when the network is overloaded or the service is beyond the allowed usage permits for these applications.

A CAC function is executed whenever there is a new connection call, to determine whether to accept or reject that request. The parameters used by CAC to make a decision are the applications' description of the required QoS, as well as the available resources in the network that can meet those QoS requirements.

Signaling protocols are associated with CAC, by checking and notifying all network resources of the expected call, making sure that the resources are available and the call request can be admitted (while network resources are reserved by the network elements). Such an example is Resource Reservation Protocol [57] that was described in Chapter 2.

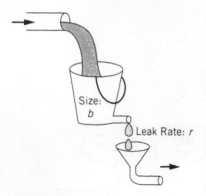

FIGURE 6.5

Leaky bucket

6.4.2 Traffic Management

As previously mentioned, traffic management is traditionally used to describe three functions: traffic measurement, traffic policing, and traffic shaping. It is also sometimes called traffic conditioning (e.g., in the DiffServ environment).

This subsection describes traffic access control mechanisms that together are called traffic management. A brief theoretical background is provided in the subsequent subsections, followed by formal descriptions of these mechanisms.

6.4.2.1 Theory

The following subsections describe a few basic concepts of traffic management and specifications.

6.4.2.1.1 Leaky and Token Bucket

Leaky and token bucket are very often used to model or to control traffic, as the following describes. This subsection describes these models, in order to better understand the traffic models and control algorithms. *Leaky bucket* [413], or Simple Leaky Bucket (SLB), is the simplest approach to specify required bandwidth and limit users to their allocation.

The principle is this: imagine a leaky bucket (as in Figure 6.5), from which a flow of fluid (or packets) drains at some fixed rate. Fluid (or packets) can be poured in (or added) to this leaky bucket at the same rate of the drained flow. From time to time, after periods of slow or no pouring, there is room in the bucket, and fluid (or packets) can be added quickly, up to the bucket's capacity. If liquid is poured (or packets are added) beyond the drained rate for long enough periods, the bucket will eventually overflow, and the liquid (or packets) will be lost.

The SLB mechanism is as follows: SLB increments a counter whenever a packet is sent (into the bucket) and decrements the counter periodically, according to the rate (r) in which packets leaves the bucket. If the counter exceeds some threshold (the "bucket size") (b), the packet is discarded. The user specifies b as well as the

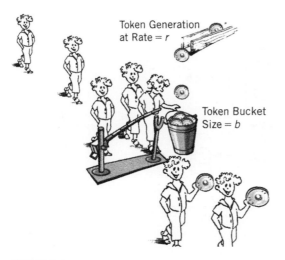

Token Generation at Rate = r

Token Bucket Size = b

FIGURE 6.6

Token bucket

rate r at which the counter is decremented. r therefore determines the average bandwidth, and b defines a measure of burstiness (how many packets can be sent at once, or very quickly, to the "bucket"). These two parameters define the required bandwidth and limit the flow of packets into the network.

The resulting (output) traffic after the SLB is a flow that is bounded by r, even when bursts of traffic enter the SLB. *Token Bucket* (TB) is based on the principle of SLB and is very similar to it, except in the terms of the resulting output traffic, which can be bursty (not bounded) as described below.

The principle of the TB is to allow packets into the line when there are enough "tokens" (credentials) accumulated at a controlled rate (r) in a "bucket" of size b. Imagine a bucket of tokens at the entrance to a subway station, where tokens are falling into the bucket (size b) at the rate r (see Figure 6.6). If there are enough tokens in the bucket, people take one when they arrive, and use it to pass through the stile and into the subway. Thus, for each outgoing packet, one token is removed from the bucket. If there aren't enough tokens for all arriving passengers (or packets), then those who can't get a token simply "overflow" and vanish. During idle periods when no passengers are arriving, the bucket will start to fill again.

Like the SLB, the TB allows incoming bursts of packets that can build up during idle periods or periods of low-rate packet arrivals. But unlike the SLB, the rate of the output traffic is not bounded always to r, and follows the input traffic rate (up to the overflow point, when there are no more tokens, and then the traffic is clipped and leaves the TB at a rate of r).

In some circumstances, multiple buckets are used in tandem (concatenated) to model or to control traffic. When buckets are concatenated to control multiple rates, the resulting rate will be the minimal rate across all buckets (where

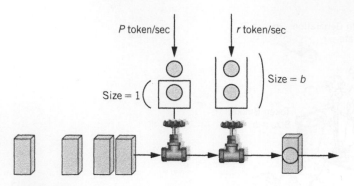

FIGURE 6.7

Dual leaky bucket

the bucket sizes should be in increasing order). The most common use of tandem buckets is the Dual Leaky Bucket (DLB)—also called the Dual Token Bucket—which is used mainly to control the incoming burst rate (p) when the second bucket has vacancy. In this case, a small bucket (that can hold one token, for example), with high token generation or leak rate p, is concatenated to the "normal" bucket of size b, which runs at rate r. The (b, r) bucket controls the average rate, whereas the $(p, 1)$ bucket controls the peak rate, as depicted in Figure 6.7.

The resulting effect of DLB is that users or source applications are allowed, over the long term, to transmit at a rate r while preserving their right to burst at a higher rate p for limited periods—up to $b/(p - r)$ time units when the bucket of size b is initially empty. This can be seen in Figure 6.8 (since at the maximum burst duration t, the number of packets that enter and exit the buckets are equal, or $pt = b + rt$).

When it is not a 1-byte traffic, but a packet nature of traffic, then at the maximum burst duration t we have $M + pt = b + rt$, where M is the maximum packet size and p is the peak rate of the burst. The maximum burst duration is therefore $(b - M)/(p - r)$ time units, as can be seen from Figure 6.9.

The maximum amount of traffic (in bytes) is expressed in terms of a traffic envelope $A(t)$. $A(t)$ indicates the upper bound traffic allowed in time t, and maintain $A(t) \leq min(M + pt, b + rt)$.

6.4.2.1.2 Traffic Specification

Traffic specification is commonly defined by TSpec [384]. TSpec takes the form of a token bucket where the bucket rate is r and the bucket size is b, plus an additional peak rate (p). There is also a minimal policed unit (m), in which any packet size less than m will be considered as of size m with regards to resource allocation

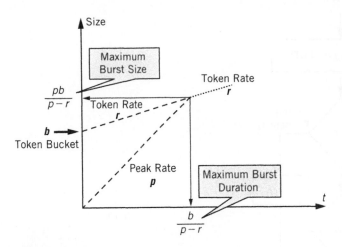

FIGURE 6.8

Burst rate and period

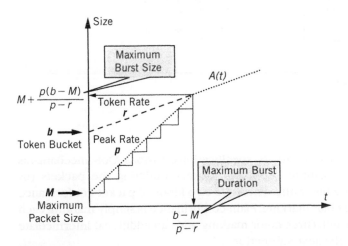

FIGURE 6.9

Token bucket rate

and policing, as well as a maximal packet size (M). All rates are in units of bytes per seconds and all sizes are in bytes. Traffic specification is used to characterize traffic enforcement (policing) and traffic shaping, as described in the following subsections.

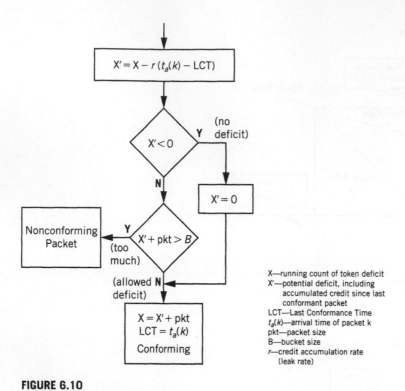

FIGURE 6.10

"Continuous state" leaky bucket

6.4.2.1.3 Packet Conformance

When traffic does not conform to its specifications, various QoS mechanisms have to respond by, for example, dropping packets, buffering the packets (or some of them), or shaping the traffic. There are two kinds of packet conformance methods—plain or three-color marking. Plain conformance is simply marking each packet as conforming or not. Three-color marking adds an additional intermediate marking, as described in the next subsection.

The continuous-state leaky bucket algorithm is an example of a simple conforming method. The algorithm is depicted in Figure 6.10. In this algorithm, the level (X) of the bucket decreases at a rate of r bytes per "time-unit." When a packet arrives at time t_a, the bucket has already lost $r(t_a - LCT)$ from its level (where LCT, the Last Conformance Time, is the last time the bucket level was updated). If, at that time, there is enough space in the bucket for this packet (that has size pkt), the packet is conformed, and the level of the bucket is raised by this packet's size. If there is no room—the packet is nonconformed, and the bucket's level keeps decreasing.

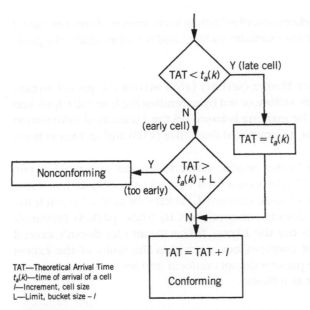

TAT—Theoretical Arrival Time
$t_a(k)$—time of arrival of a cell
I—Increment, cell size
L—Limit, bucket size – I

FIGURE 6.11

Virtual scheduling

Another algorithm, which is not based on buckets, is the Virtual Scheduling Algorithm (VSA), and is described in Figure 6.11 (where I is Increment, the packet size equivalence, and L is Limit, equivalent to the bucket size minus the packet size). Here, it is assumed that at a sustained rate, the arriving packets (or cells, in the original algorithm), are evenly spaced, and coming in at every I units of time. Obviously, if a packet appears after it was expected, then it will be confirmed. If it appeared L unit-times before it was expected (which reflects the tolerance dictated by the "buffer size" L), then it is also confirmed. Once the packet is confirmed, then the next expectation time (the Theoretical Arrival Time) is updated by an additional I time-units.

These two equivalent algorithms are used in the definition of the Generic Cell Rate Algorithm (GCRA) defined in Annex A of ITU-T I.371 [223], where I is the inter-arrival time of a packet (or a cell in GCRA case), and L is the time tolerance.

6.4.2.1.4 Three Color Markers

Three Color Markers (TCM) use green, yellow, and red to indicate packets' conformance. A "green" packet means a conforming packet, a "yellow" packet means a nonconforming packet that might eventually be handled according to "best-effort" treatment, and a "red" packet means a nonconforming packet that probably cannot be handled.

Each of the Three Color Markers described below is composed of two functional modules: one meters the traffic (or estimates its rate), and the other marks the packets according to some criteria.

srTCM Single-Rate Three Color Marker (srTCM) [168] meters the packet stream, and marks[3] the packets as green, yellow, or red by extending the burst size limit and allowing two priority levels. The marking is based on the Committed Information Rate (CIR) and two burst levels, a Committed Burst Size (CBS) and an Excess Burst Size (EBS).

The srTCM uses the token bucket model, where tokens are generated at rate CIR. When packets enter the TB, the Committed Token Count (Tc) and, in some cases, the Excess Token Count (Te), are examined. Packets are marked green if the Committed Token Count (Tc) does not exceed the CBS (these packets conform), yellow if they exceed the CBS but the Excess Token Count (Te) doesn't exceed the EBS (these packets do not conform, but are within the limits of the excess burst), or red otherwise (these packets do not conform, and are beyond the excess burst). The srTCM algorithm is as follows:

```
Two token buckets C and E are initially (at time 0) full, i.e., the committed
token count Tc(0)=CBS and the excess token count Te(0)=EBS.

Thereafter, the token counts Tc and Te are updated CIR times per second
as follows:
if Tc < CBS, Tc is incremented by one, else
     if Te < EBS, Te is incremented by one, else
          neither Tc nor Te is incremented.

When a packet of size s bytes arrives at time t, the following happens:
if Tc(t)-s >= 0, the packet is green and Tc is decremented by s down to 0, else
     if Te(t)-s >= 0, packet is yellow & Te decrements by s down to 0, else
          the packet is red and neither Tc nor Te is decremented.
```

trTCM Two-Rate Three Color Marker (trTCM) [169] meters the packet stream, and marks the packets as green, yellow, or red by separating control for peak rate and committed rate with individual burst sizes. The marking is based on Peak Information Rate (PIR) and CIR, as well as on two burst levels: CBS and Peak Burst Size (PBS).

The trTCM also uses the token bucket model, where tokens are generated at rates CIR and PIR into two buckets. When packets enter the TB, the Peak Token Count (Tp) and in some cases also the Committed Token Count (Tc) are examined. Packets are marked red if the Peak Token Count (Tp) exceed the PBS

---

[3]In [168] there are actually two marking schemes, the Color-Blind mode and the Color-Aware mode. Only the Color-Blind mode, which we describe here, is important for our purposes; the Color-Aware mode, which recolors already marked packets, is another variant.

(nonconformant packets), yellow if the Committed Token Count (Tc) exceeds the CBS but the Peak Token Count (Tp) does not exceed the PBS (nonconformant packets, but within limits of the peak burst), or green otherwise (conformant packets). The trTCM algorithm [169] is as follows:

```
Two token buckets C and P are initially (at time 0) full, i.e., the committed
token count Tc(0)=CBS and the peak token count Tp(0)=PBS.

Thereafter, the token counts Tc and Tp are updated CIR and PIR times per
second, respectively, as follows:
if Tc < CBS, Tc is incremented by one
if Tp < PBS, Te is incremented by one

When a packet of size s bytes arrives at time t, the following happens:
if Tp(t)−s < 0, the packet is red, else
    if Tc(t)−s < 0, the packet is yellow and Tp is decremented by s, else
            the packet is green and both Tc and Tp are decremented by s.
```

A modified trTCM was proposed [6] to handle in-profile traffic differently, that is, to save conformance tests for the green packets (and sometimes to save these packets from being marked red despite their eligibility for green). Furthermore, the modified scheme does not impose peak-rate shaping requirements on edge devices, but rather checks the Excess Information Rate (EIR) for in-profile traffic. The modified trTCM algorithm [6, 169] is as follows:

```
Two token buckets C and E are initially (at time 0) full, i.e., the
committed token count Tc(0)=CBS and the excess token count Te(0)=EBS.

Thereafter, the token counts Tc and Te are updated CIR and EIR times per
second, respectively, as follows:
if Tc < CBS, Tc is incremented by one
if Te < EBS, Te is incremented by one

When a packet of size s bytes arrives at time t, the following happens:
if Tc(t)−s > 0, the packet is green and Tc(t) is decremented by s, else
    if Te(t)−s > 0, the packet is yellow and Te(t) is decremented by s, else
            the packet is red and neither Tc nor Te is decremented.
```

tswTCM The Time Sliding Window Three Color Marker (tswTCM) [119] is different from the srTCM or trTCM in that it does not use the token bucket model at all, nor does it use the two functional modules that srTCM and trTCM are composed of—metering and marking. Instead of metering, tswTCM uses rate estimation, and instead of marking according to deterministic criterion, it marks probabilistically.

tswTCM estimates the packet stream rate based on simple control theory principles of proportionally regulated feedback control. Then, it compares the traffic stream against the Committed Target Rate (CTR) and the Peak Target Rate (PTR).

tswTCM uses a rate estimator that provides an estimate of the running average bandwidth. The rate estimator takes into account burstiness and smoothes out its estimate to approximate the longer-term measured sending rate of the traffic stream.

tswTCM marks with green the packets that contribute to a sending rate not above the CTR. Packets that contribute to the portion of the rate between the CTR and PTR are marked either with green or with yellow according to a probability that is based on the fraction of packets contributing to the measured rate beyond the CTR. Packets causing the rate to exceed PTR are marked red according to a probability that is based on the fraction of packets contributing to the measured rate beyond the PTR, or yellow or green in the complementary probabilities.

Using a probabilistic function in the marker is beneficial to TCP flows, as it reduces the likelihood of dropping multiple packets within a TCP window. An example of one realization of tswTCM [119], using one form of rate estimator (time based) is as follows:

```
Initially assign:
AVG_INTERVAL = a constant; // Time window over which history is kept
Avg_rate = CTR; // Avg_rate averages the Arrival Rate of traffic stream
                      (initially it is the Committed Target Rate)
t_front = 0;
```

Thereafter, when a packet of size *s* bytes arrives at time *t*, the following happens:

```
Rate estimation:
Bytes_in_win = Avg_rate * AVG_INTERVAL;
New_bytes = Bytes_in_win + s;
Avg_rate = New_bytes/( t - t_front + AVG_INTERVAL);
t_front = t;
```

```
Coloring:
if Avg_rate <= CTR the packet is green; else
      if Avg_rate <= PTR, then
                  {
            p0=(Avg_rate - CTR)/Avg_rate;
            packet is yellow with probability p0;
            packet is green with probability (1 - p0);
            } else
                  {
                  p1=(Avg_rate - PTR)/Avg_rate;
                  p2=(PTR - CTR)/Avg_rate;
                  packet is red with probability p1;
                  packet is yellow with probability p2;
                  packet is green with probability [1-(p1+p2)];
                  }
```

6.4.2.2 Measurement and Metering

There are basically three simple ways to measure traffic, which were used in the preceding subsections. More particularly, traffic measurement can be executed by one of the three mechanisms:

- *Point samples*, which are measures taken periodically or sporadically, and which are simply sampling the traffic at specific moments.
- *Time-window samples*, which means that several measures are taken within a time window, and traffic representation is computed according to either the average or the maximum measure.
- *Exponential averaging* is done when averaging a time window is not enough, and instead the average is computed based on weighted previous measures and current measures. The weight influences the adaptive nature of the resulted measure and its smoothness.

6.4.2.3 Policing

The theoretical background in the previous subsections should clarify both what traffic policing is and how it works. When traffic can be described and specified (e.g., peak and sustained rates, required delay bound and variations, or loss tolerance), then violating or conforming packets, as well as the extent of the violation or conformance, can be identified and treated. Policing (and shaping, as described below) can be regarded as an open-loop control mechanism on the network traffic.

The policing function is to monitor the traffic and to take corrective steps when the measured traffic deviates from its specification. Policing means allowing or disallowing packets (traffic) into the network (links), or prioritizing their right to enter the network. Policing can be executed by actually dropping packets, delaying packets or marking packets according to priority for other mechanisms in the traffic handling modules that will decide what to do with these packets later on.

Policing is sometimes defined as either simple policing or as a reshaping mechanism [384]. In this subsection we refer to just simple policing, and in the next subsection we refer to shaping mechanisms. Generally speaking, it is assumed that policing is done in the network edge, whereas shaping is executed in all network branches and merging points.

Policing mechanisms can be classified according to three of the mechanisms previously described, Leaky Bucket, Virtual Scheduling, and Windows (i.e., the moving window that was described in tswTCM, exponentially weighted moving average, etc.). Policing can be done by the TCM methods or by simple conforming tests (i.e., marking whether a packet conforms or not), as described earlier.

6.4.2.4 Shaping

Traffic shaping is essentially another way of looking at policing, that is, it means arranging traffic to meet its specification. Shaping is done by reducing rates or burstiness, smoothing excessive peaks, delaying packets until they can enter the

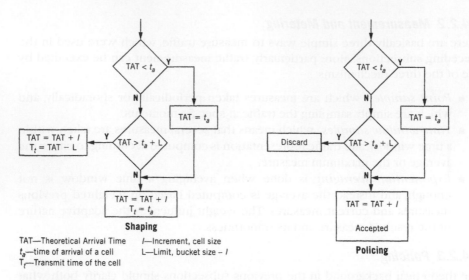

FIGURE 6.12

I.371 Generic cell rate algorithm (GCRA)

line, or generally making the packets conform to the expectations of other QoS mechanisms in the network.

Shaping mechanisms usually use buffers to hold or delay excessive packets until they are allowed into the network (link), and decisions about when to allow the packets in are very similar to the policing mechanisms. For example, the most common shaper is a buffer in front of a token bucket, so that packets are buffered until they are allowed into the network, when tokens are available, according to the traffic specification. Another implementation of a shaper, which is not based on the leaky bucket but on the virtual scheduling algorithm, demonstrates how every policing mechanism can also act as a shaper, as shown in Figure 6.12.

6.4.3 Packet Queueing

Buffers (in the network, in equipment, or in chips) play a crucial role in packet behavior and traffic nature in terms of packet loss, shaping capabilities, delays, and delay variations. In the next two subsections, two aspects of queueing will be discussed: (1) handling packets on their way into the buffer and inside the buffer, and (2) policies of how these queues—or actually the packet inside these queues—are served or scheduled.

To design and implement a system (or equipment, or a chip) with adequate buffers to provide the required QoS, some queueing theory knowledge and usage is recommended; however, such theory goes beyond the scope of this book. For our purposes, it is sufficient to know that queueing theory models arriving packets with random inter-arrival periods; it also models how packets are queued and serviced according to a service discipline that has either fixed or random characteristics. The resulting queueing theory models can then be used to evaluate delays, buffer sizes, and the relation between packet loss, buffer size, and traffic character. For more on this subject, the reader is referred to the vast literature of queueing theory.

6.4.4 Buffer Management

Buffer management techniques determine how to deal with buffer overflows in cases of overrun packets or congestion (buffers in the network). They drop or discard packets in order to solve the congestion in the network, equipment, or chips.

Along with simply having to treat overflow situations, buffer management techniques take different approaches to solving them. They must also try to achieve [123, 125]: (1) minimal packet loss and queueing delay, (2) the avoidance of global synchronization of packets' sources, and (3) high link utilization.

There are two primary methods for managing buffers: passive and active. In passive queue management, simple techniques are used to avoid queue overflow, and the queue is handled regardless of the packets' source, traffic, fairness, network status, and so on. In Active Queue Management (AQM), queue overflows are avoided while such factors as packets' source, traffic and network flows, fairness, network status and congestion level in the network are also considered in order to minimize packet losses and network congestion. AQM signals packets' sources by selectively marking or dropping packets before congestion happens, thus optimizing packet losses and delays, as well as network throughput and utilization.

One way to signal congestion to packets' sources in AQM is by using the Explicit Congestion Notification (ECN) mechanism [362], and the other way is through packet drops. ECN is used by applications that are sensitive to delays or packet losses, in order to avoid using packet drop as an indication of network congestion. ECN uses two bits in the ECN field of the IP header (the TOS field in IPv4 contains the ECN field). One bit is the ECN-capable transport (ECT) bit, and the other is the congestion experience (CE) bit. ECT is set to indicate that the end-nodes of the link are ECT-capable, and the CE bit is set in the network' buffers to indicate congestion to the end-nodes.

Most of the algorithms described in the following are AQM techniques, apart from the tail drop and drop on full techniques, which are usually not adequate to current networks' QoS requirements. Most of the algorithms described are targeted for IP networks, or even more specifically with TCP congestion considerations in mind; however, some ATM-based algorithms are also described.

In this subsection, basic techniques are described briefly. There are many studies on this subject, many more algorithms, simulations and performance evaluations that cover all sorts of specific conditions or traffic types, such as, for example, Random Exponential Marking [27], Proportional-Integral (PI) controller [174], Proportional-Differential controller [398], Smith Predictor-based PI-controller for AQM [281], Dynamic Random Early Detection (RED) [39], Gentle RED [371], scalable control [345], or Adaptive Virtual Queue (AVQ) [266]. More detailed description of buffer management can be found in many textbooks and research papers—for example, [76, 128, 130].

6.4.4.1 Tail Drop

This is the simplest way to handle buffer overflow: Drop all the incoming packets when the buffer is full. Simple as it might be, it is obviously not fair, nor is it efficient in terms of solving congestion situations [56]. It is not fair since the buffer could be dominated by one—possibly even low-priority—stream of packets, which send many packets to the queue. This phenomenon is called lockout. It is not efficient, since it causes queues to be completely full for long period of times, and many packets can be lost before the sources of the packets recognize the full queues. At that point, the sources will reduce the rate of sending packets, causing periods of unloaded queues and links. The problem is amplified in networks of buffers, when many packets will be dropped at some point after using buffers along the path. Until the sources understand that there is a buffer-full situation (through a back-propagation of a signal or otherwise), many other packets will be dropped after occupying resources, and then the network will be drained for a while. This will eventually cause low link and queue utilization of the network.

6.4.4.2 Random Drop on Full

Random drop on full is one solution to the lockout problem described before [56]. Instead of dropping the arriving overflowed packets (tail), randomly selected packets in the queues are dropped. The full queues problem is not handled in this method.

6.4.4.3 Drop Front on Full

Drop Front on full is another solution to the lockout problem described in the tail drop method [56]. Packets at the head of the line (HOL) are dropped rather than the arriving ones (the tail packets). This method was shown to improve performance in specific circumstances (TCP load, for example, [268]), by reducing the duration of congestion episodes, since the packet loss indicating congestion reaches the sender earlier during a time of full queue. The full queues problem, however, is also not handled by this method.

6.4.4.4 Random Early Detection

Random Early Detection (RED) is an AQM algorithm, in contrast to the Tail Drop or the Drop on Full methods [46]. RED drops the arriving packets randomly, not those in the queue, as Random Drop on Full does. In principle, as the queue begins to build up, arriving packets are dropped at a low but increasing probability.

In its simplest form, RED discards (drops) arriving packets in a probability that is determined by the queue size at the instant at which the incoming packets should enter. This probability increases as the buffer gets fuller, starting from some threshold, and the rate of increase depends on several buffer occupancy thresholds, as shown in Figure 6.13. The resulting curve creates a RED profile that can be used for specific applications or network environments.

A more sophisticated RED algorithm [128] drops arriving packets in a probability that increases as the estimated *average* queue size grows (rather than one measured at the precise instant that the packet arrives). This means that RED assigns low probability (or none) for dropping an incoming packet when the queue was relatively free for some time before this packet arrives, or high probability if the queue was relatively full (see Figure 6.14).

The RED algorithm has two phases. In the first phase, for each packet arrival, RED calculates the new average queue size (avg) using a low-pass filter with an exponential weighted moving average (explained in the following). Then, in the second phase, it decides whether or not to drop the incoming packet based on calculated probability (p_a), which is a function of the average queue size (avg), defined minimum and maximum thresholds (min_{th} and max_{th}, respectively), and load (whether the last packet arrived into a loaded or free queue).

When computing the average per each arriving packet, it is incremented or decremented by a weighted difference between the existing queue size when the packet arrives (q_{size}) and the known average at that time (avg):

$$avg = avg + w\,(q_{size} - avg),$$

where w is the weighting factor. If $q_{size} = 0$—that is, the packet arrived when the queue was empty—RED reduces the average exponentially by the idle time

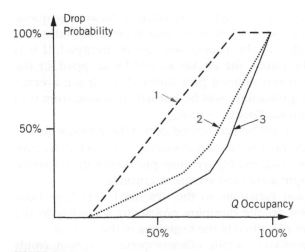

FIGURE 6.13

RED profiles

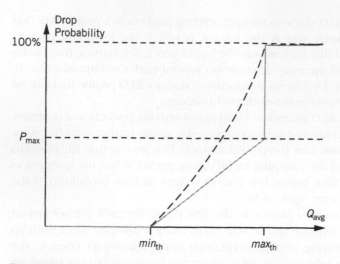

FIGURE 6.14

Random Early Detection

of the queue, m. This means that $avg = (1 - w)^m avg$, where m represents the (typically small) number of packets that the queue could have received during its idle time. Each of these imaginary packets that could arrive to an empty queue reduces avg by a factor of $(1 - w)$. That is, when the packet arrives, the average is computed as if m small packets arrived at an empty queue during the idle period, just before this packet. The average can be estimated in units of bytes or of packets.

The probability of dropping a packet (p_b) is derived from the average queue size with regard to the thresholds; if it is below the minimum threshold, min_{th}, the probability will be zero, that is, the packet will not be dropped. If it is above the maximum threshold, max_{th}, the packet should be dropped, or the probability is 1 (or some maximum defined probability P_{max}). If the average is between the thresholds, the probability will be linearly increased from 0 to P_{max}; that is, $P_b = P_{max} (avg - min_{th})/(max_{th} - min_{th})$.

The final drop probability (p_a) will be accelerated only if the previous packet was also above the minimum threshold. It means that successive packets that arrive to relatively full queues will be assigned higher drop probability than a single packet that arrives at the same queue size and average occupancy.

Many implementations of and variations to the described RED have been suggested and used; the simplest set the dropping probability according to the queue size and the thresholds, as described in the beginning of this section.

Random Early Detection algorithm maintains efficient queue utilization, avoids a full-queue phenomenon, and handles bursts of packets without loss. Using randomness in RED algorithm eliminates synchronized packet streams that cause the lockout phenomena described in tail drop method.

6.4.4.5 Weighted Random Early Detection

Weighted RED is used for discarding prioritized packets for buffer management, and is simply achieved by using different RED profiles for each of the incoming packets, according to the packets' priority. An example of using Weighted Random Early Detection (WRED) for two priority packet flows is shown in Figure 6.15.

6.4.4.6 RIO (Differential Dropping)

Random Early Detection In/Out (RIO) is a variation of RED and WRED algorithms [81]. Basically, it is a WRED with two priorities for incoming packets, called in RIO *in*-packets (high priority) and *out*-packets (lower priority). In-packets are packets that are in their "service allocation profile," whereas out-packets are those that are out of their "service allocation profile." Nevertheless, the concept of RIO is general enough to be described for the two-priority case.

The "out" RED algorithm drops the out-packets much more aggressively than the "in" algorithm, and starts dropping them at a much smaller queue size. In addition, in-packets get priority over the out-packets not only by choosing proper thresholds, but also by using just the in-packets average queue occupancy for calculating the in-packets dropping probability, whereas the average queue

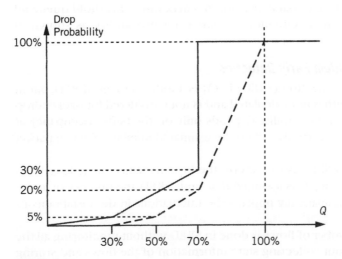

FIGURE 6.15

Weighted Random Early Detection

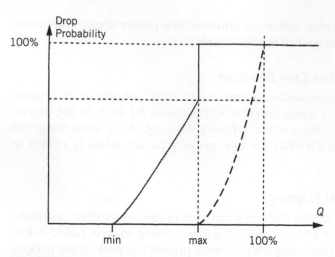

FIGURE 6.16

Random Early Detection In/Out

occupancy for the out-packets is calculated based on *all* packets. The result is that the out-packets can be controlled before the queue reaches a point where in-packets might be dropped.

Another way to look at RIO is according to two RED profiles. One drops the out-packets from some minimum threshold queue-fill percentage to a maximum threshold, and then tail-drops all incoming out-packets, and the other profile begins to drop in-packets from this maximum threshold queue-fill percentage to 100% queue occupancy in increasing probability, as shown in Figure 6.16.

6.4.4.7 Stabilized Random Early Detection

Stabilized Random Early Detection (SRED) [343] is another variant of RED, but in SRED average queue length is not calculated and is not considered for packet drop decisions. The packet drop probability depends only on the buffer occupancy at the moment that the packet arrives and on the estimated number of active packet sources (flows).

The main idea behind SRED is to stabilize the instantaneous queue length (its length at any instant in time). It is assumed that under RED, the number of active packet sources (flows) dramatically impacts the fluctuations in the instantaneous queue length.

The estimation of number of flows is done in SRED without monitoring all the input packets, and without collecting state information of the flows and storing them. When a packet arrives, an estimation of the effective number of flows [343] can be made based on the frequency of recently buffered packets that arrived from the same flow, that is, the more packets from the same flow arrive at the queue, the fewer flows are assumed. This estimation is achieved by maintaining a zombie list that records recent flows. When a packet arrives, it is compared to a packet that is randomly chosen from the zombie list. The result of a hit or a miss is used to estimate the number of active flows, as well as to detect misbehaving flows for more aggressive packet dropping. Then, the probability of packet drop is calculated based on the buffer occupancy (grouped into three categories—up to one-sixth of the queue, one-sixth to one-third of the queue, or more than one-third), multiplied by a factored number of the estimated active flows, as described above. The mechanism described above for calculating the packet drop probability is called "simple SRED."

A better packet drop probability calculation is suggested ("full SRED," or "SRED," for simplicity) in [343], which modifies the "simple SRED" by a factor that accounts for the misbehaved flows. These flows are assumed to flood the queue, causing a bias in the estimated number of active flows, so the suggested factor simply increases the dropping probability in intensive flows (those flows that send large number of packets into the queue, or more than the average flows).

The details, assumptions and formulas are given in [343], for those who would like to read more on this approach.

6.4.4.8 Flow Random Early Detection

Flow Random Early Detection (FRED) [283] (sometimes called Fair RED) is also based on RED, however, it is a per-flow buffer management technique. FRED tries to solve the issue of fairness in dropping randomly selected packets, so that packets coming from slow rate streams (flows) will not disappear in the randomly chosen packets to be dropped. In other words, instead of dropping packets proportionally, FRED drops packets from a filtered set of those flows that send larger number of packets into the queue than the average flows.

The entire algorithm is not detailed here (the reader is referred to [283]), but its principal concept is as follows: FRED works like RED, but monitors every incoming flow for its packets, making sure some packets will be queued, regardless of the flow intensity (its packet's frequency in the entire number of incoming packets). In order to do this, FRED maintains a parameter $qlen_i$ that counts packets in the queue for each flow i, and defines a global parameter min_q for the minimum number of packets any flow is allowed to buffer unconditionally. FRED makes sure that any flow will have at least min_q packets in the queue, and starts to drop packets from that flow randomly, as RED does, when the queue reaches min_{th} occupancy.

To cope with nonadaptive packet sources (those packet sources that are not sensitive to the congestion indication created by packet loss), FRED uses another mechanism. FRED maintains an additional parameter $strike_i$ that counts the number of times a packet is received from source i when there are max_q packets in the queue, and where max_q is a global parameter denoting the maximum number of packets any flow (from a packet source) is allowed to queue in the buffer. FRED makes sure that no flow will have more than max_q packets in the queue, and in cases where the flow's $strike_i$ is high, FRED reduces the number of the allowed queued packets from that packet's source (flow) to that of the average flow.

Flow Random Early Detection originally assigned values to all parameters, an algorithm to modify them while working, and a modified way to calculate the average queue length, but these details are not relevant to the understanding of FRED's principle.

6.4.4.9 Balanced Random Early Detection

Balanced Random Early Detection (BRED) [23] is a close variant of FRED. It is also a per-flow buffer management technique, and a very simple algorithm.

BRED maintains per-flow accounting for the active flows in the queue, and drops packets based on the queue occupancy for those flows. Calculation complexity is limited by the queue size (state information is maintained only for flows having packets in the queue).

Like FRED, BRED maintains a parameter $qlen_i$ which counts packets in the queue for each flow i, and, in addition, it defines three global parameters: (1) min_q, for the minimum number of packets any flow is allowed to buffer unconditionally; (2) thr_q, for the number of packets any flow can have in the queue before its packets start being dropped more aggressively; and (3) max_q for the maximal number of packets any flow is allowed to buffer in the queue. Two additional parameters, p_1 and p_2, denote the dropping probabilities of packets from a flow, when the number of packets from that flow exceeds min_q (where p_1 applies) or exceeds thr_q, respectively (where p_2 applies), and $p_2 > p_1$.

BRED makes sure that any flow will be allowed for at least min_q packets in the queue uninterruptedly, and starts to drop packets from that flow randomly, at probability p_1, when the queue contains more packets from that flow than min_q. BRED drops the packets of a flow if the number of packets from that flow exceeds thr_q in the queue with probability p_2, and drops all packets of a flow if the number of packets in that flow exceeds max_q.

6.4.4.10 Longest Queue Drop and Random Dropping

Longest Queue Drop (LQD) [401] is also a per-flow buffer management method, like FRED and BRED. The difference between LQD and FRED or BRED is that LQD drops packets within the queue according to some algorithm, whereas FRED and BRED calculate the dropping probability of arriving packets and drop them accordingly *before* they enter the queue.

This mechanism works by soft-partitioning the queue, so that every incoming flow gets a partition for its use. The nominal buffer allocation, or partition (b_i), that flow i gets is simply B/n, where B is the entire queue size, and n is the number of flows (packet sources) that send packets to the queue. When flow i requires more than b_i at some point of time, it simply uses the unused queue if there is enough room in it (if the total occupancy is less than B). Once a packet from flow j arrives to a full queue (i.e., the total current occupancy of the queue is B), and this flow still has room for its packets in "its" partition (i.e., the flow's current occupancy q_j is less than b_j), the LQD algorithm clears room for this packet by dropping, or pushing out, the HOL packet belonging to the flow that most exceeds its quota in the queue (i.e., the flow k for which $q_k - b_k$ is the largest of all flows).

Simply put, the entire queue is partitioned for the use of each of the incoming flows. Overflowed packets from some flow may be queued on other partitions' expense as long there is room. When a packet arrives that belongs to a flow that has not yet queued to its full quota, another queued packet that is overquota is dropped, thereby allowing the entitled packet to be queued.

Dynamic soft partition with Random Drop [401] is identical to LQD, with one very small modification: When an incoming packet finds the queue full (the queue's total occupancy is B), then LQD randomly selects such a flow from those that exceed their allocated partitions (i.e., their $q_k > b_k$), instead of searching for the flow that most exceeds its quota in the queue (i.e., the flow k that $q_k - b_k$ is the largest of all flows). It then drops its HOL packet, thus clearing room for the incoming packet.

LQD maintains a partition and a queue status for each flow, and requires running a search among all the queued flows to find out which is the longest overflow whenever there is a full queue. This calls for $O(\log n)$ complexity, and $O(n)$ space requirements. Approximated LQD (ALQD) is proposed in [401], similar to a scheme called Quasi-Pushout [285], to simplify the time and space complexity of this buffer management scheme to $O(1)$.

6.4.4.11 BLUE

Blue [124, 125] is a simple AQM algorithm that tries to improve RED with regards to network utilization (along with its queues), and to packet loss.

Blue calculates the incoming packet drop (or mark) probability (p) according to four predefined parameters: L, a threshold to the queue length at the instant that the packet arrives, *inc_delta* and *dec_delta*, the increased or decreased change in p, respectively, and *freeze_time*, a minimal time interval in which modifications to p cannot be made. The calculating algorithm is as follows:

Start with some arbitrary dropping probability p. Each arriving packet is dropped (or marked) at probability p. If the packet is dropped, or the instantaneous queue length exceeds L, then p is increased by *inc_delta*, provided, however, that the time interval since p was last updated exceeds *freeze_time*. In other words, p is updated according to the packet loss and queue length, but not too often, as defined by the minimum interval *freeze_time*. In other words, the changes in p are not too aggressive. Conversely, if the incoming packet arrives into an empty queue, then p is decreased by *dec_delta* (provided, again, that the time interval since p was last updated exceeds *freeze_time*).

As a result, instantaneous queue length, packet loss, and link utilization (empty queue) are used as the indices of traffic load to determine the dropping probability, rather than just queue occupancy, as other AQM techniques usually use.

A variation of Blue [124, 125] ignores L and the instantaneous queue length, and increases p just when the packet is dropped, and if it exceeds the permitted update frequency. However, considering the instantaneous queue length in calculating p makes it possible to leave space in the queue for transient bursts, as well as to control the queuing delay in large-size queues.

6.4.4.12 Partial Packet Discard

Partial Packet Discard (PPD) is a technique that originated with the concept of ATM, where packets are partitioned among several cells, and, once a cell is dropped from the queue, the rest of the packet's cells become useless. The purpose of PPD is to identify, from the rest of the incoming cells, those belonging to the same packet from which a cell was dropped. Once identified, these cells can be discarded, so as better to utilize the buffer.

This principle can be applied to any sequence of packets that can be identified as belonging to an entity that is worthless if not received in whole (e.g., at the IP level, it can be packet segments).

PPD algorithm specifically describes the mechanism of identifying the packet cells in ATM, as defined in the AAL5 (ATM Adaptation Layer 5) and the VC (virtual channel) of the cells. Any dropped cell results in the addition of its VC into a drop list, and any subsequent cell belonging to the same VC is dropped, until the last cell of this packet arrives, and its VC is removed from the drop list. The last cell of a packet is identified as an EOM (end of message) cell by the payload type identifier (3 bits) of the cell header.

Some inefficiency still exists, since part of the packet is still handled by the queue, wasting buffer space. Early Packet Discard solves this inefficiency.

6.4.4.13 Early Packet Discard

Early Packet Discard (EPD) is also a technique originating with ATM, but it improves on the PPD just described. The main idea is not to wait until the queue is full and only then to start to drop cells that belong to the same packet, but rather to begin doing so before the queue is full. The EPD mechanism sets a buffer threshold, and when the buffer occupancy exceeds this threshold, EPD begins to drop cells that belong to new packets (identified by the first cell of a new VC), and add their VCs to the drop list. The rest of the mechanism is as described in PPD, and until the EOM cell arrives, cells that carry VCs that are in the drop list are dropped (so that the entire packet is discarded) until an EOM cell arrives and the VC is removed from the drop list.

Both PPD and EPD are insensitive to the intensity of the flows, that is, to packet sources that might dominate the traffic and consume all resources and buffers due to their intense packet transmit activity. To solve the issue of unfairness, Early Selective Packet Discard was introduced.

6.4.4.14 Early Selective Packet Discard

Early Selective Packet Discard (ESPD) is the last technique described here that originated from the ATM world. This technique is an enhancement of the EPD and PPD, which aim to drop all the cells belonging to a packet once there is congestion, while at the same time preserving some fairness by dropping more packets from intense flows than from moderate ones.

ESPD simply adds a queue threshold that is lower than the EPD's threshold, and removes VCs from the drop list in a different way. Instead of removing the VC from the drop list when the packet ends (when the EOM cell arrives), ESPD leaves the VC in the drop list until the lower threshold is reached, thereby allowing new VCs to enter the queue instead of those that are listed in the drop list.

The drop list itself is limited, so that just a few VCs will be discarded (unless the queue is full, and VCs are always added to the drop list). Additionally, there is a drop timer ticking when there are VCs in the drop list that makes sure that the drop list is refreshed from time to time, when it is expired, to maintain fairness across sessions.

6.4.4.15 Buffer Management Summary

We discussed in this section various buffer management schemes, and we can summarize them by making distinctions between them along three dimensions [143]. Each of these dimensions is concerned with the question of detecting or anticipating a congestion situation, handling such a situation, and (optionally) notifying the end nodes about it. These dimensions are (1) performance objectives, (2) parameters used, and (3) policies regarding dropping or marking packets.

Performance objectives can be (1) achieving fairness, or (2) maximizing utilization while minimizing packet delay (latency that directly affects queue size) and packet loss, or (3) stabilizing queues in the network. The parameter used by these schemes is usually queue length, but input rate or idle link events are used as well, usually as congestion indices.

6.4.5 Packet Scheduling and Service Discipline

Service disciplines are based on various scheduling techniques, methods, and algorithms. Scheduling methods are critical elements in packet networking. Scheduling algorithms determine how to solve packet contentions, like which queue (or packet) to serve (or transmit) next. Priority is often attached to scheduling algorithms for handling various QoS parameters associated with different queues (or packets).

There are many scheduling methods, schemes, and algorithms, for various network configurations, implementation complexity, performance, and QoS requirements. The scheduling methods can be categorized in many ways. One way to distinguish between the schedulers [361] is between time stamp-based (or deadline-based) algorithms and Round-Robin-based algorithms. Time stamp-based algorithms achieve good fairness and delay bounds, but suffer from computational complexity, whereas the Round-Robin-based algorithms are usually simpler, but suffer from large delay bounds.

Another way to categorize schedulers [440] is by their work-conserving nature. A work-conserving scheduler is never idle when there is a nonempty queue, whereas the nonconserving scheduler can be idle even when there are packets in its queues, due to its scheduling algorithms (e.g., each packet has an eligible time to be served, and even when the server is idle, packets will not be served if they are not eligible).

Some schedulers are coupled with traffic shapers, to provide end-to-end delay guarantees on a per flow basis or other QoS requirements. These combinations are termed Rate-Controlled Service schedulers. This category of schedulers will not be described in this subsection, and the interested reader is referred to [441], for example.

There is a vast amount of research and practitioners' literature that studies scheduling methods and proposes many kinds of schedulers and variations of schedulers to meet certain packet or cell networks, interconnection networks, tandem queues, and so on. This subsection describes only those scheduling techniques that we would likely find in network processors, traffic managers, and network requirements environments. It does not discuss many other scheduling algorithms such as Stop-and-Go [150], Elastic Round-Robin [244], Prioritized Elastic Round-Robin [243], Smoothed Round-Robin [80], Stratified Round-Robin [361], Stochastic Fair Queueing [311], Bin Sort Fair Queuing [78], Leap Forward Virtual Clock [399], eligibility-based Round-Robin [279], and Rate-Controlled Static Priority [441], just to mention some.

Another important category of scheduling policies is hierarchical or aggregating policies, which are also not described in this subsection. These policies are also called Class-Based Queueing [130]. Hierarchical Round-Robin [242] and Hierarchical Packet Fair Queueing (H-PFQ) [49] are examples of policies that belong to this category. In this group of schedulers, several scheduling policies can be used for various levels of the hierarchy, or even in-between classes at the same level (i.e., interactive classes may be serviced by priority-driven schedulers, while noninteractive classes may use rate-based schedulers). Scheduling algorithms that involve multiple nodes are also not treated in this subsection.

Finally, it should be emphasized that we simplify these schedulers for the sake of clear understanding of what is required. A more detailed description of scheduling can be found in [76], for example.

6.4.5.1 Scheduling Parameters and Fairness

Each scheduling mechanism can be characterized by "performance," or behavior that can be evaluated according to several parameters, such as delay, utilization (in terms of network usage and in terms of overhead), complexity, robustness, fairness, and starvation. Many of these parameters are trade-offs and the different scheduling schemes sometimes improve one parameter at the expense of another. In the trade-off among delay, complexity, and fairness, it can be argued that complexity and delay are most important, since fairness affects only the short-term distribution of service. If complexity's importance is downgraded when using network processors that handle all scheduling internally and transparently, delay becomes the most important issue, since it affects burstiness, and as a result, affects buffer sizes that limit packet losses. Ideal delay behavior of a scheduler will isolate queues in terms of their mutual sensitivity, and will be bounded independently of number of queues. Utilization and delay can be represented quantitatively, in absolute or in relative terms.

Starvation means that in the case of higher priority packets, there is a chance that low-priority packets will not receive any attention at all. Priority can be achieved by giving 100% attention to any packet of higher priority, or, to decrease the likelihood of consequent starvation, by giving that packet a higher chance of service than all other packets (say, twice).

Fairness means that all packets (of the same class or importance) will receive equal attention from the scheduler, regardless of history of scheduling, place in the queue, and so on. This is basically a qualitative measure, and several attempts to measure and quantify it are presented in the following.

Jain's fairness index [233] computes the relation between all queues' service; it is 1 if it is perfectly fair, and approaches 0 if it is not:

$$FI = \frac{\left(\sum_n x_i\right)^2}{n \sum_n x_i^2}$$

where n is number of sources, and x_i is the normalized service for each of the i queues. Normalized service can be calculated by dividing the actual service by any criterion to find the fair service.

Another way to obtain fairness is by a Fair Index (FI) (termed "normalized service" in [151]) and Service Fair Index (SFI) [396], which are obtained by the ratio of the service that the scheduler gave to queue i, and the assigned minimal, guaranteed service defined for that queue (GS_i):

$$FI_i = \frac{S_i(\tau_1, \tau_2)}{GS_i}$$

where $S_i(\tau_1, \tau_2)$ is the service received by queue i between τ_1 and τ_2. Then, the SFI is defined by: $SFI = |FI_i - FI_j|$. A scheduler is perfectly fair if it always satisfies $SFI = 0$, for any two queues i and j.

A worst-case packet fair [50, 347] is another fairness measure that compares packet flow to ideal fair scheduling (fluid flow). In a fair scheduler, any packet is not delayed in its queue more than its queue size (at its arrival time) divided by the rate the queue is being served, Q_i/r_i. A worst-case fair scheduler is one that maintains for all its queues an upper bound of additional worst case delay C_i, expressed in absolute time (i.e., $Q_i/r_i + C_i$). This time can be normalized by the relative queues' weight (i.e., $C_i = C_i r_i/r$), and the normalized worst-case fair index (WFI) is defined to be the maximal c_i among all queues (i.e., $WFI = \max_i \{C_i\}$).

6.4.5.2 Scheduling Methods

The following subsections describe the most important or common scheduling methods that are used in traffic managers or in network processors.

6.4.5.2.1 First In, First Out

Also known as FCFS (first come, first served), this is the simplest scheduling mechanism, and works exactly as its name implies: the first packet to arrive is handled first. In other words, packets are handled by their order of appearance. There are many deficiencies in this mechanism, such as there being no way to prioritize incoming packets or to treat various classes of incoming packets, as well as to suffer a stall where a very long packet is treated and the other incoming packets are waiting.

In designing FIFO queues, one must consider the nature of the incoming traffic, and prepare sufficient buffer size to accommodate the incoming packets. In case of overflow in the incoming queues, packets must be dropped. When used in cascade queues, the uncorrelated queueing delays in the chained queues creates a severe problem of jitter. An example of FIFO scheduler, which handles three packet sources, is depicted in Figure 6.17, where t_a is the packet arrival time and t is the scheduling time.

6.4.5.2.2 Strict Priority

Also known as Priority Queueing (PQ), this is the second simplest scheduling algorithm. Taking into account the priority of packets, packets are simply reordered in the queue so that the highest priority packet is placed at the head of the queue. There are many deficiencies in this mechanism, such as a potential starvation of low-priority packets. An example of Strict Priority (SP) scheduler, which handles three prioritized packet sources, with W = 3 the highest priority, is depicted in Figure 6.18.

6.4.5.2.3 Round-Robin

This mechanism assumes several queues in which each receives a specific type of packets (for instance, different classes of packets, different sources, etc.). The Round-Robin (RR) scheduler simply chooses one queue at a time, in a cyclical order.

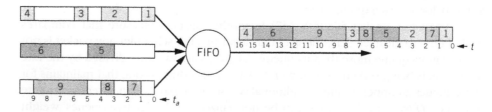

FIGURE 6.17

First In, First Out

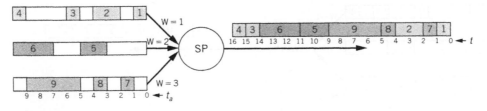

FIGURE 6.18

Strict Priority

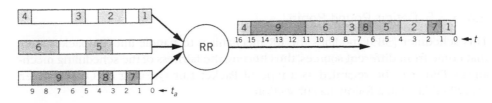

FIGURE 6.19

Round-Robin

It picks the first packet in that queue, if there is one, and then goes to the next queue, and so forth (see Figure 6.19). In this way no starvation can happen, and each source of a class of packets is responsible for its queue behavior. If all sources are equal in priority and average packet size, the scheduler will perform reasonably well. Still, very long packets by one source might cause uneven service utilization.

For cases where sources have priorities, the Weighted Round-Robin (WRR) scheduler, described in the following, is a better answer. For cases where sources have variable packet sizes, the Deficit Round-Robin (DRR) scheduler, described in the following, provides the best solution.

6.4.5.2.4 Weighted Round-Robin

This scheduler is based on the RR mechanism, and addresses the issue of handling different priorities of the incoming packets. Each of the queues gets a weight proportional to the priority of the packets it receives and contains, or related to this priority by some rule or formula. Then, the WRR scheduler picks packets from the queues relative to their weight rather than one by one, as with the regular RR scheduler. This means, for example, that in a three-queues system (see Figure 6.20), where queues A, B, and C have weights of $w_a = 1$, $w_b = 2$, and $w_c = 3$, respectively, the scheduler will approach queue A 17% of the time ($= w_a/(w_a + w_b + w_c)$), queue B 33%, and queue C 50%. This can be achieved by approaching the queues in many ways, for example, through a repeated sequence of AABBBBCCCCCC, or of ACBCBCACBCBC. Obviously, the latter is better for fairness purposes.

FIGURE 6.20

Weighted Round-Robin

6.4.5.2.5 Deficit Round-Robin

This scheduler [385] modifies the RR mechanism to handle uneven packet sizes that come from different sources, thus hurting the fairness of the scheduling mechanism. DRR can be regarded as a type of Packet Fair Queuing (PFQ) scheduler, described later in a following subsection.

Each queue, if it has packets to be served and in its turn, is served if a "quota" of service it has accumulated up until that point is sufficient to handle its head-of-line (HOL) packet. If the queue cannot be served, its quota for the next round is increased, and if the queue is served, its quota is decreased by the size of the packet the queue just released. An algorithm that will implement DRR is as follows (see also Figure 6.21).

Q_i is the allocated quota for queue i for each of the RR iterations.

DQ_i is a variable describing the deficit that queue i has before each of the RR iterations.

AQ is a list of queues (by indices) that contains at least one packet to be served.

$S_i(k)$ is the size of the head-of-line (HOL) packet in queue i, at the RR iteration k.

Upon packet arrival to a queue i, and when queue i is not in AQ, then its index (i) is added to the tail of the AQ.

In every RR iteration k, the queue i in the head of the AQ is checked for potential service:

    a) $DQ_i \leftarrow DQ_i + Q_i$
    b) if $S_i(k) \leq DQ_i$ then
        1. $DQ_i \leftarrow DQ_i - S_i(k)$
        2. Serve HOL packet of queue i
    c) if queue i still has more packets to be served, i is moved to the tail of AQ.
    d) if queue i is empty, then i is removed from AQ, and $DQ_i \leftarrow 0$.

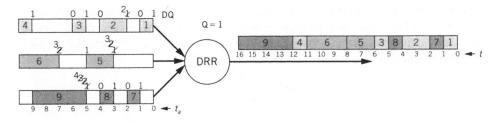

FIGURE 6.21

Deficit Round-Robin

A DRR scheduler can be implemented on a regular RR scheduler (without any priority considerations) or on a WRR scheduler, where in each RR iteration, higher priority queues are approached more than lower priority queues.

6.4.5.2.6 Earliest Due Date

Also known as Earliest Deadline First (EDF) [287], EDD assigns a deadline by which each packet must be served, in order to meet latency requirements, for example. The scheduler is supposed to arrange the packets so as to serve the packets within their deadlines; however, meeting these deadlines is not always feasible due to overbooking, and packets then start to miss their deadlines.

There is a fairness problem with this mechanism, since packets that are received last with some deadline will suffer from a longer delay than packets with the same deadline that were received before them.

Several extensions exist to the EDD scheduler that reduce or limit delay or jitter. Delay-EDD [127] is a work-conserving algorithm, intended to guarantee bounded delay, provided the source of the packets obeys an agreed-on average and peak packet rates that it sends to the scheduler. Jitter-EDD [420], on the other hand, is a nonwork conserving algorithm intended to guarantee a bounded delay-jitter (i.e., a maximum delay difference between two packets) by "holding" packets until they are eligible in terms of the jitter.

6.4.5.2.7 Processor Sharing and Generalized Processor Sharing

Processor Sharing (PS) is a technique [257, 258] that serves N FIFO queues simultaneously by dividing the entire service capacity equally among all nonempty queues (each gets $1/M$ of the service capacity, if M queues are nonempty, $M \leq N$). Processor Sharing is not practical for packet processing, because it splits service attention infinitesimally between all queues. In other words, PS services each of the queues for whatever duration it is available for servicing, whereas for packet processing, this duration must be at least as long as the time required to handle an entire packet. Generalized Processor Sharing (GPS) is a generalization of PS, to allow non-even requirements (priorities) from each of the queues.

GPS is the ideal scheduling mechanism but, unfortunately, it is also impractical for packet handling. Nevertheless, it is the basis for a whole class of important schedulers called Packet Fair Queueing (PFQ).[4] In short, the GPS scheduler [348] will grant queue i at time t with a service rate $g_i(t)$ of:

$$g_i(t) = \frac{w_i}{\sum_{j \in B(t)} w_j} \, r$$

where r is the total available service rate, $B(t)$ denotes the set of nonempty queues at time t, and w_i is the relative weight of queue i (or the minimum, or guaranteed allocated service rate for queue i, given $\sum_i w_i \leq r$). An example of GPS with weights of 1/2, 1/4, and 1/4 is shown in Figure 6.22.

A GPS scheduler is perfectly fair, but, as was mentioned above, it is a "fluid-input" scheduler that can take from the input as much as it requires to fill in available service space. This is not appropriate for packet input because of the packet's structure, which cannot be subdivided as the scheduler is available (i.e., the scheduler either has available space for an entire packet or it does not). The next subsections describe various approaches for approximating the GPS for the packet case.

6.4.5.2.8 Packet Fair Queueing, Packet Approximation of GPS

The problem in using GPS for packets is that packets are quantum units of data, and either the scheduler services the queue or it does not. In cases where the queue deserves service but the packet has not yet been received, a problem is created in which the server is unemployed while it waits for the packet. The decision to wait itself depends on knowledge that does not yet exist; that is, what will happen in the queue and for how long it should wait.

The basic idea in handling packets is to emulate the GPS continuous case by ensuring that packets will receive the same treatment as if they were continuous

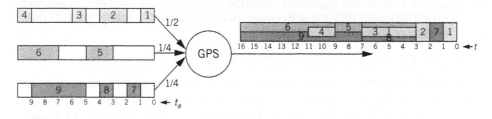

FIGURE 6.22

Generalized Processor Sharing

---

[4]Fair Queueing was introduced by Demers [100] and later its weighted version [348] was introduced, describing the WFQ scheme (described in the following); here we use PFQ as a general concept, not just WFQ.

streams of bits. PFQ or Packet Approximation of GPS (PGPS)[5] [348] techniques are based on GPS in that they maintain a function (the "system's virtual time," $V(t)$), that follows the progress of the GPS server and tries to stay as close to it as possible. This function is used to determine the virtual finish time of all the HOL packets in all queues, which represent the time that these packets must leave the system, and to time stamp these packets accordingly. Packets are then served according to the increasing order of their time stamps (virtual finish time).

Serving the packets according to their virtual finish time, by increasing order, is a policy called "Smallest virtual Finish time First" (SFF) [49], and is very clear and intuitive.

There is another policy that maximizes the schedule fairness not only by choosing packets according to their time stamps SFF, but that also chooses among the eligible packets. The eligible packets are those that would have been served at the same time as their equivalents in the GPS server. This policy is called "Smallest Eligible virtual Finish time First" (SEFF) [49]. In other words, this PFQ subcategory, using the SEFF policy, also considers the packets' virtual start time. This can be achieved or emulated, among other techniques, by a rate-controller or a shaper [396] positioned just before the scheduler, or integrated with the scheduler; thus, this PFQ subcategory is sometimes called shaper schedulers.

Regardless of the policy used, the end result is that the scheduler picks up packets from queues not only by their priority, but also by their length, that is, a short lower-priority packet might be picked up by the PFQ scheduler before a longer, higher-priority packet.

Let us begin to explain the PFQ techniques by assuming, for simplicity, a theoretical case where the queues are never empty. After doing this, we will move to the more general case, in which the queues do not fulfill this assumption; that is, they are sometimes idle or empty. The treatment in this general case differs among the PFQ techniques, and is described later.

Under the first assumption in which the queues are never empty, we can emulate the GPS scheduling mechanism by assuming that each packet receives service immediately after the packet that arrived prior to it in its queue. Packet k receives the service rate (r_i) of the queue (i) to which it belongs, and the finish time ($F_{i,k}$) can be calculated by:

$$F_{i,k} = F_{i,k-1} + \frac{L_{i,k}}{r_i}$$

where $L_{i,k}$ is the length of arrived packet k in queue i, and let $F_{i,0} = 0$ for all queues (i). Now, packets will be served according to the increased value of $F_{i,k}$ (if several packets will have the same finish time, they will be picked arbitrarily).[6]

---

[5] Here we refer to PFQ and PGPS as a general class of schedulers that tries to "packetize" the fluid-based GPS. PFQ and PGPS sometimes refer to a specific scheme of this class of schedulers (as described in the following subsection).

[6] In a busy server, when all its queues are nonempty always, all packets are eligible; hence, it is right to use SFF policy.

The end result will be that the server will be fully employed, and packets will be served packet-by-packet at the server rate r. Moreover, each packet will terminate its service time at $F_{i,k}$, as if it were receiving the GS rate of the queue it belongs to (r_i), exactly like in the GPS case.

In the real world, however, queues are not always busy. In order to prevent the server from being nonproductive (if the service rate of the empty queue will be unused), the other queues get increased service rates, at the expense of the empty queues. On the average, and for long enough periods, all the queues will receive the reserved service rate. The virtual finish time ($F_{i,k}$) can now be calculated by using the system's virtual time $V(t)$, which takes into account these increased service times by being a representation of the progress of the work done by the system:

$$F_{i,k} = \max \{F_{i,k-1}, V(a_{i,k})\} + \frac{L_{i,k}}{r_i}$$

where $a_{i,k}$ is the arrival time of packet k into queue i, and $V(a_{i,k})$ adjusts the work done by the server for the nonempty queues, while queue i was waiting for this packet k.

The following subsections describe various algorithms that compute the system's virtual time in various ways, as well as using the SFF or the FESS policies.

Weighted Fair Queueing (WFQ) is a PFQ scheduler [348] that uses the SFF policy described in the previous PFQ subsection. WFQ uses a system's virtual time function that represents the progress of the work done by the server, defined by:

$$V(0) = 0$$

$$V\left(t_{j-1} + \tau\right) = V\left(t_{j-1}\right) + \tau \frac{r}{\sum_{i \in B_j} r_i}$$

where B_j is the set of queues that are nonempty in the interval (t_{j-1}, t_j), t_i is the time when a packet arrives or departs the system, and $t_{j-1} \leq \tau \leq t_j$. The rate of the progress achieved by the server is $\partial V(t)/\partial \tau = r/\sum_{i \in B_j} r_i$, and since $\sum_i r_i \leq r$, this rate is always bigger than or equal to 1 (i.e., when there are empty queues, the system's virtual time runs faster than time τ). In the corresponding GPS scheduler, the non-empty queue receives more service, and more progress can be achieved in this queue (at the expense of the empty queues); thus, the virtual time of the entire system seems to fly more quickly. Moreover, the reader can verify that the service granted to each nonempty queue i at time t is the rate of progress achieved by the server, multiplied by the relative weight of this queue, that is, $g_i(t) = r_i \, r/\sum_{j \in B(t)} r_j$, as defined in the GPS schedule.

The resulting finish virtual time is: $F_{i,k} = \max\{F_{i,k-1}, V(a_{i,k})\} + L_{i,k}/r_i$. The example shown in Figure 6.23 clarifies this important scheduler.

A packet can depart the WFQ server at a maximum delay of the maximal packet length served (L_{max}) divided by the server's service rate (r), relative to the equivalent GPS server [348]; that is, $d\_i,k^{WFQ} - d\_i,k^{GPS} \leq L_{max}/r$, where d_i,k^{WFQ} and $d_{i,k}^{GPS}$ are the departure times of the kth packet on queue i under WFQ and GPS, respectively. However, a packet can depart the WFQ server before the departure

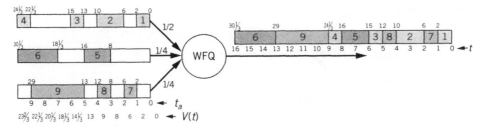

FIGURE 6.23

WFQ scheduler

time in an equivalent GPS server. This means that the worst-case packet fair of WFQ can be more than a packet, and is found to be (see [50]) equal or greater than $((N-1)/2)\, L_{max}/r = O(N)$, where N is number of queues.

Weighted Fair Queueing scheduler has a complexity of $O(N)$, where N is the maximum number of nonempty queues, due to the required calculation at each packet arrival or departure (the system's virtual time and the set B_j). Other PFQ algorithms have been developed in order to reduce this complexity, since this is a major issue in implementing WFQ. Another alternative is, of course, to use network processors.

Virtual Clock (VC) is also a PFQ scheduler [442], and it also uses the SFF policy described in the PFQ subsection above. VC tries to ease the complexity involved in calculating and maintaining the system's virtual time for each packet arrival or departure. VC uses real time to approximate the system's virtual time function, as defined by:

$$V(t) = t$$

so,

$$F_{i,k} = \max\{F_{i,k-1}, a_{i,k}\} + L_{i,k}/r_i$$

Since real time is always less than or equal to the system's virtual time of WFQ, a VC scheduler will pick packets arriving to empty queues more quickly than WFQ will pick them. This is due to the gap created between the virtual finish time of the working queues (and the time stamps their packets get), and the slower real time that will be assigned to the long-idled queues when they finally get packets. As a result, and because of ignoring the work progress done by the server, queues that have more "busy periods" may receive poorer service from the VC scheduler, even if those queues did not overburden the server at the expense of the other queues. In other words, packets from one queue that use the server when there are no packets in the other queues will be stalled when packets arrive to the other queues; these later packets will get unfair priority before the packets of the first queue. This is shown in the example in Figure 6.24.

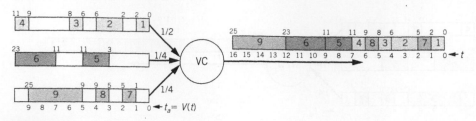

FIGURE 6.24

Virtual Clock

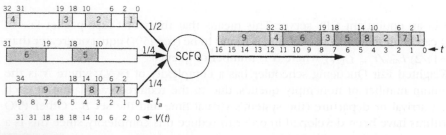

FIGURE 6.25

Self-Clocked Fair Queueing

The normalized WFI of a VC algorithm can be arbitrarily large, even in a simple two queue case [348, 397]. The complexity of VC, however, is relatively small, $O(\log N)$, where N is the maximal number of nonempty queues.

Self-Clocked Fair Queueing (SCFQ) is another PFQ scheduler [151] that uses the SFF policy. It also simplifies the maintenance of the system's virtual time by assigning the finish time of every processed packet to the system's virtual time. The most important aspect of SCFQ is that it is better than a VC scheduler in terms of fairness because it keeps the system's virtual time closer to the work progress done (the virtual finish time of the processed packets). It avoids the time gap in the case of a long-idled queue that receives packets and time stamps based on an "old" system's virtual time that makes them prioritized. This is shown in the example in Figure 6.25.

While this scheduler is simpler than most PFQ schedulers, it still suffers from larger delay bounds than WFQ [440], although its WFI is similar to WFQ.

Worst-Case Fair Weighted Fair Queueing (W²FQ) As was explained for the WFQ, the maximum delay between the times a WFQ scheduler will serve a packet and the time a GPS will serve it is equal to the processing time of the largest packet. However, a packet can finish service in a WFQ system much *earlier* than in an equivalent GPS one, and create an unfair situation for some of the packets (see for example in [50]). W²FQ implements the SEFF policy and is a shaper scheduler subcategory of PFQ.

As in the WFQ case, a packet can depart the W²FQ server at a maximum delay of the maximal packet length served (L_{max}) divided by the server's service rate (r), relative to the equivalent GPS server [50]; that is, $d_{i,k}^{W^2FQ} - d_{i,k}^{GPS} \leq L_{max}/r$, where $d_{i,k}^{W^2FQ}$ and $d_{i,k}^{GPS}$ are the departure time of the kth packet on queue i under W²FQ and GPS, respectively. However, the SEFF policy reduces the time difference between packet departures in the W²FQ server and the equivalent GPS server. This means that the WFI of W²FQ is better than that of WFQ, and found to be [50] equal to L_{max}/r. The WF²Q algorithm has the same WFQ complexity ($O(N)$), and is also not easy to implement.

Starting Potential-based Fair Queue, WF²Q+ Despite the name, WF²Q+ [49] is very different from WF²Q, and the common things they share are: first, the fact that both are PFQ algorithms; second, the fact that both are trying to improve the WFI; and third, that they both implement SEFF policy, albeit in a very different way. Starting Potential-based Fair Queue (SPFQ) [397] was suggested in parallel with WF²Q, and proposes the very same algorithm.

WF²Q+ and SPFQ offer low implementation complexity, and, at the same time, have the same bounded delay of a packet compared to the GPS scheduler and the WFI of the WF²Q. The system's virtual time in these algorithms is defined as:

$$V(t + \tau) = \max(V(t) + \tau, \min_{i \in B(t)}(S_i^{b_i(t)}))$$

where $b_i(t)$ is the sequence number of the HOL packet in queue i, $S_i^{b_i(t)}$ is the virtual start time of that packet, and $B(t)$ is the set of nonempty queues at time t.

This definition of the system's virtual time guarantees that it is equal to or bigger than the minimum of all virtual start times of all HOL packets. Any arriving packet will have a virtual starting time at least equal to one of the queued packets. This definition also makes sure that at least one packet in the system has a virtual starting time that is lower than the system's current virtual time, thus making this packet eligible to be served and enabling the SEFF policy to be implemented and to be work-conserving.

To decrease implementation complexity, the start and finish virtual times are redefined. Instead of calculating start and finish virtual times for each packet, a pair of start (S_i) and finish (F_i) virtual times is calculated for each queue i, whenever a packet k arrives to the HOL according to:

$$S_i = \begin{cases} F_i & \textit{if packet k was in a nonempty queue} \\ \max\{F_i, V(t)\} & \textit{if packet arrived to empty queue} \end{cases}$$

$$F_i = S_i + \frac{L_i^k}{r_i}$$

where L_i^k is the length of packet k in queue i (the one that became the end-of-line packet). Maintaining the set of eligible queues sorted by their virtual finish time and the system's virtual time can be accomplished with $O(\log N)$ complexity [49].

6.4.5.3 Packet Scheduling Summary

A taxonomy of service disciplines was proposed [440] that summarizes the different kinds of schedulers. According to this taxonomy, there are two dimensions by which schedulers can be categorized: one axis is the work conserving property (either work conserving or nonwork conserving), and the other axis is whether the server contains a rate controller or not, in addition to a scheduling scheme.

Many packet schedulers were described in this subsection, covering two main classes of schedulers: simple schedule schemes, and more complicated schemes that aim to provide fairness by trying to approximate the GPS "classical" algorithm for continuous, nonquantified, resource sharing. Although some of these schemes might look complicated, and the number and the variety of them might look redundant, all of them are used by many applications, depending on the network condition and the requirements. Many address TCP optimization, where packet loss (of particular flow) might cause performance issues; others address end-to-end delay and jitter, and so forth.

In packet flow handling, these complicated algorithms can be addressed by using the processing power of the network equipment (this method is difficult and not flexible), or by using traffic managers. These traffic managers can be either part of the network processors (or other processing elements), or network processors already implement traffic management as an integral part of them. These traffic managers can cope with the network and application requirements in a transparent way, although they differ according to the algorithms they implement.

6.5 SUMMARY

This chapter is the second to discuss the theory behind network processors. Network constraints and packet flow handling mechanisms are described, in order to understand how network devices should treat packets and flows, to achieve QoS differentiation.

The chapter first discusses service differentiation definitions and descriptions, and then it describes the QoS mechanisms in more detail. These mechanisms are also subdivided into three groups, according to their roles and place in the network devices: *admission control, traffic management*, and *packet queueing* (which itself is subdivided into *buffer management* functions and *packet scheduling* functions).

The next chapter continues the theoretical discussion, but it comes closer to the network processors themselves, and discusses possible architectures for implementing these devices.

Architecture

In the previous two chapters, we described what each packet goes through in networks, networking equipment, and chips. In this chapter, we describe the basic architectures and definitions of network processors that process and handle those packets in the network. The main idea in this chapter (and in network processors) is that when it comes to high-speed networking environments, a single processing unit is not sufficient for carrying out the processing tasks of packets, and various ways of parallel processing and multiprocessor architectures must be considered.

This chapter describes various computation schemes, as well as network processing architecture in general. Beside the processing element, other architectural components (e.g., input/output [I/O] and memories) as well as interface standards are described at the end of this chapter in order to provide a comprehensive understanding of network processors design and interface, both at the system level (board and equipment), and at the networking level.

This chapter also outlines advantages and drawbacks of programmable and Application Specific Integrated Circuit (ASIC) devices, which are used for networking devices, and discusses the trends and technologies that are available for designing network processing devices. It is important to note, however, that as technology moves quickly, these trends, available technologies, and tools are rapidly changing, and will inevitably offer new ways to solve the networking challenges. The next chapter provides a description of the implications of multi-processing for programming (what is called "*programming models*").

The knowledgeable reader who knows basic computer architectures can skip the background section (Section 7.2). On the other hand, those who require more information than is provided in this background section are encouraged to turn to other references such as [104, 171] or [94].

7.1 INTRODUCTION

Network processors can be categorized according to their use and the way they have evolved. These categories can significantly impact the architecture of the network processors, or sometimes, inversely, are characterized *by* the architecture. The three main categories of network processors are:

Entry-level network processors, or access network processors, which process streams of up to 1 to 2 Gbps packets, and are sometimes used for enterprise equipment. Applications for such access network processors include telephony systems (e.g., voice gateways and soft switches), xDSL[1] access, cable modems, wireless access networks such as cellular and WiMAX, other narrowband and broadband wireless networks, and optical networks (in the access network, that is, FTTx,[2] PONs, etc.). A few examples of such network processors are EZchip's NPA, Wintegra's WinPath, Agere, PMC sierra, and Intel's IXP2300.

Mid-level network processors (2–5 Gbps) contain two subgroups of network processors (NPs): legacy NPs and multiservice NPs, which usually are used for service cards of communication equipment, data center equipment and Layer 7 applications (security, storage, compression, etc.). In the legacy subgroup, one can include the classical, multipurpose NPs like AMCC, Intel's IXP 1200 and 2400/2800 NPs, C-port, Agere, Vitesse, and IBM's NP (which was sold to Hifn). Examples of multiservice and application (Layer 7) network processors are Cavium, RMI, Broadcom, Silverback, and Chelsio.

High-end network processors (10–100 Gbps) are used mainly for core networking and metro networking, usually on the line cards of the equipment. Examples of such network processors are EZchip's NPs, Xelerated, Sandburst (who was bought by Broadcom), and Bay Microsystems.

Each of these categories was either designed for specific applications and thus has specific architecture patterns that support these applications (e.g., matched accelerating functional units), or was designed using almost general-purpose processing elements in their cores. Legacy network processors, designed around 2000, are an example of the latter case, and include the three major processor manufacturers—Intel, IBM, and Motorola.

The architecture of network processors can be described in many ways, and the three ingredients common to computers (processing unit, memory, and I/O) are not sufficient to comprehend it fully. An extended general framework for classifying network processors was suggested in [381], which includes the following five dimensions.

---

[1] A reminder from Chapter 4: Digital Subscriber Loop (DSL) interfaces include many variants for speed and range, such as Asymmetric DSL (ADSL), High speed DSL (HDSL), Very high speed DSL (VDSL), and more. As a group, these are commonly notated as xDSL.
[2] A reminder from Chapter 4: FTTx refers to Fiber to the Home (FTTH), or to the Zone (FTTZ), or to the Curb (FTTC), or any place in the access area.

- Parallel processing approach
  - processing element level
  - instruction-set level
  - bit level
- Hardware assistance
  - coprocessors
  - functional units
- Memory architectures
- Network processor interconnection mechanisms (i.e., on-chip communications)
- Peripherals

This chapter, as well as the rest of the book's chapters, is based on the preceding dimensions. For network processors, the first point—parallel processing—is crucial. We saw in previous chapters how many tasks have to be achieved during the packet processing period. It means that much work—analysis, decisions, modifications and data movements—has to be executed within nanoseconds. A single "conventional" processor cannot meet these requirements. Even when considering Moore's law[3] and the clear trend of dramatic increase in processor power over the years (a trend that is not likely to change), networking requirements are increasing even faster. Therefore, the necessity of basing networking equipment on network processors that use multi-processing or parallel architectures is clear.

The next section, therefore, provides a background and definitions for computing models, mainly parallel processing. Sections 7.3 through 7.5 describe the processing level: first, the general question of hardwired processing versus programmable devices; second, parallel processing topologies; and finally, the instruction set architecture (ISA) level. The chapter ends with a detailed description of network processors' components—memories, internal busses, and external interfaces.

7.2 BACKGROUND AND DEFINITIONS

This section provides a very basic introduction to processor architectures, dealing mostly with parallel processing and multiprocessors. Microarchitecture (which goes into the processing element level) is hardly treated here, as it is extraneous to an understanding of network processors' architecture for those who need to evaluate the various alternatives.

---

[3]Moore's law is an interpretation of the 1965 statement of Gordon Moore of Fairchild and a founder of Intel, which relates to the doubling of the number of transistors on a chip, or the chip performance, every 18 to 24 months.

7.2.1 Computation Models

Before we discuss the various options of network processor architectures, some very basic definitions of processing architectures are required. The common architecture we use for General-Purpose Processors (GPPs) is the *control flow* (or *instruction flow*) architecture, also called Von-Neumann architecture. According to this architecture, instructions are fetched sequentially from the memory to the Central Processing Unit (CPU), in which they are executed, followed by some loads and stores of data from and to the memory.

There are also some parallel, multiprocessing architectures that are based on the basic multiple Von-Neumann architecture, as described next. Another architecture, the *data-flow*, is an exotic architecture that seems to fit network processing requirements and sometimes is used for network processors, but it is usually used for marketing purposes more than as an accurate architecture description. According to this architecture, flow of processing is determined by data availability; in other words, this is a data-driven model of computation where processing is enabled only when data for all input operands become available by predecessor processing. Packet and network processing can be seen as having a dataflow nature; thus, dataflow architecture can fit network processing.

7.2.2 Multiprocessing—Flynn Typology

Using parallelism is an obvious way to achieve speedup when it is required, and multiprocessing is used to achieve parallelism. Network processors use many forms of multiprocessing and parallelism. In order to cover the parallelism issue systematically, we will use Flynn's terminology and taxonomy [131, 132]. This taxonomy is based on two axes that are simultaneously allowed in the architecture: the number of instruction (or control) streams and the number of data streams (see Table 7.1). The term *stream* refers to a sequence of either instructions or data that the processor uses for operation. The four basic types of architectures are as follows:

1. SISD architecture (Figure 7.1), in which a single instruction stream controls a single data stream—that is, classical Von Neumann architecture in which instructions are fetched from the memory to the CPU and are executed. Some of these instructions may load and store data from and to the memory.

Table 7.1 Flynn Taxonomy

	Single instruction stream	Multiple instruction stream
Single data stream	SISD	MISD
Multiple data stream	SIMD	MIMD

2. SIMD architecture (Figure 7.2), in which a single instruction stream controls multiple data streams. This architecture includes vector and array processors. The idea is that one control unit manages multiple processing elements (arithmetic function units, for example), simultaneously carrying many data operations from and to the memory.

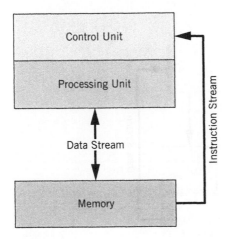

FIGURE 7.1

SISD architecture

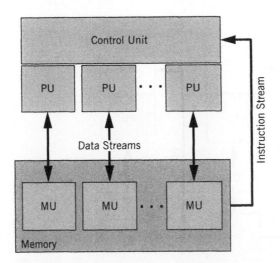

FIGURE 7.2

SIMD architecture

3. MISD architecture (Figure 7.3), in which multiple instruction streams control a single data stream. Although commonly recognized as an experimental, nonpractical or exotic category, some references include systolic arrays, pipelined vector machines, and data-flow architectures in this category.

4. MIMD architecture (Figure 7.4), in which multiple instruction streams control multiple data streams. Most multiprocessor systems and multicomputer systems use this architectural category.

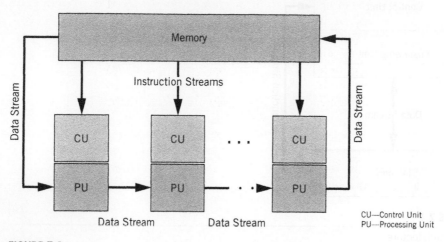

FIGURE 7.3

MISD architecture

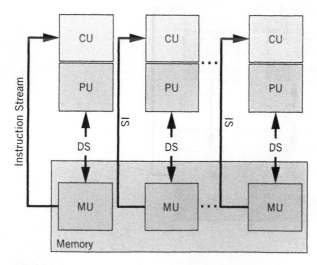

FIGURE 7.4

MIMD architecture

The MIMD category is subdivided into two major classes:

- *Shared memory multiprocessors*—architectures that share a common memory (see below) and are said to be tightly coupled.
- *Distributed memory* (or *message passing*) *multicomputers*—architectures that use distributed memory between the processing modules, and are said to be loosely coupled.

Shared memory can be implemented in various ways, although usually it is done according to either the Uniform Memory Access (UMA), the Nonuniform Memory Access (NUMA), or the Cache Only Memory Architecture (COMA) models. Under the UMA model, the physical memory is shared equally and uniformly by all the processors (although the processors can possess local cache memories). Tightly coupled architecture usually refers to UMA implementations. NUMA implies the opposite; it means that some processors share their local memory, and the collection of all local memories create the global memory. It also means that access to memory is not equally shared (local processors have the advantage). A subclass of NUMA is COMA, which means that the collection of all local caches creates the global memory. Some variants of these models are in use, like the Cache Coherent NUMA (CC-NUMA) or Cache Coherent UMA (CC-UMA) which means that all processors are aware of any update of any memory by any processor.

Multicomputer systems, usually loosely coupled MIMD architectures and sometimes called Multiple SISD (M-SISD), use distributed or private memory, and include Massively parallel processors that are composed of grids and hypercubes, and Clusters of Workstations or Network of Workstations that are beyond our scope.

Multiprocessor systems can be further subdivided according to various other criteria, but the most notable subcategory is the Symmetrical Multiprocessors (SMP). The SMP is a tightly coupled architecture of similar processors that have the same priority and share the same memory. Another subcategory is the asymmetric processors and coprocessing, which have a tightly coupled architecture of specific, dissimilar processors, and that usually share the same memory. These subcategories are further detailed in the next subsections.

Other categories were defined after Flynn's taxonomy, among them being the single program over multiple data streams (SPMD), and the multi-program over multiple data streams (MPMD). Some position SPMD between MIMD and SIMD architectures, since SPMD can imply similar programs running on several computing nodes. These categories actually refer more to parallel granularity, as described later, and more to computing style or programming models (detailed in the next chapter) than architectures.

Modern implementations of MIMD are hybrid with respect to shared memory and distributed memory, that is, there are several distributed nodes of multiprocessors sharing common memory, as shown in Figure 7.5. This is a very common practice when it comes to some network processor architecture implementations, as described in the following.

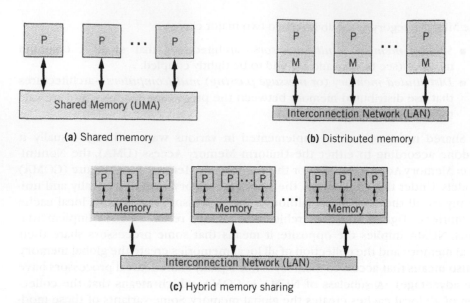

(a) Shared memory

(b) Distributed memory

(c) Hybrid memory sharing

FIGURE 7.5

MIMD memory usage

7.2.3 Multiprocessing—Practical Implementations

There are many other issues in multiprocessing that are not covered by the plain Flynn typology (or implied by this typology), and which are very much relevant to network processing. This subsection relates to the main considerations in the implementation of *chip multiprocessors* (CMP, which are chip-level multi-processors), *multicore* processors (MCP, or multiple processors cores that share caches and the chip interconnect network) or multiprocessors systems on a chip (MPSoC).

We start with some definitions of terms and descriptions of parallelism levels, continue with a discussion of basic symmetrical and asymmetrical configurations as well as of types of the processing elements themselves, and conclude with an observation summary.

7.2.3.1 Some Basic Definitions

A *scalar processor* is the simplest form of CPU, which is SISD. It can consist of many internal functional units, for example, an Arithmetic Logic Unit (ALU), or it can be assisted by external hardware assists or coprocessors that are attached to the CPU. Coprocessors execute their instructions after they are triggered or dispatched from the CPU, and work asynchronously with the CPU. These coprocessors can be *vector processors* (i.e., SIMD architectures) for various vector arithmetic, imaging or signal processing or other applications (like security encoding/decoding, pattern and key searching, and so on, in the case of network processors). Coprocessors usually cannot use I/O (unless they are I/O coprocessors), and are restricted to using the main memory (i.e., they have to be synchronized with the CPU).

When dealing with processor performance, some metrics are used. The commonly used terms are outlined here:

- The number of instructions issued (i.e., commence execution) in one cycle is called the *issue width* (IW), and in scalar processors is always 1. IW is called also *instruction issue rate*, or the *degree* of the processor.
- The number of executing *instructions per cycle* (IPC), and its reciprocal—*cycles per instruction* (CPI)—measures the complexity of the instruction set and the parallelism in its execution.
- Performance is correlated to the IPC multiplied by the clock rate of the processor, and today is proportional to $IW^{0.5}$.

To understand what multiprocessors can parallelize, a brief description of the "processed material" units might be required.

At the highest level, processors execute *programs*, or actually instances of the programs (called *processes*) that reside in the processors' memories. Parts of programs, for example, subprograms, procedures, subroutines, functions or methods can be further identified, and are relatively independent parts of the larger program, performing some specific task. The processes can be partitioned into multiple interrelated *tasks*, which are sets of instructions residing in the processor's memory (sometimes, "tasks" imply real time processes). *Threads* are lightweight streams of processes (or tasks); they are dependent splits that run simultaneously with other threads, and are handled differently than processes and tasks that are run by the processors' operating system. There are several ways for threads to be scheduled on a processor (i.e., to create a *context switching*,[4] deciding how and when to switch between them), but the most efficient way is the *Simultaneous Multithreading* (SMT).[5] SMT refers to the processor's or the multiprocessor's execution of instructions from a number of threads simultaneously (in Intel's case, it is called *hyper-threading* rather than multithreading). The lowest level of the "processed material" is the instruction level. A parallelism level, based on these levels of "processing material," is described in the next subsection.

---

[4]Context switching usually refers to process switching, changing one context to another, and it means storing and loading the states of the processor in the current and the switched processes.

[5]Context switching in threads can be done at every cycle, which is called *fine grained multithreading* (FGMT) or *interleaving multithreading*, and the processor that uses it is called a *barrel processor*. This guarantees real time properties, but is not efficient in cache utilization and single thread performance. The other extreme is *coarse grained multithreading* (CGMT), which lets a thread run until some stall occurs that yields a context switching.

7.2.3.2 Parallelism Levels

There are three basic mechanisms that are relevant to implementations of parallelism in processing elements: *Instruction level parallelism* (ILP), *thread-level parallelism* (TLP), and *data-level parallelism* (DLP). ILP means that more than one instruction is executing at the very same time, in various technologies, as described briefly below. TLP implies multiple threads that are running simultaneously or pseudo-simultaneously, as described in the previous subsection. MCP exploit TLP techniques, and achieve high performance in a simpler way than ILP. Obviously, this is efficient when the applications are threaded, or can be threaded. DLP means that the same instruction is repeatedly used for multiple data instances, for example, loops. DLP is less common and less general than ILP or even TLP. DLP works for very specific applications that can benefit from its implementation, for example, imaging or scientific applications, where SIMD (vector and array) architectures are efficient for these applications.

Fine grained parallelism usually refers to ILP in a single processing element, and is achieved mainly in *superscalar*, *pipelining*, *Very Long Instruction Word* (VLIW), and *Explicitly Parallel Instruction Computing* (EPIC), as described in the next subsections. *Medium grained parallelism* refers to multithreading and multitasking levels and usually refers to TLP, and *coarse grained parallelism* refers to the program level parallelism. TLP, medium-, and coarse-grained parallelisms are achieved mainly in multiprocessors and multicomputers. It is worth noting though, that SMT actually brings the TLP to the ILP, as several instructions (from different threads) are running simultaneously.

The finer the grain of the parallelism, the higher will be the level of parallelism achieved, but it is by far more complex in terms of the mechanisms, communication, scheduling, and overhead circuitry requirements.

7.2.3.3 Symmetric Parallelism

Symmetric multiprocessing, as described before, refers to systems in which their processors have equal access rights to the systems' components, memory, peripherals, and so on. Usually, in SMP we mean also multiprocessors of the same kind, and furthermore, these processors are general-purpose processors. Most traditional parallel processing is done in symmetric parallelism, that is, using SMP, and there are many network processors (or their core processing units) that are basically composed of symmetric parallel processing units. The emerging CMP (or the MCP) are mainly TLP engines that obviously also work in symmetrical parallelism. The main advantage of SMP is that it is easy to write code, since all processors run the same instruction set. The main drawback is that synchronization must be accounted for, meaning adding specific means and instructions to handle resource sharing in a proper way (mainly memory).

Symmetric multiprocessing can be done by tightly coupled architecture, sharing one memory, or it can come in a loosely coupled architecture. In both cases synchronization has to be watched, but it can be done on various levels. Sharing a memory requires strict synchronization, but allows more parallelism on the same data by separate threads. Several programming models are described in the next chapter.

7.2.3.4 Asymmetric Parallelism, Functional Units and Coprocessors

Asymmetrical or heterogeneous processors are usually found when there are very different processing needs in various stages of the application, for example, the use of a graphic or video processor in a personal computer or a gaming platform, side by side with the GPP. The purpose of asymmetric processing is to increase performance by matching application requirements, or parts of the applications, to a suitable processor. Such a processor should be architecturally optimized to handle the tasks that the applications require, for example, number crunching, vector processing, signal processing, video raster graphics, and so on.

In network processing, where distinct tasks can be defined during packet processing, heterogeneous processing can contribute very much to increasing processing performance. Heterogeneity can be expressed in many ways; for example:

- Very dissimilar parallel processors, each responsible for a part of the packet processing phases, which have different architectures, instruction sets, and programs;
- Additional, supportive coprocessors, executing internal procedures, written in their own instruction sets, which are used for compression, security, look-ups, and so on;
- Various functional units that are attached to the processors and are triggered by special purpose instructions (that are part of the processors' instruction sets, in contrast to the coprocessor case), and that are executed by the processors. These functional units are considered as part of the processors as far as programming is concerned.

7.2.3.5 Processor Families

Network processors use the entire range of major processor families: Complex Instruction Set Computing (CISC), Reduced Instruction Set Computing (RISC), pipeline, superscalar, systolic, VLIW, data-flow, and combinations of the above. In this subsection, a short description of each of the main processor families is described.

7.2.3.5.1 Complex Instruction Set Computing

Complex Instruction Set Computing is based on an ISA that allows multiple and complex addressing modes for hundreds of complex instructions, using variable instruction formats, and usually executed by a microprogram (low-level operations) control in the processor. The term was coined only after RISC computing was introduced, to distinguish between the two families of architectures. CISC was originally designed to allow programmers and compilers to produce small-sized programs, using many options of coding. The CPI for CISC is dependant on the instruction,

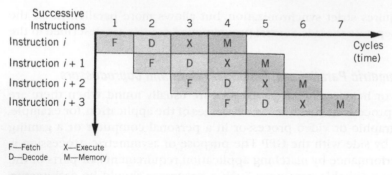

F—Fetch   X—Execute
D—Decode   M—Memory

FIGURE 7.6

Instruction pipelining

and is very long compared to RISC ISA (up to 20) since it involves lots of decoding and executing cycles. It has proved to be inefficient, since most of the complexities are not even used in regular coding, but the complexity of the processor microarchitecture nevertheless remains high.

7.2.3.5.2 Reduced Instruction Set Computing

In RISC, lots of the complexities of instructions and addressing modes were abandoned for the sake of simplicity, higher processor clock, and much smaller CPI (up to 4). RISC instructions are usually executed by *hardwired control*, in a fixed number of cycles and fixed instruction format for all instructions, which allows simple architectural improvements, like superscalars and so on (described later).

7.2.3.5.3 Pipelining

Pipelining is a mechanism that allows ILP by overlapping processing stages of different instructions (usually sequential). If we assume an instruction that has four stages (i.e., Fetch, Decode, Execute, and Memory), then in a four-stage pipeline, when the first instruction enters its second phase (decode), the successive instruction enters its first phase (Fetch), and so forth, as can be seen in Figure 7.6.

The number of stages in an instruction pipeline can be as small as a few in some architectures, to as many as tens of stages in others. The advantage in long (or deep) pipelines is that a higher ILP, or IPC, can be achieved, but one problem lies in overhead. An even bigger drawback to long pipelines is the potential for resource conflicts and stalls in the pipeline, which can happen when specific instruction sequences or instances that are dependent cause some parts of the pipeline to stop functioning and to enter a wait state for several cycles. These stalls can be relieved by *out-of-order* instruction issuing,[6] or other avoidance techniques.

---

[6]Instructions that are issued not according to their order in the program (what is called *dynamic scheduling*). It is possible when the instructions are totally independent (they use independent data and have no impact on the flow of the program). This is in contrast to *in-order* instruction issuing (or *static scheduling*).

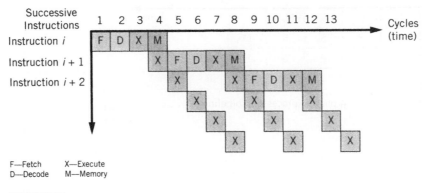

FIGURE 7.7

Vector processor pipelining

Also, every branching instruction causes the pipeline to be flushed and to restart in the new program location, unless a branch prediction or other mechanisms are used to alleviate this flushing inefficiency. The end result is decreasing the IPC. In pipelining, these instruction and data dependencies are called *hazards*. These hazards can be categorized as follows:

- Structural hazards that occur when two instructions are about to use the same structure (unit).
- Data hazards that are caused when two instructions are trying to use the same memory location.
- Control hazards, when an instruction affects whether another instruction will be executed.

Unless the hardware is handling these hazards, it is the programmer's (or the compiler's) concern. Handling hazards is described in the next chapter, when dealing with programming models. Without any stalls, the speedup[7] of k-stage pipelining is k.

Finally, pipelining can be used not just for ILP. It can be used conceptually for processor pipelining for processes and tasks. Vector processors are an example of pipelining arithmetic functional units that compute one vector operation, as depicted in Figure 7.7.

We can categorize the aforementioned pipelining concept by various levels of pipelining, ranging from (a) instruction pipelining within a single processor, to (b) processor pipelining in a multiprocessor system, a vector processor, or an array processor, to (c) program or task pipelining, which divides the programs or tasks between various stages, as can be seen in Figure 7.8.

---

[7]Speedup is the ratio of the time to execute N instructions on a simple implementation of a scalar processor and the time to execute these N instructions on the processor in hand (in this case, a k-stage pipeline).

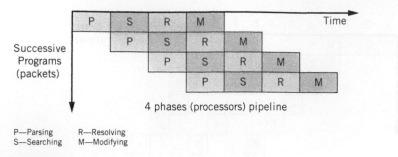

P—Parsing   R—Resolving
S—Searching M—Modifying

FIGURE 7.8

Program pipelining

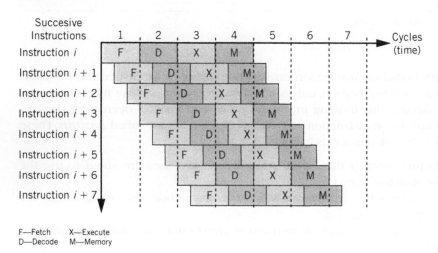

F—Fetch   X—Execute
D—Decode M—Memory

FIGURE 7.9

Super pipeline of degree 3

There are several variants of pipelining (linear and nonlinear, synchronous and asynchronous, etc.), that are beyond the scope of this book, but the most important one is the super-pipelining. In super-pipelining of degree n, the k-stage pipeline starts a cycle every $1/n$ of the pipeline cycle, as shown in Figure 7.9.

It means that an instruction runs in the pipeline for k cycles, as in the regular pipeline, but in each cycle there are n pipelines commencing a stage, in a time shift equal to $1/n$. This produce n times more instructions in the super-pipeline at any stage, and ILP equal to n. The total speedup of k-stage super-pipeline of degree n, is therefore kn.

7.2.3.5.4 Superscalar

Superscalar is a mechanism that allows ILP by issuing and running multiple instructions at each stage simultaneously (see Figures 7.10 and 7.11). Superscalar is said

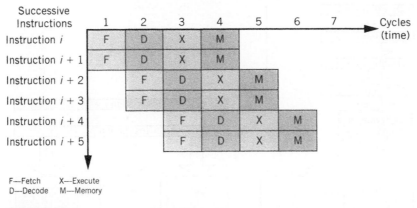

F—Fetch   X—Execute
D—Decode   M—Memory

FIGURE 7.10

Superscalar of degree 2

to be of degree m, when m instructions are issued at the same time to m k-stage pipelines (or, in other words, the issue width is m). The speedup of superscalar of degree m compared to scalar k-stage pipeline is m (or mk compared to nonpipeline scalar processor).

When superscalar implementations use super-pipelining techniques, the end result is k-stage, m-degree superscalar, n-degree super-pipeline, which yields very high ILP (and a speedup of kmn).

In practice, however, due to cache misses, data dependencies, and mispredicted branches,[8] superscalars are not fully utilized, and it is quite common to achieve only 50% of these theoretical speedups. Multiplexing threads on the various pipelines (in multithreading techniques) improves these speedups, due to the reduced potential of data conflicts.

7.2.3.5.5 Very Long Instruction Word

Another method of achieving parallelism is the VLIW architecture, where more functional units are usually employed than the number used in superscalar architecture. VLIW is based on "horizontal *microcoding*," and thus uses *microprogrammed control* for its instructions that can be hundreds of bits long (in contrast to common RISC architectures, which use hardwired control). The VLIW instruction groups several instructions for each of the functional units in the processor, that is, for the ALU, memory operations, branching instructions, and so on. In other words, the VLIW instruction is quite fixed in its format, and typically is

---

[8] As mentioned before, pipelining inefficiency can result from pipeline flushes caused by branch instructions. One mechanism that reduces this inefficiency is to feed the pipeline with instructions that are fetched from the predicted branch location. Mispredicted branches cause pipeline flushes, as is the situation with unpredicted branches.

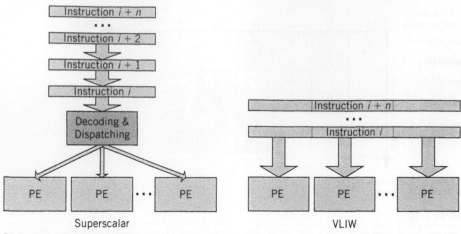

PE—Processing Element (Functional Unit, Execution Unit, Pipeline, etc.)

FIGURE 7.11

VLIW and superscalar architectures

"slotted" such that the first instruction must be ALU instruction, the second one is a movement instruction, and so on.

Very Long Instruction Word architecture is comparable to superscalar architecture, since the two employ several processing units in parallel, albeit in very different ways, as shown in Figure 7.11. VLIW is much simpler to implement in hardware terms, uses static scheduling of instructions and worst case scheduling, that is, it must wait until all instructions are done before issuing the next instruction, in contrast to superscalar, where dynamic scheduling is feasible, as is coping with variable execution latencies of instructions.

Very Long Instruction Word cannot run more than one instruction that relates to a specific functional unit in the same cycle (the same VLIW instruction), that is, there cannot be two ALU instructions in one VLIW instruction. Moreover, different functional units that are employed during one cycle must be used so that no data conflicts may arise. The end result of all this is that sometimes the functional units are not utilized, or extra *nops* (no-operation instructions) are used, wasting memory bandwidth and space.

Conventional programs written with "normal" instructions must be compacted efficiently into the VLIW instructions, without resource conflicts, data or control hazards, and with maximum utilization of the functional units.

It is important to stress that in the *pure VLIW* architecture, the programmer or the compiler must guarantee that instructions within a VLIW group are independent (or slots are filled with "no-operations"). Since it runs many instructions at one time, it has a very low theoretic CPI. VLIW architecture is also very simple to implement in hardware, and higher clock rates are possible as a result.

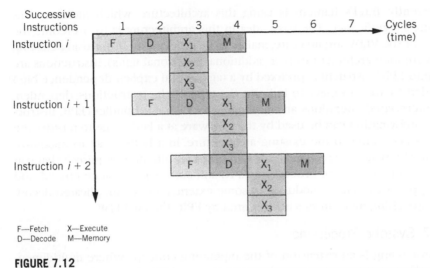

F—Fetch   X—Execute
D—Decode   M—Memory

FIGURE 7.12

VLIW architecture

Current VLIW implementations are based also on superscalar processing, so the end result can be seen in see Figure 7.12.

As can be understood from the description of this architecture, VLIW instructions and architecture force very strict programming and compiling that cannot be used except for the very specific VLIW architecture they were written for. Any modification of the architecture, for example, adding functional units, causes a rewrite of the entire code. This means that pure VLIW is a software centric architecture, and that the programmer, or the compiler, or both, are responsible for parallelism and data movements and dependencies. Dynamic, run-time resource scheduling is impossible in VLIW architecture.

Pure VLIW developed into *tainted VLIW* architectures, which are more realistic. In tainted VLIW, intergroup checks are also done by hardware, and the programmer or the compiler does not schedule stall cycles. Pipelines of varied numbers of stages can be used by the tainted VLIW architecture. This tainted VLIW architecture, however, evolved to what is called Explicitly Parallel Instruction Computing, which is described next.

Very Long Instruction Word architectures were abandoned for decades due to their problematic implementation and usage, and despite their academic soundness. However, since it is now being used again in some network processors, it is described here briefly.

7.2.3.5.6 Explicitly Parallel Instruction Computing

Explicitly Parallel Instruction Computing (EPIC) was derived from VLIW, or at least its roots are from VLIW concepts. It can be regarded as *Variable Length*

VLIW. Currently, Intel's Itanium is using this architecture, which indicates this architecture's significance. The basic idea in this architecture is to overcome the drawbacks of the VLIW architecture, mainly in its inflexibility about using existing code for extended architectures (e.g., additional functional units). Instructions are bundled into EPIC instruction, prefixed by a tag (several explicit dependence bits) that is added by the compiler to indicate the type of the instructions, dependencies and interrelated operations, and dependencies between bundles (EPIC instructions). This information can be used by the hardware at a later stage to reorder the instructions according to the existing architecture, in a better way, to maximize parallelism. The result is a reduced code size (nops are no more required), that can be executed on succeeding processors. EPIC still suffers from cache missed, and since it supports dynamic scheduling to some extent, a more complicated decoding and dispatching mechanisms are required by EPIC than in VLIW.

7.2.3.5.7 Systolic Processing

Systolic processing is an extension of the pipelining concept, where multidimensional pipelines are connected to form a *pipenet* [177] that implements fixed algorithms. The processing is done in the interconnecting processing elements, on multidimensional flows of data streams. The layout of the interconnected network can be fixed or programmable, but at any rate it calls for strict programming and is not used commonly.

7.2.3.5.8 Dataflow Processing

Data-flow processing is an entirely different computing paradigm, as noted at the beginning of this section. Instead of program (or control) flow, processing is dictated according to data and results availability. This data-driven processing is achieved by tagging the data, specifying its "belonging" to instructions, and once all required data for some instruction are ready the instruction is triggered and executed. In this way, *data tokens* (results) are roaming between the instructions, duplicated according to the number of instructions that require them for execution, and vanishing once the instruction "consumes" them. The control-processing is done by a mechanism that matches ready data tokens with starving instructions, in contrast to the Von-Neumann control-processing that runs instructions fetched from memory as pointed to by the program counter.

An example of the data-flow computing paradigm at the application level (rather than the architectural level) is a spreadsheet, where any change of data is reflected in many data items (cells) and causes a recalculation of the entire workspace.

It is very tempting to think of packet processing as a kind of data flow, since packets can flow from one processing element to another, and intermediate processing tasks (searches, for example) are also finished and everything is ready to continue processing in the next processing phase. However, it is very hard to implement data-flow architecture and there are no current implementations of network processors based on data-flow architectures. Nevertheless, there are some marketing efforts to present some network processors as such.

A high degree of parallelism can be achieved in this kind of architecture, since data-flow execution is driven by the availability of the data (equivalent to the operands in the Von-Neumann architecture's instructions), limited by data dependencies and the number of functional units or processing elements.

7.2.4 Basic Chip Design Alternatives

In the previous subsections, computing models, processors, and multiprocessing taxonomy was briefly described, and multiprocessing and parallelism were presented as a means of getting the high performance architecture required for contemporary packet processing. However, there is an alternative option for getting the required performance; that is, with special purpose hardware, hardwired circuitry, or "custom silicon." In this subsection a very short description of this option is described. It is important to understand that eventually, a mixed solution based on both processors and customized silicon circuitries, as well as customized processors, is probably the answer for high-speed networking requirements.

In order to understand the impact and significance of choosing processing elements for network processing equipment, we must first delve into chip design and characterization. In a nutshell, chips are Integrated Circuit semiconductor devices, which began as a few simple logic circuits that were integrated on one piece of silicon (called a "die"), and progressed to Large Scale Integration (LSI) and to Very Large Scale Integration (VLSI) on a die. Application Specific Integrated Circuits (ASIC) is a customized chip for particular use. Complex programmable logic devices and Field Programmable Gate Arrays (FPGA) are programmable logic devices of different mechanisms; both include many programmable logic circuits and programmable interconnections. ASICs are designed and FPGAs are programmed using hardware description languages like VHDL and Verilog, and require later verification, simulation, and many other design processes to ensure their proper functioning before fabrication.

To ease the design of ASICs and FPGAs, there are many hardware libraries of common functions (usually called IP-blocks, or intellectual property blocks) as well as many tools and utilities (commonly referred to as electronic design automation, or EDA). ASICs and FPGAs were used in the 1980s as small to medium "glue-logic" chips (integrating logic circuitry between onboard components), and they evolved into very complex, huge chips that can include several full-scale embedded processors (in the form of IP-blocks). In fact, there are many GPPs and special purpose processors (SPPs) that are designed and implemented using ASICs. A more recent type of customized chip/processor is the configurable processor. A configurable processor is a predefined processor core with design tools that enable adaptation of parts from the processor core with additional instructions that highly customize the processor and its ISA for a specific application. The result is usually an IP-block that is used by FPGA or ASIC developers to create an optimized embedded processor for some application.

Since ASIC is specifically designed and optimized, and its circuits, modules, and interconnect are aimed at a specific task, it is generally faster than FPGA, and faster than GPPs. However, designing ASIC is a very long and costly process, with no flexibility to make changes that may be required during the long designing phase. Moreover, ASIC can almost never be reused in further revisions or generations of the products, which makes it very specialized.

A chip platform is an abstraction used to provide a comprehensive, multilevel perspective of a chip that incorporates its modules, modules interconnection, design methodologies, design tools (EDA), and library of hardware and software common functions (IP-blocks). A platform-oriented approach provides a system perspective, and develops in isolation from the VLSI or logic levels of the chips.

A programmable platform is a chip platform that can be modified, configured or programmed at various times and rates, and at various granularity levels, both in terms of hardware components and in terms of changes. Changes may include the chip's modules (like logic circuits for encoding or ciphering, or ALUs or other functional units for varied algorithms), interconnects and data paths (like cascading or paralleling functional units, modules or processors, for varied algorithms or tasks), or control circuitry (like configurable processors for varied applications or tasks). Changes can be made at many points in time: at the design phase only (e.g., in fixed devices such as ASIC or a configurable processor), at fabrication time or when beginning to use the device (e.g., in configurable devices such as FPGA), between various application periods during the lifetime of the device (in reconfigurable devices), or as the application runs, at every cycle, as in any traditional processor. Like chip platform abstraction, programmable platforms include all the programming languages, tools, models, and so on as part of the abstraction.

During recent years, many technologies for speeding up design and get-to-market have been suggested, and some of them have had significant impact on the architecture of network processors, or network processors-to-be. The most notable trend is the System-on-Chip (SoC), which started in the mid-1990s. SoC is a significant portion of an end application that integrates many diverse functional units (processors, memory, analog and digital circuits, and peripherals), all on one chip, single die, including the software (*firmware*) that runs it. SoC became a cornerstone in embedded systems and, like ASICs that were used for product differentiation at the chip level, SoC is used for product differentiation at the level of the embedded system. Obviously, SoC has economic and performance advantages that allow this differentiation. MPSoC followed SoCs, to enable better performance and functionality by using more processors, whether they be homogenous or heterogeneous [240].

Current network processors are basically MPSoC, and as MPSoC design methodologies continue to improve, we can expect more and better network processors to carry network processing tasks in all parts of the network, starting from the telecom core, through the metro networks, edge networks, enterprise edges and cores, and even in SOHO (small office and home office) equipment.

A short reference to Network on Chip (NoC) [48] is in place here, for two reasons: (a) NoCs are *not* networking chips, that is, they are not the type of chips that replace network processors, switches, or any other networking tasks, and (b) NoC offers a technology of designing chips, particularly SoCs and MPSoCs, that uses structured inter-module interconnection between the chip's subsystems (this is sort of micro-LAN with all required "equipment," that is, switches and routers, and is usually based on packet switching). NoCs allow complex chip designs without being sensitive to signal propagation, clock synchronization, bit-errors, or cross-talks of high-speed parallel busses. They also enable easy next-generation designs and scalability and save a great deal of die space by using serial networking between the chip's subsystems.

7.2.5 Summary

There are various ways to increase performance in network processors; most, if not all, are achieved by using multiprocessing through different techniques. An illustration of this trend can be found in [254, 281]; their findings are summarized in Figure 7.13.

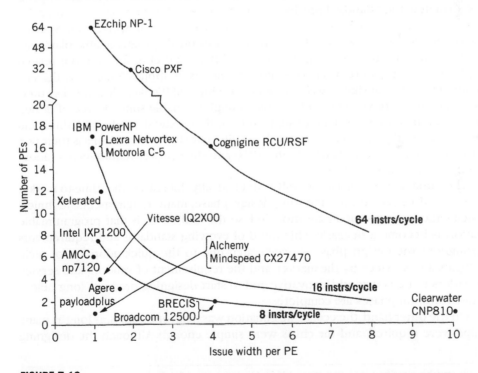

FIGURE 7.13

Trade-offs between number of PEs and issue width [254, 381]. (Courtesy of K. Keutzer et al., ICCD '02, © 2002 IEEE.)

7.3 EQUIPMENT DESIGN ALTERNATIVES: ASICS VERSUS NP

Before turning to network processor architectures, this subsection provides a short discussion of processing element choices, and justifies using network processors in the first place. As it turns out, in terms of application fitness and performance, there are two main alternatives that are used to design and build communication equipment: ASICs or network processors, which we compare briefly here. Although this chapter deals primarily with architectures, some economic considerations are appropriate here, since, at the bottom line, they dictate not only how solutions are adopted, but also how architecture is developed.

ASICs governed product design from the 1980s through the mid-1990s, when other alternatives appeared. The justifications for ASICs were the strong demand for product differentiation, high performance, and inexpensive chip price, using per-application silicon. As ASIC design became tremendously complex, expensive, and long, as described in the preceding section, it became crucial to attempt to reuse the same design for as many chips as could be used for similar applications. Where an ASIC was used by one vendor, Application Specific Standard Products (ASSPs) are ASICs that were developed for sale to other vendors for similar applications (hence, the "standard product"), saving these other vendors the expensive design phase. A revolution came with the development of Application Specific Instruction Set Processors (ASIPs),[9] which are essentially programmable platforms. ASIPs enable the use of a larger quantity of the same chip for multiple related applications and different versions and generations of applications, [254] thereby dividing the huge design costs between many chips. ASIPs also allow quicker time-to-market, they are versatile and flexible enough to enable some degree of design modification during the long design process, they extend product lifetime, and they are generally significantly faster than GPPs. The downside of ASIPs is their performance (several times slower than ASICs) and their power requirements (about ten times more than ASICs).

The long design phase of ASICs dramatically increases the time-to-market pressure. Moreover, during the long design phase, many requirements, technical specifications, or protocols are modified, so the end result is that programmable platforms become a necessity. This trend of evolving standards and requirements throughout the design phase is accelerated due to the quick adoption of technologies and services by the market, and the requirement of vendors and service providers to be constantly innovative and to start designing products long before standardization phases are complete.

On the other hand, the economic equation was in favor of ASICs, as long as many chips were required, and the chips were simple enough. Although the designing

---

[9]ASIPs can be classified into two types [254]: (A) Instruction Set Architecture (ISA) based, originating from classic ISA processors, and (B) Programmable hardware based, originating from programmable logic, like FPGA. This classification is quite fuzzy, as many implementations use both characterizations.

was extremely expensive, the chips were very cheap. This is opposed to the case of network processors, where the designing was (and still is) very cheap, but the network processors' chips were expensive. However, in recent years, due to advances in production, the overall trade-off between these two alternatives is shifting to favor network processors. While ASICs complexity went up, the required die size became smaller and smaller, due to the fact that chips today are based on sub-100 nm technology (or even 65 nm), which also results in a reduction in chip manufacturing costs. However, since both network processors and ASICs must continuously provide increasing throughput, these two types of devices become I/O bounded (or pad limited). In other words, the die sizes become dependent on the number of interfaces and their nature (parallel or serial), as these interfaces may require many hundreds of pins and larger pads area on the die for these I/Os. Thus, the pin count becomes a primary factor in the size of the die size in both types of devices, which brings them to about the same die size, allowing more and more "free" extra logic on the silicon that is very useful for network processors, and eventually the price of the two devices becomes quite similar in production. The difference in designing remains strongly in favor of network processors, and their flexibility of design, even during the design phase, clearly makes the network processor solution a better design alternative.

Recently, FPGA technology, which has traditionally been slower than ASIC, progressed to a stage where it can be considered an option for network processors vendors as a platform for manufacturing network processors, instead of using ASICs (at least for some applications). In the future, communication equipment vendors, may consider using FPGA as an alternative to network processors (surely to ASICs).

It should be noted here, though, that the question of ASICs, FPGAs or network processors is subject to dispute in the industry among practitioners, and it is often just a question of "religion" or taste.

7.4 NETWORK PROCESSORS BASIC ARCHITECTURES

Previous sections detailed the need for multiple processing elements in network processors and basic processing element technologies. This section addresses basic architectures of network processors in terms of how processing elements are arranged in the network processors.

The focus here is on the parallel processing dimension of network processor architectures, specifically the processing element level. One way to classify network processors [152] is by what is called the topologies of the processing elements, that is, looking at the way they are arranged either in parallel pools, in the shape of a pipeline, or in some kind of mixed topology, as can be seen in see Figure 7.14. In the following, a finer resolution of this classification is used, in order to distinguish between the various types of network processors.

As was described in the introduction, there are three categories of network processors: entry-level, or access NPs; mid-level, or legacy and multiservice NPs; and

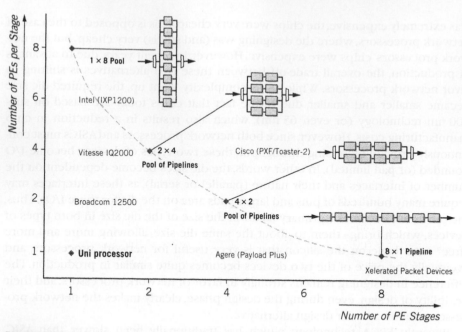

FIGURE 7.14

Design space for PE topologies annotated with existing NPUs (from Gries, Kulkarni, Sauer, & Keutzer [152])

high-end, line-card NPs. It is important to note that, practically, the two extremes—the parallel and pipeline configurations in Figure 7.14—represent two typical applications: parallel configuration is usually used for networking applications of the higher layers (L5–L7, e.g., security, storage, accounting), whereas pipeline configurations are usually used in line cards that require ultra-high processing speed at the lower layers of networking (L2–L4).

7.4.1 Homogenous Parallelized Processors

In this type of topology, a pool of parallel homogenous processing elements (called micro-engines or pico-engines) is used to process packets in one stage; each packet is treated completely by one processing element. Within this type of topology, one can count the "classical," first-generation network processors; for example, those of IBM (which was sold to Hifn), Intel, C-Port (now Freescale). More important, this topology includes implementations of multicores and CMP that target L4–L7 network processing, such as Deep Packet Inspection (DPI) and security in the access and aggregation networks.

Examples include access processors, such as Wintegra's; security and DPI services processors, such as Raza Microelectronics' (RMI); Cavium's, and Cisco's

Quantum Flow Processor (QFP) with 40 cores. (Each of these cores, called Packet Processors Engines [PPE], is a 32-bit RISC that runs 4 threads.) Alcatel-Lucent's FP2, which targets core networks (L2–L3), is also multicore with an array of 112 cores.

In this typology, all processing elements are identical, and a balancing and scheduling mechanism assigns incoming packets into the unutilized processing elements. This means that it is a pure SMP set-up; however, some additional SPPs, or hardware assists (coprocessors) may be present to assist the processing elements with specific tasks, on an equal basis (e.g., search engines or traffic managers). One clear advantage of using a homogeneous processor is that it requires only one programming language and one ISA to program the network processor. Network processors of this topology use high-level programming language and enable each processor to have a processing period that can vary as required. In the next chapter, we refer to this as a *run-to-completion* style of programming. If parallelism is hidden from the programmer by hardware circuitry, mechanisms and schedulers, it is easy for the programmer to use homogeneous processors, and at the same time to gain the performance advantage that this parallelism provides.

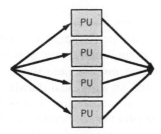

7.4.2 **Pipeline of Homogenous Processors**

The pipeline stands at the other extreme of the topology of processing elements. In a pipeline of homogeneous processors, the processing of packets is conducted in many stages; at each stage, some of the processing is done by a separate, but identical processing element. The primary disadvantage of similar processors in a pipeline, compared to parallel processors, is poor efficiency if the subprocesses done by each of the processing units in the pipe is not precisely the same. Arranging all of these similar processors in a pool, like the simple parallel topology described above, is more efficient in terms of power (work done per time unit), since all processors are utilized when there is work to be done, whereas in a pipeline one busy processor can stall the pipe and cause idle processors even when there is work to be done. One solution is having buffers of intermediate results between the pipe stages, to improve the smooth execution of the workload. The primary advantage of pipelining with similar processors comes when subprocessing in each of the processing elements can be made precisely equal, so that one can make sure that the rate of execution is predefined and equals that of a pool of similar processors.

Xelerated network processors implement this basic type of topology, that is, chains of similar processing units, although they use several pipelines in a sequence, to enable some work to be done in between segments of pipelines, to overcome the primary disadvantage of homogenous pipes.

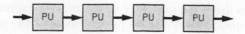

7.4.3 Pipeline of Heterogeneous Processors

More efficient pipeline setups use heterogeneous processing elements, where each of the processing elements is a specialized processor, optimized for executing a particular stage of the process. It is similar to the instruction pipelining principle, in the sense that instruction pipelining allows different, special purpose hardware (decoders, ALUs, memory interfaces, etc.), to perform each phase of the instruction. A pipeline of heterogeneous processing elements executes various packet processing subtasks in matched processing elements for these subtasks along the pipeline.

This topology has clear advantages over homogenous processors, paralleled or pipelined, since various stages of the processing are done quicker by special purpose, optimized processors. The main disadvantage of this topology is the requirement to use different programming models (and tools, languages, etc.) for each of the different kind of processors. An additional drawback is the requirement to balance the stages, that is, to smooth processing, as it is required in the homogeneous pipeline. Since each of the processing elements is different, and the subprocessing required for different packets is varied for the same stage and processing element, there is a need for interstage buffers that will hold the intermediate results in order to prevent pipeline stalls. Agere's network processors are an example of this kind of topology in network processors.

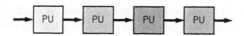

7.4.4 Parallel Pipelines of Processors

A pool of parallel pipelines is the compromise between strict parallel processors and strict pipeline of processors, in an attempt to benefit from both extremes. This topology can be compared to superscalar architecture. This topology means that at the entrance to the processing, a packet is directed to one of the pipes, and is processed inside this pipe until it is finished. Practical implementations of parallel pipelines include mostly homogeneous processing elements in the pipelines. Cisco's Toaster (PFX) is an example of this topology, where modules of homogeneous processing elements, each containing four pipelines with four stages in each pipeline, can be arranged together to have a pool of many pipelines with many stages each. Intel's network processors, as well as Freescale's C-Port network processors can be configured to work as clusters of pipelines, with few stages in each of the pipelines.

Obviously, each of the processors in the pipelines can execute different subtasks along the pipeline, and stalling can happen at the pipeline level. Stalling, however, is reduced in the overall system due to the use of multiple pipes, that is, when one of the pipelines is busy (even if not fully utilized), another pipeline can be selected for processing the incoming packet. All the advantages and disadvantages related to homogeneous versus heterogeneous processing elements, as described earlier, are valid for this type of topology.

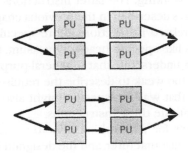

7.4.5 Pipeline of Parallel Processors

The other way to combine the benefits of parallel pools and pipelines, comprised of heterogeneous processing elements, is to arrange all of the processing elements in a pipeline of parallel processors. This arrangement is similar to the strict pipeline, and takes advantage of matching processing elements to the tasks they need to do at various stages, but instead of having one processing element in each stage, there are parallel pools of optimized processors for that stage's work. Interstage buffers keep intermediate results, and when all required data is ready for the next phase of processing, a scheduler picks up an available (idle) processing element of the next phase, and triggers this processor with the required inputs.

This topology enjoys the benefits of both the pipeline and parallel topologies, and targets carrier switches and routers that require extremely high throughput and L2–L3 processing. Like the parallelized topology, this mechanism also allows the processing period to vary as required, at each processor and at each stage (*run-to-completion*), depending on specific packets instances. It also reduces the chances of processes waiting for service, and the need to synchronize, smooth, and buffer processes and results waiting for a free processing element. EZchip's network processors are an example of such a topology.

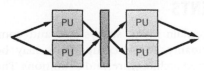

7.5 INSTRUCTION SET (SCALABILITY; PROCESSING SPEED)

Instruction sets have a major impact on performance, and they are designed very carefully by network processor manufacturers. However, the instruction set also impacts the usability of the network processor, its flexibility, and at the bottom line its overall capabilities. There are some examples of network processors that practically do not use their own programming language due to the difficulty of using the instruction set, and many development environments were developed to assist in programming.

At the two extremes of the instruction set range are, at one end, general purpose CISC or RISC instruction, arranged as VLIW or not, and at the other end, instructions that are designed for networking. The latter instructions are specifically designed for the kinds of procedures described in the previous chapters, such as check-summing, scanning, or QoS related instructions. More specific instructions are protocol-oriented instructions like specific header checksum, payload calculations, and so on. It is important to understand that a general-purpose RISC instruction set might be too limiting or too weak to describe the required operations on packets, while an instruction set that was too specific might also be wrong, as changes in the protocols would invalidate these instructions.

Optimized instructions for networking operations are therefore the goal of every network processor manufacturer, and the designed instruction set should be examined for its efficiency and flexibility. This flexibility can be determined by trying to see how well the instruction set performs in changed networking tasks and requirements, and of course, changed protocols.

Another dimension in instruction set design depends on the network processor configuration: in homogeneous processing elements there is just one instruction set, which simplifies programming and code reuse. When heterogeneous processing elements are used, then there might be several instruction sets, one for each type of processing element in the network processor, which make programming a more complex task. A solution to this problem is the use of a common set of similar instructions for all processing elements types, with additional specific instructions that these processing elements require for their unique operation.

The last issue in instruction sets is the amount of "work" that each instruction can actually do (the power of the instructions). This means for example, that an instruction can move data depending on a condition determined by a flag and, while moving the data, can update flags, increment pointers, update checksum counters, and so forth. Some network processors achieve similar capabilities by using simple CISC instructions in VLIW architecture.

7.6 NP COMPONENTS

Network processors contain many subsystems and components that together comprise their functionality. These components may be internal or external, depending on the selected architecture, and definitions. The following subsections detail the internal components of network processors that affect its functionality and performance the most: memory, interfaces (external and internal), and control and synchronization mechanisms. The external, peripheral components, such as coprocessors and hardware assists, host processors, switch fabrics, and traffic managers are described in Chapter 9.

7.6.1 Memory

Memories are used inside the network processor at several levels, as well as outside the network processor. In network processors, memory serves many functions, like storing the program and registers' content, buffering packets, keeping intermediate results, storing data that is produced by the processors while working, holding, and maintaining potentially huge tables and trees for look-ups, maintaining statistical tables, and so forth. In addition, memory access time should match the wire speed at which packets are processed; thus, network processors require very fast memories. These two factors, size and speed, are the main constraints that memory imposes on the network processor. Additionally, one expects to have cheap memory chips that are not power consumers. Another dimension to consider in memories is their technology, which influence all the above factors. Very often, memory structure and architecture makes one network processor fit for an application while another is incapable of executing the required application. This may result from inadequate table sizes or program sizes, or frame buffers that are not enough.

Using internal memory is obviously the best solution from the point of view of speed, since no time-consuming interfacing circuitries are used, very wide memory access is possible, and memory can run at the internal processor clock. On the other hand, using internal memory severely limits the possibility of using a large amount of memory, or scale in memory size (not to mention it might be a "waste" of processing silicon space). Therefore, as always, a compromise is used: some memory is located internally, while external memory is interfaced. This creates a memory hierarchy in terms of speed and size, as well as in terms of technology.

In order to understand what can be achieved from memories, a short overview of memory technology and architecture is required. We already described Content Addressable Memory (CAM) in Chapter 5. The more common memory is the Random Access Memory (RAM), a memory that returns the value stored in a given address (or stores a value in a given address). There are two basic implementations of RAM: Static RAM (SRAM) and Dynamic RAM (DRAM). The static and dynamic notations refer to the ability of the RAM to sustain its stored values, that is, in SRAM, contents remain as long as the SRAM receives power. In DRAM, on the other hand, the contents of the DRAM elements decay over time, and in order to maintain them, the DRAM must be refreshed (by pseudo-block-read operations) every few milliseconds (usually up to 32 or 64 ms).

Static RAM reads contents faster than DRAM, because (a) DRAM requires periodical refreshes that might collide with a read operation, (b) SRAM internal cell structure allows quicker reading than the DRAM's cell structure, and (c) commodity DRAM requires additional time to access the required cell, because the cell address is multiplexed (i.e., first the higher address bits are given to the DRAM and then the lower bits), while SRAM accepts all its address bits at once. SRAM also required less power than DRAM (depending on the clock frequency), and is easier to work with (simpler control, addressing, and interfacing in general). The drawbacks of SRAM are that it is a bit more expensive than DRAM, and it contains far fewer data cells than DRAM of the same die size. This is because in a DRAM cell only one transistor and one capacitor are required, compared to about 6 transistors in SRAM cells. The main trade-off therefore between DRAM and SRAM is density and speed.

7.6.1.1 Static RAM Technologies

There are several types of SRAM chips that can be classified according to the architecture they use. First of all, there are Synchronous SRAM and Asynchronous SRAM; where in synchronous SRAM, a clock signal synchronizes all address, data, and control lines, in asynchronous SRAM, the data lines are synchronized with address transition. Then, there is Zero Bus Turnaround architecture (also known as No Bus Latency), in which the switching from read to write access or vice-versa has no latency. Synchronized burst architecture means that synchronous bursts of read or write operations can be executed by a sequence of addresses that are determined by the SRAM (in which the address bus is not used for accessing the subsequent data content). This can be done either linearly, that is, in increasing order of addresses, or in an interleaved mode (which some processor may find useful). Pipeline architecture means that data is latched in pipeline operation, so that a subsequent read operation can initiate while data is being read from the pipelined SRAM simultaneously. This enables a bit of higher throughput, at the expense of one extra clock cycle latency. Flow-through architecture is a bit slower (in terms of throughput), but with no read latencies (see Figure 7.15).

High-speed SRAMs use Double Data Rate (DDR) technique, which means that data is transferred both on the rising edge of the clock and on its falling edge, thereby doubling the data bandwidth at the same clock rate. Quad Data Rate (QDR) is used for even higher bandwidth rates, and is achieved by having the SRAM use two separate ports to access the memory, one for data input and the other for data output. Each uses a DDR interface.

As of this writing, SRAM can reach tens of Mbits per chip, at an access time of a few nanoseconds. Data access width is from 8 to 32 bits wide and some SRAMs support 18, 36, and even 72 bits wide data buses for specific uses.

7.6.1.2 Dynamic RAM Technologies

There are many DRAM technologies, and only the most common are described here. The most common DRAMs for networking applications are Synchronous DRAM (SDRAM), Double Data Rate (DDR SDRAM), Quad Data Rate (QDR SDRAM), and

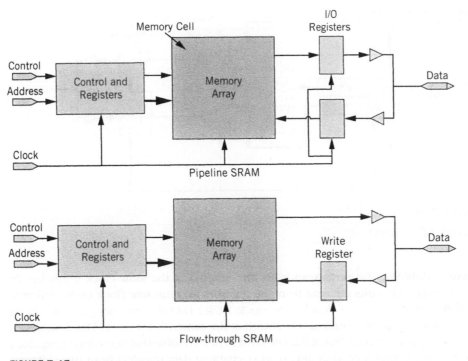

FIGURE 7.15

Pipeline and flow-through SRAM

Reduced Latency DRAM (RLDRAM). Some of the technological terms for rate interfaces are also used in SRAM, and were described above. DRAM is architecturally different from SRAM, not only due to its electronic basis; DRAM's memory array is arranged in banks (typically 4 or 8), and each cell is addressed by its bank, row, and column position. Row and column are determined by splitting the address bus into two—half of it used to determine the row address and half of it for the column address (see Figure 7.16).

Synchronous DRAM simply means that all data, address, and control signals are synchronized by the clock signal (where asynchronous DRAM reacts as fast as it can, following the control signals).

Using a pipeline mechanism in DRAM means that a couple of instructions can be processed simultaneously by the DRAM (provided it is synchronous DRAM). In other words, a read or write request can be accepted in the middle of executing the previous read or write. For example, a read can be done from one address, and while the data is presented on the output bus, the next address for read can be presented on the address bus.

Double Data Rate (DDR) DRAM (sometimes called SDRAM2), are SDRAMs that use the rising edge of the clock and its falling edge for data transfer, like DDR SRAM,

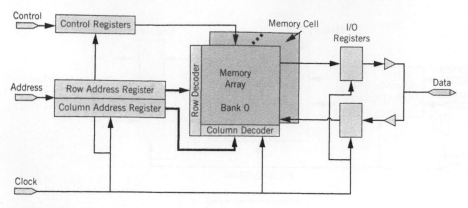

FIGURE 7.16

Pipeline DRAM structure

thereby doubling the performance (in throughput) at the same clock frequency. In other words, two bits are read from the memory array at one clock cycle, and multiplexed on the I/O at a double rate clock. DDR2 DRAM is another jump in DRAM performance, that is, doubling bandwidth at the same DRAM clock. This is achieved by clocking the internal bus of the DRAM at a speed double that of the incoming clock (at which the memory array also works), enabling data transfers from two different cells to be done at one cell cycle. This and using rising and falling edges of the clock increase the bandwidth by four, compared to a SDRAM at the same clock. In other words, four bits are read from the memory array at one clock cycle, and multiplexed by the busses and the I/O at quadruple rate. However, this is achieved with a serious drawback—the latency increases by a few clock cycles. It means that rate increases as well as latency, and in some scenarios (particularly in low clock frequencies), DDR may have greater throughput than DDR2. DDR3 DRAMs are considered the generation after DDR2 DRAMs, working at higher clock rates than that of the DDR2 DRAMs, by reading eight bits (a byte) from the memory array at each clock period, and multiplexing them on the busses and I/O at eight times the clock rate. Obviously, the latency issue gets worse.

The Joint Electron Device Engineering Council (JEDEC, the organization that standardizes all memories) standardized DDR-200 to DDR-400 (for a memory clock of 100 to 200 MHz that produces 200 to 400 million data transfers per second), DDR2-400 to DDR2-800 (for a memory clock of 100 to 200 MHz and a bus clock of 200 to 400 MHz, that produce 400 to 800 million data transfers per second) [237], and DDR3-800 to DDR3-1600 (for a memory clock of 100 to 200 MHz) [238].

Another family of DRAM is Direct Rambus DRAM (DRDRAM) and its follower, XDR DRAM. DRDRAM has extremely high bandwidth, but suffers from very long latency and very high power consumption, which XDR DRAM solved. XDR DRAM is targeted to the high-end graphics market, thus it is beyond the scope of this book.

Specific DRAMs that fit networking are the Fast Cycle RAM (FCRAM) and the Reduced Latency DRAM (RLDRAM). FCRAM is a DRAM with a memory array, which is not typical to DRAMs, and which enables the use of addressing the cell like in SRAMs. This means that column and row addresses of the accessed cell are provided together, unlike in the regular DRAM, in which row addressing[10] precedes column addressing[11] by a couple of clock cycles. FCRAM also uses pipeline architecture, and has DDR interface.

As its name implies, RLDRAM has low latency compared to other DRAMs, while still working at high rate (being a DDR device). This is achieved due to two factors: (a) using more data banks in the memory array (memory array is organized in 8 banks, which is twice the number of banks in regular DRAM), thus decreasing the probability of hitting an inaccessible bank (which happens for a few clock cycles after accessing a bank), and (b) addressing is like in SRAM or FCRAM described before. There is QDR like RLDRAM, which is termed separated I/O (SIO), that is, it uses separate read and write data busses and allows simultaneous read and write operations, just like the QDR SRAM operation.

Due to the DRAM cell structure, cells may flip, which makes DRAM more sensitive to bit errors. Though this is quite rare (happening several times a year in a 256-mbyte size memory), it is recommended to use some sort of mechanism to check, or correct, such errors. The most common ways to check or correct are by using a parity bit (a 9th bit for every byte, for example), or Error Correcting Code (ECC). There are several schemes for ECC, and the most frequently used is Single Error Correcting, Double Error Detection (SECDED). For a 64-bit word, eight more bits are required for SECDED,[12] and for a 128-bit word, an additional nine bits are required. Every time a word is accessed for read or write, a SECDED ECC mechanism computes the ECC bits, and either stores them side-by-side with the word, corrects the retrieved word, or sets off an alarm, as the case may be. This mechanism is usually part of the DRAM controller, and can be turned on or off, according the requirements.

As of this writing, DRAM can reach several Gbits per chip (ten times more than SRAMs), at an access time of about 15 nanoseconds (which is slower than SRAM). Data access width is from 4 to 36 bits wide, with bandwidth reaching several Gigabytes per second.

7.6.1.3 Conclusion

Choosing the right memory type is a matter of how it is intended to be used, since in some usage patterns, one technology results in better throughputs than another technology, and it can be the other way around in other cases. Among the criteria to consider in choosing the right memory is the read/write interleaving ratio, the

---

[10]Row addressing is done by using the Row Address Strobe (RAS).

[11]Column addressing is done by using the Column Address Strobe (CAS).

[12]SECDES requires $n + 2$ ECC bits for a 2^n bits word. For 64 (2^6) bits words, $n = 6$, so we need 8 ECC bits. However, just seven bits are enough to correct a single bit; it is not enough to detect a second bit failure. For correcting two bits in a 64 bits word, 12 bits are required.

sequential/randomness of addressing, burstiness of read/write operations, simultaneous read and write operations, size (density of the memory), power, and so on.

7.6.2 Internal Connections (Busses)

Due to the many network processor's processing elements, functional units, on-chip memories, I/O interfaces, schedulers and so on, there is a need for a vast number of internal busses that convey data from and to the various units. In order to move as much data as possible in one clock cycle between the internal units, these units are connected by parallel busses. The architecture of these internal busses is one of the major factors that determines network processors' performance. Therefore, internal busses not only have a functional role, but they are designed to enable the high bandwidth of data movements inside the network processor. Busses of 128, 256, or even of 512 parallel bits are quite common, and there are many such busses. The end result is that the aggregated bandwidth of the internal main busses exceeds 1T bytes/s (10^{13} bps). For the sake of example, a bus of 512 bits, clocked at 250 MHz, is capable of moving 128 Gbps, and there are many such busses. Three major problems emerge with these busses: (a) much of the chip area is "wasted" by busses, (b) there is a lot of potential "noise" that has to be considered by the chip designer, and (c) due to the long paths resulting from the large chip size, synchronization problems are exacerbated (signal competition, delays, and clock synchronization).

One technology that potentially can ease network processors' design is NoC. In a nutshell, instead of using a huge amount of very large busses, NoC uses a network inside the chip to interconnect all the internal units, that is, a sort of LAN (where the local is very local—the die itself). NoC has switching and routing elements to enable the transmission of data packets between the chip's units. The problem with this approach, however, is that the serial links between the units may not be sufficient to stand up to the wire-speed processing of packets by the network processor. However, NoC can reduce the number of busses, where performance is less of an issue.

7.6.3 External Interfaces

External interfaces are very important in the network processor, since they practically dictate its maximum performance, whether through ingress or egress rates, interfaces with essential memories, coprocessors, or other external circuitry. The most important interfaces are the packet input and output interfaces in terms of aggregated rate, and memory interfaces in terms of aggregated rate, types, and addressable memory. External memory is required for holding statistics, counters, and lookup tables.

External interfaces are made by internal circuitry or external chips (that interface with the network processor). Usually, when standard interfaces are involved, they can be achieved by using IP blocks internally (as described in Section 7.2.4), or as attached chips (ASSPs) externally. There are two basic interfaces—serial and parallel. Serial interfaces are usually used for I/O, whereas parallel interfaces are used for memory interfaces.

Serialization de-Serialization (SerDes) devices may be used around the network processor and inside it, to drive interfaces across distances and reduce the number of signals (and pins) that need to be routed across the system's boards and backplane. SerDes chips are transceivers that convert serial to parallel and vice versa at rates of above 10 Gbps, and most of them are capable of full-duplex operation. SerDes chips (which come as ASSPs) are used often in SONET/SDH, XAUI, Fiber Channel (FC1, FC2, and GbFC), and Gigabit Ethernet applications. SerDes circuitries are usually used in some of the chip interfaces, mainly in packet interface, PHYs, mappers, and so on.

External interfaces can be grouped into two functional groups: (1) interconnection interface between network processing devices (which can be further subdivided into those that interface with the network, e.g., PHY devices, and those that interface with the switch fabric), and (2) interfaces with nonnetwork processing devices that contain interfaces with external memories, coprocessors, and host computers. The first type of interface is called a *streaming interface* by the Network Processing Forum,[13] and the second type is called a *look-aside interface* (see Figure 7.17). The most commonly used interfaces between IP networks run at 1Gbps and above, and the network processors are XAUI, GMII/XGMII, and System Packet Interface (SPI); the most commonly used interfaces between switch fabrics and network processors are SPI and XGMII (and also a legacy CSIX, described in the following). Other common interfaces are those that interconnect external memories (SRAM and DRAM) with network processors, such as the DDR and QDR. In the following subsections all these interfaces are described, as well as several other less commonly used interfaces. Many interfaces that are rarely used with network processors (as of this writing) are not included here (e.g., InfiniBand, Fiber Channel, Serial ATA, RapidIO,[14] just to name few familiar standards) although some processors may use these or other, less familiar interfaces (storage coprocessors, for example).

7.6.3.1 Packet Interfaces

Packet interfaces are interconnection interfaces between any network processing devices, such as switch fabrics, other network processors and dedicated network coprocessors (e.g., classification coprocessor), or line (PHY) interfaces and framers. As mentioned above, the Network Processing Forum refers to this

---

[13]The Network Processing Forum (NPF) had merged into the Optical Internetworking Forum (OIF) by mid-2006.

[14]FreeScale (Motorola at the time) intended to use RapidIO for its next generation C-Port network processor, the C-10, which was eventually not developed. RapidIO was also a candidate for the Network Processing Forum's next version of its Look Aside interface (LA-2); as of the writing of this book, this also did not happen.

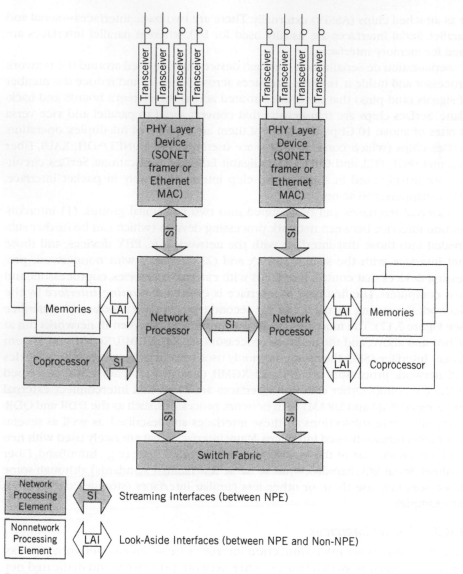

FIGURE 7.17

The Network Processing Forum reference model for external interfaces

type of interface as streaming interfaces. They also standardized the entire group by creating subgroups according to their usage (as described in the following).

When the interfaces are between the network processors and the line (PHY), the network processors usually contain all the packet logic down to the data-link layer, and the PHY interfaces are done externally. Sometimes MAC layer circuitry can be found also outside the network processor. At any rate, all the interfaces are usually done by using IP blocks (inside the network processor), or specific ASSPs that handle the required interfaces.

Since packet interfacing is of utmost important in network processing (which is packet centric), packet interfaces will be described in somewhat more detail than the average user may need. This, however, will enable the reader to understand network processors specifications, and evaluate their limitations in packet I/O.

7.6.3.1.1 Network Processing Streaming Interface and Scalable System Packet Interface

There are two standards that describe packet interfaces: (1) the Network Processing Streaming Interface (NPSI)[15] [332], and (2) the Scalable System Packet Interface (SPI-S)[16] [341], which is actually based on and is the successor to the first. Both define a framework specification of point-to-point data path interfaces with the support of addressing and flow control (for multiple framer channels, switch fabrics and/or coprocessor destinations and classes). NPSI defines the interconnection between network processing devices (e.g., framers, switch fabrics, network processors, classification coprocessors, encryption coprocessors, traffic manager coprocessors, or any other network coprocessors), targeting data transfer rate of 10 Gbps (OC192). It contains three main modes of operation (or group of interfaces): switch fabric interfacing, framer interfacing, and network processor-to-coprocessor interfacing.

Network Processing Streaming Interface is a unidirectional, 16-bit-wide data path (with additional clock and control signals, and flow control that is two-bits wide and a clock in a reverse bus), running at 622 Mbps[17] per bit lane, and is based on SPI-4.2 (or SPI-5, as described in the following), with protocol concepts taken from CSIX-L1 (also described in the following). The NPSI switch fabric mode supports 4096 egress ports, with 256 classes, and network processor (or coprocessor) interfacing mode that supports up to 256 channels. The NPSI framer interface mode is simply the SPI-4.2, as described in the following.

Scalable System Packet Interface was introduced after NPSI, and actually replaced it, and is also a successor of the SPI interfaces (described below). It is a generic, point-to-point interface, not specific to some physical interface or bandwidth that is capable of addressing multi-framer channels, switch fabrics, or network processors. SPI-S defines the data link level requirements in terms of packet definition, framing, addressing, error detection, and flow control. It is used for Ethernet packets ($\geq$ 64 bytes), ATM cells (48 or 53 bytes) or IP packets ($\geq$ 40 bytes), in a bidirectional, unidirectional or asymmetric manner.

---

[15]NPSI was copyrighted by the Network Processor Forum in 2002.

[16]SPI-S was copyrighted by the Optical Internetworking Forum in 2006.

[17]The bus clock is actually 311 MHz, but since it is double-edge clocking (both on the rising edge and the falling edge of the clock, which is called Double Data Rate, or DDR) it runs 622 Mps per bit lane, where bit lane means a signal in the bus, carrying one data bit (in NPSI case it is one Low Voltage Differential Signaling (LVDS) pair, see footnote 23). Usually, for busses with single edge clocking, data rates (in Mbps) will be the bus clock (in MHz) multiplied by its bits-width.

7.6.3.1.2 UTOPIA

Universal Test & Operations PHY Interface for ATM (UTOPIA) bus is part of the ATM world, and is mentioned here briefly, since there are still some interfaces to this bus in a few traffic managers or other line card chips. UTOPIA [31] describes the data path interface between an ATM link and PHY devices, that is, it describes how cells are actually sent from the data link (ATM layer) to the physical layer (ATM PHY) chips, and vice versa. UTOPIA can use Synchronous Optical Network/Synchronous Digital Hierarchy (SONET/SDH), fiber or STP (shielded twisted-pair) with 8B/10B[18] encoding or other physical medium. There are four versions of UTOPIA, marked 1 to 4, that appeared sequentially as line interface rates increase. UTOPIA level 1 covers line rates up to 155 Mbps (OC-3), by using an 8-bit-wide data path and a maximum clock rate of 25 MHz. UTOPIA level 2 [32] supports up to 622 Mbps data rates (OC-12), using a 16-bit-wide data path and a clock rate of 33 MHz (to meet the PCI bus, described below) and 50 MHz (to reach 622 Mbps). UTOPIA level 2 also allows interface between the ATM layer to multiple PHY ports (MPHY), more than the one ATM PHY chip that level 1 allowed. UTOPIA level 3 [33] covers rates of 1.2, 2.4, and 3.2 Gbps, by using a 32-bit-wide data path and a 104 MHz clock. UTOPIA level 4 [34] answers the 10 Gbps (OC-192) requirements by using the same 32-bit-wide data path clocked at up to 415 MHz (eight times the OC-1 clock rate). UTOPIA is not a peered interface, which means that there should be one end in a master mode and the other working in slave mode.

7.6.3.1.3 System Packet Interface and Packet-Over-
SONET-Physical Layer

System Packet Interface (SPI) defines an interface that "channelizes" packets and cells between the physical layer and the data link layer, as standardized by the Optical Internetworking Forum (OIF). SPI is a Packet-over-SONET/SDH (POS) framer[19] interface, or Ethernet aggregator interface. It is a full-duplex interface, which has two familiar levels of rates, level 3 (SPI-3) and level 4 (SPI-4), following the POS-PHY level 3 and POS-PHY level 4, respectively.[20]

POS-PHY level 3 interface (which is SPI-3) supports Packet over SONET applications using a framer, at 2.5Gbps (OC-48). The interface consists of 8 or 32 signals for data, eight control signals, eight optional multichannel status signals,

---

[18]8B/10B encoding came from the FiberChannel standards, and is used to minimize error by "balancing" (averaging) serialized sequences of bits. In other words, in case of 8 bits all "0" (or all "1"), 8B/10B encodes (by a simple lookup table) the 8 bits to 10 bits made of "1" and "0" that are said to be "DC-balanced" (i.e., the resulted code almost always has 5 bits of "1" and 5 bits of "0"). There is frequency bandwidth overhead that is required now for sending 25% more bits, but the system has fewer errors; thus, the overall performance increases (in terms of throughput).
[19]The POS framer is at the physical layer, and originates from [377, 378] and is also known as POS-PHY.
[20]Both POS-PHY levels 3 and 4 originated from the Saturn development group.

and one clock signal, all per each direction, and all are TTL[21] signals, running at 104 MHz.

Packet-Over-SONET-Physical Layer level 4 interface (which is identical to SPI level 4 phase 2, or SPI-4.2)[22] supports multi-Gigabit applications (e.g., 1 × 10 GbE or 16 × 1GbE) using an Ethernet aggregator, or Packet over SONET applications using a framer; for example, 1 × OC-192 (i.e., 10 Gpbs), 4 × OC-48 (i.e., 4 × 2.5 Gbps), or 16 × OC-12 (i.e., 16 × 622 Mbps). It supports 256 channels (different addresses, each with its flow-control) over its point-to-point connection. The interface consists of 16 LVDS[23] pairs for data and a few signals for control, clock, and status, per each of the directions, and runs at about 700 MHz. SPI-4.2 was designed for interfacing between the physical layer (PHY) and the data link devices (PHY-link), but it is also used for PHY-PHY and link-link connections. SPI-3 and SPI-4.2 emerged from PL-3 and PL-4 respectively; both originated at the ATM forum (PL stands for POS Layer).

Two other interfaces are the POS-PHY level 2 operated up to 832 Mbps, which is no longer common, and the SPI-5, which targets the 40 Gbps (OC-768) interface. SPI-5.1 is based on a 16-bit-wide bus per direction, running at 2.5 GHz, and is very similar to SPI-4.2 in all other aspects.

There is one more complementary standard of the OIF that is associated with SPI, and should not be confused with SPI, namely the SerDes to Framer Interface (SFI). This standard defines the interconnection in the "line side" (as opposed to the "system, processing side" that SPI defines), that is, between the optical module and the processing chips, or between two different processing chips. In other words, SFI defines the point-to-point interfaces between SerDes components, the Forward Error Correction (FEC) processor, and the framer. The aggregated rate supported by SFI level 4 (SFI-4) is 10 Gbps (OC-192), for use in POS, ATM, and 10 GbE, and it is bidirectional. SFI-4.1 uses 16 channels per direction, each clocked at 622 MHz, while SFI-4.2 uses a 4-bit-wide bus per direction running at 3.125 GHz. SFI-5.1 reaches 40/50 Gbps by using two (bidirectional) 16 channels with 2.5/3.125 Gbps per channel, for a total of 40/50 Gbps, while SFI-5.2 reaches 44.4 Gbps per direction by using a 4-bit-wide bus per direction, running at 11.1 GHz [342].

Figure 7.18 depicts a simple reference model (on the left side of the Figure 7.18), in comparison with SONET/SDH (on the right) and IEEE 802.3 [186] 10 and 1 Gbps Ethernet (in the middle of the Figure 7.18). Note that SPI is a boundary interface of system/physical, as well as being the SPI.

---

[21]Transistor-Transistor-Logic (TTL) is a 0-0.8 V or 2-5V interface for logic "0" or "1." TTL used to be very common in discrete logic circuitry, but suffered from power, speed, and distance issues.
[22]SPI-4.1 is the first phase of SPI-4, targeted at 10 Gbps (OC-192), but works slower than SPI-4.2. SPI-4.1 defines unidirectional interface that works with a 64-bit-wide bus, running at 200 MHz, whereas SPI-4.2 is a bidirectional interface that uses a 16-bit-wide bus per direction, running at 622 MHz.
[23]Low Voltage Differential Signaling (LVDS) was introduced in the mid-1990s to support very high-speed interfaces over cheap, twisted-pair copper wires. It is based on transmitting low current (3.5 mA) in one of the wires according to the logic level ("0" or "1") through a 100 Ω resistor, creating a 0.35 mV difference—hence, the Low Voltage.

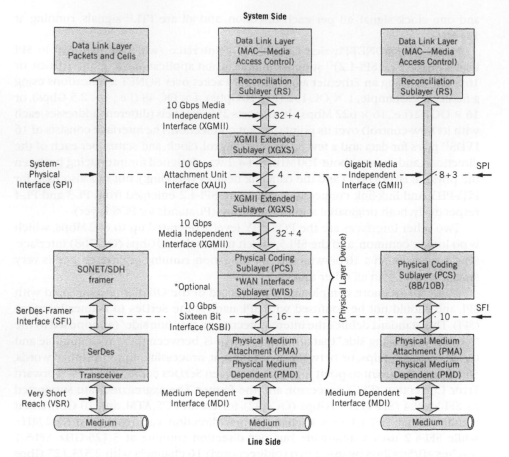

FIGURE 7.18

SPI-SFI reference model

7.6.3.1.4 IEEE 802.3 Ethernet Interfaces

IEEE 802.3 [186] defined several Ethernet interfaces, ranging from 1 Mbps to 10 Gbps rates (as was described in more detail in Chapter 2). In network processors, the 100 Mbps ("fast Ethernet") is rarely used (and lower rates are totally irrelevant), so these interfaces are not detailed here. More practical and common are the 1 Gbps Ethernet (GbE) and the 10 Gbps Ethernet. Figure 7.18 provides a reference model of the interfaces, from the MAC (Media Access Control) layer down to the media (a more detailed description is provided in Chapter 2). The upper layers of the Ethernet are handled by the network processor software. In this subsection, we discuss the interfaces of the network processors with regard to Ethernet packets. These interfaces may include: Media Independent Interface (MII) that is used for 10 and 100 Mbps Ethernet (and is not described here), Gigabit Media Independent

Interface (GMII) that is used for GbE, and 10 Gigabit Media Independent Interface (XGMII) that is used for 10 Gbps Ethernet.

Gigabit Media Independent Interface and Gigabit Media Independent Interface The GMII interface bus is full-duplex, with 8 data bits, 2 control bits, and a 125 MHz clock per directions and two more control signal bits, as shown in see Figure 7.19. The XGMII interface is also full-duplex, with 32 data bits, four control bits, and a 156.25 MHz clock per direction, which is used in its rising and falling edges (i.e., DDR, which means 312.5 Mbps per bit lane), as shown in Figure 7.19.

XAUI For XGMII there is an extension block, called XGMII eXtended Sublayer (XGXS), which permits the extension of the operational distance of the XGMII and the reduction of the number of interface signals. By using an XGXS at the data link layer side, that is, the RS end (DTE XGXS), and an XGXS at the PHY end (PHY

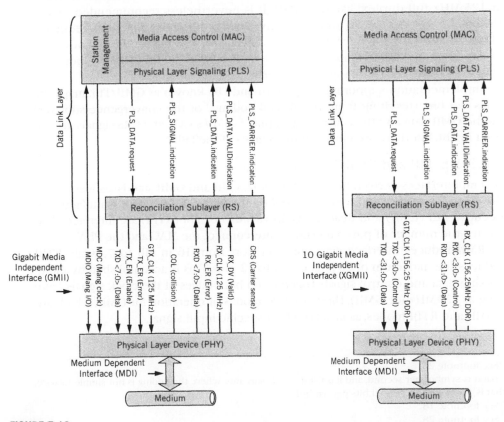

FIGURE 7.19

GMII and XGMII interfaces

XGXS), it is possible, for example, to extend the physical separation between the MAC and the PHY components across a circuit board. The interface between these two XGXS blocks is called 10G Attachment Unit Interface (XAUI) and is reminiscent of the old 1 Mbps Attachment Unit Interface (AUI), which was used to connect the system by its data link layer to a separated medium to which the PHY was attached. At any rate, XAUI is important in itself, and is used in network processing interfaces for 10 Gbps Ethernet PHYs instead of the XGMII. This is because XAUI is simpler, targeted for chip-to-chip interface implemented with traces on a printed circuit board, and contains four pairs of Low Voltage Differential Signaling (LVDS)[24] per directions, each running at 3.125 MBaud.[25]

TBI Ten Bit Interface (TBI) is used quite frequently for GbE connections. Contrary to the GMII, which is an interface between the MAC layer and the PHY devices, TBI is an interface *inside* the PHY, between the Physical Coding Sublayer (PCS), the upper PHY sublayer, and the Physical Medium Attachment (PMA). TBI is thus a physical instantiation of the PMA service interface, working at 125 MHz full duplex, like the GMII. It is nevertheless recommended by the IEEE802.3 for interfacing since it conveniently separates the high-frequency circuitry from the logic functions of the MAC and PCS sublayers, it is intended to be used as chip-to-chip interface, and it provides compatibility between chips designed by various manufacturers.

Implementation supporting both GMII and TBI, known as GMII/TBI, maps the TBI's ten bits (resulting from the 8B/10B coding[26] of the convergence sublayer) on the GMII data and the control signal pins.[27] GMII's COL and CRS signals are not used in TBI, but there are other control signals of the TBI.[28]

7.6.3.1.5 MII and GMII Derivatives

Several interfaces derived from the IEEE 802.3 MII and GMII standards were introduced to the market and were accepted by the network processor manufacturers. Reduced MII (RMII) and Reduced GMII (RGMII) or TBI (RTBI) are interfaces that reduce the number of pins required to interconnect the MAC and the PHY.

RMII reduces interface pin count per port from 16 in MII to 7, while RTBI reduces the maximum of 28 pins down to 12 pins. This is accomplished by multiplexing data and control signals together on both edges of the clock signals (running at 125 MHz for GMII). There are two modes of operation for GMII, which are RGMII and RTBI modes, as indicated by a specific input signal.

---

[24]See footnote 23.

[25]Baud is symbol per second, and it is used to measure rate where the coding is not simple bitwise, that is, there may be several bits per symbol.

[26]See footnote 18.

[27]See footnote 28.

[28]Detailed description of these signals and their function is beyond the scope of the book, and the interested reader is referred to the standard, IEEE 802.3 [186, clause 36.3].

Serial MII and GMII (SMII and SGMII) are specifications published by Cisco Inc. (ENG-46080 and ENG-46158, respectively). The basic idea is to use two pins per port, and, optionally, to connect several PHY devices to one MAC layer, synchronized with one clock. SMII is composed of a global synchronization signal and a global 125 MHz reference clock, in addition to the two signals per port (receive and transmit). If a long trace delay is used (more than 1 ns), than four optional signals may be used instead of the synchronization signal (two per direction; one is clock and the other is synchronization). SGMII uses two data signals (receive and transmit) and two clock signals (for receive and for transmit) to connect the PHY to the MAC. The data signals run at 1.25 Gbaud, using the clock signals at 625 MHz (which is used in its rising and falling edges, i.e., a DDR interface). Each of these signals uses a differential pair LVDS.

7.6.3.1.6 Common Switch Interface Specification

Common Switch Interface Specification (CSIX) was defined by the CSIX group, which later became the Network Processing Forum (NPF), together with the Common Programming Interface Forum (CPIX) [331]. The first standard (and the only one, to date), called CSIX-Level 1 (CSIX-L1), defines a full-duplex interface between traffic managers (and network processors) and switch fabrics for ATM, IP, MPLS, Ethernet, and so on. It discusses CSIX-Level 2 (CSIX-L2) for carrying information between traffic managers (or network processors) but this information was not standardized.

Common Switch Interface Specification-Level 1 defines a physical interface and a "directive" header (called CSIX frame, or CFRAME), which carries information as to how to treat the frame (i.e., to control the behavior of the interface between the traffic manager and the switch fabric). Each received frame is classified and forwarded across the switch fabric (that works according to the CSIX), according to the CFRAME.

Common Switch Interface Specification-Level 1 physical interface is intended to be used for board-level connectivity (i.e., up to 20 cm trace length), and supports data rates of up to 32 Gbps. The data path is 32-, 64-, 96-, or 128-bits-wide per direction, clocked at up to 250 MHz. In addition, several control lines exist (for each 32 bits of data, and for each of the directions—transmit and receive—there are parity, start of frame, and clock control pins).

A CSIX-L1 header consists of 2 to 6 bytes, followed by optional payload (up to 256 bytes) and 2 bytes of vertical parity. The first 2 bytes contain the payload length, frame type, and ready bits (for link level flow control). The frame type indicates whether the frame is idle, unicast, multicast (or carries the multicast mask), broadcast, or is a flow-control frame.

Common Switch Interface Specification-Level 1 supports up to 4096 switching ports (12 bits destination addresses) for the connected traffic managers or network processors. In addition, there are up to 256 distinguished classes, defined by the traffic manager (or network processor) and by the switch fabric vendor, that is, the 8-bit class provides a mechanism to handle priorities and QoS services within the switch fabric. This means that the switch fabric can use the class to differentiate between CFRAMEs that are addressing the same switch fabric port, and handle them differently, according to their classes.

7.6.3.1.7 Access Interfaces

Access interfaces, in the context of network processors, are sub-1 Gbps interfaces. They include many types of interfaces, and are mainly used by the access network processors (hence, their separate classification). Packet-based data (e.g., Ethernet), cell-based data (e.g., ATM), and continuous data (which is often referred to as Time Division Multiplexing [TDM], e.g., digital telephony) have various interfaces. These interfaces are described very briefly in this subsection, since they are beyond the scope of the book.

Ethernet interfaces are done with the MII, GMII, RMII, and SMII definitions described above. Utopia level 2 (described earlier) is used frequently to interface cell-based data (e.g., ATM OC-3 and ATM OC-12—i.e., 155 Mbps and 622 Mbps, respectively).

Packet-Over-SONET-Physical Layer level 2[29] [378] is used for packets over SONET/SDH, to carry IP packets over Point-to-Point Protocol [387], over HDLC protocol [388], and over SONET [386]. POS-PHY can be also used to carry Frame-Relay data or even Ethernet packets (although it is quite rare).

Time Division Multiplexing interfaces include the serial T1/J1 and the serial E1 interfaces,[30] which are connected to Plesiochronous digital hierarchy (PDH)[31] networks. T1/J1/E1 can also be interfaced to the network processors by using xDSL.[32] Higher speeds PDH connections are also used, like the T3/E3.[33]

7.6.3.2 Host Interfaces

Almost all network processors have interfaces with host processors for various tasks, starting from initializations, configuration, slow-path packet processing, management tasks, as well as reporting, maintenance, and other functions that the

---

[29]POS-PHY level 2 emerged from UTOPIA level 2 and SCI-PHY, for variable size packets. The interface is 16-bits-wide per direction, clocked at up to 50 MHz, and, hence, capable of full-duplex, 800 Mbps connection. It supports multiple PHY interfaces (by a 5-bits address bus), and has seven additional control signals and three other optional control signals per direction.

[30]A reminder from Chapter 2: T1/J1 are the North American and Japanese standards of serial PDH interfacing and framing, that are running at 1.544 Mbps (T1/J1) and 2 Mbps (E1). These interfaces multiplex (in time division technique) basic voice (Digital Signal) channels of 64 kbps (which is called DS-0). T1 is also known as DS-1, as DS-1 is the protocol name that the physical T1 uses. T1 framing can be of Super Frame (SF) scheme or Extended Super Frame (ESF); both synchronize the T1 frames by 8 kbps framing channels that are used differently and are not compatible.

[31]A reminder from Chapter 2: PDH are networks that are almost synchronous, which means that they use the same clock rate but from different clock sources.

[32]A reminder from Chapter 4: See footnote 1. Additionally, it is important to note that the DSL modem, at the CPE (Customer Premises Equipment) end, and the DSLAM (Digital Subscriber Line Access Multiplexer) at the other end of the line, that is, at the CO (Central Office), are working on an ATM or Ethernet infrastructure, which carries Point-to-Point Protocol (PPP), which in its turn encapsulates IP packets.

[33]A reminder from Chapter 2: E3 runs at 34.368 Mbps, whereas T3 runs at 44.736 Mbps. T3 is also known as DS-3.

network processor is not optimized to perform (or not capable of doing at all). Some network processors have an embedded processor that runs some of the slow-path tasks, management, or maintenance, but even then it is typical to find an external host processor.

When a host processor is attached, it must be interfaced with the network processor; this can be done either with some proprietary interface, memory-type interface, or a standard interface used by host systems, thereby enabling the network processor to be attached to a system and not just glued to some CPU.

The standard way of achieving such interface is through the use of a PCI bus, or its followers, PCI-X and PCI express. Peripheral Component Interconnect, or PCI, was introduced originally by Intel around 1990 at the component level (PCI 1.0), as a way to interconnect devices on motherboards of personal computers and servers. The purpose was to allow various devices, including memory, to interconnect directly without CPU intervention. It evolved also to specifying a connector (for "cards" to be attached to the motherboard) and a bus that supported up to five devices. Since it supported bridges, each of these devices could have been used to connect more devices. These advantages—coupled with software features that enabled auto configuration (Plug and Play, or PNP) that appeared simultaneously—turned the market around, and most computer vendors adopted PCI and abandoned previous Industry Standard Architecture (ISA) and Extended ISA (EISA) busses. The PCI Special Interest Group (PCISIG) was organized in 1992, and became responsible for the standard and its derivatives.

The original PCI standard (version 2.0) was a 32-bit-wide bus, running at 33 MHz, which limits the theoretical throughput to 133 Mbytes/s. Later revisions went up to 66 MHz (PCI 2.2) while PCI-X supports 133 MHz, and a 64-bit-wide bus that can reach just above 1 GBytes throughput. PCI-X version 2.0 enables bus clocks of up to 533 MHz, hence supporting up to 4.3 GBytes/s throughput. PCI-X is also a parallel bus, and is backward pin compatible with the 5 and 3.3 V cards that PCI 2.2 (and higher versions) supported. Both PCI and PCI-X are bidirectional busses, but work in half-duplex mode.

PCI Express, or PCIe, was also introduced by Intel (in 2004), and is currently the fastest bus interface supported by PCISIG. It is an entirely different bus architecture than the PCI and PCI-X (and not compatible with these standards). It is a serial interface, full-duplex, point-to-point bus. While in PCI and PCI-X, the slowest device dictated the entire bus speed, PCIe enables speed negotiation by each of the attached devices. PCIe has multiple lanes for data transfers, and each is a serial channel capable of reaching 250 MBytes/s.[34] PCIe runs a layered protocol stack, and is capable of using a PCIe switch for

---

[34]PCIe standards define 1, 2, 4, 8, 12, 16, or 32 lanes, and the respective busses are known as PCI-Express/ x*N*, where *N* is number of lanes, for example, PCIe/x16 means 16 lanes PCI-Express.

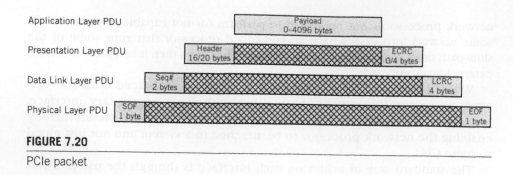

FIGURE 7.20

PCIe packet

interconnection and routing between several PCIe interfaces.[35] Its lowest layer, the physical layer, has an electrical sublayer that uses two unidirectional serial Low Voltage Differential Signaling (LVDS)[36] pairs, one per direction, running at 2.5 Gbps, creating a lane of four wires. PCIe 2.0 doubles this bandwidth to 5 Gbps, and the forthcoming PCIe 3.0 brings it to 8 Gbps. At the upper physical layer, data is framed (one start-of-frame byte, and one end-of-frame byte), interleaved on all used lanes, and 8 B/10B[37] encoded (PCIe 3.0 is not using the 8 B/10 B encoding). The data-link layer of PCIe adds a sequence number to the packet header (two bytes) and a CRC (4 bytes, called Local CRC, or LCRC) as a packet trailer, and uses acknowledgments to maintain data integrity, as well as flow-control protocol, credit-based,[38] to make sure that the receiving party is capable of receiving the sent packets. The upper layer is the transaction layer, which is used for interfacing with the application. This layer is responsible for addressing devices and memory and also maintains a handshake with the peered layer, as well as with the data-link layer. This layer adds a header (16 or 20 bytes) and an optional CRC (4 bytes, called end-to-end CRC, or ECRC) as a packet trailer. The payload itself is 0 to 4 K bytes. The PCIe packet is shown in Figure 7.20.

PCIe has the potential to impact network processors, not only in interfacing with hosts (it is a better interface since it requires fewer I/O pins on the network processor interface), but also as a platform for data transfers between devices, since it can reach 8 GBytes/s with 32 serial channels (lanes). PCIe 2.0 doubles this bandwidth, having 5 Gbps rate on each of its lanes, and PCIe 3.0 is expected to bring it to 256 Gbps.

---

[35]In 2004, the Advanced Switching Interconnect Special Interest Group (ASI-SIG) even published a standard for switching PCIe busses, called ASI. ASI dissolved into PICMG at 2007, but PCIe switches are commercially offered as chips for onboard implementations, and as systems for inter-computer and peripherals PCIe interconnection.

[36]See footnote 23.

[37]See footnote 18.

[38]A device advertises its capacity to receive (i.e., credit), and the sending party knows its limits in sending packets, while it gets the updated line credit from the receiving party as packets are processed. This saves latency and increases throughput.

7.6.3.3 *Memory and Coprocessors Interfaces*

Practically, most network processors interface memory chips and coprocessors according to the requirements and the pin layout and functions of the memory chips or the coprocessors. However, there is one standard—the Look-Aside (LA) interface of the Network Processing Forum (NPF), now the Optical Interfacing Forum (OIF)—that is used mainly by search engine coprocessors and CAM memories that are used for searching. A few comments on interfacing memories are required before describing the LA interface.

7.6.3.3.1 Memory Interfaces

Naturally, all SRAM and DRAM technologies that were described in the memory subsection above should be considered in interfacing memories with network processors, that is, QDR, DDR, and so on. This means that some of these interfaces may have to be multiplexed internally according to the memory interface, in order to work with the double rate clocking (DDR), for example. Other kinds of interfaces may require other functions in the network processor internal interface circuitry. For example, bursting also has to be considered, when it is applicable, as well as bus turnaround.

Additionally, a network processor can interface several memories in parallel, applying the same address bus and control signals to all of them, and create a data bus of any width, which is the sum of all the widths of the connected memory chips data busses. In other words, to create a 64-bit-wide data bus, for example, the network processor can use four parallel memory chips, each with a 16-bit-wide bus. This is very useful in increasing the overall memory throughput, and the only limitation is the extent of memory width that the network processor really requires, or can have in terms of internal busses, amount of pins, and so on. Memory size, on the other hand, is determined by the chips sizes and their amount, as well as by the width of the address bus used by the network processor (a 28-bit-wide address bus can handle up to 256 M addresses, for example). When Error Correction Code (ECC) is used in DRAMs, extra memory chips might be required to store the ECC bits, and the data bus should be affixed with the ECC bits that have to be interfaced in addition to the data bits (that is, if the ECC mechanism is inside the network processor, in a DRAM controller, and is not a part of an external memory circuitry).

It should be noted that SRAM memory chips have similar interface in terms of pin functions, and are simpler to interface than DRAMs. As mentioned above, DRAMs usually require a DRAM controller, which might be inside the network processor or external to it, and DRAM memory chips also have quite a similar interface to the DRAM controller.

7.6.3.3.2 LA-1

The Look Aside Interface (LA-1)[39] is the Network Processor Forum's first generation of look aside interface. Its objective is to interface memories (mainly CAMs)

---

[39]LA-1 was copyrighted by the Network Processing Forum in 2002.

and coprocessors, or any adjacent device to a network processor that is off-loading network processing tasks. LA-1 interface is basically a SRAM interface, memory mapped (8–512M address space), with some modifications that are meant to incorporate coprocessors, like variable access latency, out-of-order responses, and so on. The address bus is used for register addressing when appropriate, when interfacing coprocessors, and an in-band handshake occurs for transferring the required control and results to and from the coprocessor. LA-1 uses separated read and write 16-bit-wide data busses (SIO), and with 200 MHz DDR interface, it transfers 32 bits, yielding 800 Mbytes/s bidirectional data transfers. LA-1 enables bursts of two read or write operations in the DDR scheme, that is, two bytes are transferred from consecutive addresses in the first operation, followed by an additional two bytes from consecutive addresses (the two least significant address bits are not used). There are four additional parity bits per direction (with two lines per direction using DDR), which are used for even-byte parity.

A following modification to the standard, known as LA-1B, enhances the LA-1 in speed (x2.5 bandwidth, that is, 2 GBytes per direction), allows multiple operations simultaneously (e.g., lookups), and enables bursts of four consecutive operations (up to 8 byte data transfers from or to consecutive addresses).

LA-2 specification has not been released (although it has been discussed), and it considered adopting RapidIO, HyperTransport, or derivations of SPI-4 or SPI-5.

7.6.4 Control and Synchronization Mechanism

This mechanism is what makes the network processor work, and, more than that, determines how it will work. More abstractly, this mechanism is also responsible for the way the user of the network processor (the programmer who implements the network processor in its system designs) actually uses it. It can allow the user to ignore all synchronization tasks (questions such as: Where is the packet? Where are its results? Can it be sent to the search engine? Are the results ready?), parallelism (What processing element is free? What about packet ordering?), buffer management (Is this place available? How big is the search data base?), and so forth.

Control and synchronization circuitry is designed first and foremost to make the processor function properly. In a pipeline configuration, the control mechanism handles all the intermediate results, in a way that smooths the pipeline operation and benefits from the pipeline even in situations of imbalanced use of its stages. In a parallel configuration, it keeps track of packet ordering and proper use of common network processor resources (memories, functional units, etc.).

7.6.5 Functional Units

Search engines and traffic managers are the main processing elements that typically reside within the network processor. There are, however, many implementations of search engines and traffic managers that are external to the network processor and are working as either a coprocessor or attached as a memory-mapped device.

The integration of these two types of processors in the network processor adds substantial advantage, in terms of the rate of data movements and requires fewer components at the system level.

Any network processor, as any other GPP, must have functional units to enhance its capabilities. The most obvious one is the ALU. Additional network-related functional units are those that handle the packet checksums (IP and TCP), converters that assist in converting packet content from some representation to another one, scanners that assist in identifying specific contents in the payload, look-aside managers, memory managers, and so on. These functional units have many materializations, and detailing them is beyond the scope of this book.

7.6.6 Embedded Processors

Some network processors contain embedded processors such as ARM, PowerPC, MIPS and the like, or even configurable processor cores (such as Tensilica), which assist the network processor with control functions, in addition to taking some of the slow-path operations onto themselves.

7.7 SUMMARY

In this chapter, we discussed the basics of network processor architecture from a theoretical point of view, describing what technologies are available to use, and how they all integrate in the design of network processor. In addition, the interfaces used for connecting the network processor to its working environment (from both the network side and the system side) are described, to enable the reader to understand how to plan, design, and build a system with network processors.

The next chapter discusses the programming model, and together, these two chapters provide an answer to the challenge of how to cope with the mission of packet processing in a high-speed, demanding, and complex networking environment.

Software

In the previous chapter, we discussed network processor architecture in terms of the processing element configuration, other internal units in the network processor, and interfaces to the network processor. At several points during this discussion of the architecture, we mentioned possible impacts on and by programming models. This chapter describes programming models of network processors, as well as some important principles that are relevant to their programming, and concludes by describing the typical programming environment of network processors.

Programming a network processor is very different from programming any other processor; some see it as notoriously difficult, while others find it very strange. Some network processors are fairly easy to program, almost like general purpose processors, while other network processors require specific tools to assist with their programming, which is very complicated. At any rate, developing software for a network processor is a complex task that requires very careful attention to multiple areas: the networking environment, demanding performance constraints within a typical heterogeneous architecture that has many types of processing elements, search engines, traffic managers, and other functional units.

This chapter sheds light on the peculiar programming world of network processors. It is not the intention of this chapter to provide a general programming concept (surely not describing a specific programming language), but rather to discuss the relation between network processor architecture and programming, as well as the typical programming tools and utilities required to program a network processor. In addition, it should be noted that the wide coverage of programming issues in this chapter, which touches on the entire spectrum of network processors as well as conventional equipment, might result in the feeling that programming a network processor is harder than it really is. Once a network processor is chosen, most of the details in this chapter become irrelevant. However, choosing the right network processor from among all possibilities, and understanding the relevant programming issues and tasks, requires the broad overview provided in this chapter.

In the introduction to the previous chapter, Architecture, we categorized the network processors according to three groups: entry-level or access network processors, mid-level processors (legacy and multiservice network processors), and high-end network processors that are targeted for core and metro networking equipment, usually on line cards. This categorization is obviously important in describing the software of network processors, although all three of these groups share many common attributes. It is important to note, however, that parallel processing and programming, mainly on symmetrical multiprocessor (SMP), chip multiprocessor (CMP), and multi-core processor (MCP) architectures, is usually used for networking applications of the higher layers (L5–L7, for example, security, storage, and accounting). Pipelining architectures are usually used for ultra-high speed packet processing on line cards, at the lower layers of networking (L2–L4), and performance is a key issue that impacts the software and programming of network processors for these applications.

One last remark on the software of network processors: there is usually a host processor—either embedded in the network processor or interfaced to it—that is responsible for managing the network processor's packet processing elements (the data path[1] processing). This is the control path processing, which is not the focus of this chapter, although in some implementations the two are combined. Writing software for the host processor (particularly if it is external host) is "usual programming," which is briefly described in Section 8.2.

8.1 INTRODUCTION

Designing and programming equipment for networking applications can be done in a traditional way, with traditional architecture, which means that both the data and control path tasks are processed by a general purpose processor that uses an operating system (OS) or a kernel, with the applications (the control-path tasks) running on top of it. Designing and programming networking equipment can also be based on network processors, with a minimal amount of highly-efficient specific software executing the required data path processing. It can also be a combination of the two, by assigning the control path processing to a host processor that is organized in the traditional way, and executing the data path processing in the network processor.

There are several terms used when referring to "programming," with each describing a different level of abstraction or implementation. These terms are often vaguely defined, and some even overlap, or are used interchangeably: programming models, programming paradigms, programming languages, programming (or software) architecture, as well as the development environment terms that include programming environment, programming tools, and programming utilities.

---

[1]As previously defined, the data path refers to packet manipulations by the equipment, (primarily packet forwarding), and the control path refers to control tasks that manipulate the equipment and its functionality.

In essence, a programming model provides an answer to how a programmer views and uses the hardware architecture; a programming paradigm provides an answer to how a programmer views the execution of a program; a programming language provides an answer to how a programmer describes the required algorithm or task; programming architecture provides an answer to how the software elements are organized; and a programming environment provides an answer to how a programmer actually codes his program.

8.1.1 Programming Models

A programming model, for our purposes, is the programmer's view of the computer architecture, which includes the specifications of the communications and the synchronization requirements that are addressed at the programming stage. In other words, a programming model is what a programmer uses while coding an application. This means that the communications and the synchronizations primitives are exposed to the user-level, and this is the realization of the programming model. Having saying that, it is important to understand that a programming model has to fulfill two contradictory requirements: on the one hand, it should be a general, broad abstraction of the computer architecture, and on the other hand, it should highlight and provide access to the specific attributes and capabilities of the specific computer architecture. A good programming model should provide the right balance, so as to exploit the maximal benefits, power and efficient implementation of the platform it relates to, while offering simple programming [382, 404, 405].

There are several other descriptions and definitions of programming models, where the most simplistic one is portraying a layered or modular model of the various components or the programming levels of an application, a utility or a tool, for example, Java programming model, C++ programming model, visual Basic programming model, or XML programming model. We are not using these definitions here.

Programming models are important for understanding how network processors' architectures are used by higher programming layers, or how programmers must use the network processors during design, or when programming. Usually, programming models are offered with specific programming languages (when they exist), and distinguishing between the two is not always clear. Programming models are offered by manufacturers of network processors (e.g., Intel's micro-ACE) [192], by software vendors (e.g., IP-Fabric's Packet Programming Language, PPL [327], Teja's Teja-NP), and by academia (NP-Click [383] and NetVM [41]).

8.1.2 Programming Paradigms

Programming paradigms are patterns, or models, that serve as disciplines for programming computers. They should not be confused with programming styles (i.e., how a program is written), or programming techniques (e.g., stepwise progress of a program or "divide and conquer" progress). There are four major programming paradigms: imperative, functional, object-oriented, and logic, as well as many other paradigms that are described in the vast literature of programming, such as: declarative, procedural, event-driven or flow-driven, structured or unstructured, visual, constraint based, and data-flow programming. Some of these paradigms have common attributes and can be combined, and some are known by their "taboos" (e.g., never use "go-to" or "jump" instruction in a procedure, never use states or variables, etc.).

For our purposes, there are two main programming paradigms that are used in network processors: imperative programming and declarative (mainly functional) programming.[2]

Imperative programming means that the program maintains states, which are modified by a sequence of commands over time. In the simplest terms, it means a directive concept, that is, "do this, and then do that." It is the oldest programming paradigm, and most computer hardware (von Neumann architecture based) and machine-languages (as described in the following subsection) support it naturally. Procedural programming is very similar to imperative programming, but it combines imperative programming with procedures, methods, functions, routines, or subprograms (pieces of a program that define a course of action for doing something).

Functional programming means that the programs evaluate functions and expressions, and use the results for further evaluations.[3] Although this is theoretically in contrast to imperative programming, where the emphasis is the sequence of commands, in practice, some mixture happens. The main cause for impure functional programming is side-effects,[4] that is, function evaluations that have some side effects that result in some state modifications, rather than just returning function values.

One of the advantages of the functional programming paradigm for network processors (which are usually made of multiprocessors), is that it is inherently capable of running concurrent functions. This is due to the fact that no state is

---

[2]Interestingly, these two paradigms are rooted back in the 1930s, when a group of scholars at Princeton University, among them Alan Turing and Alonzo Church (Turing's guide), described the Turing Machine and the Lambda Calculus (the basis of the imperative and functional programming paradigms, respectively). Ten years later, a third scholar of this group, John von Neumann, implemented the imperative paradigm in computer architecture, which became known as the von Neumann architecture. One important thing to note here is that one formulation of the Church-Turing thesis (there are various equivalent formulations), is that these two programming paradigms are equivalent in power [256, p. 232].

[3]Functional programming tends to be considered an impractical "academic paradigm," with lots of odd terminologies, requiring at least a PhD to work with it or explain it. The truth is that many of us use functional programming without even noticing it. Network processors definitely use this programming paradigm, as outlined in this chapter and detailed in its appendix.

[4]Side-effect is a term that is also used for processing elements, for example, arithmetic logic unit's (ALU's) operation that result in some flag modifications. Data and control hazards may occur in pipelining architectures due to these kinds of side-effects, as detailed in Chapter 7.

changed (no side-effects), therefore function execution can be done in parallel, independently, and to some extent—in any order (this has to be examined and scheduled though). Also, testing and debugging programs written in functional programming languages become easy, again, due to the lack of state changes, and the sufficient unit-test examination.

Declarative programming sometimes refers to functional, logic, or constrained based programming. However, declarative programming is actually the opposite of imperative programming, because it describes what is to be achieved, and not how to achieve it, like in imperative programming. Various implementations of declarative programming imply how programming happens eventually (sometimes there are imperative notions encapsulated in the declarative framework, to execute the required declarative statements). For our purposes, however, declarative programming refers to functional programming.

A very popular use of declarative programming is in "domain-specific" environments (e.g., simulations, mathematics, or packet-processing). In networking environments there are several languages that fall into this category, as well as being even more specific, for example, packet classification languages. Some examples of these languages are described in the following subsection and in the appendices of this chapter.

8.1.3 Programming Languages

We assume that most of the readers are familiar with programming language, from low-level, machine languages (assemblers), to high-level languages such as C, Java, and so on. For our purposes, there are also many functional languages that are domain specific rather than general purpose, which are sometimes known as very-high-level languages, and among these are several network-processing languages. However, all kinds of languages are used in network processing. Due to the critical nature and importance of performance, machine languages (assemblers) are used very often, at least in the critical packet data-path processing. After assembler languages, the second most commonly used language is C, or variants of C. For specific tasks, like classification, several functional languages have been offered by network processor manufacturers (e.g., Intel's NCL [192] and Agere's FPL [86]), software vendors (e.g., IP-Fabric's PPL [327]), and academia (e.g., NP-Click [383] and NetVM [41]).

8.1.4 Programming Architectures

Programming architecture usually refers to the way software is designed and built, for example, layers, structure, modules, procedures, macros, interface in general, application program interfaces (APIs), and so on. We shall not deal with this at all, as it is entirely out of the scope of this book; however, readers should be familiar with proper software coding and programming architectures.

8.1.5 Development Environments

When we talk about development environments (sometimes called Integrated Development Environments, or IDEs, or System Development Kits, or SDKs), we refer to programming environments and frameworks, including utilities and tools that aid the programmer in designing, coding (editing the source code), compiling, linking, simulating, debugging, and implementing the required software. A typical environment works like the programming environment we use for Windows (e.g., Visual Studio) or for UNIX (e.g., KDevelop or Eclipse). Most development environments also contain dependent modules of the network processors architecture, for example, hardware abstraction modules in source code, including files of run-time libraries. Intel's Internet eXchange Architecture (IXA), packaged in Intel's SDKs for their Internet eXchange Processor (IXP) architecture, is an example of such a system.

There are many development environments, and usually every network processor provides his own. In addition, there are many others offered by software vendors (e.g., Teja) to provide better tools to ease the programming of network processors, or to provide a tool that will be agnostic to any hardware that might be chosen later. These later types of development environments are called cross IDEs.

Development environments are a critical part of network processor programming. As mentioned before, network processors are often hard to program, and program-writing is difficult without the aid of a very functional development environment that can assist in writing, using macros, definitions, and later in visualizing the use of memories, registers, and all processing elements. Moreover, since all development environments for network processors deal with networking issues, a very handy functionality is to offer assistance in referring to specific networking terminologies, packet structures, header structures and protocols, as the programmer usually needs to refer to these as his raw data.

8.1.6 Summary

In this introductory section we briefly described how programming models are relevant for network processors, what programming languages are used, and what programming environments offer. While this discussion related to network processing programming in general, we must also distinguish between data-path and control-path programming, since they are very different in all respects. The next sections provide more details on how to handle each.

8.2 CONVENTIONAL SYSTEMS

Although we are dealing with network processors and with data- and control-path processing, it is more than likely that the system into which network processors are integrated contains many "conventional" subsystems, at least the control-path processing subsystem. Therefore, a short overview of software of "conventional" systems can illuminate the entire picture and assist in better design and implementation of networking equipment based on, or containing, network processors.

8.2.1 Introduction

By conventional systems, we mean all kinds of networking equipment that include general-purpose-processor-based equipment, and that comprises the majority of the networking equipment in use today. This includes a very wide variety of systems, from tiny Small Office and Home Office (SOHO) equipment (gateways, interfaces, adopters, etc.), Network Interface Cards (NICs) in Personal Computers (PCs) to peripheral equipment (printers, storage devices, etc.), branch-office interface equipment, data-center equipment, and up to huge core routers and telecom switches. These systems use various types of microprocessors, multiple processor systems, and many types of programming models, paradigms, and languages.

In network processor-based equipment, the control-path processing, as well as the equipment management and maintenance processing, are done by "conventional" subsystems. The software architecture of a conventional system, in a nutshell, may take the following layered structure, as shown in Figure 8.1.

The *operating system* (OS) is the main layer that manages, synchronizes, and utilizes all hardware and software resources available on the managed platform (the hardware). Some parts of the OS must reside at all time in the memory, while other parts can be stored in a storage device and fetched when required. The OS usually runs in a privileged mode, uninterrupted, and at highest priority. The OS provides services (like memory management, task scheduling, I/O, access and security management, file systems and networking, etc.) to the applications, and interconnects

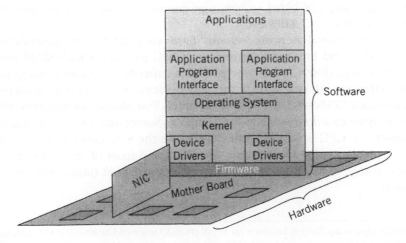

FIGURE 8.1

Software architecture

the user (through the I/O) with all applications and the other I/O devices. There are several types of OSs, and the relevant type for most networking equipment is Real-Time OS (RTOS). The core of the OS, the part that always resides in the memory, is called the *kernel*.

There are two kinds of kernels—microkernels and monolithic kernels; microkernels contain the most essential components, such as memory and task (and thread) management, whereas monolithic kernels include everything, including networking, file systems, and so on (e.g., Linux). The kernel is loaded into memory at the time of initialization (in what is called *boot* or *bootstrapping*), and it loads the rest of the required system. The kernel manages the CPU, the memory and the I/O, and interacts with the applications and the other parts of the OS by system calls or inter-process communications (IPC), and with the hardware through the *device drivers* by interrupts or timely polling schemes. At the lowest layer, there is the *firmware*, which is software that is embedded in the hardware components— either in their Read Only Memory (ROM), Erasable and Programmable ROM (EPROM), or flash memory. The device drivers, the kernel, and the OS interact with the hardware by using the firmware.

Network interfaces for these hardware platforms are either: (1) embedded interfaces, (2) NICs, or (3) line cards in "heavy" network equipment. NICs and line-cards contain software, and the entire description given above on software is relevant to them as well.

Some of the conventional network systems' hardware platforms are just general purpose computers that are equipped with network interfaces (usually NICs) and that run some application on top of an OS. This OS can be either a general purpose type (Linux, for example, is quite popular for this type of system), or a dedicated type, sometimes RTOS, for example, QNX or VxWorks. Applications interface with the OS by an Application Program Interface (API)[5] or by using system calls. These systems usually use x86-based architecture microprocessors, or some RISC-based architectures (PowerPC, MIPS, etc.).

Other conventional network systems' hardware platforms are embedded systems (e.g., NICs and line-cards, printers or SOHO gateways). Embedded systems are designed to perform mainly one predefined specific application, using microprocessors that are integral parts of the devices they control. These systems use 32, 16 or 8 bits data CPUs such as ARM, Xscale, Intel's 960, Motorola's 860, PowerPC, or MIPS, or simpler examples such as 8051 and PIC. Sometimes, as noted in the previous chapter, the CPU is further embedded inside the SoC, either in ASIC or FPGA designs. Typically, embedded systems use RTOS, and most of the software, called *firmware,* is stored in a ROM or in other type of permanent (nonvolatile) memory (e.g., flash).

---

[5]The Portable Operating System Interface for UNIX (POSIX) is a well-known open API standard that emerged from UNIX, was adopted by the IEEE, and is now supported by many OSs.

8.2.2 **Principles**

Operating Systems are used in some of the systems; other systems may use a thin kernel, or a customized, special-purpose OS, or one based on a developing platform that eases the application development. Some systems use specific, one-layer software design that includes device-drivers, kernel functions and the application itself. Others may use a general-purpose OS that has some middleware (infrastructure software services) above it, with applications on top. This broad spectrum of systems makes it difficult to categorize the conventional programming model. Embedded systems, as well as other networking applications that use general-purpose computers, use various programming models to perform the required networking applications.

Interrupt-driven (or *event-driven*) software, is used by the kernel (in the case of layered software architecture, particularly when OS is used), or by an application (in the case of one homogeneous block of code that does everything). In interrupt-driven software, I/O events, set-up timers, and other events create interrupts, which cause the processor to call a corresponding module in the program (that matches the interrupt). Software interrupts can be created to have the same effect; these interrupts are caused by protocol software or events created or calculated by procedures, so as to prioritize some urgent (or insignificant) tasks that the software wishes to prioritize. Interrupt-driven processing is typically asynchronous in nature, which means that the program is written in blocks, and events dictate the flow of execution.

Another scheme for running a control program (usually a simple application used by an embedded system) is to loop indefinitely, during which time software states and events are polled, and hardware is polled for status, information, and events. Tasks are then called according to these events or states to perform some timely processes. This scheme can be applied in conjunction to a state machine concept, whereas at every loop, states are examined and procedures are triggered and executed according to the states, causing the states themselves to be modified according to the process results and the events that occurred. This scheme can be viewed as synchronous, since one can synchronize processing and event handling by the order of the flow of the program. This scheme, however, can also use interrupt-driven processing combined with the polling, to ensure some critical real-time event handling, or to prioritize the tasks within the control loop.

Some systems use an object-oriented programming paradigm, where the application is designed and implemented accordingly. This, however, is outside the scope of this book.

Other, more advanced schemes execute tasks concurrently, either using multiprocessing when it is possible, or using techniques like multiprogramming, multitasking, or multithreading. These were discussed briefly in Chapter 7 and described further in the following.

8.2.3 Multiprogramming, Multitasking, and Multithreading

Programs, or applications, are copied into memory (what we call *processes*) and executed by the processors. *Tasks* are relatively independent parts of the programs, for example, subprograms, subroutines, procedures, functions, methods, and so on. *Threads* are dependent splits of processes and tasks, and are mainly used for seemingly simultaneous, inter-correlated execution.

By multiprogramming, multitasking, and multithreading, we usually mean how to run programs simultaneously, whether from the user perspective (when we use only one processor) or in actual parallel execution, when we use some MIMD architecture.

Multiprogramming is quite an old scheme, used primarily on mainframes in the past, and it simply indicates that several programs run on a uniprocessor interchangeably, that is, the programs are switched in execution for every time slot, interrupt, I/O request, virtual memory management operation, and so on. In multiprogramming, there is no intercommunication or synchronization at the process level, and multiprogramming is therefore not relevant to network processors. From here on, we will focus on multitasking and multithreading.

When more than one process is executed on a single processor, the processor picks a process to execute for every short time period (slot), and switches to another process in the following slot. These slots can be fixed or varied, and can be the results of time-outs, interrupts, or resource constraints. Whatever the reason is for process switching, each replacement of a process in these slots is called *context switching*, since the processor changes the context it executes, and requires some handling, such as storing and retrieving process status and register contents. These switch contexts create overhead on the system, but nevertheless are worth doing (for example, if the alternative is to wait for an I/O to be completed); moreover, without switch contexts, there cannot be any multitasking or multithreading. The mechanism of context switching described above, in which every assigned time slot is preemptive, means that a scheduler triggers and decides which task or thread is replacing what and when. *Preemptive multitasking* is done in most OSs that schedule the tasks and the threads to be executed. Nonpreemptive multitasking (or *cooperative multitasking*) is more common in control programs (described before), where during the infinite loop, each task schedules the one that follows it when it finds it appropriate to do so (either by limiting itself, recognizing the importance of another task given a change of state, or simply calling another task it requires).

Multitasking and multithreading can be applied to uniprocessors as well as to multiprocessors. In multiprocessors' architectures, multitasking and multithreading makes the system much more efficient, and actually, without multitasking or multithreading, it is even hard to exploit multiprocessors. Some view the main difference between multitasking and multithreading as memory usage: Threads always share the same memory (or address space), whereas multitasking can be done also in separated (or distributed) memories, by using specific mechanisms such as the Remote Procedure Call (RPC).

Multithreading refers to the ability to run many threads simultaneously, as described previously. It is implemented at two levels: the processor level (described in Chapter 7), and the software level. The processor level is concerned with how the processor (the hardware) does multithreading, regardless of the software, and the software level is concerned with how programmers call for the multithreading capabilities of the processor, and how they write programs that make use of multithreading.

At the processor level, multithreading can be done in *fine-grained multithreading* (FGMT), *coarse-grained multithreading* (CGMT), or *Simultaneous Multithreading* (SMT). FGMT, or *interleaving multithreading*, switches between threads every machine cycle, and the processor that does it is called a *barrel processor*. This scheme guarantees real time properties, but is not efficient in cache utilization and single thread performance. In pipeline architectures it wastes many cycles due to cache misses, mispredicted branches, and requires many threads to be effective and many register files to handle all context switching. CGMT, on the other extreme, lets a thread run until a stall occurs that yields a context switching. It means that in pipelining, the processor can work as efficiently as possible, and no thread's performance is sacrificed, although long latencies occur in context switching. SMT means that the processor (or multiprocessor) executes instructions from a number of threads simultaneously, and it is like CGMT with an out-of-ordering[6] execution that makes the threads' execution more efficient (in Intel's case, this is called *hyper-threading*).

At the software implementation level, we have to distinguish between two sublevels: one is how to use the hardware multithreading architecture of the processor, which depends on how multithreading is actually done, and the other is how to write multithreading programs. The second question, how to write multithreading programs, is answered in the next section. As for the first question—it is done by specific calls to the OS or kernel module that handles the processor scheduling, with a request to create a thread, kill one, schedule it, and so on. Usually threads relate to each other (and communicate) as part of a parent-task, for example, one thread fetches and organizes data, a second thread is responsible for processing the data and putting the results in a buffer, while the third thread format the results and outputs them. Threads can also be independent, and the only common thing between them is that they use a shared memory that makes context switching faster.

---

[6]As explained in the previous chapter, instructions can be executed in a different order than they appear in the program, or out-of-order, if they are totally independent (they use independent data and have no impact on the flow of the program). This is also called *dynamic scheduling*, and is in contrast to *in-order* instruction execution (or *static scheduling*).

Although we are discussing conventional systems at this point, an example from network processors can clarify how multithreading is implemented: In Intel's IXP's architecture of network processors, every micro-engine (the basic processing element that runs in parallel with other micro-engines and executes data-path processing) runs several identified threads,[7] each connected to predefined ports, and the threads are defined and used by specific software calls at the upper software layer (the application). Another example is Cisco's Quantum Flow Processor (QFP) with 40 cores, called Packet Processor Engines (PPE); each is a 32-bit RISC that runs four threads.

8.3 PROGRAMMING MODELS CLASSIFICATION

Programming models can be classified according to the underlying architectures, primarily uniprocessors, multiprocessors, and data flow. The various processors' configurations and processing methods that we described in the previous chapter lead to the following subclassifications:

- Uniprocessors/sequential programming.
  - SISD, or von Neumann architectures.
  - Multiprogramming.
  - Pipelining (at the instructions level).
- Multiprocessors/parallel and concurrent programming.
  - Parallel (SMP and not SMP).
    i) Distributed memory.
    ii) Shared memory.
  - Pipelines (at the processors level).
    i) Systolic arrays.
- Data-flow processing/data-driven programming.

As described in the previous chapters, the last classification, data flow, is not really implemented in network processors, although some vendors claim to do so.

Since most network processors are multiprocessors, or have many processing elements in their architecture, the two most important classifications that are relevant to network processors are *parallel programming* and *pipelining*. However, pipelining can be done at the processors level (i.e., task-pipelining, when we have multiprocessors) or it can be instruction-level pipelining (in each of the processors). As was mentioned in the previous chapter, instruction pipelining is also very common in high-performance processor architectures, and therefore is also very common in network processors. Although they are programmed very differently, they have the same principles, and we shall refer mainly to instruction pipeline in the following discussion of pipeline programming.

---

[7]IXP1200 uses four hardware threads and IXP2xxx uses eight.

8.4 PARALLEL PROGRAMMING

Parallel programming has been researched and used for many decades now, mainly for high-performance financial and scientific applications that use very complex calculations and simulations (such as geophysics, weather and climate, etc.) and that may even use super-computers. Parallel programming models provide a bridge, or a mapping, between software and the multiprocessing hardware, as noted above. For our purposes, we need to understand the principles of parallel programming models that apply to network processors and to multiprocessors in networking applications, as these are an important category of network processors. Moreover, an important architectural candidate for executing specific kinds of networking applications, is parallel, general-purpose processors, grouped together on a chip (CMPs), or MCPs. These are used primarily for high-level applications that handle layers 5 to 7 (e.g., security, storage, accounting) and it is possible that they will be used for next generation network processor architectures.

8.4.1 Definitions

Many definitions exist for parallel programming models: some are biased toward computation models, others describe coding styles and practices, and still others relate to actual programming techniques and models. All three of these types are described as follows.

8.4.1.1 Parallel Computation Models

The main parallel computation models are Parallel Random Access Machines (PRAM) [133], Bulk Synchronous Parallel (BSP) [416], and a model based on Latency, overhead, gap, and Processors parameters (the LogP model) [133]. Each of these models has many variants, but for our purposes, it is sufficient to know the main principles.

Parallel Random Access Machine is essentially a shared memory MIMD programming model, and it is an ideal parallel computer, according to which there are p parallel processors that run one application, and use a shared memory with synchronous, simultaneous access. Several subcategories of PRAM describe the way data is accessed, among them the Exclusive Read Exclusive Write and Concurrent Read Exclusive Write models. The PRAM model is widely considered to be unrealistic for describing multiprocessor architecture, as it lacks, for example, the communications overhead and memory contention [165].

Bulk Synchronous Parallel is a distributed memory MIMD programming model used for synchronized multiprocessing, according to which there are p nodes that are each composed of a processor and its local memory. The nodes are all connected by an interconnection network, with a barrier synchronizing facility that enables the processors to interact. It models parallelism by multiple (p)

sequences of supersteps, which each consist of three phases: local computation, communications, and barrier synchronization. By the end of the synchronization, the exchanged data is available in all p local memories. The supersteps are parameterized by L, the synchronization period, and g, the ratio between computation and communications throughput.

LogP is a more detailed distributed memory MIMD programming model that emphasizes the interconnection network, with p processors doing the computing and an interconnect network over which they communicate by point-to-point messages. It is a more constrained message passing mechanism than BSP (message passing is described in the next section), and it lacks explicit synchronization [165]. LogP is parameterized by latency (L) of the message from a source processor to a sink processor, overhead (o) time in which the processor is engaged in I/O, gap (g) time between consecutive messages, and P, the number of processors. The interconnection network is considered to be a pipeline of length L, having a processor overhead o at each side, and a packet initiation rate g.

8.4.1.2 Parallel Coding Models

There is one important distinction between the various parallel programming models that impacts the coding effort significantly, and is expressed in the programming languages and development platforms. This is the way synchronization is defined and achieved between the parallel processes accessing shared data and results (which are not necessarily in the same memory). *Explicit parallelism* (in contrast to automatic, *implied parallelism*) offers full control over parallelism during the execution, but with the high penalty of having to define specifically when and where each module or variable can work and interact with another. This is done by using specific run-time libraries that enable the programmer to call the required system functions (create threads, synchronize data, etc.) that eventually facilitate parallelism. It can be eased by having compiler directives imbedded in the C/C++ or assembler source code that advise the compiler about these interrelations for creating an executable code that is as efficient and optimal as possible. *Implied parallelism* stands at the other extreme, where parallelism is done either as a result of hardware-assist or architecture, and the programmer actually thinks and works on a scalar problem.

In network processing, this distinction comes into effect when many processors are working on many packets simultaneously. Parallelism can be accomplished when each packet has to be assigned manually to a processor and may be also to a thread, and intermediate results and data have to be synchronized with other threads, processes, processors, ports, and packets. In addition, the programmer has to be aware of resource sharing, load balancing, and so on. This is the explicit parallel programming penalty. Alternatively, parallelism can be also accomplished when the programmer has one copy (single image) of the program to consider, working on one packet. Actually, the hardware (or a scheduler) takes care of executing multiple copies of this program on many packets simultaneously, takes care of resource sharing, and takes care of load balancing, critical timing, and so on.

8.4.1.3 Parallel Programming Models

Initially, there were just two types of parallel programming models: data parallelism and message passing. These were generalized, and more finely tuned models are commonly used today. Today we talk about:

- Data parallelism
- Message passing
- Shared memory
- Thread-based parallelism

The programming models and architectures are not always the same, as, for example, message passing can be implemented in shared-memory architectures. The following subsections detail these models.

8.4.2 Data Parallelism

Data parallelism is not so relevant to network processors, since it is usually applied to large data arrays (e.g., matrices representations). Data parallelism is a highly abstracted approach that has a single-instruction control structure, a global name space (variables are globally accessible), and synchronous parallel execution on data variable and arrays. An example of a language extension that uses data parallelism is High-Performance Fortran [137, 261], which is based on the Single Program Multiple Data (SPMD) model described in the previous chapter.

8.4.3 Message Passing

Message Passing (MP) programming means that each node has its local memory, and many processes or threads, residing in this node or in other nodes, can access the memory by receiving and sending messages among themselves (Figure 8.2). The processes and threads can use their own names for variables, and the variables have to be sent and received among the processes and threads by some communications means (e.g., bus or interconnect network). Synchronization is created by the need for communication, but is also required in each of the processors, to ensure data integrity and validity.

In order to enable message passing, specific protocols are required, especially the definition and structure of the exchanged data, as well as the preparation of the data that is to be exchanged (loading, formatting, storing, etc.). Although the message passing paradigm is a simple concept for achieving parallelism, the programming that is required to enable it is a bit more complicated than simply sharing variables in common memory, which is described in the following. The first implementations of the message passing concept appeared with the parallel machines of the mid-1980s, as sets of library calls for the Fortran and C languages that were used to connect processors working simultaneously on partitioned data

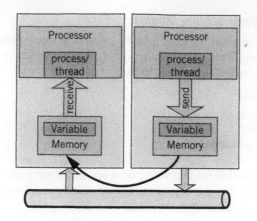

FIGURE 8.2

Message passing

over a communications network. Later, in the beginning of the 1990s, a de facto standard for message passing known as the Message Passing Interface (MPI) [316] was adopted to allow code portability. At about the same time, another paradigm, the Parallel Virtual Machine (PVM) [146,359] was offered mainly for interoperability. Based on these standards and languages, today many scientific, engineering, medical, industrial, Peer-to-Peer (P2P), and grid computing applications use clusters or network of workstations or other parallel (distributed) computing used by many computers.

Remote Procedure Call (RPC) is a computing scheme that is used for parallel, distributed memory computing, and which uses message passing [391]. RPC enables one procedure to call another procedure with a message (parameters or information, as described by an external data representation standard, XDR [113, 392]), and to wait for its results (data). Equivalent paradigms are the Open Software Foundation's Distributed Computing Environment, Microsoft's Distributed Component Object Model and .NET, and Sun's Java Remote Method Invocation.

For our purposes, message passing is important in parallel implementations of multiprocessors in network processors that are not using a shared memory. Its principles can be applied to the network processor interface to the host-processor, that is, interfacing the fast-path processing with the slow-path processing. However, MPI or its derivatives are not used in network processors, and each vendor uses its own schemes—some of which are software-based, as described above, some of which are hardware-assisted. Hardware-assisted message passing means that the processors prepare the messages that carry variables, state information, packet segments, processing results, instructions, and so on, and once a scheduler assigns a processor to work on a packet, it attaches the relevant message as well. In the case of host-NP communications, hardware assists in making the relevant buffer or pointer available to the processors, and interrupts or otherwise triggers the message passing.

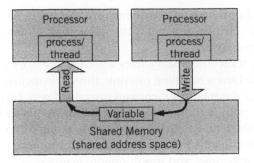

FIGURE 8.3

Shared memory

8.4.4 Shared Memory

The Shared Address Space (SAS) programming model allows many processes (or tasks, or threads) to load and store contents using the same variable, whether it has a different or a common name, as if these activities were done in their own local memory (Figure 8.3). Data and resource synchronization becomes critical, however, especially between writes and reads. Various algorithms, programming structures, and programming means are used to enable this synchronization, by using locking mechanisms, semaphores, mutually exclusive objects,[8] and so on.

Network processors use shared memory very often, and the usage can be classified into two main patterns: networking data (packets) that are accessed in parallel by multiple processors working on the same packet (usually in pipelining configurations), and state and control shared data (e.g., search tables, statistical tables, state variables) used by multiple processors working on different packets. Synchronization of access to the shared data (packets or other data) is done either by software means (as outlined briefly earlier) or by hardware means (network processor circuitry, creating mutually exclusive access schemes).

8.4.5 Thread-Based Parallelism

Threads have been described several times in this book from different aspects: first, in Chapter 7, from the hardware viewpoint, as part of the processors architectures (i.e., how multithreading is implemented by the CPUs), and then briefly from the perspective of the OS (or the kernel) (i.e., how threads are physically mapped on the processors by the kernel or the OS by specific system calls

---

[8]A mutually exclusive object is called a *mutex*, and is a very common way to manage concurrent use of a shared resource, which can be a variable or a piece of code. The shared resource may have to be restricted from use at some points of time or states (e.g., interrupts cannot interrupt themselves endlessly, or even once if they are not written in a recursive way, that is, they might override variables that the calling code is using).

that use the specific processing elements appropriately). In this subsection, we concentrate on the third layer of multithreading, that is, how to use multithreading in the application layer.

Thread-based parallelism usually uses a shared memory model. The focus in thread-based parallelism is handling multiple threads concurrently, so that the task will be executed in the most efficient and organized way. Obviously, synchronization is as critical a point in thread-based parallelism as it is in the shared memory model, but beside the data synchronization that is required to maintain data integrity and consistency, thread synchronization is also important to ensure that tasks are done in the proper order, and in a complete manner.

A common scheme that is used for thread-based parallelism is the Fork–Join model. In Fork–Join, there is a master thread running in a sequential region of the process, and at some point in the beginning of the parallel region, a fork creates a team of parallel threads that run concurrently and join at the end of the parallel region. The join occurs after the threads synchronously completed their missions, leaving the master thread to continue.

There are several standardization efforts for thread-based parallelism, among them the Unified Parallel C (UPC), POSIX threads (Pthreads), and OpenMP [298]. In UPC, the programmer views multiple threads, each of which seems to have its own memory and affinity with some shared memory. But in fact, the threads can use either SAS or distributed memory, or even the SPMD model (described in the previous chapter). POSIX threads are C language UNIX-type API for handling threads, thread attributes, mutexes (which synchronize threads based on access to data), and conditional variables (which synchronize threads based on data values). OpenMP stands for open multiprocessing, and is a C/C++/Fortan language API that is based in a multiple, nested Fork–Join model.[9] It is a scalable, portable model for multithreading, shared memory parallelism. OpenMP defines compiler directives, run-time libraries, and environment variables (that control the execution of parallel code).

8.4.5.1 Threads in Network Applications

Thread-based processing in networking applications can be done by having different threads assigned to handle different layers, different protocols within a layer, or different packets. It means, for example, that all tasks belonging to a particular layer (e.g., networking) that usually have the same priority as all other tasks at this layer, and higher priority than those in the layers above, will be handled by one or more high-priority threads that are dedicated to this layer. This simplifies software design, as threads call each other when interlayer treatments are required (e.g., threads of the data-link layers call lower-priority threads of the networking

---

[9]Nested parallelism means that a thread can be split into several threads at some point. A nested Fork–Join model means that several threads can undergo the Fork–Join structure in the parallel region, that is, fork threads that are joined and terminated before the sub-master threads join again to the master thread.

layer, which call even lower-priority threads in the transmission layer, which call the lowest-priority threads at the application layer). In addition to simplicity, this scheme has the advantage of taking care of priorities; however, it suffers from the high context-switching overhead that is required when a packet is handled throughout the layers.

Another simplification is when specific threads are assigned to specific protocols (e.g., TCP or UPD in the transmission layer, or FTP or Telnet in the application layer). Again, the major drawback is context-switching overhead. A better way of dealing with packet processing is to assign a thread, or a couple of threads to a packet, from its entry in the system to its emergence out of it, and to conduct the entire process using these threads perhaps in several stages (see task pipelining in Section 8.5).

8.4.5.2 Threads in Network Processors

As was mentioned at the end of subsection 8.2.3, Intel's IXP and Cisco's QFP architectures of network processors make use of threads. However, this is quite unique among current network processors' architectures, despite the fact that multicore architectures and multiprocessors systems on a chip (MPSoC) that potentially can be used for network processing may well be suitable for multithreading [91, 437]. Other network processors (including Intel's IXP architecture) use either pipelining or other parallel computing models. The next section describes pipeline programming in more detail.

It is important to note that threads at the software level should be exercised with great care, if they are used at all [275]. In many cases, threads are non-deterministic and make programs unreadable and very hard to use. Furthermore, embedded computing generally avoids using threads [275].

8.5 PIPELINING

As was mentioned in the previous chapter, pipeline architectures and pipeline programming are often used in the line-cards of networking equipment, where ultra processing speed is a must. As efficiency in pipeline programming is a key requirement, high-level languages, programming models or any programming environment that cannot exploit the maximal performance of the pipeline hardware will probably not be used. Therefore, some pipelining practices are required to exploit the maximal performance of the pipelining architectures, as described in the following.

Although there is no formal pipeline programming model, or even pipelining API, many of the parallel synchronization issues described in the parallel programming model exist in pipelining. For example, in instruction pipelining—when we talk about a uniprocessor running one or several threads interchangeably during several machine cycles (time)—data hazards can be considered synchronization problems. Synchronization is more of an issue when we discuss process pipelining by several processors, where data synchronization can be problematic in exactly the same ways as in parallel programming. Pipelining, however, poses additional problems, such as control hazards that have to be considered.

We introduced data and control hazards in Chapter 7 on architecture; in this section, we extend the discussion to the software implications. We will focus here on instruction pipelining, since it involves programming considerations and is common to most network processors. However, it is also worth saying a few words about task pipelining.

Task pipelining appears in several network processors, and implies cross-references to the same data, for example, packet, look-up tables, statistical tables, and so on. These problems are usually handled by the applications (and those who write the applications must be aware of them, and implement some synchronization mechanism). Some network processors avoid synchronization problems with packet data by "carrying" it across processors along the pipeline; however, the cost is that data is copied and cannot be cross-referenced or handled. Still, look-up tables, statistics, and other common data should be synchronized.

Instruction pipelining in processors can be handled by schemes that resolve hazards; for example, like automatic flushing of instructions from the pipe once there is a control hazard, or insertion of no-operation cycles, stall the pipe, or reorder instructions in cases of data or control hazards. These mechanisms (except instruction reordering) generally waste machine cycles, and in the case of network processors, when every machine cycle counts, it might not be a satisfactory solution. When high-level languages like C/C++ are used, there is always some waste of machine cycles due to the compiler that cannot take advantage of the program logic flow in order to reorder instructions and use pipelining instructions wisely. Very high-level languages, such as functional languages, are potentially capable of reordering instructions, since they are usually designed around the data-flow logic.

Using as an example the case of a four-stage instruction pipeline, the next subsection describes how to write efficient code in assembly.

8.5.1 Instruction Pipeline Overview

For the sake of this example, let's assume a four-stage pipelined processor, with each stage lasting one clock cycle. In the first stage (F in Figure 8.4), an instruction is fetched from the instruction memory block and stored. The instruction is decoded in the second stage (D in Figure 8.4). The third and fourth stages (E1 and E2, respectively, in Figure 8.4) are the two execution stages, where E1 is generally used for reading the instruction source data and E2 is generally used for writing the destination data.

With each clock cycle, another instruction is fetched and begins its four-cycle execution. Since source is read at E1 in each instruction, the source data must be valid at that stage. The destination data is written at E2, and is valid only after E2.

Program Execution	Clock Cycles						
	CC1	CC2	CC3	CC4	CC5	CC6	CC7
INST#1	F	D	E1	E2			
INST#2		F	D	E1	E2		
INST#3			F	D	E1	E2	
INST#4				F	D	E1	E2

FIGURE 8.4

Instruction pipeline execution

Therefore, the source resource for an instruction cannot be the same as the destination resource for the previous instruction.

8.5.2 Data and Control Hazards

There are pipeline events where the next instruction cannot execute properly in the next clock cycle(s). These events are called hazards. With a data hazard, an instruction depends on the results of a previous instruction still in the pipeline. A control hazard arises from the delay in determining the appropriate branch instruction to fetch.

8.5.2.1 Data Hazards

An instruction is actually executed in the two clock cycles of E1 and E2 stages. As described above, in each of the instructions, the destination data is always written in the fourth cycle (E2), and cannot be read correctly until the next clock cycle. As long as the INST#1 destination resource write cycle has not been completed, the source resource of INST#2 cannot be the same as the destination resource of INST#1 (see Figure 8.4). Only INST#3 can use the destination resource from INST#1. Most registers (used as destination resources) are undefined for the duration of one clock after being written to.

Instructions must be organized (reordered) so that a nondependent instruction, which does not read the resource, is added between instructions. A NOP (No OPeration) command must be inserted if no useful instruction can be found to insert between the two resource-dependent instructions, but this is, of course, a waste of one cycle. Most general purpose pipeline processors are capable of doing reordering or stalling as required, and the programmer is not bothered by data hazards. In network processors, this is usually not the case, since every clock cycle counts, and the programmer is left with the responsibility of dealing with data hazards. The following sample code is illegal, as it contains a data hazard.

Example of bas code:

```
MOV  REGA, REGB, 2;
MOV  REGC, REGA, 2;          // REGA is not ready
```

8.5.2.2 Control Hazards

Control hazards can be the result of pipeline processing, and arise during branching. Execution occurs in the third and fourth stages of the four-stage pipeline (see Figure 8.5). In E1, the branch instruction checks the condition code operand, decides whether or not to branch, and computes the next address. (If the branch is not taken, the pipeline continues with the PC+1 address.) No execution occurs in E2 for branching.

Since the branch is not executed until the E2 stage, two other consequent instructions (from addresses PC+1 and PC+2) enter the pipeline. Only the instruction in the fourth cycle (INST#4) fetches the instruction from the branched address. The result is that all four instructions are executed, despite the fact that the first instruction was a jump, and the next two instructions were not supposed to be executed. If these two instructions, following the branch, are nondependent (and avoid data hazard), one possibility for solving this data hazard is to reorder the instructions, as explained in the following.

General-purpose pipeline processors usually cope with control hazards by either flushing the pipeline in case of a branch (thus losing cycles), or trying to predict if and whereto the branch will take place. Then, the processor fetches instructions from the predicted place immediately following the branch instruction into the pipeline (and thus, if the prediction was right, no cycles are lost). Some processors even fetch instructions from several potential branching addresses to a number of pipelines, and pick the right pipeline after the branch takes place. At any rate, the programmer is unaware of control hazards in general-purpose pipeline processors, and can program by ignoring their potential possibility.

In network processors, where each cycle is crucial, programmers are supposed to deal with the control hazard by either "creating" flushing (by adding NOPs) or by reordering instructions. To make use of the control hazard, instructions should be organized so that two instructions relating to the program logic are added *after* INST#1, before INST#4 is executed, and are executed whether the branch is taken

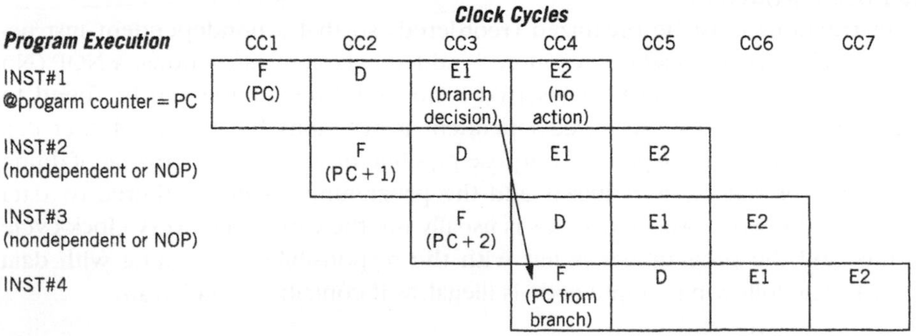

Program Execution	Clock Cycles						
	CC1	CC2	CC3	CC4	CC5	CC6	CC7
INST#1 @progarm counter = PC	F (PC)	D	E1 (branch decision)	E2 (no action)			
INST#2 (nondependent or NOP)		F (PC + 1)	D	E1	E2		
INST#3 (nondependent or NOP)			F (PC + 2)	D	E1	E2	
INST#4				F (PC from branch)	D	E1	E2

FIGURE 8.5

Pipeline execution of a branch

or not, without influencing the branching. If no useful instructions can be added, NOP commands must be inserted. When no NOPs are inserted, the next two instructions are executed, as just described. If one NOP is inserted, then just one instruction, which should be a nondependent instruction, is executed. Otherwise, two NOPs should be inserted.

Care should be exercised when several jump instructions are sequenced, as the first jump will happen, but immediately after executing one instruction from the branched place, the next jump will occur, and the next instruction to be executed will come from the second branched address.

Example: How branching happens in a four-stage pipeline:

```
JMP LABELA;
ADD REGB, 4;
MOV REGC, 0x8000;
....
....
LABELA:
SUB REGD, REGC;
MOV REGA, 5;
....
```

Description: The first jump (to LABELA) will be done, but the two instructions following the Jmp LABELA will be executed also (i.e., ADD REGB, 4 and MOV REGC, 0x8000) before the instructions at LABELA (i.e., SUB REGD, REGC, etc.)

8.6 NETWORK PROCESSOR PROGRAMMING

In this section we integrate all the issues discussed theoretically in the previous sections, and describe programming models, software practices, languages, and development environments that are relevant to network processors. We begin by discussing two major issues that dominate the way programming is done on network processors and relate to pipelining and parallelism. We then discuss programming languages, functional programming, and programming models, as far as they are relevant to using network processors.

As we have mentioned, we can identify two kinds of network processors applications (high- and low-layer networking applications), and their corresponding architectures (parallel and pipelining, respectively). Each also has corresponding programming practices: high-level applications use high-level languages and programming models that simplify and shorten the programming process, ensuring reliability and expressiveness at the expense of performance; and low-level packet processing (at the line-cards level) rely on low-level assembly language, probably without programming models or any other overhead that might jeopardize performance.

Although network processors use simple processing elements, it is very complicated to program them due to their architectural heterogeneity, complexity, and strict performance requirements. However, some common characteristics of network processors' programming can be identified:

- Real-time programming.
- Event-driven programming (in which incoming packets are the events).
- Multiprocessing; parallel programming, pipeline programming, or both.
- Interupt-driven[10] programming is most unlikely to be found.
- Preemptive multitasking is most unlikely to be supported. Most of the network processors work in "cooperative multitasking," that is, one task calls another one.

Additionally, most network processors' program structures, or program-flows, are of "rule/pattern and action" schemes. This scheme is underneath the complex assembly code that eventually describes the rules, patterns, and actions. This scheme can also be directly expressed in functional programming languages, as detailed in the following. This scheme, "rule/pattern and action," is also the basis for most of the existing programming models, as well as those that are now being developed.

Finally, most network processors come with some development environments that include many examples for network applications, and allow coding, debugging, simulation, in-circuit emulation, and software uploading.

8.6.1 Run-to-Complete versus Run-to-Timer

As we have mentioned repeatedly throughout this book, network processors have to process packets at wire speed. There are two fundamental approaches to achieving this:

- Run-to-timer, hard-pipelining, task-level parallelism,[11] or fixed-processing rate (which all produce the same rate of performance), or
- Run-to-completion

Run-to-timer, hard-pipelining, task-level parallelism, or fixed-processing rate of a packet guarantees a predefined packet processing rate, and although latency might be relatively high, this mechanism ensures that packets can be processed at wire-speed. According to this method, there is a processor, a task, or a thread assigned to every micro-phase processing of the packet. They can work in parallel, but the common characteristic of these mechanisms is that they are bound by

---

[10]In this discussion, we exclude a category of CMPs and MCPs that are composed of several general-purpose processors grouped together and that are used for high-level networking applications (e.g., security). These processors are also sometimes called "network processors."

[11]Task-level parallelism does not refer here to general or parallel multitasking, but rather to many tasks that are running in parallel and have to be synchronized in their execution time.

either by time or number of instructions, and programming has to consider these constraints. Since the amount of work that can be done on a packet is bounded, this scheme cannot be used if there is a requirement to manipulate some packets according to more complex, variable, or flexible algorithms. Moreover, software that is written in this way can pose a challenge in later modifications, as the timely execution of each phase might call for large portions of the software to be rewritten.

Run-to-complete is basically a nonpreemptive mechanism, under which every task is executed according to its requirements until it is completed, and wire-speed is achieved by parallelizing the processors that are doing the run-to-complete tasks. In this method, every packet takes its time, as required, and therefore, packets that have very simple processing requirements finish quickly, enabling other packets to be processed. Packets that require complex treatment can receive it and still maintain wire speed processing, by taking advantage of the processing time of those packets with lighter requirements. In this method, however, latencies are variable, and should be taken care of either by limiting the processing when writing the application or measuring processing time. On the average, the processing time should be such that all packets will be processed at wire-speed. It is important to understand that run-to-completion actually allows achieving a fixed processing rate by simply writing the number of instructions that will create a fixed latency, fixed processing time, and therefore fixed packet processing rate.

8.6.2 Explicit Parallelism

Programming languages and models can be distinguished by having either explicit parallelism, or implied parallelism. Explicit parallelism can also refer to pipelining considerations, as described above, and is relevant when writing parallel instructions in a Very Long Instruction Word (VLIW) assembly line, or, to a lesser degree, in Explicitly Parallel Instruction Computing. It should be clear that there is often a trade-off between the effort involved in writing each parallelism explicitly and the achieved performance. However, some hardware mechanisms can be used to ease and automate the parallelism effort and save it from the programmer. For example, having a scheduler that picks up a free processing unit, allocates all the required data and resources, and lets it run in a transparent parallel mode can be of great assistance and a helpful performance tool.

Parallel programming is also eased by programming models and languages—specifically functional programming, as described in the following sections.

8.6.3 Functional Programming and Programming Languages

Most network processors are programmed by assembly languages. Many network processor manufacturers are also providing C/C++-like compilers, but these are rarely used in performance-demanding environment, as discussed above. In the entry and access network processors category, it is more common to use C/C++ and libraries of functions, while such use is rarely found in the high-end, core networking line-card network processors.

Many research studies have shown that performance decreases by 10–30% when using high-level (or functional) languages, compared to using assembly languages. However, it obviously depends on the application, the network processor, and the language used.

Most functional languages are based on a model of the processing scheme of network processors. This processing scheme assumes a task (or multiple tasks) that examines every incoming packet and potentially uses a data table to look for a rule or a predefined pattern. Following a match, the task then executes an assigned action on the packet for this rule or pattern. This model, "rule/pattern-action," has many syntaxes, parallelism levels, and conditions that are specific for each functional programming language. An example of such "event-action" rules is the software architecture for programmable lightweight stream handling [144, 145].

Intel's Network Classification Language (NCL) is an example of a functional programming language that is described in Appendix A of this chapter. Other examples of functional languages that are part of more general programming models are provided in the descriptions given in this chapter's Appendices B and C. These examples make the concept of functional programming clear, and give some insight to the capabilities of functional programming for network processors.

8.6.4 Programming Models

To begin with, there are no practical programming models for network processors. Most network processors are quite complex, with many, various, and specific functional units (e.g., search engines), many memory types (e.g., frame buffers, look up tables, external memories), different degrees of synchronicity, and so on. On the one hand, this complexity raises the difficulty of creating a generic, effective, and efficient programming model, and on the other hand, it raises the need for it, if portable, simple, readable applications are to be written. As performance, reliability and predictability are critical, most implementations sacrifice expressiveness and readability that can be offered by programming models. This, however, might change as hardware performance will increase by using more multiprocessing (CMP), multicores (MCP), MPSoC, and other means. After all, performance was sacrificed somehow in custom ASICs in favor of network processors, for the sake of gaining time-to-market and flexibility advantages.

There are also no OSs that run on network processors,[12] nor are there kernels, device drivers, or any conventional software. As mentioned previously,

---

[12]Actually, some researchers have proposed operating systems for network processors; for example, Linux-based OS (PromethOS) for network processors nodes [372, 373].

network processors usually do not support interrupts or preemptive multitasking, which makes OSs impractical anyhow. This lack of support, however, is not uncommon in embedded systems; embedded computing usually exploits concurrency without using threads, and programmers tend to use low-level assembly code. Despite considerable innovative research, in practice, programming models for embedded computing—network processors included—remain primitive [275].

However, several generic software, development, and programming environments have been offered, and some are even called programming models. These include commercial systems such as Teja-NP (acquired by ARC), or IP Fabric's PPL [327]; research systems such as NP-Click [382]; NetBind [70]; OpenCom [276]; VERA [245, 390]; or more general frameworks such as NetVM [41].[13] Most of these programming environments originated from router abstraction and implementation on generic hardware platforms, sometimes by network processors. According to these programming environments, a network processing application is written in a high-level functional or imperative language, and is compiled to be executed in the most efficient way, synchronizing all resources it utilizes, performing the required IO and memory interface, executing RPCs, and controlling all hardware assists and functional units.

NP-Click and IP Fabric's PPL are described in Appendices B and C, respectively, of this chapter. These examples provide some insight into the concept of programming models, as well as their benefits and disadvantages for network processors.

8.7 SUMMARY

In this chapter, we discussed the software aspects of network processors, starting in programming paradigms, moving through programming languages, and ending with various aspects of programming models.

A concluding remark should emphasize the fact that in software, there is no right or wrong: software definitions are not absolutely sharp, and software practices are diversified. Nevertheless, some common attributes and practices do exist, and this chapter covers their relevance for network processors. In the next chapter, we concentrate on various components that surround network processors, that is, switch fabrics, coprocessors, and host-processors.

---

[13]Although designed for any packet processing element or system rather than specifically for network processors, this framework can definitely be used for network processors.

APPENDIX A
PARSING AND CLASSIFICATION LANGUAGES

There are two well-known parsing and classification languages in the industry, namely, Intel's NCL (now part of Intel's IXA), and Agere's FPL (now LSI logic's). In this appendix, we describe NCL, as it is more commonly used.

Network Classification Language runs (NCL) on the Strong ARM processor, which is used in Intel's network processors. It is not part of the fast path, that is, it is compiled by a specific compiler prior to running the NP, and the generated code is used for the classification process itself. NCL is a high-level language, which mixes declarative (nonprocedural) and imperative (procedural) paradigms, and is optimized for fixed-sized header packets. NCL defines the classification rules and their appropriate actions, that is, a function or sequence of steps that are invoked when a classification is performed. NCL can also be used to define sets (data tables that associate defined data with packets) and searches (to determine if a packet has a matching element in a set, based on values of specified fields).

Network Classification Language statements are composed of protocol definitions, followed by rules. Predicates (Boolean expressions that can be used in the conditional part of rules), sets and searches can be defined between the protocol definitions and the rules. NCL uses unique names to identify entities like protocols, protocol fields, predicates, sets, searches, and rules. For the purpose of definitions (and not for run-time), NCL allows arithmetic as well as logical and relational operations such as addition, shifting, bitwise or logical "and," greater than, and so on.

The following section attempts to describe the essence of NCL, leaving out all details of syntax, operations, operators, and reserve words, which can be found in [192] and, to some extent, in [86]. The goal of the following description is to provide the reader with the ability to read and understand NCL language for packet processing definitions, and to use it sufficiently for most purposes. In the following, *italics* are used to represent variable names in NCL.

PROTOCOL (HEADER) DEFINITION

The protocol statement is the declarative part of NCL, and is used to define naming conventions used later by the classification process (the imperative part). It defines field names of headers (or protocols), and set Boolean and value expressions. The syntax of the protocol statement is defined in the following example, which clarifies both the protocol statement and the syntax:

```
protocol p_name {
size {keyword bytes}
field f_name {f_description}
intrinsic i_name {}
predicate c_name {c_description}
```

```
demux {
        Boolean_exp {np_name at offset}
        default {np_name at offset}
    }
}
```

The protocol statement defines a named protocol entity at a defined size that can include many declared fields of that protocol, as well as nested protocol elements that are demultiplexed (the demultiplexing is detailed by an expression). It may also include some protocol based intrinsic functions (like TCP/IP's checksum).

Fields and predicates can be accessed by using the protocol name, a dot, and the field or predicate name, for example, *p_name.f_name* or *p_name.c_name*.

Size Statement

Bytes specifies the protocol element size in bytes, where *keyword* indicates the way size is defined (e.g., constant, aligned, etc.). For example, size declaration can be written as follows: `size {const 20}`.

Field Statement

Each field in NCL is specified by *f_description*, which is a tuple composed of an offset from the same base (label or symbol, usually based on the protocol entity *p_name*) and size. Both offset and size are expressed by default by bytes. An example of *f_description* of the first four bytes in the TCP payload (after the 20-byte TCP header) is: `TCP [20:4]`. If bit addressing and size is required, then the offset and size are prefixed by a `"<"` and terminated by a `">"`, for example, the specification *f_description* of the first four bits in the destination port of the TCP header is: `TCP[<16>:<4>]` (the destination port starts at offset of two bytes, i.e., 16 bits). The entire destination port field of a TCP header is defined, for example, by: `field DPORT {TCP[2:2]}`.

Intrinsic Statement

Intrinsic functions in NCL can be either *chksumvalid* or *genchksum*, and can be used only in IP, TCP, or UDP headers. *Chksumvalid* returns "true" if the checksum is OK, while *genchksum* calculates the checksum of the IP, TCP or the UDP packet, as required. An example of defining the checking of the TCP header (assuming the protocol describes a TCP header) is: `intrinsic chksumvalid {}`, which, by referring to it at a later stage, for example, `tcp.chksumvalid`, returns either "true" or "false."

Predicate Statement

Predicates are logic "variables" (Boolean expressions) that can be used later by rules; an example of a predicate is: `predicate FIX_PORT {DPORT < 0x1000}` (where DPORT is a previously defined field in the protocol statement). If the TCP destination port is less than 1000 h, then when using the FIX_PORT predicate, it returns a "true." Predicates can use other predicates as part of their Boolean expression.

Demux Statement

The last primitive in the preceding definition of NCL (and it must be the last in the protocol definition) is the demux statement, which comes from the word demultiplexing. The demux statement allows the inclusion of nested protocols (or headers), that is, protocols (or headers) that are contained or wrapped in the main protocol (or header), forming a logical protocol (or header) tree. An example is an IP header that may be followed by a TCP, UDP, or ICMP header, depending on the value of the Protocol field in the TCP header. The demux statement chooses only one protocol (*np_name*) from any number of the included protocols in the demux statement, according to the Boolean expressions (*Boolean_exp*) associated with each of the included protocols. If no Boolean expression is "true," then the default protocol is selected (the one that is defined by a *default* instead of a Boolean expression). It works as follows: as packets flow in, the parser checks the expressions according to their order in the demux statement, and in the first "true" occurrence, the parser uses the nested protocol as defined in the corresponding statement, and continues working (i.e., parsing the incoming packet) by those nested-protocol definitions. An example of an IP header definition is given in the next section, and it clarifies the demux statement, as well as the entire protocol statement.

Protocol Statement—Summary

An example of the IP header clarifies the usage of the protocol statement as follows:

```
protocol ip {
size { const 20 }
field version { (ip[0:<4>] }
field IHL { ip[<4>:<4>] }
field h_length { IHL << 2 }
field tos { ip[1:1] }
field total_length { ip[2:2] }
field id { ip[4:2] }
field flags { ip[6:<3>] }
field frag_offset { ip[6:2] & 0x1fff }
field ttl { ip[8:1] }
field protocol { ip[9:1] }
field chksum { ip[10:2] }
field source { ip[12:4] }
field destination { ip[16:4] }
```

```
intrinsic chksumvalid {}
predicate broadcast { destination == 255.255.255.255 }
demux {
      ( protocol == 6 ) { tcp at h_length }
      ( protocol == 17 ) { udp at h_length }
      default { otherProtocol at h_length }
      }

}
```

Comments are allowed in NCL, like in the C/C++ style, that is, /*...*/ or any line starting with //...

By using the naming convention described before, protocols may be redefined and extended by adding fields or predicate statements; that is, defined_protocol_name.added__field_name or defined_protocol_name. added_predicate_name. **An example of adding a field is:** field TCP.SPORT {TCP[0:2]}.

RULE DEFINITION

The rules are the imperative part of NCL and are actually the core of the classification language. The rules use the data structure, predicates and intrinsic functions defined in the protocol statements (the declarative part) and, depending on specific values of the analyzed packets, the rules do something with these packets. The syntax of the rule statement is defined as follows:

rule *r_name* { *predicate* } { *action* }

The rule statement defines a named rule (*r_name*) that triggers *action* when the *predicate* is "true" in the processed packet. The *action* is done by invoking an *action_function* with parameters (*arg1, arg2, ...*). The predicate can be any Boolean expression composed of defined predicates combined with Boolean subexpressions.

Conditional execution of a group of rules is defined by a clause prefixed by "with *Boolean_exp*", as follows:

with *Boolean_exp* {
predicate *c_name* { *c_description* }
rule *r_name* { *predicate* } { *action* }
}

Conditional rules are used when some analysis or validation processes are required during the classification, before creating the rule; for example, process TCP ports and flags only if the TCP packet has a valid checksum (as shown in the following example). An example for using a rule (*CountGoodPackets*) and a conditional rule (*OpenTelnet*) is as follows:

```
predicate ipValid {ip && ip.chksumvalid && (ip.version == 4)}
predicate tcpValid {tcp && tcp.chksumvalid}
```

```
rule CountGoodPackets {ipValid} {add_to_ip_count()}
with (tcpValid) {
predicate NewTelnet {(tcp.flags & 0x02) && (tcp.dport == 23)}
rule OpenTelnet {NewTelnet} {start_telnet(tcp.dport)}
}
```

SET AND SEARCH DEFINITION

By using predicates and the protocol definition, packets could be grouped (i.e., classified) by their common *structure*. Sets and searches are additional mechanisms for classification, which enable grouping of packets that are associated by their *content*. Sets and searches use data tables that are required either for comparison with some predefined fields in the incoming packets, or for storing data from the analyzed packets, mainly to carry stateful classifications (cross and interpacket references). An example of stateful classification, that is, using interpacket content references, is storing the session identifications (source IP address, destination IP address, source port, and destination port) of some specific packets for further classification of similar packets (e.g., assigning some priority, egress port, etc.)

The set statement is used to define data entries in a table; each entry is identified by one or more key values. The syntax of the set statement is as follows:

set *t_name*
< *number_of_keys* > {
size_hint {*expected_population*}
}

The set statement defines a named set (*t_name*) that contains *number_of_keys* (in the range of one to seven) per entry, and has about *expected_ population* entries at the beginning (it may rise as the table grow, but initially should be power of two). The content of the set is provided either at initialization time, or by action references (called by rules).

A search statement identifies a unique correspondence between some fields of the analyzed packet (according to the protocol statement) and key values of a given set. The syntax of the search statement is as follows:

search *t_name.s_name*
(*key1, key2, ... keyn*) {
requires {*Boolean_expression*}
}

The search statement defines a named search (*s_name*) that uses set *t_name*. For each packet, a search is done in the *t_name* set, using fields *key1, key2,* and so on ... *keyn*. Each key is expressed in the form of a protocol name, a dot, and a field name (i.e., *p_name.f_name*). The optional "requires" clause causes the search either to be done (if the *Boolean_expression* is "true"), or to be skipped.

The search *t_name.s_name* returns a "true" if the search resulted in a match between the packet's fields and the set's contents, or a "false" if there was either no match, or the search was not done (due to the false condition in the "required" clause). An example for using a searchable set is given in the following:

```
set addresses <1> {
size_hint {1024}
}
search addresses.ip_src ( ip.source )
search addresses.ip_dst ( ip.destination )

predicate red_address { addresses.ip_src || addresses.ip_dst }
```

The predicate *red_address* will become "true" if the analyzed packet has an IP address (source or destination) that is stored in the data table set *addresses*.

SUMMARY

Network Classification Language provides a rich declarative part that enables the user to define protocols, headers, and even payloads (with fixed, predefined offsets), and an imperative part that enables stateless and stateful classification, based on packet content and structure. A more complex, summarizing example is given as follows:

```
#define TFTP 21 /* port number for tFTP */
set ip_udp <3> {
size_hint { 256 }
}

search ip_udp.udp_dport ( ip.source, ip.destination, udp.dport ) {
requires { ip && udp }
}

predicate ipValid {
ip && ip.chksumvalid && (ip.h_length > 10) && (ip.version == 4)
}

rule CountGoodPackets { ipValid } { add_to_ip_count() }

with ipValid {
predicate tftp { (udp.dport == TFTP) && ip_udp.udp_dport }
rule tftCount { tftp } { Count_tftp_packets() }
}
```

In this example, one data set is defined, *ip_udp*, that contains an IP source address, an IP destination address, and the UDP destination port. A search is defined such that every UDP packet searches for both its IP addresses and its UDP destination port in the *ip_udp* table. The first rule, *CountGoodPackets*, always executes, and if the packet is a valid IP packet (determined by the *ipValid* predicate), then a function is activated (*add_to_ip_count*).

If the packet is a valid IP packet, then, and only then, the rule *tftCount* is executed. It checks the *tftp* predicate, that is, if the packet is with a destination port that contains TFTP, and the packet has both its IP addresses stored in the table *ip_udp* as well as with the TFTP port in this entry of the table. If this is the case, then a *Count_tftp_packets* function is called.

APPENDIX B
CLICK AND NP-CLICK LANGUAGE
AND PROGRAMMING MODEL

Click was developed in about 2000 [262, 263, 322], and was originally intended to describe routers functionalities and to build flexible, configurable routers. It is a declarative language that supports user-defined abstractions, a software architecture for building configurable routers, and it can be used as a programming model when amended with proper components that bridge the gap between the software and the hardware. Click is used for many networking applications, including building Linux-based networking kernels that are based on it. NP-Click, offered in about 2003 [382], was based on Click's functionality and concepts, and was designed to build applications on network processors in an easy and efficient way. NP-Click is a programming model that was written for the Intel IXP1200 as an alternative to using its own programming model.

Click and NP-Click are described briefly in the following sections in order to give the reader some insight into the various approaches of programming languages, architectures, and models. This description is not intended to enable the reader to start programming in these languages, and it might not be relevant to all readers, depending on the network processor architecture and application they may choose. However, in discussing programming models and concepts for network processors, these specific examples may help familiarize the reader.

CLICK

Click is based on the terminology, abstraction, and components of networking environments. Networking tasks—for example, queuing, forwarding, classifications, look-up, and any other packet processing modules—are the basic building blocks of Click and are termed *elements*. Each element, or task, may have inputs and outputs, termed *ports*. These ports carry the packets between the elements (the tasks) along the *connections*, or edges that interconnect these ports. The result can be described as a directed graph of elements at the vertices, and packets travel on the connections. There are two types of ports that describe the way the two connected elements at each end of the connections transfer packets: push or pull. For example, a push output port of a source element can be connected to a push input port of a sink element, which would mean that the source element initiates packets to the sink element's input port, which later acknowledges the receipt of the packets to the source element. In Figure B8.1 push ports are notated by a black rectangle for the output push port and a black triangle, for the input push port.

Pull communications means that the sink element pulls packets from the source element, for example, the scheduler empties a queue, as shown in Figure B8.2 (where pull ports are notated by the empty rectangle or triangle). An agnostic port

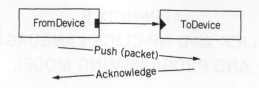

FIGURE B8.1

Push communications

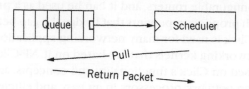

FIGURE B8.2

Pull communications

is a port that complies with the other port's type, at the other end of the connection (either push or pull), and it is notated by a double lined port.

In Click, each element is defined by a class, which specifies its data layout and behavior, that is, how many ports are defined for this type of element, what handlers it supports, how it processes packets, and so on. Each element may have some initialization parameters, called a configuration string, which is supplied to the class at initialization time (like constructor arguments in objects), that is, IP address, state of the element, and so on. Method interfaces (a list of methods that other elements can access) and handlers (methods that the user can access) are defined for the elements. Click offers both language syntax, which describes the elements, methods and handlers, ports, and connections, as well as schematic representation of the elements, their ports, and the interconnections, as defined in Figure B8.3.

A Click design is composed of a network of various types of elements and interconnections. Click uses a library of such elements that represents about 150 common network components that perform some task or algorithm (e.g., Counter, Classifier, Queue, Priority Scheduler, Random Early Detection block, Paint, Shaper, etc.). An example of a simple, unidirectional IP firewall is shown in Figure B8.4. Elements and their connections are declared as follows:

```
name:: class (config-string);          //declaration
name1 [port1] -> [port2] name2;        //connection
```

The IP Firewall in the example would then be defined as follows (this description is stored in a configuration file that Click processes):

```
// Declare the Elements
src :: FromDevice(Eth0);            //Read packets from eth0 interface
check0 :: Classifier (...);         //Screen out non-IP packets
check1 :: CheckIPHeader (...);      //Validate IP packets
filter :: IPFilter (...);           //Do the actual firewall function
```

```
sink :: ToDevice(Eth1);        //Put OKed packets to et1 interface
drop :: Discard;               //Get rid of non-IP packets
// and define the connections
src -> check0;
check0[0] -> check1;
check0[1] -> drop;
check1 -> filter;
filter -> Queue -> sink;
```

One element that is particularly important to network processors is the *classifier* generic class. At initialization, the classifier receives a configuration string

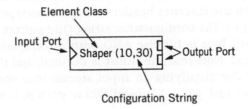

FIGURE B8.3

Sample element

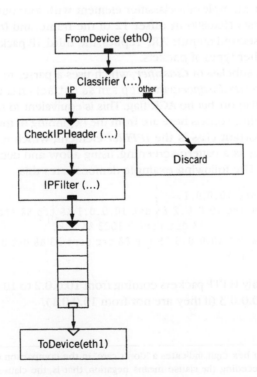

FIGURE B8.4

A simple IP firewall example

according to which it creates a directed acyclic graph—a "decision" tree—that is used later, at run time, to classify incoming packets. Each incoming packet's data is compared to this tree, one word from the packet at a time with the branching-node content of the decision tree and down the tree nodes, until a leaf is reached that points to an output port of the classifier.

In other words, the classifier has one input (in which it receives the incoming packets) and several outputs, according to the number of pattern clauses in the configuration string. A pattern clause is either "offset/value" or "offset/value%mask," and several clauses can be used in a pattern (separated by space). Examples may help clarify: The pattern clause 12/0800 in the configuration string means a simple comparison between the 12th and the 13th bytes of the incoming packet (in the Ethernet header, it is the Ethertype field) and 0x0800 (which is the IP type). The configuration string (the pattern clause) 33/02%12 means that the 33rd byte is bitwise AND with a 0x12 masking, and then is compared with 0x02. Note that the offset is decimal, and the value and mask are hexadecimal.[14] For classifying an input stream into several output ports, the syntax *Classifier* (*p1, p2, ...*) is used, where each p_i is a pattern matching clause.

```
Classifier(12/0806 20/0001, 12/0806 20/0002, 12/0800, -)
```

This classifier is an example of a classifier element with four output ports: the first outputs ARP requests (0x0806 in offset 12 of the frame, and 0x0001 in offset 20 of the frame); the second outputs ARP replies; the third, IP packets; and the fourth port outputs all other types of packets.

IPClassifier is a subclass of *Classifier*, which uses a parser to build this decision tree. For example, the *IPClassifier* clause tcp syn && not ack means classify TCP packets that have SYN flag on, but no ACK flag. This is equivalent to the 9/0c 33/02%12 *Classifier* clause (where offsets here are from the beginning of the IP header).

A similarly important class is the *IPFilter* element, which receives a friendlier configuration string as a pattern-screening, using allow and deny (or drop) operators and a pattern. The following example demonstrates this:

```
IPFilter(drop src 10.0.0.1,
        allow src 10.0.0.2 && dst 10.0.0.1 && tcp && src port www
              && dst port > 1023 && ack,
        allow dst 10.0.0.3 && tcp && src port 23 && dst port > 1023 && ack,
        deny all)
```

This filter allows only HTTP packets coming from 10.0.0.2 to 10.0.0.1, and TELNET packets going to 10.0.0.3 (if they are not from 10.0.0.1).

---

[14]A "?" in the value, as a hex digit, indicates a "don't care" in the comparison of the relevant nibble of the packet. A "!" preceding the clause means negation, that is, the clause must not match the packet. A "-" matches all packets.

Click is extendable by adding C++ classes that define elements.[15] Any additional abstractions, elements, modifications, and so on are added to the Click software, compiled again, and executed with a proper configuration file.[16] The compiled Click application loads the configuration file, and starts executing it by running a scheduler in the background that invokes all elements in push and pull paths as required, in a cyclical order. The scheduler follows each path until stopped in a queued packet or a discarder packet, and keep invoking all relevant elements along this path. In multithreaded architecture, several such paths can be executed concurrently, and due to packet independence, it can work in a consistent way (although out-of-order processing may result).

NP-CLICK

NP-Click is a programming model that was implemented on Intel's IXP1200. It was used to ease the three major problems in this architecture: exploiting parallelism, arbitrating resources, and improving data layout efficiency [382]. In order to achieve this, NP-Click uses Intel's IXP-C language (instead of C++ used by Click), to cope algorithmically with the major issues that IPX1200, like any paralleled-based network processor, has to deal with: (a) it must control threads balancing and allocation (a key issue is assigning the right threads in the right processing elements); (b) it must map data to the various memories, each with its own characteristics; and (c) it must handle resource sharing in an optimized way, apart from the programmer's use of these resources (in a multithreaded environment, resource contention might be a severe cause of stalls).

NP-Click is used either to implement elements (like Click elements, but written in IXP-C language in a simpler way), or to implement the application by assembling and using these elements (creating the "configuration file" equivalence). An NP-Click *element* is a functional block, like in Click, that has a *type,* which defines its functionalities and the semantics of its ports [382]. Multiple elements of the same type can be used in the application design. An *instance* refers to an implementation of these elements in the design, according to the application's mapping onto the architecture, and there may be multiple instances of an element in order to exploit parallelism. The programmer can map application elements onto architecture components; that is, by assigning some tasks to threads on the architecture, or by defining memory allocation for some data variables. At any rate, the application implementation is separate from the network processor's architectural considerations (assigning tasks to processing elements, memories, or threads).

---

[15]See the detailed explanations in the Click site [82].

[16]There are some differences when Click is used in User-Level of Linux (usually for debugging), and when it is used in Kernel-level of Linux (used for production, actually replaces the Linux as an Operating System).

The algorithmic assignment of software (Click) elements to hardware elements (processing elements, memories, and threads) is a very complex problem,[17] which is solved in NP-Click by 0-1 integer linear program formulation, since the problem is not so big in the current technology. Among other constraints, the assignment has to consider the processing elements, the run-time limitations of each of the tasks, and the instruction memories so that any task length, in instructions, will not exceed the program memory.

NP-Click was tested in several applications (forwarding and QoS) on IXP1200, and found to perform approximately 90% of the same applications written manually with IXP-C. On the other hand, it was three or four times faster to write these applications using NP-Click than to write them using IPX-C [383].

It should be noted, however, that NP-Click is primarily used to solve the issues of simplifying the tedious work of programming parallel processing elements, resource sharing, and arbitration. It fits very well for any parallel architecture, provided that the relevant NP architecture is defined to NP-Click. Pipelining architecture is very different, and although NP-Click might ease the programming by implementing it on an assembly of elements, it would be meaningless to assign software (Click) elements to hardware elements when they are pipelined processors. One way to approach pipelining by NP-Click would be to organize tasks into levels in order to map packet processing stages to processor element pipelines [383].

---

[17]These assignment problems are known to be the hardest in complexity, which are NP (nondeterministic polynomial time decision) problems of NP-complete class.

APPENDIX C
PPL LANGUAGE AND PROGRAMMING MODEL

Packet Processing Language (PPL) is IP-Fabric's [327] language for generic packet processing; it is hardware independent, and targets any network processor or any custom designed ASIC or SoC for packet processing (although it was started and functions on Intel's IXP architecture). PPL is a very high-level language, or programming model, that enables definitions of how packet processing is to be done and translates it for specific hardware without mixing the processing description with the underlying processor's complexities. PPL allows the programmer to describe concepts like pattern searching, encryption, queueing, and so on in a packet-centric manner (in which the only existing data is packets, and the only valid operations are packet manipulations), rather than writing assembler or C/C++ instructions.

Packet Processing Language can interact with external code if required, by forwarding packets or using a RPC mechanism. Some external, proprietary, libraries that provide extensions for some packet operations exist, and can also be used for optimized algorithms, for example, searches. After writing the PPL code and compiling it to a software virtual machine, it is mapped onto specific network processor architecture by runtime environment libraries. The entire development environment is Eclipse-based user interface.

Using the PPL instead of the native assembler instructions in the most efficient way may decrease the performance by about 20 to 30%, as found on the Intel 2xxx network processor, but at about the same performance as writing with C, which is more difficult and not as portable as the PPL. However, like all programming models that were described in this chapter, PPL also has the benefit of using a simpler language that requires a shorter development time, thus saving time-to-market, ease on debugging and, later, adaptation to varying requirements.

The PPL language is composed of two main primitives: the rule and the event. Rules list one or more optional conditions (*condition_exp_list*) that, if met, trigger sets of actions (*action_list*) in the following syntax:

```
Label:  Rule <condition_exp_list> action_list
```

For example, a rule that if the upper 16 bits of the IP destination address match entry 1 in array *iptable*, and if the packet is a TCP packet with only the SYN flag set, then a policy named *tcpconn* (and defined in the following) should be applied:

```
Rule EQ(IP_DEST/16,iptable(1)) EQ(TCP_SYNONLY,1) APPLY(tcpconn)
```

Events identify the occurrences that cause a defined group of rules to be processed, such as packet arrival. For example, if a packet arrives on port 1, then execute a rule according to which the packet should be forwarded unconditionally, as follows:

```
Event(1)
Rule FORWARD
```

Packet Processing Language has additional primitives, like Policy, Run, Wait, Array, and DeviceMap. Additionally, the Define statement, allows substitutions of strings in the PPL program during the compile time.

The most important additional primitive is Policy: Policies are labeled functions, usually complex packet oriented tasks, often with an internal state, that are invoked by the rules statement (by the APPLY action of the rule), and are defined by:

```
Label: Policy function
```

If a table does not already exist, a policy can be written, for example, that creates a table of at least 100 entries, searches associatively for a key composed of IP_SOURCE and L4_SPORT in this table and, if not found, stores this searched information associatively in the table:

```
Traffic: Policy ASSOCIATE NUMBER(100)
         SEARCHKEYS(IP_SOURCE,L4_SPORT)
```

Run defines a group of statements that can have their actions performed concurrently with other groups. Wait is used mainly to denote the end of concurrent actions defined in a group of rules. Array is used to define indexable tables, and DeviceMap is used to define a mapping between the PPL environment and the specific hardware used.

Values in PPL are also packet-centric, as PPL "understands" IP, TCP, and UDP headers and fields, and can use the named packet fields. It can also use dynamic fields in the packets, for example, PFIELD(*n*).b defines the *n*-th byte of the packet. It also recognizes some state variables, that describe the packet's state; for example, PS_FRAGMENT is a Boolean indicator of the current packet being fragmented.

The PPL program is structured in groups of rules triggered by events, exceptions or "start of program," and runs in parallel unless there are some synchronizations declared. It is a quite flexible language; however, it is restricted to the list of functions allowed in the policies and rules library. At any rate, this language is a good example that gives good insight into what can be achieved with functional language used as a programming mode.

NP Peripherals

Most of this part of the book has described packet processing algorithms and functions as well as network processors (hardware and software). Implementing network processors in network systems, however, requires several additional components, which we call network processors' (NPs) peripherals. Some are for adjunct processing (e.g., security functions), some are used for higher-level processing (e.g., control plane processing), some are for data storage (e.g., memories), and some are used for interconnecting all of the network system's components (of which the network processor is only one).

In this chapter, we describe the two important categories of network processors' peripherals that are unique to networking: switch fabrics (the interconnection functions), and coprocessors (for adjunct functions). Network processors' peripherals are changing rapidly, and some of their functions are being absorbed into the network processors or integrated into the host processors as they become more powerful. This chapter ends the part on network processing and network processors with a brief discussion of the network processors' peripherals.

Again, as with the case of the network processor architecture, no implementation details are provided, since the actual implementation is of secondary importance to the goal of this book. More importantly, the reader should be familiar with the general concepts and behavior of the NP peripherals in order to make efficient use of them in the design or evaluation phase.

9.1 SWITCH FABRICS

Switch fabrics have been used for many decades now for many applications; they are based on many technologies, and have many interpretations. A generic definition of a switch fabric includes hardware and software components that enable data moves from an incoming port to one or more output ports. Although a switch seems to be a simple hardware function, a switching fabric contains many such simple hardware switching units in integrated circuits, coupled with the software that controls the switching paths. This may involve buffer management, queueing, or other traffic management aspects. The term "switch fabric" uses the metaphor of woven material to portray a complex structure of switching paths and ports. In general, the term can refer to a wide spectrum of switching applications, starting from inside the integrated circuit, to interconnection chips that switch data among board components, to all of the switching nodes, collectively, in a network. In this section, however, we refer only to a subset of switching fabric devices that are relevant to network processors.

After defining what switch fabrics are for, we describe switch characterization according to how efficiently they perform switching, which becomes important for distinguishing and choosing switch fabrics. The main attributes of how they function have to do with throughput, delay, jitter, loss, and fairness. Other attributes that relate to the implementation of switch fabrics include amount of buffering, implementation complexity, and interfaces. Obviously, the goal is to maximize throughput with minimum loss and delay, with a minimum number of buffers and a minimum degree of interface complexity, as well as a low power consumption and cost.

9.1.1 Introduction and Motivation

Network processors receive packets from their input ports, process them packet by packet, and either forward the processed packets throughout their output ports or discard them. Packets that are sent to a line (i.e., the "network side") are simply sent over the network processor interface to the destination network. Packets that are forwarded to the "system side," either for further processing or for another network interface (that is not directly attached to the same network processor), must go through some kind of switching to access their destination. This switching can be done through the backplane of the system (possibly by a specific switching card attached to it). Switching may be required even on a single line card or a system board, when packets are transmitted between several network processors and other components on the same board.

Switching is the basic underlying technology that enables packets to be transferred simultaneously and at the required ultra-high speeds among the processing elements. Switching, in our context, is therefore the function that connects many packet-outputs to many packet-inputs, in a variety of configurations and techniques, which enable packets to transverse the switching devices. These techniques may include some traffic management, priority considerations, and other flow handling, to the extent that traffic managers (as coprocessors or otherwise) may sometimes be integrated or treated together with switch fabrics. However, we shall defer the discussion of traffic managers to a dedicated section later, and ignore the traffic management capability in this subsection, despite its availability in many commercial switch fabrics.

The switch fabric is the chip or chipset that is responsible for the switching function. It can be as many as 15 to 30 chips; some are subswitch modules that are interconnected to provide the required switching bandwidth (up to several Tbps), while other chips may be controllers, and so on.

Switching theory and switch fabric technology are as old as the days of telephony, and switch fabrics are fundamental to telecommunications (e.g., telephone exchanges). Switch fabrics also became essential to data communications (e.g., in Ethernet switches, IP routers, and other network systems). Therefore, a vast body of research exists on switch fabrics that includes countless algorithms and performance evaluations of the algorithms, the configurations, the traffic patterns, and the switching requirements. We shall not enter this field, as using switch fabrics does not require the knowledge of designing one; however, some configurations and parameters that are used for switch fabrics are detailed in the following for better utilization of switch fabrics with network processors and network systems.

In order to define the switching requirements of network systems that are relevant to network processors, it is worth noting that usually network systems deal with 24 to 96 interconnected channels, running real-time traffic at 10 Gbps each. This requires a switch fabric that should be capable of handling 300 Gbps to 1 Tbps aggregated traffic, with minimal frame delays and frame losses. The next generation switch capacities should be 4 to 10 times larger than that.

9.1.2 Models and Operation

Traffic types and patterns are key factors in determining switch technology, architecture, and performance. Basically, we distinguish between packet switches (usually for data-networks and computer interconnection applications) and circuit switches (usually for telecom networks). Packet switches can be subdivided into variable length packets (e.g., Ethernet and IP) and fixed length packets (e.g., ATM cells).

Switch fabrics are used for interconnecting traffic from N input (ingress) ports to M output (egress) ports. Such a switch fabric is denoted by $N \times M$. Traffic interconnection is enabled by traffic-paths that are created in the switch fabrics for each flow, from ingress to egress. If all possible paths are used, then the switch operates in its maximum throughput. However, since these paths sometimes overlap (i.e., have joint segments or use shared elements in the switch fabric), blocking may occur, and not all of the paths can be simultaneously active, and this degrades the switch throughput. These blocking situations are the main reason for performance degradation in switch fabrics; they have been widely investigated for decades, and are discussed in the next subsection. Generally speaking, buffers are used to avoid or ease blocking situations, as well as scheduling and arbitration algorithms for synchronizing the traffic flows between the ingress ports, buffers, switch fabrics paths, and egress ports.

We can therefore model a switch fabric as a switching core coupled with input and output buffers, and several mechanisms that map packets to ports, and control, calculate, route, synchronize or configure paths and places in the switching core

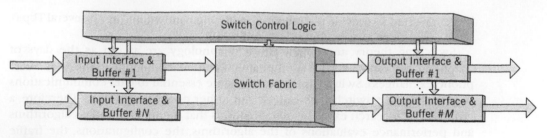

FIGURE 9.1

Switch model

(see Figure 9.1). The control mechanism can use either in-band (part of the traffic flow) or out-of-band signaling for flow control and configuration, based on the traffic destination and the switch core's status.

The input interfaces are used for traffic reception, error checking, segmentation, and buffering as required, while routing, switching, and configuration decisions are made and executed. The output interfaces buffer, assemble, reorder, prioritize, and schedule the output traffic as required. The ingress traffic may be delayed for enabling the control logic to allocate a free path for it, or possibly to dictate the required switching function in the switching core to create such a free path. Attaching buffers to the ingress ports and/or egress ports results in several models of switches, namely Input Queued (IQ), Output Queued (OQ), or Combined Input Output Queued (CIOQ) switches, which are described next.

9.1.2.1 Blocking

There are three main reasons for the blocking condition: (a) *output blocking* happens when two traffic flows (or frames) try simultaneously to access an egress port; (b) *internal blocking* happens when two traffic flows try simultaneously to use an internal path or a switching element, or they run into some other internal bottleneck, and (c) *input blocking*, or *head of line (HOL) blocking*, which happens when a frame at an ingress port is prohibited from crossing the switch (possibly due to internal or output blocking), and that causes all other ingress frames following it to be stalled, even if there are clear paths for them to their destinations in the switch.

A switch that operates without internal blocking is called a *nonblocking* switch. This nonblocking operation can result inherently from the switch architecture, regardless of the traffic flow (or frame) that has to be switched. This switch is called *Strictly Nonblocking* (SNB) switch. This operation, however, can also be guaranteed for each traffic flow entering the switch, if the switch controller assigns or reconfigures a free path for this traffic flow, connection, or frame. If the rearrangement includes the reassignment of already active connections, then this switch is called *Rearrangeable Nonblocking* (RNB) switch. If, on the other hand, this rearrangement is just the setup of a new connection, regardless of the current switching state and without interrupting existing connections, this switch is called *Wide-sense Nonblocking* (WNB).

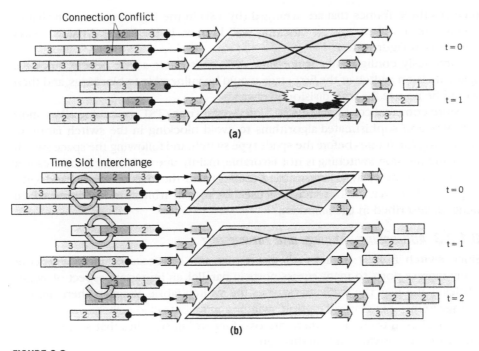

FIGURE 9.2

Time-Space switch

9.1.2.2 Time- and Space-Division Switching

Switching can be performed either by one stage of switches or by multiple stages of switching. At each stage, switching can be done either in the space domain (using separated and dedicated paths for carrying traffic flows), or in the time domain (using time division multiplexing [TDM] techniques). Time-Slot Interchange (TSI) is a time division switching technique according to which data frames are placed in time slots so that two goals can be achieved: (a) the order of the time demultiplexing defines the destination of each of the data segments,[1] and (b) rearranging the data segments in the "right" order can minimize blocking of the frames when they are transmitted to a blocking, space division type switch. A configuration of the latter case is called Time-Space (TS) switching, and can be demonstrated as follows: Assume three streams of frames, each of which is time division multiplexed with each frame destined to one of three egress ports (see Figure 9.2a). The left side (the first stage) is a time division switch, and the right side (the second stage) is a space division 3 × 3 switch fabric. At every time slot, the space division stage

---

[1]For example, if the frame destinations are 3,1,2,4, a TSI can reorder the frames in the ingress port according to the destinations. In actual implementations, the frames enter a sequential queue, from which a control mechanism "picks" the frames according to their destinations.

receives three frames that are arranged (by TSI) in the ingress ports such that no connection conflicts and no blocking will happen in the space division switch fabric (as in Figure 9.2b).

Obviously, configuring a Space-Time (ST) switch has a little benefit, since the space division switch at the first stage would be vulnerable to blockings, and there would be no use for time division afterwards.

More complex configurations of Time-Space-Time (TST) switching allow more complex and sophisticated algorithms to avoid blocking in the switch fabric by reordering the frames before the space type switch, and following the space switching. Time division switching is not favorable, mainly due to latency issues (caused by the inherent delay in transmitting frames) and the high memory bandwidth required. However, the TS principle is used for switch arbitration and frame scheduling, as described in the following.

9.1.2.3 Addressing, Routing, and Forwarding

Since switch fabrics for network processors are networks for interconnecting many ingress ports to many egress ports in parallel, an important aspect of switch fabrics is the way the traffic transverses the switch fabric. Basically there are two main schemes for that: self-routing switches and controlled switches. In addition, for frame-based traffic flow, there are two forwarding schemes that switch fabrics use: store-and-forward and cut-through.

A self-routing switch is an inherent result of the switch architecture. It means that the switching elements of the switching fabric behave (forward the frame) based on the frame's header contents. Self-routing switches usually refer to multi-staged, space division switches that are described in the following (e.g., Banyan networks). In these networks, each switching element, at every stage, changes its state (and forwarding path) according to the content of the frames' header that it has to forward. For example, in the first stage of switching elements, a specific 1-bit in the frame's header may contain "0" to indicate one egress port of that switching element, or "1" to indicate another. In the subsequent stage, the switching element "examines" the next bit of the frame's header, and if it's "0," forwards the frame to one port, otherwise to the other port, and so forth. This is explained and demonstrated again when we describe multistage switch fabrics in the following. In single stage switches, "self-routing" is an outcome of how the switched traffic is introduced to the switch fabric; for example, in time division, it is the slot location in the frame, and in space division, it refers to the assignment of the traffic to a specific port and path. In other words, self-routing fabrics use the architectural principles of the switch fabric core to handle frame forwarding; for example, a bit in the frame's header dictates the state of a switching element along the path, or the location (in time or in space) of the switched traffic.

Cut-through and store-and-forward schemes of switching were described in Chapter 2 when we discussed a network's bridges and switches. The same principle applies to switch fabrics; that is, switching can be done after the entire frame is queued, analyzed, switching decisions are made, and finally the frame is switched. Or, switching can be done "on the fly" by examining the header of the incoming frame, making all forwarding decisions in time ("at wire speed"), and allowing the frame to flow throughout the switch fabric without queueing it.

The switch fabric controller maps the ingress port of an incoming frame to its egress port and tags the frame appropriately. Then the frame transverses the switch fabric core to its destination port, after the controller ensures that the path is available (or reconfigures the core to have such a path, if possible). It can be done either in store-and-forward, or in cut-through manner, as described above. The frame can be forwarded in the self-routing switching core, for example, according to the attached tag (or its header); a specific 1-bit in the tag or header may contain "0" to indicate one egress port, or "1" to indicate another. Lastly, the switch fabric controller strips off the tags that were used by the frames to transverse the switch fabric core (it can be done at the egress port or throughout the switch core path).

9.1.2.4 Avoiding Blocking

There are several ways, architecturally and algorithmically, to reduce blocking and increase throughput of switch fabrics, some of which apply only to frame-based traffic flows, and some to all kinds of traffic. We focus on frame-based traffic [13], as this is most relevant to network processors. Some of these ways include [13]:

- Increasing the internal switch fabric core speed to accommodate several input traffic flows simultaneously (in time division schemes).
- Using buffers in the input ports, in the switch fabric core, and in the output ports, in order to reorder the transverse frames in various locations of the switch fabric for avoiding contentions.
- Running a back-pressure, arbitration, or handshake mechanism between the various stages, buffers, and ports, in order to synchronize traffic flows in an optimal way.
- Configuring a number of switch cores to work together in parallel, to increase the potential free paths for traffic flows.
- Load-balancing the traffic flows before the switch fabric, to reduce the probability of blocking caused by bursts with similar patterns.

Some of the techniques are described in more detail in Section 9.1.4. Others, which relate to switch fabric models and architectures, are described in the following subsections.

9.1.2.4.1 Output Queued Switches

Even in a SNB switch fabric, blocking might still happen at the egress port—for example, when two or more traffic flows are forwarded to the same egress port—and no switch fabric core can solve this. The easiest solution is to place a buffer in the output port, and to run the switch fabric core faster than the aggregated speeds of all its incoming traffic flows. In other words, the switching core runs at speedup N (where there are N inputs), assuming all frames at the N ingress ports might be destined to the same egress port, and each of the output ports uses a queue to receive the N potential frames arriving faster than its speed. The output port will then transmit the traffic further on, from the output queue. (OQ) thus, has the highest throughput and lowest guaranteed and average delay, and is therefore sometimes called an Ideal Output Queued Switch (IOQS). However, it is quite an impractical solution to use such a high speedup, and it is used just for modeling and as the highest boundary of performance in comparison to other schemes.

9.1.2.4.2 Input Queued Switches

Another way to reduce blocking probability is by queueing all ingress traffic in a buffer *before* the ingress port, so that when there is a clear path from the ingress port to the desired egress port, the frame at the head of the queue will transverse the switch. However, due to HOL blocking, it is not an efficient solution. The HOL blocking, which is the input blocking, happens when successive frames must be held just because the current frame at the head of the queue is waiting for a free path, thereby causing a decrease in performance. In IQ, the upper limit to throughput is 58.6% for large N, when independent and identically distributed (IID) fixed-length frames enter the switch at uniform distribution (when the frames are served from the input buffer by First In First Out—FIFO scheduling) [246]. It can be slightly higher (63.2%), when a frame-dropping mechanism is applied, and only when the input ports are highly utilized (>88.1%).

One way to solve the HOL blocking issue is to queue the incoming traffic in separate queues in each of the input buffers, according to the traffic destination. This way, a frame that is destined to an egress port with a clear path to it can be transmitted to the switch fabric core, even if a frame before it is waiting in another queue for another egress port. In other words, several *Virtual Output Queues* (VOQs) are maintained at each of the *input* ports (see Figure 9.3), such that separated buffers are reserved for each egress port and frames can be chosen according to path availability (and other Quality of Service considerations, such as priority).

Input Queued (IQ) requires either self-routing fabrics or some arbitration module to handle the injection of ingress frames into the switch fabric core, and VOQ requires it even more. Such arbitration algorithms are described in Section 9.1.4.

9.1.2.4.3 Combined Input Output Queued Switches

Once there are buffers at the ingress and egress ports of the switch fabric, there can be algorithms that can increase switch throughput, reduce frame delay, or dictate some desired property of the switch. Some of these algorithms are described in Section 9.1.4.

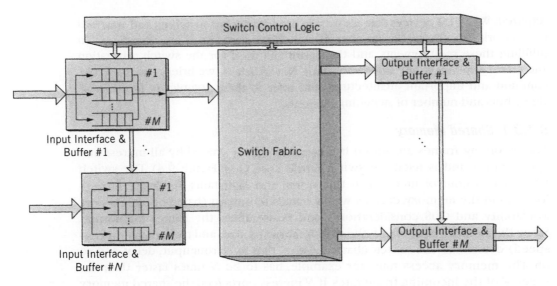

FIGURE 9.3

Input Queued with Virtual Output Queue

9.1.3 **Architectures**

Switch fabrics were introduced first for the telecom industry (telephone exchanges) and then for parallel computers (interconnection networks) and for data communications (and ATM boosted the research on switch fabrics). Therefore, an enormous number of switch fabric architectures were described for a vast number of cases, with various traffic patterns, frame types, applications, complexity, speed, and so on. These were modified over the years as technology evolved. The most common architectures are described in this subsection.

Switch fabrics can be classified into three main architecture categories [406]: shared memory, shared medium (or shared bus), and space division. Shared memory and shared medium architectures are time division fabrics, and they work pretty much according to what their names imply; that is, they use a shared memory or bus to switch the data from its source port to its destination port in different time slots. Space division means transmitting the data over a physical, identified, and reserved path from the source port to the destination port for the duration of the packet transverse. Space division switch fabrics can be further subclassified into crossbar fabrics, multistage interconnection networks, and other space division fabrics that have N^2 disjoint paths for N input (ingress) ports and N output (egress) ports (termed $N \times N$ switch).

Although we cover several architectures, we again mention that for the network processor and network systems needs (particularly the number and the speed of the ports), a fast switch with minimum delay and more than 1Tbps capacity is required. With VLSI devices that are usually used for packet switching and attached to network processors, the number of switching elements is less significant in fulfilling these requirements, and the algorithms used for the switch arbitration and frame scheduling are more important. Nevertheless, we briefly overview the common and important architectures, and refer to their complexity in terms of algorithms and number of switching elements.

9.1.3.1 Shared Memory

The incoming frames are stored in a memory that is shared by all ingress and egress ports, and is itself the switch fabric core (see Figure 9.4). The switch fabric has a control mechanism that stores and maintains the frame buffers location in the memory, decides which frames to output (based on egress port availability and QoS considerations), and reassembles the frames for output. Since the memory is, in effect, the switch fabric, its size, and technology (access speed) determine the switch characteristics, that is, throughput, delay, and so on. The memory access time, for example, has to be N times faster than the speed R of the incoming frame rates, if N ingress ports feed the shared memory switch.

This architecture is quite simple and cheap, and requires no complex scheduling. Multicasting and broadcasting capabilities are inherently achieved by this architecture, since shared memory allows the frame to be available to all required ports simultaneously. This architecture can reach 100% throughput with minimal buffering to achieve a specified level of packet loss. Scalability of shared memory is also simple, by using multiple shared memory cores in parallel, and feeding each one of them with a slice of the incoming frames. This technique is called stripping or byte slicing. This scalability requires no additional control mechanism, as it uses one control mechanism for all memory cores, similar to the single shared memory case described before. The main drawback of this architecture lies in its sensitivity to faults. If one of the slicing lanes fails, the entire frame is useless, while it still draws resources from the switch. For these situations, a backup memory is architected, so that when such a failure occurs, the frame slice or the entire port's traffic is switched over to the backup memory for switching.

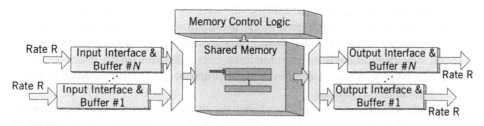

FIGURE 9.4

Shared memory

Shared memory can either have static queues or dynamic queues (in size), that is, it maintains a buffer for each egress port so that its size is always $\leq M/N$ (M is the memory size, and N is the number of egress port) in static queues. This type of shared memory operates like the OQ switch, as described in the following.

Memory speed should be at least $2NR$ when R is the incoming and outgoing frame rates in bps, and N is the number of ports. For example, with 128 bits wide bus access to a 1 GHz memory, a 128 Gbps throughput can be achieved. Practically, shared memory's throughput is about half that. If interfaces support TDM, then although the shared memory is inherently a single stage time switch, it can become a three-stage, TST switch.

9.1.3.2 Shared Medium (or Bus)

Shared medium is a similar approach to shared memory, where all ingress ports and egress ports are connected through a shared medium or a bus (which may be a ring, a hypercube, a mesh, a torus, etc.). This architecture is simple, and multicasting and broadcasting are simple to implement, but only an extremely fast shared medium or bus, or a complex arbitration scheme (bus controller), can be used efficiently while preserving QoS demands (see Figure 9.5). Scalability can be achieved by adding more busses that work in parallel, for increasing the aggregated bus speed, and the traffic is split between the busses.

Shared medium is the simplest interconnection, using the standard interface and protocols, and multicasting and broadcasting are simple. It limits, however, the scalability due to bus contention, requires an ultra fast bus, and is limited to aggregated 100 Gbps.

Like in shared memory, if interfaces support TDM, then although the shared medium is inherently a single stage switch, it can become three-stage, TST switch. This architecture, however, is relevant more to network systems (such as configuration of the systems' line cards) than to the switch fabrics that network processors work with.

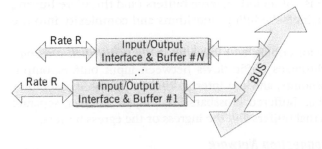

FIGURE 9.5

Shared bus

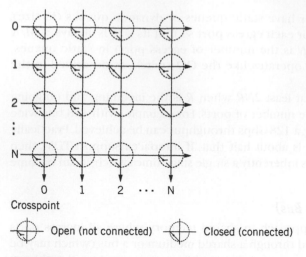

Crosspoint

$\ominus$ Open (not connected) $\oplus$ Closed (connected)

FIGURE 9.6

Crossbar switches

9.1.3.3 Crossbar

The simplest space division architecture is the crossbar switch. Crossbar, crosspoint, or crossover switches are various names that refer to this basic architecture, where many horizontal inputs may be switched over one or more of many vertical outputs (see Figure 9.6).

This is the most popular and efficient architecture, as it enables high throughput, QoS, low latency, multicast and broadcast, and it is strict nonblocking, modular, and still very simple. It is not so scalable, as the switch grows in $O(N^2)$ for N ingress ports and N egress ports, has no fault tolerance (once a switching element fails, the entire crossbar is useless), and it is subject to output blocking.

To cope with the output blocking, one can either run the crossbar core at a higher speed (N times the rate of the ingress ports), or a *buffered crossbar* can be used (see Figure 9.7). Buffered crossbars answer most of the crossbar blocking issues, at the expense of including large buffers (and therefore hurting the scalability) and introducing scheduling algorithms and complexity into the crossbar core.

An additional alternative for coping with blocking is the *arbitrated crossbar*, which schedules and synchronizes traffic flows between input buffers, output buffers, the crossbar's crosspoints, and the internal buffers (at the crosspoints), as described in the following. Buffered crossbars may require simpler, separate arbitration between the internal buffers and the ingress or the egress buffers.

9.1.3.4 Multistage Interconnection Network

Using one monolithic switch fabric core has the advantage of simplicity and speed, at the expense of scalability and fault tolerance. In order to improve scalability and

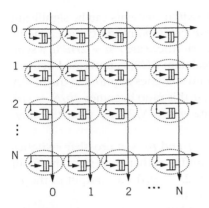

FIGURE 9.7

Buffered crossbar

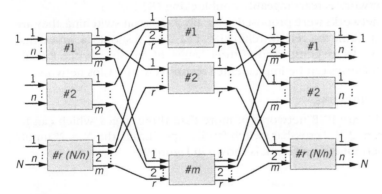

FIGURE 9.8

Clos network

fault tolerance of switch fabrics, many configurations of networks of small switch fabrics are offered, structured in multistage or otherwise. These networks can offer various degrees of scalability (in terms of size, e.g., number of ports) and fault tolerance, usually at the expense of switching delay.

9.1.3.4.1 Clos Networks

A Clos network [83] is a network composed of three stages of crossbar switch fabrics (see Figure 9.8). Symmetric Clos networks are networks in which the last stage is similar to the first one, and the Clos network has $n \times m$, $r \times r$, and $m \times n$ crossbar switch fabrics, with r crossbar switches at the first and last stage, and m crossbar switches at the middle stage. These three integers (n, m, and r) define the symmetric Clos network. Clos networks were proposed in order to reduce the number of switch elements in the $N \times N$ crossbar switch fabric ($N = rn$), while maintaining nonblocking operation. The saving of crosspoint switch elements can be significant for large crossbar switch fabrics; instead of the r^2n^2 crossspoint switching elements that are required in the $N \times N$ crossbar switch fabric, there are $r^2n + 2rn^2$ in a symmetric Clos network, in the case of $m = n$ (to maintain RNB, as described below). This is just $(r + 2n)/rn$ of the number of crosspoint switches required in the $N \times N$ crossbar switch fabric. When substituting $r = N/n$, this ratio becomes $2n/N + 1/n$, which can be optimized (minimized) by setting $n = \sqrt{N/2}$, and then the ratio becomes as low as $2\sqrt{2/N}$.

In a symmetric Clos network, the $n \times m$, $r \times r$ and $m \times n$ crossbar switches are connected between the stages by a perfect shuffle exchange. Perfect shuffle in this case means that for $x = 1, 2, ..., p$ and $y = 1, 2, ..., q$, the xth output of the yth switch is connected to the yth input of the xth switch.

This architecture supports either a SNB or a RNB operation, at the expense of latency and complex management techniques. When $m \geq 2n - 1$, then the symmetric Clos network is strict nonblocking (SNB).[2] When $m \geq n$, then the symmetric Clos network is rearrangeable nonblocking (RNB).[3]

Although Clos networks were proposed originally for circuit switching, they are used also for packet, frame and cell switching [88, 112, 280]. It should be noted that although Clos networks are composed of three stages of crossbar switch fabrics, any of these switch fabrics can be implemented recursively by a Clos network.

9.1.3.4.2 Benes Networks

Benes networks [47] are RNB networks of more than three stages, which can be built recursively from 2×2 crossbar switch fabrics. Specifically, the $N = 2^n$ input/output Benes(n) permutation network is shown in Figure 9.9.

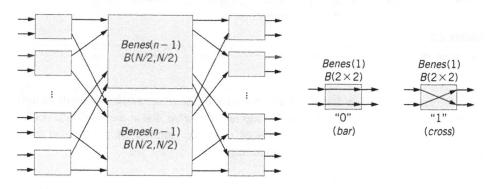

FIGURE 9.9

Benes(n)—permutation network of $N \times N$, $N = 2^n$

---

[2]In the case of nonsymmetric Clos, the condition for SNB is that $m \geq$ *number of inputs in the ingress crossbar + number of outputs in the egress crossbar* -1.

[3]In the case of nonsymmetric Clos, the condition for RNB is that $m \geq$ *the largest number between the input ports of the ingress crossbar and the output ports of the egress crossbar*.

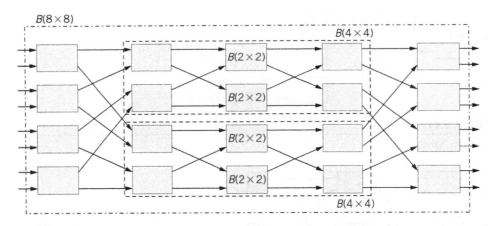

FIGURE 9.10

The 8 × 8 Benes(3) network

Benes(n) has $2n$-1 stages (or $2log_2N$-1 stages), and each contains $N/2$ Benes(1) (2 × 2 crossbar switches). For example, Benes(3) is an 8 × 8 network that has 5 stages, and is shown in Figure 9.10.

The number of 2 × 2 crossbar switches required in Benes(n) is $N(log_2N$-1/2), and if we consider each 2 × 2 crossbar switch as having four switching elements (crosspoints), then the Benes(n) requires $4N(log_2N$-1/2) crosspoints.

9.1.3.4.3 Cantor Networks

Parallelizing $m \geq log_2N$ Benes networks together by multiplexing and demultiplexing (creating a TST switch configuration) results in a Cantor network, which is a SNB network. Cantor networks have $O(Nlog_2^2N)$ 2 × 2 crossbar switches.

9.1.3.4.4 Banyan Switching Networks

The principle of Banyan networks [149] comes from a binary tree of 2 × 2 crossbar switches. A Banyan network is identified by having exactly one path from the ingress port to the egress port, and the fact that it is a self-routing network. At each of its stages, the crossbar routes the incoming frame to one of the two crossbar outputs; at stage i, the i'th bit of the incoming frame determines to which output of the crossbar the frame is to be routed. In other words, if the first bit value is 0, then the frame will be switched in the first stage to output "0" of this crossbar, and if this bit is 1, the frame will be switched to output "1" of this crossbar.

An example is shown in Figure 9.11. A rectangular $N \times N$ Banyan network is constructed from identical 2 × 2 crossbar switches in log_2N stages. A Banyan switching network interconnects the two inputs of each of the 2 × 2 crossbar switches in a variety of topologies, one of which is shown in Figure 9.12. The

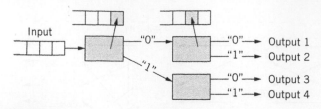

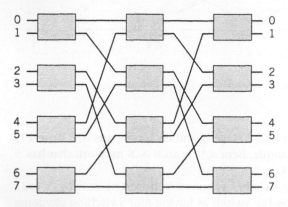

FIGURE 9.11

Banyan network

FIGURE 9.12

Omega network

most interesting ones, for our needs, are the rectangular SW-Banyans, which have the same number of ingress and egress ports, and are constructed from identical switching elements. Banyan networks are not fault tolerant, as only one path exists between any input to any output. Although this path is not used for other connections (so blocking cannot happen on the paths), the 2×2 crossbar switches in the network can be blocked if both input ports of these 2×2 crossbars try to use the same output port.

One way to avoid blocking is to speedup the Banyan crossbars and to add buffers to the internal crossbar switches. Another way is to use multiple Banyan networks in tandem (Tandem Banyan Switching Fabric), so that blocked frames are routed to the following Banyan network, while unblocked frames are routed to the output port of the network [407]. The third option is to sort the incoming traffic, as explained in the next subsection.

Banyan networks are very efficient, and introduce *O(logN)* delay in the frame routing due to the *log N* stages of the network. They are easily scalable, since they have *N/2* switching elements at each stage, or a total of *(N/2)logN* 2×2 crossbar switches.

There are several implementations of the Banyan network. An example of one is the Omega network shown in Figure 9.12, which connects each stage to the following one in a perfect shuffle[4]; that is, each output port is connected cyclically to the next switching element in the next stage.

Other subclasses include the Delta networks [349], Flip networks, Cube networks, Baseline networks, and Modified Data Manipulator networks, which are beyond the scope of this short overview. They all are rectangular $N \times N$ networks which are constructed from $b \times b$ crossbar switches arranged in $log_b N$ stages. Although these networks were originally designed for circuit switched fabrics, they have been considered also for packet switching fabrics, and they all have the same performance for packet switching. They are self-routing networks, modular, and can operate in synchronous and asynchronous modes [13, 101, 411].

9.1.3.4.5 Batcher–Banyan Switching Network

Since Banyan switches are internal blocking, other alternatives have been proposed. Based on the observation that internal blocking may be saved by sorting the frames according to their destinations, and introducing the sorted frames in a continuous sequence (without "holes") to the Banyan ingress ports, a Banyan network is preceded with a sorter. The Batcher network is such a sorter [43] (see Figure 9.13), and, when combined with Banyan, results in the nonblocking Batcher–Banyan network. Output blocking is still possible, however.

Each 2×2 comparing element in the Batcher sorter compares the two complete destination port addresses of the frames, and sends the frame with the higher address to its "up" output port and the frame with the lower to its "down" output port. Eventually, after $logN(1 + logN)/2$ stages, the Batcher sorter actually

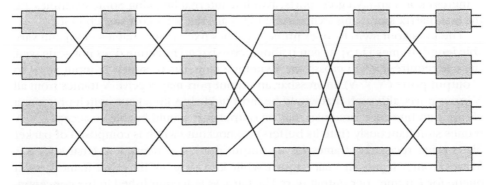

FIGURE 9.13

An 8 × 8 Batcher sorter

---

[4]Perfect shuffle in this case means that each output port is connected to the input port that has the number created by cyclically rotating left the output port number; for example, in 8 ports, port 1 (001_2) will be connected to port 2 (010_2), and port 6 (110_2) will be connected to port 5 (101_2). The ports are numbered sequentially, from the first crossbar input to the last input of the last crossbar in the stage.

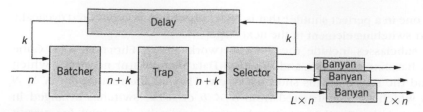

FIGURE 9.14

Sunshine switch

completes a hardware sorting of all frames, according to their destination addresses. Since at each stage there are $N/2$ comparing elements, there are $NlogN(1 + logN)/4$ comparing elements, which, even with the $(N/2)logN$ switching elements of the Banyan network, are still fewer than the N^2 switching elements required in the simple crossbar.

To cope with output blocking, various techniques were proposed; one, for example, is the Sunshine switch (see Figure 9.14). The idea is to parallelize L Banyan switches, so that no more than L frames to the same destination can be handled simultaneously (with an output port logic that can handle it). In the case of more than L frames, some of them are trapped and recirculated through a delay buffer, while the others are sent by the selector to different Banyan switches.

9.1.3.5 Other Space Division Architecture

Many other architectures were proposed for providing N^2 disjoint paths among N ingress ports and N egress ports, avoiding internal blocking. Some examples are the bus-matrix switching architecture [337] and the knockout switch [436].

The knockout switch, for example, switches fixed length frames [436] and variable length frames [115]. As noted above, crossbar architecture does not scale well, since it requires $O(N^2)$ crosspoint switching elements to switch N input ports to N output ports ($N \times N$). In crossbar, an output port may receive N frames from all N input ports, although that would be most unlikely. A knockout switch provides a cheaper architecture, assuming that an output port would be able to accept $L < N$ frames simultaneously (into its buffers). A knockout switch is composed of packet filters (that recognize frames for an output port), a concentrator (that selects L frames from N inputs in a fair manner, while knocking out the other frames), and a queue for L frames per output port. The fairness is accomplished in the concentrator by pairing all inputs and randomly choosing one of each pair, and running this contest several times among the "winners" and the "losers" until L frames remain "winners" eventually.

9.1.3.6 Parallel Switching Architectures

Parallelism is used not just for processing, as described extensively in the previous chapter on network processor architectures, but also for switching fabrics. The principle is simple: by using multiple "planes" of switching, each plane is a switch fabric core, and by splitting the incoming traffic among the switching planes, the same effect as speeding up the switch fabric core is achieved. There are several architectures for parallelizing the switch fabric, and we already describe one, the Cantor switch, which is a parallel Benes network.

Parallelism is used mainly to scale switch capacity, although it can also serve as a fault tolerance measure. Parallelism can be based on bit or byte slicing (every bit or byte is sent to a parallel switching plane) with one scheduler that operates on all switching planes identically. Parallelism can also be done on a cell or flow splitting, based on time slicing (each switching plane treats its frame) with a scheduler that calculates for each cell or flow its switching plane, path, and schedule.

This concept of parallelism has many implementations and derived architectures, such as the Parallel Packet Switch [230], Parallel Buffered Packet Switches [191], Parallel Shared Memory, or Distributed Shared Memory [231], just to name a few.

9.1.4 Algorithms and Performance

The main purpose of switch fabric algorithms is to avoid blocking, reduce packet loss and packet delay, while keeping the switch fabric as simple as possible, with a minimal number and size of buffers, switching elements, and switching speed. As noted previously, speeding up the switching core (relative to the frames' input rate, to allow more frames to use the core's paths), using TSI methods and buffers (either IQ, OQ or buffered switch) can reduce significantly the blocking probability [250, 313]. A redundant configuration of multiple and parallel switch fabric cores can also reduce the blocking probability, as well as providing other advantages, such as load balancing and fault tolerance, at the expense of size, cost, and power consumption.

A buffered switch can result in 100% throughput, for example, in a crossbar switch fabric (as $N \to \infty$), when each crosspoint can buffer one frame [236]. This is achieved without speedup or VOQ, under the assumption of Bernouli IID or uniform distribution of the incoming traffic [284]. A buffered crossbar can provide guaranteed performance (rate and delay, and 100% throughput) with a switch speedup of 2 [291, 414], and when using scalable scheduling algorithms that operate in parallel on the ingress and the egress ports [79].

Ideal OQ switches provide the highest throughput at the lowest possible and guaranteed delay. In order to avoid the required high speedup in OQ architectures (which is N), IQ is used. However, the maximal throughput for non-VOQ IQ is 58.6% for large N (for IID fixed-length frames entering the switch at uniform distribution) [246]. To solve this HOL performance degradation, VOQ and CIOQ architectures were introduced to emulate the OQ. This leads to synchronization, scheduling, and arbitration algorithms for connecting ingress ports to egress ports and buffers.

Many synchronization and scheduling algorithms have been proposed to match buffers, ingress ports and egress ports, that is, to find a free path for a frame to transverse the switch toward the egress port while maximizing the throughput at a guaranteed packet delay and loss, at an acceptable switch fabric core's speedup. Centralized schedulers, which are arbiters, are used to determine the configuration of the switch fabric, and to maintain frame order [252]. Some algorithms use in-band or out-of-band "backpressure" signaling to inform the previous switch stages (or input buffers) of congestion at the egress port or at internal junctions in the switch fabric core, and to cause these previous stages to queue frames rather then to forward them. Other algorithms may use variations of credit-based flow control, according to which the ingress port accumulates "system" credits that it uses when it forwards frames. Backpressure can be used for controlling the rate at which credits are generated in the credit-based flow control.

Such algorithms are based on matching algorithms in bipartite graphs and the marriage theorem [139, 164]. Matching algorithms find edges in a graph such that each edge has no common end-nodes with other edges. A CIOQ switch, for example, can be considered a bipartite graph, with ingress ports on one side of the graph and egress ports on the other, and a matching algorithm can be applied to calculate the match. Perfect matching happens where every ingress port has a mate among the egress ports, and vice versa. The marriage theorem states that if the number of ingress ports equals the number of egress ports and, if for every subgroup of ingress ports, there is at least the same number of potential egress ports, a perfect matching is possible.

Maximum Matching algorithms try to match as many ingress ports to egress ports as they can, and their complexity is $O(n^{3/2} + (m/log\ n)^{1/2})$ [16], where n is the number of nodes and m is the number of possible paths.

Maximum Weight Matching (MWM) [314] algorithms try to find the "best" matching among ingress ports and egress ports based on various weighting criteria, such as queue priority, queue length, waiting time, and so on. The complexity of such algorithms is typically $O(n^3)$ [236], although it can be as low as $O(mn + n\ log\ n)$ [137], where n is the number of nodes and m is the number of possible paths.

The problem with matching algorithms is that they have to be calculated online, which becomes quite impractical in high-speed switching. Scheduling thus can be a bottleneck in itself, so speeding up the schedulers becomes an issue, and various solutions to that have been proposed (such as [158]). Simpler scheduling algorithms, such as the Weighted Round Robin, fail to provide uniform rate guarantees in nonuniform traffic. Some examples of scheduling algorithms include Parallel Iterative Matching (PIM) [17] and its simplified version, Iterative Round-Robin Matching [315]; iSLIP [312, 313]; Dual Round-Robin Matching [74]; iterative Longest Queue First (iLFQ) and Oldest Cell First (iOCF), which are two MWM algorithms [314]; Reservation with Preemption and Acknowledgment (RPA) [300]; Batch [103]; and FIRM [379]. iSLIP,[5] for example, which is very common, achieves 100% throughput

---

[5]iSLIP is an iterative version of SLIP, which is a variation of Round-Robin matching and is based on PIM (which suffers from being complex and unfair).

for uniform traffic; it is used to configure the crossbar switch and decide the order in which frames will be served. Many of the above examples are distributed scheduling algorithms, which ease the online calculations.

An interesting approach for relatively low online scheduling complexity is Birkhoff von Neumann (BvN), which can be used when the input rate matrix is known *apriori* [72, 73]. The main idea in this scheduling algorithm is to use a capacity decomposition approach [52, 423], which has a computational complexity of $O(N^{4.5})$, but the online scheduling algorithm[6] is $O(logN)$. It provides 100% throughput, and uniform rate guarantees for each input-output pair, without internal speedup.

Most of these algorithms are designed for cell-based switching, that is, they work in fixed-length time slots. Packet mode schedulers bridge the packet/cell difference by either: (a) segmenting the frame, switching the segments, buffering them at the output port and reassembling them when all segments comprising the packet are collected, or (b) by continuous delivery of the frame over the switch fabric, while blocking the ingress and egress ports of the switching fabrics, as well as internal junctions, to the frame [36, 156]. It is important to note that packet-mode CIOQ cannot emulate OQ, whatever the speedup (while the cell-base CIOQ can). However, a packet mode CIOQ switch can perform like a cell-based CIOQ (mimicking OQ) with a speedup of 1, when additional, bounded packet delay is allowed [36]. Packet mode scheduling of IQ crossbar switch fabric, with a speedup of 1, can achieve 100% throughput in several kinds of arrival traffic patterns [141, 142]. Similar results on packet-based switching were shown [299] by modifying the cell-based MWM algorithm, and achieving 100% throughput for ingress traffic with IID packet length (having finite mean and variance), and that is distributed as Bernoulli IID.

Another category of algorithms has adopted the principle of load-balancing ahead of the switching fabric [417] so as to spread the traffic flow uniformly and to enable smoother operation of the switch fabric. Load balancing is used for the BvN switch, which decreases online complexity to $O(1)$ [73] while maintaining 100% throughput, guaranteed rate, bounded delays, and packet ordering (using multistage buffering). Another example is a scheme, in which a frame is sent twice on a full mesh interconnection switch. The first time it is sent into a VOQ according to uniform distribution; therefore, it is load-balancing the IQ. Then it is sent again, from the VOQ to its real destination. In other words, logically, this is a two-stage mesh switch, where the first mesh load balances packets across all VOQs by sending 1/Nth of the traffic to each VOQ, and then the second mesh redirects the traffic to its destination by servicing each VOQ at fixed rate. It turns out that without any scheduling, it performs extremely well—better than any other interconnection topology (hypercube, torus or ring), and guarantees 100% throughput [251, 253].

---

[6]This algorithm can be Packet-Generalized Processor Sharing (PGPS), or Weighted Fair Queueing (WFQ), both of which are described in Chapter 6.

9.1.5 Implementation and Usage

There are three basic requirements for switch fabrics: interconnecting line cards across the backplane, packet switching between ports (on a line card and on different line cards), and QoS support while switching the packets.

We already touched on the issue of interconnect networks, or switch fabrics in Chapter 7, when we discussed interfaces and specifically PCIe (for interchip or C2C), Advanced Switching Interconnect[7] (ASI, for inter-board or backplane), and InfiniBand (for interchassis or rack). Advanced Telecom Computing Architecture (ATCA), which is a part of the PCI Industrial Computer Manufacturers Group (PICMG) standardized the carrier grade communication equipment, mainly high-speed interconnect technologies. ATCA proposed mechanical, electrical, and some functional (e.g., management) interfaces between boards, backplanes, and systems, including a switch fabric interface (in the data transfer part of the interface, called "Zone 2"). There are 16 slots in the ATCA chasis; each supports a switch fabric interface of 15 full-duplex channels to the other slots, each 4 10 Gbps, yielding a 640 Gbps chassis capacity.

The generic line card contains a switch fabric interface, through which the line card is connected to the chassis and to the other line cards through the switch fabric. Line cards may use backplane connections that are multi-2.5, 3.125, or 10 Gbps serial lines ("lanes").

Today interfaces to switch fabrics are based on SPI-4.2 (10 Gbps) or SPI-5.2 (40 Gbps), and future interfaces are expected to rely on Ethernet. Older interfaces are based on the Network Processor Forum's (NPS) CSIX-L1 or NPSI (the streaming interface).

There are many possible configurations for using the switch fabric: it can reside on the backplane and be connected directly with the line cards, or it can reside on a separate card (or cards) attached to the backplane and be connected through a mesh to all line cards (see Figure 9.15). It can even be distributed, that is, used in every line card, and the line cards can be interconnected by a mesh (or shared bus) through the backplane.

Switch fabric implementations support redundancy that is either passive (1:1), load-sharing ($N+1$) or active ($1+1$). Passive redundancy is the simplest, and it means that additional switch fabrics are configured for backup purposes, that is, in case of failure, the backup switch fabric becomes active automatically. The major drawbacks in this configuration are the inefficient utilization of the hardware, and the loss of data in the active switch fabric in case of failure switchover. Load-sharing is the cheapest and most efficient, since all switch fabrics are active, though in the case of a failure, the performance degrades as one of the switch fabrics is disabled. Using enough redundant switch fabrics ensures that in the worst

---

[7]ASI was defined by ASI-SIG, a Special Interest Group that was formed for developing a switch fabric architecture based on PCI Express technology, and later merged with the PCI Industrial Computer Manufacturers Group (PICMG) in 2007.

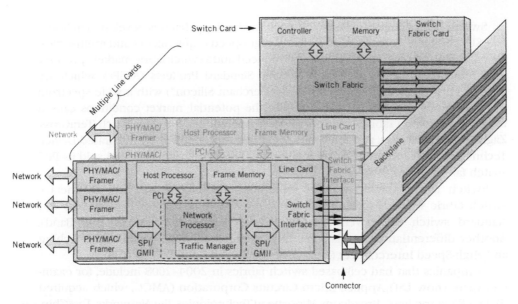

FIGURE 9.15

Centralized switch fabric and backplane

case, there will be enough switching capacity left in the switch fabric. Active redundancy is the most expensive and inefficient configuration, but guarantees no data loss, since the two switch fabrics work in parallel, with just one of them connected to the output ports. In case of a failure of the connected switch fabric, an automatic switchover disconnects the faulty switch fabric and connects the backup one, with no loss of data.

9.1.6 Commercial Aspects

Switch fabric chipsets are used for enterprise systems, multiservice provisioning platforms (MSPPs), or carrier-class Ethernet applications. Enterprise systems contain relatively low-cost and low-scale switch fabrics, mainly for the enterprise data networks (L2, Ethernet, or L3 IP routers). Some storage networks require these kinds of switch fabrics, where the requirement here is large packets and low latency requirements, supporting not only Ethernet and IP, but also iSCSI, Fiber Channel, and InfiniBand. MSPP applications require a protocol agnostic switch fabric, which are transport-based (TDM-like) or data-based (fixed and variable length packets). Carrier-class Ethernet applications require scalable, redundant, and extremely high-speed switch fabrics.

Switch fabric chipsets usually include multiple chips to scale to the required capacity. These are usually arranged as multistage switch fabrics, with several chips for memory, arbitration, and control.

Switch fabrics chipsets used to be a core technology that was developed in-house by ASIC designers. Due to the complexity, high-speed requirements, and multiservice functionality, in-house development was reduced, and a switch fabric market appeared. This trend resulted in Application Specific Standard Products (ASSPs), which are general purpose switch fabric chipsets ("Merchant Silicon") with a wide spectrum of applications and technologies. Despite the potential, market conditions caused many of the switch fabrics vendors to cease operations (e.g., PetaSwitch, TeraCross, Zagross, Tau networks, and TeraChip; ZettaCom was acquired by Integrated Device Technology [IDT], Sandburst was acquired by Broadcom, and even IBM sold its PRS switch fabric line to AMCC).

Switch fabric market analysts sometimes distinguish between proprietary switch fabric vendors (cell, frame, and TDM switching using SPI interfaces) and standard switch fabric vendors (RapidIO, Ethernet, PCI Express, InfiniBand). Another differentiation that is sometimes made is between backplane switching and High-Speed Interconnect (HSI), used mainly for Chip to Chip (C2C).

Companies that had cell-based switch fabrics in 2004–2008 include, for example, Agere (now LSI), Applied Micro Circuits Corporation (AMCC, which acquired IBM's PRS fabric line), Broadcom, Mindspeed Technologies, Tau Networks, TeraChip, Vitesse Semiconductors, and ZettaCom (now IDT). They supported up to 2.5 Tbps switching using 10 Gbps interfaces, in <4 ms latency (some with 1 ms), up to 16 (64 in some cases) subports and up to 1K flows (usually tens). In these systems, frame payloads are fixed, can be selected, and are up to 160 bytes. All chips use 8b/10b encoding, hence reducing bandwidth on the serial links; the link over speed is usually two times, resulting 2 to 3 Gbps link speed on the backplane, and for 10 Gbps ports 8 links are required. Most of these chipsets require 15W per 10 Gbps, and their price has been around $300 per 10 Gbps. Most 160 Gbps configurations require 15 to 25 chips.

Companies that had packet-based switch fabrics between 2004 and 2008 include, for example, Broadcom, Dune Networks, Enigma, Erlang Technologies, Marvell, and Sandburst. These systems support up to 20 or even 40 Tbps, using 20 Gbps line interfaces. All chips require four to eight links on the backplane per 10 Gbps port, consume 10 to 15W per 10 Gbps, and have cost approximately $300 per 10 Gbps. Most 160 Gbps configurations require 15 to 25 chips (like the cell-based configurations) with some exceptions (<10).

Other companies that have been in the business of switch fabrics include, for example, Fujitsu, Fulcrum, Internet Machines, Mercury Computer, Mindspeed Technologies, NextIO, PetaSwitch Solutions, PMC-Sierra, Power X Networks, StarGen, TeraCross, Transwitch, Tundra, and Xyratex.

9.1.7 Switch Summary

Switch fabrics are characterized by their levels of throughput, packet loss, and packet delay, which are always trade-offs due to switch fabric architectures. In addition, switch scalability and complexity are of prime importance in selecting a switch fabric. In this section, we described the various models and types of switches, their architectures, capacity (throughput) scalability as well as size (port) scalability, and some arbitration and scheduling algorithms.

Technical issues in switching include scheduling, routing and reconfiguration algorithms, arbitration, backpressure for indicating congestions, and so on. Some scheduling and buffering issues are very similar to those we described in Chapter 6. For more information, the interested reader may turn to many books on switch fabrics and interconnect networks [250], as well as articles on specific issues having to do with architectures, algorithms, and performance of switch fabrics.

9.2 COPROCESSORS

Coprocessors are essential network processors' peripherals for special purpose processing. Due to their importance, some of them are integrated with the network processors or with the switch fabrics (particularly in the case of traffic management). Coprocessors can be categorized into traffic managers, security processors, search processors, classifiers, storage processors, and other application accelerators.

Coprocessors can be attached to network processors in a *look-aside* scheme, or in a *flow-through* scheme. In the look-aside scheme, the network processor may share some functionalities and resources (mainly buffers and frame memory) with the coprocessor, has a specific interface to the coprocessor, and sometimes even commands the coprocessor (e.g., initiating instructions, events, or procedures). In the flow-through scheme, the coprocessor is a totally independent module, carrying specific task it is specialized for, and the combination of it with the network processor resembles the pipeline architecture.

9.2.1 Traffic Managers

Traffic flows are one of the most important issues in networking, and traffic managers are the processing units that are responsible for making sure that traffic will flow as desired. Some network processors have minimal, superficial traffic management capabilities, whereas others have comprehensive traffic management integrated into the network processing itself. Traffic management capabilities can also be found in switch fabrics, due to the complexities of the switches, the existence of control logic, and the small marginal costs in adding these capabilities to switch fabric chips, where their die size is dictated by the I/O pads, leaving plenty of room for buffers and logic required for traffic management.

Traffic management basically ties the network side, the system side, the ingress and the egress ports of the network processor to the processing unit. These connectios are controlled by the flow processing functions we discussed in detail in Chapter 6; that is, measuring, policing, queueing, scheduling, shaping, and statistical gathering.

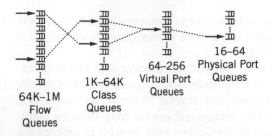

64K–1M
Flow
Queues

1K–64K
Class
Queues

64–256
Virtual Port
Queues

16–64
Physical Port
Queues

FIGURE 9.16

A multilevel hierarchical queueing system

At the core of the traffic manager there is a queuing system, and around it there are all the traffic flow handling functions. The number of queues and the ability to distinguish between the queues and the frames determines the granularity of flow handling. The result is that every frame can receive a different treatment according to the Class of Service (CoS) to which it belongs, and every CoS can be implemented by various levels and parameters of QoS.

Queues can be arranged in a flat or multilevel hierarchical organization. Flat organization is simple and intuitive, and is adequate when the traffic manager has to handle a relatively small number of traffic flows from a network with one parameter that distinguishes its flows, for example, destination, priority, and so on. Multilevel hierarchical queueing system may be required to handle traffic that originates from various networks, sources, or applications. Such cases can include access or metro networks, where traffic managers are required to treat millions of flows, coming from different sources, and each flow may represent the specific application of one user, each with its own CoS. In multilevel hierarchical queueing, queues may be arranged according to the physical ports where frames arrive, then according to the virtual ports that represent different networks, then according to the various classes (e.g., users), and finally according to the flows (see Figure 9.16).

Traffic managers can have four or more queues in the multilevel hierarchical queueing system. Sometimes the second level queueing (virtual ports, in our example) can be used for VOQ to a switch fabric.

Commercial traffic mangers have been offered by companies such as Acron Networks, Agere (now LSI), AMCC, Azanda Network Devices, Bay Microsystems, Dune Networks, EZchip, Erlang Technologies, Freescale, Internet Machines, Kichips, Marvell, PetaSwitch Solutions, Sandburst (acquired by Broadcom), Tau Networks, TeraChip, Teradiant Networks, Vitesse, Xelerated, and Zettacom.

9.2.2 Security Processors

Security is a huge area in itself, which touches on or is combined with networking in many applications. Networking applications that involve security become complex and require special purpose hardware assist to off-load security tasks from the networking devices, or even from the processing devices. However, since security involves knowledge that is beyond the scope of this book, we have to be very brief and provide just a system perspective on this subject, while referring interested readers to the vast literature on security and implementation details. A very brief description of the security aspects in network processing is provided in Chapter 5.

The two main security protocols that network processors have to deal with are IPsec [249] and Secure Sockets Layer (SSL, which was replaced by Transport Layer Security—TSL [102]). IPsec is the IETF standard for IP security, and it is composed of a suite of RFCs. It provides privacy, authentication and integrity through encryption algorithms. IPsec runs at the IP layer, whereas SSL/TSL runs above this layer, thus leaving the TCP/UDP services unsecured, while using them. IPsec is used mainly in Virtual Private Networks (VPN), whereas SSL/TSL is typically used in client server applications (such as mail, web browsing, and telnet).

Due to the enormous computation load required to handle millions of secured flows, network processors usually trigger integrated or look-aside security coprocessors, providing them with the required commands, plain text, or the cipher text and parameters. Alternatively, independent security processors work in the flow-through scheme, running the security protocols and encrypting/decrypting the data stream to and from the network processor. Most security coprocessors function as SSL accelerators or IPsec accelerators using one of these schemes.

An additional important application is content inspection, which is carried out by specific accelerators. This application is required for intrusion detection and prevention, antivirus, antispam, sophisticated firewalls, and so on. The growing complexity of the content inspection, coupled with increasing wire speed requirements and the expanding scope of these applications, makes content inspection accelerators a vital component in the network systems.

There are many kinds of security processors available commercially, which can be categorized as integrated security processors (part of the network processor), IPsec and SSL accelerators, or content inspection accelerators. Commercial integrated security processors have been offered by companies such as Cavium, Freescale, Intel, Misteltoe, P.A. Semi, Raza Microelectronics, and SafeNet. IPsec and SSL accelerators were offered by Broadcom, Cavium, Corrent, CyberGuard (which acquired NetOctave and was acquired later by Secure Computing), Hifn, Layer N (which became Vritestream Networks and was acquired by nCipher), SafeNet, and Zyfer. Content inspection accelerators were offered by NetLogic, Seaway (formerly Camelot Content Technologies, which was acquired by FreeScale), Sensory Networks, and Tatari (which was acquired by LSI).

9.2.3 Search Processors and Classifiers

In Chapter 5, we described in details the classification and searching requirements of network processing. In many cases, these tasks are performed in the network processors, but there are situations where specific hardware assists, search engines, processors, or accelerators are used in a look-aside scheme. It usually happens when the lookup capabilities of the network processor are limited, or cannot cope with the specific application they are designed for. Classification processors can also be used in a flow-through scheme, if network processor offloading is required.

Many of the search processors or classifiers are built around CAM devices, but when search or classification complexity is high, then a more sophisticated kind of processing is required, and algorithms can be carried out to perform the required search or classification. In such cases, the search processor or classifier may include a dedicated processing unit for the algorithms, although there are also hybrid CAM/algorithms search engines.

Cypress (which acquired Lara Networks), IDT, and NetLogic, for example, are vendors of search engines and classifiers, but many other vendors, including some network processor vendors, have offered various kinds of search processors and classifiers (HyWire and Silicon Access Networks, for example).

9.2.4 Storage Processors

Storage processors can be either independent processors or coprocessors. The requirements of storage devices and information flow impose specific traffic patterns and protocols that make storage processors critical to high-speed storage applications. These traffic patterns are, for example, larger blocks (frames), high speed, and very low latency and jitter. Different protocol suits exist for storage, such as the Small Computer System Interface (SCSI) and the Fiber Channel, which are disk block-level protocols. Even storage that is attached through IP networks has specific protocols that use the common networking protocols we discussed (mainly datacenter types, i.e, Ethernet, IP, and TCP). Such protocols may be at the disk-block level (similar to the SCSI and Fiber Channel protocols), or at the file system level (and are thus dependent on the operating system used). Disk-block level protocols include SCSI over IP (iSCSI) and Fiber Channel over IP (FCIP). File level protocols include UNIX's Network File System (NFS) and Microsoft's Windows Common Internet File System (CIFS).

All of these above-mentioned storage protocols are beyond the scope of this book. However, when they are used on top of networking protocols, some network services are very useful for the storage devices, such as high-speed encapsulation and decapsulation of storage datagrams, TCP termination (handling the TCP protocol in the network device, sometimes called TCP offload engines or devices), tunneling storage sessions, and so on.

Commercial storage processors of various degrees are (or have been) manufactured by companies such as Alacritech, AMCC, Aristos, Astute Networks, BigSur Communications, Broadcom, Chelsio Communications, Emulex/Aarohi, Freescale, Intel, iReady, iStor, iVivity, LightSand Communications, LSI Logic, Marvell, P.A. Semi, Platys (acquired by Adaptec), PMC-Sierra, Qlogic, Siliquent Technologies, Silverback (which was acquired by Brocade), Trebia Networks, and Vitesse.

9.2.5 **Application Accelerators**

These types of coprocessors are rarely used. However, there are some cases in which look-aside processing is required by the network processor, which is beyond its capabilities. In such cases, often an FPGA glued circuitry does the required application and accelerates it, so that a specific complex task is performed at wire speed (or at the network processor speed). Such applications may include deep packet analysis, high-level parsing (e.g., XML), look-ahead applications, or some extended policing, statistics, calculations, and so on.

9.3 **SUMMARY**

Network processors' peripherals are important devices that the designer and the user of network systems with network processors must consider and understand. We focused here on peripherals that are specific to network processing, that is, switch fabrics and special-purpose and network-oriented coprocessors.

With NP peripherals, we conclude the part on network processors theory. In the next part, we describe in detail the example of EZchip NP network processor architecture, programming models and software, and provide application samples and hands-on experience.

A Network
Processor: EZchip

3

This part of the book provides an in-depth description of a specific network processor, EZchip's NP, which dominates the metro networks markets. This NP's architecture is also used for another version of NP that is targeted for access network applications.

The requirements for metro networks are very challenging as a result of the flexibility requirements of the applications, the instability of protocols and methods used in the metro environment, and the performance issues of ultra-high-speed and low-latency requirements. These requirements are all the more challenging because of the wide range of network devices in which network processors are used; that is, core equipment, edge interfaces, access devices, or CPE equipment, all of which have line-cards and service cards.

The purpose of this part is to provide readers with a concrete example of all the network processing issues described in the first two parts of the book and to enable them to read and write network processor applications.

We begin with a description of the hardware architecture, and then continue with the software architecture and programming. Following Chapters 10 and 11, each of the processors and the functional units of the NP is described in a separate chapter. We conclude this part with a comprehensive example of how to write a program and use the network processor.

EZchip NP-1 is used here for demonstration, although several elements of the NP-x family are briefly mentioned (e.g., the embedded traffic management unit). The differences between NP-1 and its successors are also not discussed because the purpose of this part of the book is not to document the EZchip network processors, but rather to help readers gain insight and experience in implementing a real NP that is very common in the metro Ethernet market. Those readers who would later like to actually use one of the NPs (or any kind of network processor, for that matter) will find it much easier after reading this parts' general concepts and the description of the practical usage of the NP. This part contains the following chapters:

- Chapter 10—EZchip Architecture, Capabilities, and Applications
- Chapter 11—EZchip Programming
- Chapter 12—Parsing
- Chapter 13—Searching
- Chapter 14—Resolving
- Chapter 15—Modifying
- Chapter 16—Running the VLAN Example
- Chapter 17—Writing Your First High-Speed Network Application

The EZmde (Microcode Development Environment) demo program that can be used to write a code, debug it, and simulate an NP running with this code (with debugging features turned on) can be downloaded from *http://www.cse.bgu.ac.il/npbook*, as well as the EZmde design manual.

EZchip Architecture, Capabilities, and Applications

EZchip architecture is based on a pipeline of parallel heterogeneous processors, optimized for packet processing according to the various phases we discussed in the theory part of the book. Each processor along the pipeline is equipped with functional units, memory, and internal data buses that support its specific function in the pipeline as a whole.

There are many EZchip network processor types and versions, suited to various applications. They all, however, share the common architecture mentioned before, which will be described in this chapter. The emphasis is on the network processor chip architecture, in terms of features, applications, interfaces, and hardware, whereas the next chapter deals with the software conventions and programming.

There are many subsystems and functional blocks in the NP, not all of which will be described here. This chapter describes only those functions that are important to understand how a network processor is implemented by using EZchip's NP, as an example. Furthermore, for the sake of simplicity, not all features, possibilities, and benefits of the EZchip NPs are detailed. In this chapter, we focus on describing the general architecture of the NP-1, although some elements of NP-2 are mentioned very briefly.

10.1 GENERAL DESCRIPTION

EZchip's NP-1 is a 10-Gigabit full-duplex network processor providing wire-speed, seven-layer packet processing. NP-1 integrates search engines and many specific hardware assists that enable the required packet processing at wire speed. In addition, NP-2 also includes integrated traffic management units. NP-1 uses internal and optional external DRAM for all lookup tables and frame buffers, and optional external SRAM for statistical data.

The NP-1 is based on a five-stage packet processing pipeline (Figure 10.1). Each stage contains multiple, parallel processors optimized to perform the specific

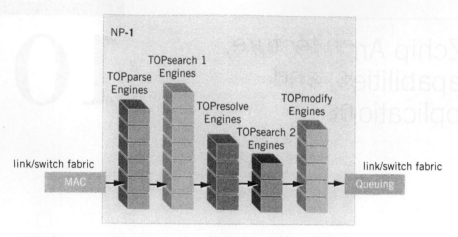

FIGURE 10.1

NP processors architecture

tasks required by that pipeline stage. There are four types of such Task-Optimized Processors (TOP engines) employed to perform the four main tasks of packet processing, that is, parsing, searching, resolving, and modifying.

Each type of TOP processor employs architecture with a customized, function-specific data path, functional units, and instruction set that supports the minimum number of machine cycles required for complex seven layer packet processing. Each TOP processor is based on a four-stage instruction pipeline that effectively results in an instruction executed every clock cycle (1 IPC). Although the TOP engines are specific for each stage, they share a common instruction set that requires only minor adjustments for operating the specific functional units and capabilities of each of the TOPs. Since there are multiple instances of the TOPs in each stage of the pipeline, the overall architecture resembles super-scalar architecture of a high degree, with the degree depending on the stage (the number of TOPs at each stage).

NP-1 uses a simple single-image programming model with no parallel programming or multithreading. Allocation of the TOP processing engines to incoming frames, passing messages between the TOPs, as well as, maintaining the ordering of frames is completely transparent to the programmer and is performed in the hardware.

NP-1 can be programmed to process packets from any source port and to any destination port (i.e., link, switch fabric or control CPU). Furthermore, processing can be divided among the ingress and egress paths in accordance with the system's architecture.

There are two stages in the NP pipeline that contain multiple instances of embedded search engines, also based on the TOPs architecture. These are used for lookups in a combination of tables with potentially millions of entries for implementing diverse applications in layer 2 to 4 switching and routing and layer 5 to 7 deep packet processing. Three types of lookup tables can be used by the NP: direct access tables, hash tables, and trees—each flexibly defined and used for various applications. Tables may be used for forwarding and routing, flow classification, access control, and so on. Numerous tables of each type can be defined, stored in embedded memory and/or external memory, and searched through for each packet. The key size, result size (i.e., associated data), and number of entries are all user-defined per table. Variable length entries, longest match as well as first match and random wildcards are supported.

Auto learning and updating of flow table entries (or other hash tables) is performed entirely by NP's TOPs with no intervention required by the host. This makes it possible for several million new or old flows to be added to or deleted from the flow table per second.

Result information, statistics counters, and per-flow rate limiters can be automatically associated with new flows and recycled when deleting old flows. Session state can be updated on-the-fly per flow, and new packets can be generated to implement various stateful functions, such as Transmission Control Protocol (TCP) session initiation/termination and dynamic TCP port allocation. Counters are available for collecting highly granular flow-based information for use in accounting and billing applications. NP chips support traffic management and Quality of Service (QoS) by various mechanisms that are not described in this book.

10.2 SYSTEM ARCHITECTURE

In this section, we briefly describe the main three components of the chip architecture, namely, the processors, the memory access, and the interfaces. We conclude with a broader discussion of the applications of the NP in network devices.

10.2.1 Task-Optimized Processors

As mentioned before, there are four types of TOP engines, each tailored to perform its respective function in the various stages of packet processing, that is, parsing, lookup and classify, forwarding and decisions, and packet modifications.

TOPparse identifies and extracts various packet headers, tags, addresses, ports, protocols, fields, patterns, and keywords throughout the frame. TOPparse can parse packets of any format, encapsulation method, proprietary tags, and so on. TOPsearch uses the parsed fields as keys for performing lookups in the relevant routing, classification, and policy tables. TOPresolve makes forwarding and QoS decisions, and updates tables and session state information. TOPsearch II optionally performs additional lookups after the TOPresolve decisions have been made. TOPmodify modifies packet contents, and performs overwrite, add

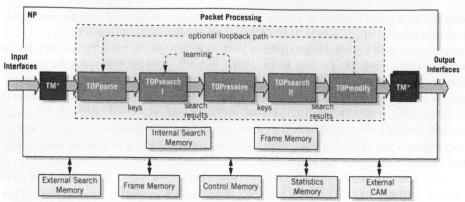

*TM is available only in NP-2 and above. location of the two traffic managers depends upon the system configuration. They may both be after the TOPs processing, or one TM may be before the TOPs processing.

FIGURE 10.2

NP architecture

or insert operations anywhere in the packet. The architecture of the NP is shown in Figure 10.2.

Multiple instances of each TOP type at the same pipeline stage enable simultaneous processing of multiple packets. Moreover, at each pipeline stage, each TOP engine provides the processing power required for seven-layer processing of that stage's tasks, at 10-Gigabit wire-speed (full duplex). Pipelining enables the passing of messages and pointers to packets from one processing stage to the next. Each TOP performs its particular task and passes its results (e.g., messages, keys, headers, and pointers) to the next TOP stage for further processing.

All TOP engines of the same type operate independently of each other and execute the same code (i.e., a simple single image programming model, with no parallel or multithreading programming) using their own program storage. All TOP engines are employed as shared resources without being tied to a physical NP port. An integrated hardware scheduler dynamically schedules the next available TOP to the next incoming packet. Ordering of packets is automatically maintained on a per-port basis.

The resulting parallel processing at each pipeline stage is completely transparent to the programmer. Furthermore, allocation of TOPs to incoming frames (ingress or egress), passing results, messages, and frame pointers from one pipeline stage to the next, as well as maintaining ordering of frames is transparent to the programmer and performed in hardware.

10.2.2 Memory Access

Multiple embedded memory cores are used for packet buffering and queuing and for storing lookup tables. These cores are accessed in parallel by the various TOPs and provide the bandwidth required for sustaining seven-layer wire-speed throughput.

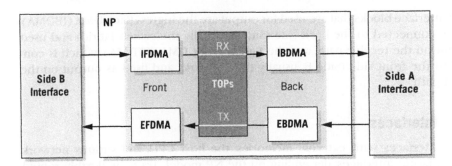

FIGURE 10.3

Processing direction

Arbitration of simultaneous accesses of TOPs to the various memory cores is controlled by an on-chip arbiter, and is completely transparent to the programmer.

Packets are stored and queued in two embedded memory cores (referred to as "*frame memory*"). The data bus width provides the required bandwidth memory accesses for each packet for receiving, classifying, modifying, and transmitting at wire speed.

Lookup tables are stored in four embedded memory cores (referred to as "*search memory*"), interfacing with the NP as described in the following section. Multiple tables across multiple memory cores can be accessed simultaneously, increasing bandwidth to speed data lookups.

10.2.3 Ingress and Egress Processing

NP schematically uses two sides, one referred to as the "*front*" and the other referred to as the "*back*." Usually, the front side is connected to the network and the line interfaces, whereas the back side is connected either to the switch fabric, the network device, the system, or to other line interface ports. One processing direction, from the front to the back, is called ingress processing, while the other, from the back to the front, is called egress processing.[1] This scheme is depicted in Figure 10.3.

The interface blocks that are used for inputs to the TOPs are the Ingress Forward Direct Memory Access (IFDMA), which is connected to the front side (Side B, usually the network), and used as input on the receive path, and the Egress Back DMA (EBDMA), which is connected to the back side (Side A, usually the switch fabric) and used as input on the transmit path.

---

[1]Ingress and egress processing traditionally referred to processing direction according to the ingress or egress packet entrance port. In other words, ingress processing referred to packets arriving from the network, while egress processing referred to packets going to the network (arriving from the switch fabric, the system, or network device side). This, however, cannot be always true, as was discussed in Chapter 5.

The interface blocks that are used for output are the Ingress Back DMA (IBDMA), which is connected to the back side (Side A, usually the switch fabric) and used as output on the receive path, and the Egress Front DMA (EFDMA), which is connected to the front side (Side B, usually the network) and used as output on the transmit path.

10.2.4 Interfaces

The NP interfaces with external memories, the host CPU, and various network or switch fabric interfaces, depending on the NP model. Traffic is flexibly routed or switched between all interfaces after processing. Generally speaking, the NP interfaces look like the schematic diagram in Figure 10.4. NP-1 is configurable to one of two link interfaces, and one of two switch fabric interfaces, as shown in Figure 10.5.

As shown in Figure 10.6, NP-2 has more and faster interfaces, and its two 10 Gbps network or switch fabric interfaces can be configured to one of three options (SPI4.2 with 192 channels for connections to SONET/DDH, SPI4.2 with 32 channels for multi-gigabit Ethernet, or Ethernet interface, either RGMII or XGMII).

A PCI interface is used for interconnection to a control CPU. Standard SDRAM or RLDRAM external memory chips can be used via NP's DDR and DDR2 interface (depending on the NP type) for external lookup tables. Standard SRAM memory chips may be employed for statistics counters via DDR or QDR interfaces (depending on the NP type).

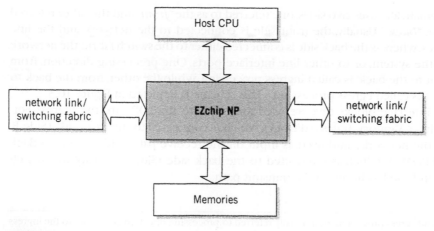

FIGURE 10.4

NP interfaces

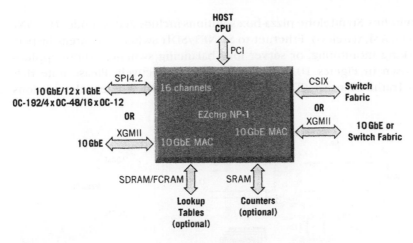

FIGURE 10.5

NP-1 interfaces

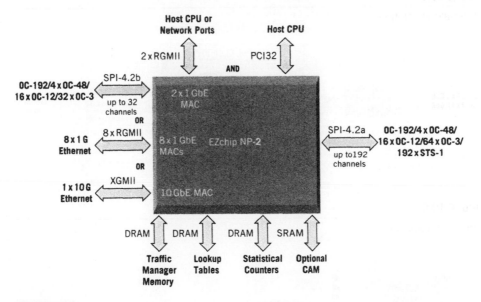

FIGURE 10.6

NP-2 interfaces

10.2.5 Using NP (Applications)

NP can be used either as a stand-alone solution ("pizza box"), in line cards, or even in service cards for some applications. Line cards in modular chassis may be used for metro switches, edge, and core routers, aggregation nodes or enterprise backbone switches. Stand-alone pizza-box solutions include access nodes (GPON/EPON OLT, DSLAM, wireless), Ethernet to SONET/SDH switches, content inspection, networking monitoring, or server load balancing switches. These applications are shown in Figures 10.7, 10.8, and 10.9, respectively. Please note that NP-1 has no Traffic Manager (TM) integrated in the chip, whereas the versions

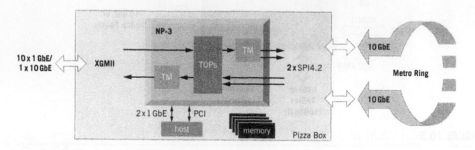

FIGURE 10.7

Stand-alone solution

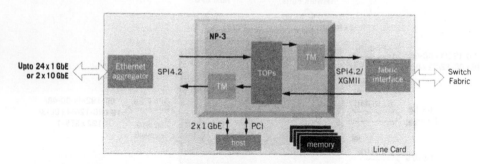

FIGURE 10.8

Line card application

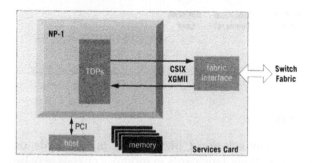

FIGURE 10.9

Service card application

NP-2 and above do. Although NP-2 is demonstrated here, it is similar to the NP-1 application in principle, and subject to NP-1 interfaces, as outlined in the previous section.

10.3 LOOKUP STRUCTURES

One of the main functions of network processors is their searching capability. Every search component (e.g., processor, CAM) requires a repository that holds all the information necessary for the required search, and is organized in a fitted data structure. EZchip NPs' search processors (TOPsearch 1 and 2) use such databases, which are all called "*structures*" collectively from here on. These structures can be organized in various ways, and they indicate any tables, trees, or data structures that are used as information sources for the search engines.

NP's embedded TOPsearch engines and search memory cores deliver lookup capabilities from embedded and external memory. These embedded search engines enable lookups in a combination of structures for implementing diverse applications involving layer 2 to 4 switching and routing as well as layer 5 to 7 deep packet processing. There are three types of lookup structures that are supported in the NP-1 architecture: direct access tables, hash tables, and trees. NP-2 supports an additional type of lookup structure that is based on M-trie, for fast IP addresses searches, but a discussion of this is beyond the scope of this book. These data structures are used for implementing various forwarding, routing, classification, policy, and state tables as required.

Multiple structures may be stored in one or more memory cores to optimize search performance. The EZchip programming environment provides optimized allocation of structures to the various memory cores; however, the NP-1 programmer may decide on any other allocation. Structure size is restricted only by the available embedded and external memory resources; hence, large hash tables and trees with more than a million entries may be defined.

10.3.1 Direct Access Tables

Direct access tables are usually used for a small number of entries (up to several thousand). The search key to a direct table entry is the pointer to the table entry.

Performance of searching in a direct access table is deterministic. A single memory access, that is, a single clock cycle, is required for accessing the information.

10.3.2 Hash Tables

Hash tables are usually used for fixed length keys, for example, DA, SA, 5-tuple IPv4/IPv6 sessions, and they can hold many entries (up to several million). In EZchip architecture, performance of searching in hash tables is deterministic. Two memory accesses are required for accessing the information. Memory utilization of hash tables using EZchip's algorithms is approximately 80%.

10.3.3 Tree Structures (Patricia)

Trees are used with variable length entries, for example, IP Longest Prefix Match and URLs, and they can hold millions of entries. EZchip implemented trees support random wildcards ("don't care" bit or byte values) anywhere in an entry, that is, in the suffix, prefix, or arbitrarily in-between. This simplifies editing policy rules and reduces the memory usage of the tree.

Wildcard examples include: IP addresses such as "10.12.*.*"; "10.12.*.7"; "*.*.5.7"; or URLs such as "www.ezchip.com*******.html"; "****.com****.mpg".

Performance of searching in trees is a function of N, the number of entries in the tree, and is directly related to the number of memory accesses (clock cycles) required to walk down a tree, beginning at the root, down to the leaf that holds the desired entry and result. The number of memory accesses is $(\log N)/3 + 2$.

For example, a tree with 256 K entries requires $18/3 + 2 = 8$ memory cycles, and a tree with 2 M entries will be searched in nine memory cycles. This is well within the cycle budget of a minimum size frame (thus the maximal frames per second can be processed) flying on a 10 Gbps wire, hence allowing sustained wire-speed searching for minimal size back-to-back frames.

For each tree, one of two search criteria may be configured: longest match or first match. Longest match can be used, for example, when searching through an IP route table, while first match is useful when searching through an access control policy table.

Longest match (or finest match) is used when there are no common wildcards among the tree entries. For example, assume these two entries in the tree structure: "10.12.*.*" with associated result X; and "10.12.15.*" with associated result Y. In this case, a key "10.12.30.11" will match the entry "10.12.*.*" and produce the result X. The key "10.12.15.11" will match the entry "10.12.15.*" and produce the result Y.

First match is used when there may be common wildcards among the tree entries. It is possible to assign a priority to each entry, to indicate a preference when multiple entries match the search key. For example, assume these two entries in the tree structure: "10.12.*.* port*" with associated result X and priority 2; and entry "10.*.*.* port1" with associated result Y and priority 1 (priority 1 is higher than 2). In this case, a key "10.12.15.20 port2" will match the entry "10.12.*.* port*" and produce the result X, while a key "10.12.15.20 port1" will match both entries, but will produce the result Y due to the priority. Memory utilization of trees using EZchip's algorithms is approximately 50% (the same as "regular" Patricia trees).

10.3.4 Learning, Updating, and Aging

NP employs several mechanisms to learn, to update, and to age entries in the structures of TOPsearch I.

10.3.4.1 Learning and Updating Mechanisms

The control CPU can update entries in all tables and trees upon initialization as well as dynamically. The hash tables can also be learned and updated by the NP-1, which provides auto-learning and can update the flow tables, Medium Access Control (MAC) tables, and so on. Millions of sessions per second can be learned, updated or deleted by NP-1 to off-load the control CPU from these tasks. NP-1 can also notify the control CPU when it updates, adds or deletes an entry in a hash.

Hash entries can be learned and updated from three sources: the host, the TOPsearch hardware learning mechanism (called "*low-learn*") and the TOPresolve state updating mechanism (referred to as "*high-learn*"):

- Layers 2 and 3 learning mechanism: TOPsearch learning mechanisms obtain Layer-2 (L2) and Layer-3 (L3) information from incoming packets and update relevant tables. The L2 mechanism learns MAC addresses and Filtering Identifiers (FIDs; mapping of the VLANs). The L3 mechanism learns IP addresses for stations in the local subnet in a manner similar to the Address Resolution Protocol (ARP) function.
- TOPresolve high learning and state updating mechanism: A programmable TOPresolve learning mechanism learns and updates hash entries on the fly. Any hash entry (e.g., 5-tuple) in any hash table can be updated flexibly.
- Host interface enables the hash entries to be updated or modified as required by the control-plane layers.

Trees and Direct tables are only updated by the host.

10.3.4.2 Aging and Refresh Mechanism

NP-1 features an aging mechanism to delete old entries from hash tables. Each time an entry is matched, the entry is refreshed. The aging process deletes all entries that have not been refreshed during a user-defined time period.

Each hash table may have aging enabled or disabled. If aging is enabled, there are two user-defined timers to choose from. Additionally, specific entries may be marked as static, so that they are not aged.

10.3.4.3 Host Notification and Messaging

The NP-1 notifies the host when TOPsearch I updates, adds or deletes an entry in a hash. Add Entry Messages are sent when the TOPresolve learning or hardware learning adds a hash table entry. A Delete Entry Message is sent when TOPresolve learning or aging deletes a hash table entry. This may be used for repository tracking or learning by the control-plane applications, run by the attached remote host processor.

10.4 COUNTERS, STATISTICS AND RATE CONTROL

The NP offers per-flow statistics and policing through the implementation of its counters. There are basically two types of counters: dedicated hardware counters and software counters that can be created and used in programming, and are described in a following subsection. We focus here on the dedicated hardware counters that are stored in external SRAM (referred to as "*statistics memory*"). Among these hardware counters types, there are Ordinary event counters, of varying lengths, and Token bucket/DiffServ counters. These counters are further divided into four groups by a programmable range register:

- Short counters—36 bits, 1 line
- Medium counters—54 bits, 1.5 lines
- Wide counters—72 bits, 2 lines
- Token bucket/DiffServ counters—72 bits, 2 lines

All internal counters are 36 bits, but may be concatenated into 54 or 72 bits. The use of the 36-bit-size is compatible with Content Addressable Memory (CAM) and Static RAM (SRAM) standards of data-bus and key widths.

The number of counters of each type depends upon the memory size in use. The counters can also read by the TOPsearch I block.

NP enables the dynamic allocation and auto association between counters and flows: when NP learns a new flow, counters are automatically allocated for the new flow and recycled when the flow is aged out.

10.4.1 Microcode Usage

Based on microcode programming, TOPparse, TOPresolve, and TOPmodify (and the Traffic Manager in NP-2's case) may add or subtract values to the counters. TOPsearch I may use the counters in either of two modes: read only or read/modify/write. The later enables NP to auto-update a counter's value. Ordering may be implemented per counter to support stateful classification. The counter values may also serve TOPresolve in its decision-making processes.

For instance, TOPparse may be microcode-instructed to build a key indicating a counter number for a particular type of frame. Using this counter number, TOPsearch I reads the counter value from the statistics block and passes it on to TOPresolve. TOPresolve can then make a forwarding decision based on the counter value, and it can possibly increment the counter.

Two methods may be used by the host to retrieve counter values: NP-push or host-poll. The host can poll all counter values, which may be time-consuming, or notification (NP-push) may be used. Each counter has a threshold that, when exceeded, causes the NP to send a message to the host via the statistics message queue. The host has the option of resetting the counters after reading them.

10.4.2 Software Counters

Programmers may implement software counters in the three embedded TOPsearch I memory cores or in the external memory. The TOPresolve learning mechanism can update hash tables by forwarding keys and results to TOPsearch I structures, according to the required application, and increment, decrement or set the result field of the relevant entries as appropriate. For example, when updating software counters, the key may be defined as the entry address or session number and its accompanying result as the desired counter value. NP contains a mechanism that ensures that the counter is incremented accurately even while frames are flowing through TOPresolve at wire speed.

The host may read these counters by accessing the search memory cores or the external memory. The number of software counters can be many millions, and is limited only by the memory resources allocated for this purpose.

10.4.3 **Token Bucket Implementation**

The NP-1 uses hardware to implement token bucket per flow (e.g., srTCM and trTCM), that is, it requires no microcode. Token bucket and related algorithms are described in Chapter 6.

Single Rate Three Color Marker (srTCM) is useful, for example, for the ingress policing of a service where only the length and not the peak rate of the burst determines service eligibility. Two Rate Three Color Marker (trTCM) is useful for the ingress policing of a service in which a peak rate needs to be enforced separately from a committed rate. In both techniques, a Meter meters each packet and passes metering results to a Marker, which (re)colors an IP packet according to the results of the Meter. This color can be coded, for example, in the DS field of the packet in a Per Hop Behavior (PHB)[2] specific manner.

Token bucket implementation enables NP to determine the token bucket reading per session. In this application, a session table is maintained in a hash structure where each hash entry identifies a session. A portion of the entry result (associated information) contains the ID of the associated Token bucket/DiffServ counters. Once a packet in the given session arrives, TOPparse creates a key for lookup in the session table. Following the lookup, the associated Token bucket/DiffServ counters are accessed to obtain the current color of the flow (i.e., green, yellow, or red). This information is then passed along to TOPresolve for discard or other decision making.

10.4.4 **Atomic Update of Counters**

This mechanism is employed in TOPsearch I to prevent instances of different resources accessing the same counter simultaneously. It essentially locks the counter during its read/modify/write stage so that it cannot be overwritten. This function is useful when back-to-back frames—each being processed by a separate TOPsearch I engine—are required to update the same counter.

---

[2]PHB is a DiffServ term, described in Chapter 2.

10.5 TRAFFIC MANAGEMENT

There is a difference in the output phase of frame interface between NP-1 and NP-2 (and above). First and foremost, the NP-2 has an integrated Traffic Management (TM) functional unit. Second, the NP-1 interfaces with the CSIX switch fabric (or XMGII), whereas NP-2 interfaces with the SPI4.2 switches rather than the CSIX. Since we focus on demonstration and practical network processing experience in this book, we describe the NP-2's TM only very briefly, and continue with the NP-1 capabilities of traffic management, programming and usage.

10.5.1 NP-2's Traffic Management

NP-2's integrated TM provides traffic management and frame queuing on both the ingress and egress paths through a full-duplex PFQ mechanism on all NP-2 interfaces. Traffic transmitted to the network links and the system switching fabric as well to the host CPU can be queued and assigned with specific QoS settings.

NP-2's TM enables provisioning of Service Level Agreements (SLAs) by supporting DiffServ and IntServ services, as well as many QoS mechanisms (as described in Chapter 6), through its queuing schemes. Flow-based bandwidth control is facilitated through programmable classification of flows, enabling Weighted Random Early Detection (WRED), priority-based and Weighted Fair Queuing (WFQ) congestion management, traffic metering, marking, and policing, and granular shaping and scheduling for thousands of queues in multilevel hierarchies.

10.5.2 NP-1's QoS, Scheduling and Congestion Management

In this subsection we describe how to use the embedded QoS features of NP-1.

10.5.2.1 Virtual Output Queuing

NP-1 implements at its output a Virtual Output Queuing (VOQ) to eliminate the Head of Line Blocking (HOLB) phenomenon often associated with switching architectures. VOQ is implemented at the NP interface to the switch fabric. Up to eight priority levels can be defined for each of the switch fabric destination ports (to other line cards). NP-1 maintains 1032 virtual output queues to the switch fabric, of which 1024 are unicast queues and eight are multicast queues. The unicast queues may be used for providing eight priority levels for 128 switch fabric destinations or four priority levels for 256 switch fabric destinations. Multicast frames are queued in the multicast queues, signifying the eight priority levels.

The VOQ block schedules the frame transmission to the switching fabric. Scheduling takes into consideration the flow control feedback received from the switch fabric. A congested switch fabric port will be masked and frames will not be forwarded to it. Scheduling has two phases: first, a priority level is selected, and then the queue is selected by, for example, applying a Weighted Round-Robin (WRR) mechanism. Prior to transmitting frames to the switch fabric, the NP-1 segments the frames to cells in accordance with the Common Switch Interface Specification (CSIX) standard.

When packets arrive from the switch fabric, NP-1 reassembles the CSIX cells to a complete frame at the egress point, and passes it for egress processing as necessary (from the switch fabric, the network device, to the line interface, the network). The frame pointer is then transferred to one of eight Egress Classify Frame Descriptor (ECFD) priority queues. This is explained in more detail in the frame walkthrough in Section 10.8.

10.5.2.2 Output Queuing

When using the 10 GbE XMGII output port, eight priority queues are provided. The channeled SPI4.2 interface can be configured to support up to eight channels with eight priorities or up to 16 channels with four priorities. Strict priority, WRR, or a combination of the two can be enforced independently at each of these ports.

10.5.2.3 Quality of Service and Random Early Detection

The NP-1 supports Strict Priority, WRR, Random Early Detection (RED), WRED, and Tail Drop QoS mechanisms.

NP-1 implements RED or WRED by tracking the status of the various queues, and through TOPresolve programming. Traffic can be selectively discarded before a port is congested to provide differentiated performance characteristics for different classes of service. TOPmodify implements tail dropping of frames to a queue that is full, as indicated by its budget expired flags.

10.6 STATEFUL CLASSIFICATION

The TOPresolve's "high-learn" mentioned before enables stateful classifying by using the learning mechanism to update entries' states in hash tables. It can be used for session state tracking by updating the resulting information associated with individual entries of TOPsearch I memory cores.

In accordance with a given state of a session, the NP can generate and transmit new frames, replies, acknowledgments, and so on. This can offload control-plane tasks from the control CPU.

10.7 MULTICAST FRAMES

NP offers several strategies for implementing frame multicast. Frames may be replicated at the ingress path before transmission to the switch fabric, or at the egress path, after traversing the switch fabric. Multiple instances of a frame can be sent to any port. These instances can be identical or modified in accordance with their destination (e.g., different destination VLAN or IP subnet).

Programmers may use the microcode in TOPparse, TOPresolve or TOPmodify to implement multiple instances of a frame on either the ingress or the egress path. Registers in each of these blocks indicate the number of copies of the frame that will be sent.

By using the Halt command (specifically `halt halt_mulc`), a different set of keys may be generated for each instance of the frame in TOPparse or TOPresolve. The same halt command in TOPmodify enables the forwarding of multiple copies of the frame to different queues. The frame pointers are recycled only after all copies of the frame have been sent.

10.8   DATA FLOW

This section describes the data flow through the NP-1 including the internal blocks of each of the TOPs. It then examines the host interface and several other functions.

The NP-1 network processor consists of the following major components, and is shown in Figure 10.10:

- Frame memory, with the frames buffered for store-and-forward operations. Additional memory cores contain search data structures (the search memory).
- Four types of TOPs to perform the four main tasks of packet processing, that is, TOPparse, TOPsearch, TOPresolve, and TOPmodify.
- Direct Memory Access (DMA) controllers and queues for input to (and output from) the chip's ports, switching fabric or host CPU. The DMA includes the Ingress Front DMA (IFDMA) and Egress Back DMA (EBDMA) on the input/receive path, and the Ingress Back DMA (IBDMA) and Egress Front DMA (EFDMA) on the output/transmit path. It also includes the Host DMA (HDMA) that serves as an output to the host interface.
- On-chip MAC controllers for 10 Gbps Ethernet.
- Other specific memory blocks and functional units that assist in various operations of the network processor (e.g., the input scheduler and the statistics block).

This section is intended to provide a high-level architectural overview of the NP-1 network processor. It first looks at the overall frame processing and then examines several of the blocks in further detail.

10.8.1   Frame Processing Walkthrough

The frame processing phases and the path of a frame in the network processor are described in this subsection. Frames may require only partial processing, depending on system configuration, the frame contents, and the microcode.

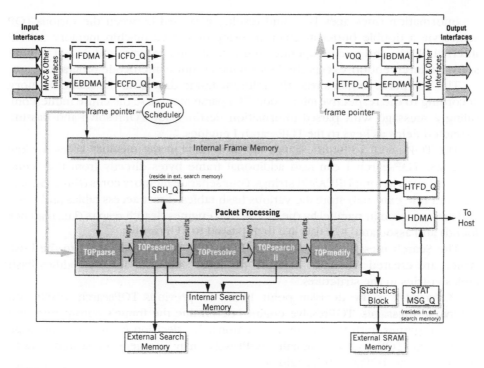

FIGURE 10.10

NP-1 data flow diagram

A frame can enter the NP-1 network processor from either the network or the switching fabric (see Figures 10.3 and 10.10). A frame is received from the network by an on-chip 10G MAC or an external framer/aggregator, and the frame is written using IFDMA to the on-chip frame memory (Rx).

Frames can be received from either a frame-based or a cell-based switching fabric. When a frame is received from a frame-based fabric, the EBDMA stores the frame in the on-chip frame memory (Tx). For a cell-based switching fabric, the EBDMA stores the cells in the frame memory (Tx) and reassembles the cells to the original frame.

Once the entire frame is stored, frame descriptors are queued in a Classify Frame Descriptor (CFD) queue, which points to the frames in the frame memory and contains additional information about the frame. Frames that arrived via the IFDMA (from the network) are pointed by frame descriptors that are stored in the Ingress CFD queue (ICFD_Q). Frames that arrived via the EBDMA (from the switching fabric) are pointed by frame descriptors that are stored in one of the eight prioritized Egress CFD queues (ECFD_Q). Weighted Round-Robin can be enforced on the ECFD_Q. The Input Scheduler is a round robin arbiter that distributes the frame pointers from the ICFD_Q (RX) and ECFD_Q (TX) queues that enter the TOPs, to the next available TOPparse engine.

Information (messages, keys, and results) is passed between the various TOP stages via a double buffer mechanism using dedicated on-chip memory. All NP internal blocks and queues operate automatically in hardware and do not require software intervention, except the TOPs which require microcode programs.

The TOPparse engines read the relevant frame data from the frame memory, according to the respective microcode. TOPparse may parse the entire frame, compiling a message with parsed information destined for TOPresolve and passing extracted fields as keys to the TOPsearch I engines.

The TOPsearch I engines search the structures in the memory cores for key matches. TOPsearch I can read additional frame bytes directly from the frame memory, such as long URLASCII strings. Four separate memory cores (three embedded and one external) store the various hash tables, direct access tables and trees, and are accessed in parallel by the TOPsearch engines. Search results (i.e., matches found) and associated key data are then passed to TOPresolve.

The Search message Queue (SRH_Q) contains updated messages to the host, which are created when the network processor adds, deletes or modifies hash tables entries in the structures.

TOPresolve is the decision point. Based on previous TOPsearch results and TOPparse messages, TOPresolve engines determine the frame's output port and queue. Together with the modify instructions, this decision is put into a message delivered to TOPmodify. Optionally, TOPresolve may also generate search keys to be used by the TOPsearch II engines.

The TOPsearch II engines search for output port-dependent data mappings, such as Virtual Local Area Network (VLAN) or Multiprotocal Label Switching (MPLS) tags, in a separate embedded memory core that stores the relevant egress information. The TOPsearch II results along with the frame's modify instructions are then forwarded to TOPmodify.

The TOPmodify engines read the frame from the frame memory and alter content in accordance with microcode and messages from previous TOPs. The frame is then rewritten to the frame memory. Frame pointers are then delivered into one or more of three queues:

The Host Transmit Frame Descriptor queue (HTFD_Q) is for frames destined for an external host CPU, via the PCI interface, and has two priority levels (high/low). This queue is mainly for frames that require a higher-level handling/control plane, such as RIP/OSPF messages. The HDMA controller fetches the frame descriptor from the HTFD_Q in order to read the frame from the frame memory, and it stores the frame in a look-ahead buffer while waiting for host access.

The Virtual Output Queue (VOQ) is used for frames destined for the switching fabric. VOQ supports multiple switch fabric destinations (channels) with multiple priority levels per destination. The Ingress Back DMA (IBDMA) reads the frame descriptor from the VOQ in order to read the frame from the frame memory. It then delivers the frame to the switching fabric. Frames are segmented into cells for cell-based fabrics. VOQ eliminates the Head-of-Line Blocking associated with several crossbar architectures in order to produce a nonblocking switching matrix. VOQ also provides an entry point for frames from the host.

The Egress Transmit Frame Descriptor queue (ETFD_Q) is used for frames destined for the network. Eight priority levels are supported for each output port. The Egress Front DMA (EFDMA) uses the frame descriptor from the ETFD_Q in order to read the frame from the frame memory. It then forwards the frame to the on-chip Ethernet MAC controller or the external POS framer/Ethernet aggregator.

The NP-1 supports frame fragmentation on the egress data path. Frames may be fragmented; for example, jumbo frames (i.e., up to 9 KB long) are segmented into standard Ethernet frames (1518 bytes long) before being transmitted to the network.

The statistics block contains counters for collecting statistical data at various stages during the packet processing. The statistical data is stored in external SRAM. The statistics message queue is an event-driven mechanism for notifying the host when statistical counters exceed a threshold, thus saving the host from polling for the counter information. A full Token Bucket counter implementation is performed inside the NP-1. A Received Frame Descriptor (RFD) table contains pointers to free buffers and to buffers containing frames in the frame memory.

All NP-1 queues are retained in the control memory (i.e., ICFD_Q, ECFD_Q, SRHQ_Q, Statistics queue, HTFD_Q, VOQ, and ETFD_Q). Frame ordering is maintained throughout the packet processing flow.

10.8.2 TOPparse Block

TOPparse decodes incoming packets and analyzes these packets. TOPparse can extract frame headers and other significant fields corresponding to all seven layers of the IP packet, for example, MAC addresses, VLAN tags, Ethernet frame types, MPLS tags, IP options, IP addresses, or ports. The extracted information is then delivered as keys to TOPsearch 1.

TOPparse is a programmable processor that executes a microcode command for each clock. Microcode downloaded to TOPparse defines the structure of the various layers, headers, and applications. New applications utilizing new data formats and protocols are supported simply by downloading new microcode to TOPparse. The TOPparse block diagram is illustrated in Figure 10.11.

Frame pointers are queued in the CFD queues of the TOPparse engines (see Figure 10.10). "In parse" dynamically distributes the frames to the available TOPparse engines. Each TOPparse engine parses an entire frame.

The hardware decoder (HW_Decoder) automatically preprocesses the frames up to the IP Options before microcode execution begins. It is designed to support IPv4 and IPv6 with or without MPLS using Ethernet or PPP encapsulation. The information

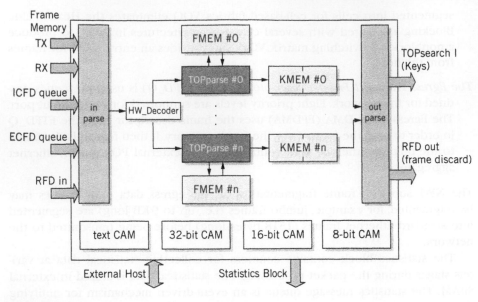

FIGURE 10.11

TOPparse block diagram

is placed in a register (HD_REG) that may be read like any other register. Using the frame offsets, the microcode program may obtain additional information from the frame by reading from the frame memory (FMEM).

In accordance with its microcode lines, the TOPparse engine classifies the entire frame, storing extracted fields as keys in the associated key memory (KMEM). The key memory is a double buffer that holds new keys from the current frame while the keys from the previous frame are being sent. Each key memory buffer contains keys that are microcode-defined.

In addition to keys, messages containing frame analysis results, such as the type of frame and its contents, are also stored in the key memories. Based on round robin arbitration and frame ordering, "out parse" selects a key memory and forwards the keys and messages to the TOPsearch I block. TOPparse may decide to discard a frame, such as one with a bad IP checksum, and free all the pointers associated with this frame from the RFD table (RFD out).

TOPparse may also perform direct table lookups in internal CAM data structures. The multiport text CAM is a shared resource containing keywords, such as GET and SERVER_PORT, for pattern searching within the frame payload. The multiport 8-bit CAM, 16-bit CAM, and 32-bit CAM are for binary pattern matching. All these CAM structures are accessed by all TOPparse engines simultaneously.

Relevant programmable accounting and statistical information pertaining to each frame and flow is collected and passed on to the statistics block. Frame ordering is maintained in the TOPparse block.

10.8.3 **TOPsearch I Block**

TOPsearch performs the various table look-ups required for the application. Layer-2 switching, Layer-3 routing, Layer-4 session switching, and Layers 5 to 7 content switching and policy enforcement are just a few of the applications possible. TOPsearch can also auto-learn and update tables with Layer-2 MAC addresses, Layer-3 directly connected IP addresses, and Layer-4 sessions.

TOPsearch receives keys from TOPparse. These can be either simple single keys or compound keys. TOPsearch matches the simple keys to entries stored in various tables, or uses the compound keys for more complex searches. TOPsearch uses a variety of search algorithms optimized for various types of searched objects.

Three primary types of data structures are supported by TOPsearch: hash tables, trees, and direct access tables. These data structure types are used to build the lookup tables and policy rules used for packet handling and the decision-making process. Unique algorithms accelerate the process of searching through the various hash tables and trees.

TOPsearch enables conditional search operations that may be applied in various applications. Search results in one data structure can trigger search in another data structure, for example, when searching for an IP encapsulation. The results of the search for the first IP address determines in which data structure the search for the second IP address will be conducted. Complex, conditional search instructions are stored in a macrocode program.

It is important to note that for simple single key searches, no coding is required. The programmer should simply transfer keys to the TOPsearch engine and expect results that are based on the found matches after the TOPsearch conducts its searches.

The NP-1 features two search stages (I and II); each with its own dedicated TOPsearch engines. TOPsearch II is useful for performing searches based on output/destination port-dependent information, such as the VLAN port, which is determined after TOPresolve.

The TOPsearch I block diagram is illustrated in Figure 10.12. "In search" distributes the keys and messages from the TOPparse engines to the available TOPsearch I engines (see Figure 10.10). "In search" identifies either single or compound keys and distributes them accordingly. Independent keys are identified as single keys and distributed to the available TOPsearch engines, enabling them to be searched in parallel. Compound keys are groups of keys that are interdependent and are loaded in the same TOPsearch engine and searched sequentially. Compound keys are input for conditional search programs where the results of a search in one data structure trigger a search in another data structure. Each TOPsearch engine can access the internal and external search memory containing the data structures being searched.

Lookup tables can be flexibly distributed among the memory cores to enable parallel searching of multiple keys. For example, MAC addresses and VLAN information may be stored in one memory core, IP addresses in a second memory core,

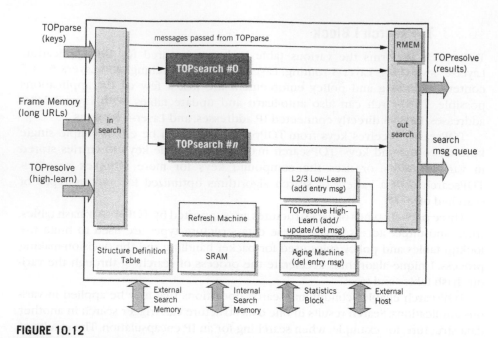

FIGURE 10.12

TOPsearch I block diagram

sessions in a third, and URLs in a fourth. All the information pertaining to a packet can then be searched and retrieved simultaneously.

The external memory is typically used to store tables (trees or hash) with many entries that otherwise would not fit into the embedded memory cores. For example, in a large router implementation, the external memory can be used to store IP subnets, while internal memory cores can be used for storing various Access Control Lists (ACLs), QoS and accounting policies. On the other hand, in a URL load-balancing switch implementation, TCP sessions can be stored in external memory, while IP addresses and various policy rules are stored in the internal memory.

All data structures can be downloaded by the host via the host interface. Layers 2 and 3 hash table entries can also be added and updated by Layers 2 and 3 low-learn, or by the TOPresolve high-learn (described next), which can update entries of Layers 4 to 7 as well.

The structure definition table indicates the structure type, attributes, and its location in the memories. Structures can be stored in one of the four memory cores, the statistics counter, internal SRAM holding small lookup tables, or ternary CAM for short field searches of up to 32 bits.

"Out search" passes all of the frame's search results (i.e., matches found or not), and key-associated data to TOPresolve. Results and/or messages can be forwarded to TOPresolve's result memory (RMEM), depending on the type of search structure and its location. Two buses are used for forwarding the results/messages in parallel. Messages from TOPparse containing frame analysis results are passed directly on to TOPresolve.

The TOPsearch I block contains several functional units/blocks that handle the structures/tables. Layers 2 and 3 low-learn is a hardware functional unit that can learn Layer 2 MAC addressees and/or Layer 3 directly connected IP addresses (e.g., stations connected to the switch's local subnets) "on the fly."

The TOPresolve high-learn functional unit/block adds, updates and deletes hash table entries based on packet tracking information from packets processed in TOPresolve.

The aging machine deletes aged entries from hash tables in the search memory. In parallel, each time that a hash table entry is used, the hash entry is refreshed by the refresh machine, if it is configured to do so. Aged entries are those that have not been refreshed during the aging period. Frame ordering is maintained in the TOPsearch block.

10.8.4 **TOPresolve Block**

TOPresolve is the decision point for handling of frames. It determines the destination, priority, format, and content changes of frames.

TOPresolve is a programmable processor and executes a microcode command per clock. Microcode downloaded to TOPresolve defines the desired resolutions for various frames and applications. The TOPresolve block diagram is illustrated in Figure 10.13.

TOPresolve determines the frame's filtering, destination port and priority based on search results, parse message, and stored learning/state information. The decisions it reaches based on these criteria, together with modify instructions, are compiled into a message that is passed to TOPmodify. TOPresolve can also generate search keys to be used by the TOPsearch II engines, if required.

"In resolve" dynamically distributes the search results to the next available buffer results memory (RMEM) (see Figure 10.13). The buffer is loaded with incoming

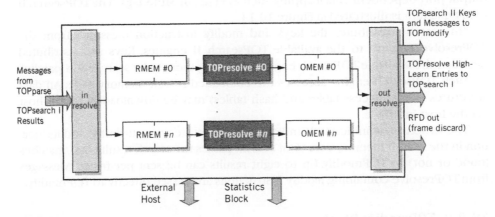

FIGURE 10.13

TOPresolve block diagram

search results and stores the search results that are currently being processed by the TOPresolve engine. Messages from TOPparse containing frame analysis results, such as the type of frame and its contents, are also stored in the RMEM.

Based on its microcode and the TOPparse message, the TOPresolve engine determines if the frame should be forwarded, the frame's destination port and queue, its priority and fields to be modified, and so on. These resolutions are formulated into keys and messages that contain modify instructions, and are stored in the out memory (OMEM). The OMEM buffer stores the keys/messages being processed and the previous keys/messages. In other words, TOPresolve decisions, along with modify instructions, are compiled into a message and passed to TOPmodify. Additionally, TOPresolve can generate search keys to be used by TOPsearch II engines. "Out resolve" sends up to eight keys/messages per frame from the OMEM to TOPsearch II.

TOPresolve also collects state information for packet tracking. The TOPresolve high-learn machine passes this information to TOPsearch I for data structure entries update. If TOPresolve decides to discard a frame, it will free all the pointers associated with this frame from the RFD table (RFD out).

Relevant programmable accounting and statistical information pertaining to each frame and flow is collected and passed on to the statistics block. TOPresolve maintains registers reflecting the status of all system queues (i.e., VOQ, TFDs, and RFDs). This information can be used to implement various QoS mechanisms (e.g., RED, WRED) through TOPresolve microcode. Frame ordering is maintained in the TOPresolve block.

10.8.5 TOPsearch II Block

TOPsearch engines in the TOPsearch II block are similar to those in the TOPsearch I block, but provide limited functionality. TOPsearch II can be used, for example, for output port-dependent data mapping such as VLAN or MPLS tags. The TOPsearch II block diagram is illustrated in Figure 10.14.

"In search" distributes the keys and modify instruction messages from the TOPresolve engines to the available TOPsearch II engines. Keys are distributed between the available TOPsearch engines, enabling them to be searched in parallel.

Each TOPsearch engine accesses the embedded search memory core. All data structures (direct access tables and hash tables) may be downloaded by the host via the host interface.

The structure definition table indicates the structure type, attributes, and its location in the search memory core. "Out search" passes the search results (i.e., matches found or not) to TOPmodify. Up to eight results can be sent per frame. Messages from TOPresolve containing modify instructions are passed directly to TOPmodify.

10.8.6 TOPmodify Block

TOPmodify's main task is to modify frame contents and forward them. TOPmodify is a programmable processor that executes a microcode command per clock.

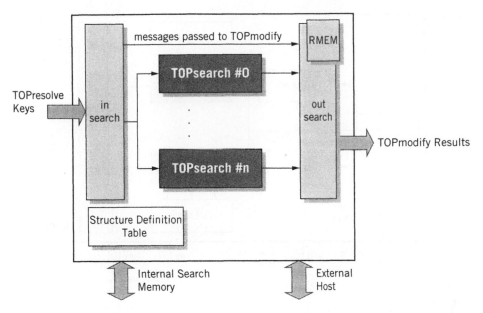

FIGURE 10.14

TOPsearch II block diagram

Microcode downloaded to TOPmodify, along with modify instructions from TOPresolve and search results, decodes the desired modifications for various frames and applications and modifies frame contents accordingly.

The TOPmodify block diagram is illustrated in Figure 10.15. "In modify" dynamically distributes the TOPsearch II results along with the frame's modify instruction messages to the next available buffer results memory (RMEM). The buffer is loaded with incoming search results while it stores the search results that are currently being processed by the TOPmodify engine.

TOPmodify reads the frame from the frame memory and stores it in its associated frame modify memory (FMEM). The FMEM has buffers that are read from the frame memory based on its RFD pointers. An incoming frame is loaded in one buffer, a second buffer holds the frame that is currently being modified, and the third buffer holds the frame that is being rewritten to the frame memory.

The TOPmodify engine overwrites, inserts or removes content in accordance with its microcode and the modify instructions associated with the frame. Once modified, the frame (in the FMEM) is rewritten to the frame memory and frame pointers are then passed to one or more of three queues based on the TOPresolve forwarding instructions.

Frames may be replicated by the TOPmodify engines; free buffers are fetched from the RFD table, and the replicated frames are stored in the frame memory while the RFD table is updated accordingly.

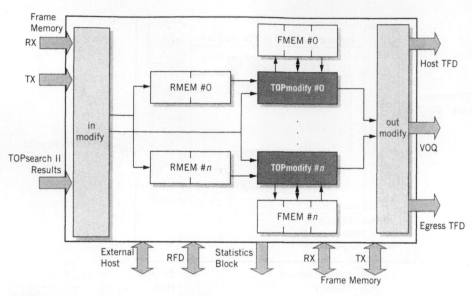

FIGURE 10.15

TOPmodify block diagram

The three output queues are: Host TFD for frames destined for an external host (via the PCI interface), VOQ for frames destined for the switching fabric, and Egress TFD for frames destined for the network.

Relevant programmable accounting and statistical information pertaining to each frame and flow is collected and passed on to the statistics block. Frame ordering is maintained in the TOPmodify block.

10.8.7 Frame Processing Ordering

The use of parallel processors provides increased processing power; however, frame ordering must be maintained. The NP has a transparent, built-in hardware mechanism to insure that frame ordering is maintained at each of the TOP blocks.

Unlike within the TOPparse engine in which entire frames are parsed, keys for the same frame may be distributed among several TOPsearch engines that are operating in parallel. Therefore, it is important not only to maintain frame ordering, but also to know how to reassemble the search results from several processor engines. For this reason, the "in search" stage assigns a tag containing the stream number, sequence number, and the total number of keys pertaining to the frame. "Out search" uses this tag to ensure that all of a frame's search results are sent to TOPresolve intact. TOPresolve and TOPmodify have similar frame ordering mechanisms.

In addition, several "lock" and "ordering" mechanisms may be activated optionally to ensure that the information related to a flow (e.g., session state information) is updated by the TOPresolve high-learn mechanism in the order in which the frames belonging to this flow are received by the NP. The same holds true for the Statistical counters corresponding to a given flow and the session hash lookup tables. Therefore, packets that belong to the same session are guaranteed to update a given state in their correct order. Subsequent packets belonging to the same session are guaranteed to be matched with the already updated state.

10.8.8 Frame Memory Interface

The NP operates in a store-and-forward mode. The frame memory stores the frame while the various TOPs perform the processing.

The frames are segmented and stored in the frame memory. The Receive Frame Descriptor (RFD) table contains pointers to all subsequent buffers belonging to the frame, with each buffer containing a segment of the frame's contents. In other words, the buffers that contain the frame's contents are linked together by the RFD table.

10.8.9 External Host Tasks and Interface

The External host has several tasks to perform in order to enable the NP to execute its application and to support it in the control plane (slow path) processing. Frames must be communicated between the NP and the External host, NP status information must be accessible to the External host, and microcode, settings, instructions, and configuration data should be sent from the External host to the NP. The main communication requirements and the hosts' tasks can be defined as follows.

Configuration—The host configures and updates NP registers, internal TOP registers and TOP resources via the register interface. Microcode is downloaded through register interfaces to each of the TOP blocks (TOPparse, TOPsearch, TOPresolve, and TOPmodify).

Search data structures—The host downloads and updates entries in the search data structures. The host is notified via the SRH_Q of any updates to these structures by the learning or aging mechanisms. The host can use this updated information as appropriate; for example, it can propagate this information to other NP devices in the system, or maintain a repository of its own.

Statistics—The host reads statistics counters collected by the network processor, either by polling counters through the register interface or by event-driven messages through the STATMSG_Q.

Receive frames—The host receives frames from the network processor via the HTFD_Q. The HTFD_Q contains a pointer to the frame as well as a message with TOP analysis information. For example, the message may contain the type of frame, search results, and forwarding information, relieving the host of the task of reanalyzing the frame. The NP acts as a preprocessor for the host.

Transmit frames—Transmit frames are posted directly into the VOQ and transmitted to the switching fabric. If required, TOPs will perform additional processing when the frame reenters from the switching fabric (in a loop-back configuration). The NP acts as an egress post-processor for the host.

The host can operate either in an interrupt-driven mode, via a PCI interrupt, or in a polling mode. For example, the host can be notified of new frames pending in the HTFD_Q by a PCI interrupt or by polling the status register. The External Host interface and tasks are handled via high-level APIs that are not discussed in this book.

NP provides a 32-bit 66-MHz PCI interface to an External Host CPU. NP uses look-ahead registers for efficient interfacing with the host. Host-destined frames are read from the buffer memory ahead of time into registers, awaiting retrieval by the host. This relieves the host of the need to access the frame memory directly. The External Host CPU can collect statistics from the network processor and download policies to the network processor.

10.9 SUMMARY

This chapter provides an overview of EZchip's NP architecture, usage and capabilities. This description should be sufficient for someone who wants to evaluate how to use a network processor and to start to program high-speed network applications. It is not the intention of this chapter to provide a reference manual or a user guide for the NP, and many details have been omitted or slightly softened due to their complexity, specific purpose, or because they have not been publicly published.

The next chapter discusses how to program the NP, and it complements the understanding of what can be achieved from network processors in general and from EZchip's NPs in particular.

EZchip Programming

In the previous chapter, we discussed EZchip's Network Processor (NP) architecture, functional units, processors, memories, interfaces, and so on. In this chapter, we will discuss how to use them (i.e., how to program them).

Programming the NP has three components: the first is the programming of the processor itself; the second is the development environment of the network processor (including compiling, debugging, simulating, etc.), and the third is running the network processor programs and applications (i.e., initializing the NP, downloading it, and working with the attached external host processor). This chapter deals mainly with programming the NP, and discusses the other two levels briefly.

EZchip's NP employs an array of Task-Optimized Processors (TOPs) to perform specific packet processing tasks with customized data paths and instruction sets. Each of the four types of TOPs performs a separate function as described in the previous chapter; they are: TOPparse, TOPsearch, TOPresolve, and TOPmodify.

The NP programming model is that of a pipelined message passing between heterogeneous multiprocessors. Although there are parallel processors at each stage of the processing pipeline, EZchip's NP programming model is not involved with parallel processing at all, as hardware circuitry handles all arbitration, ordering, and required scheduling and synchronization.

Although we have several heterogeneous processors to program (the TOPs), it is important to understand that all TOPs share a very similar programming language and style; thus, it is almost like programming a single processor architecture.

Each of the NP processor types, or the TOP engines, is a pipeline processor by itself, which means that the programmers of NP processors have to be aware of instruction pipelining, including data and control hazards (discussed in Chapter 8).

It is important to comprehend that the entire NP with all its TOPs works through pipelined processing of network packets, and that each of the TOPs works through pipelined processing of its instructions.

11.1 INSTRUCTION PIPELINE

Each of the TOPs is a four-stage pipelined processor, with each stage taking a clock cycle. In the first stage, an instruction address is fetched from the instruction memory block and stored (F stage). The instruction is decoded in the second stage (D stage) and executed in the third and fourth stages (E1 and E2 stages).

Since cycles are crucial in network processing, the TOPs are not equipped with any instruction reordering, flushing, or branch predicting capabilities; the programmer has to deal with control and data hazards by himself. As described in the programming chapter, data and control hazards can be used in such a way that no cycles are lost. This however, requires careful coding, using the options that EZchips' NP instruction set provides.

11.1.1 Data Hazards

Data hazards were discussed generally in Chapter 8, so only the implications relevant to the NP are briefly described here. An instruction is executed within the two clock cycles of the E1 and E2 stages, with read operations executed in E1, and write operations and flag settings executed in E2 (see Figure 11.1). This means that an instruction cannot use a source or a flag if the preceding instruction writes it or sets it, since these two operations happen in the same cycle and will collide.

In order to avoid data hazards, either a NOP instruction is inserted between the dependent instructions, or instruction reorder should be done, such that a non-dependent instruction—that is, one that does not read written data or count on a modified flag—is added between instructions.

The following examples demonstrate a data hazard. When the EZchip Compiler encounters such a hazard, it warns the programmer.

Examples of bad coding:

```
SUB ALU, UREG[2], REG[3], 4;
JNZ Label                              //Zero Flag is not ready

Mov UREG[4], UREG[2], 4;
Add ALU, UREG[1], UREG[4], 4;          //UREG[4] is not ready
```

11.1.2 Control Hazards

Control hazards were also described in the programming chapter, so only a few implications are discussed here. Branch instructions check the condition flags, decide whether or not to branch, and compute the next address to jump to in the E1 phase (see Figure 11.1). Control hazards result from branching that actually takes place only in the fourth cycle of the instruction (E2), where the pipeline fills up with the two instructions that follow the branch instructions. Even when the branch takes place, the two instructions that immediately followed the branch instruction and were already loaded into the pipeline will also be executed. This will cause a situation in which two more instructions will be executed after a jump instruction, as if they ignore the jump instruction.

	Clock Cycles				
Program Execution	CC1	CC2	CC3	CC4	CC5
INST#1	F	D	E1	E2	
INST#2		F	D	E1	E2

FIGURE 11.1

Instruction pipeline execution

To avoid control hazards, instructions should be organized so that two instructions of the application will be placed for execution immediately after the branch instruction, regardless of whether the branch will happen or not. Obviously, these instructions should be nondependent instructions to avoid data hazards. This will be further demonstrated in the following example. If no useful instructions can be added, NOP commands must be inserted.

NP code supports automatic insertion of NOPs for branch instructions. This is done by an operand of the branch instruction that adds zero, one or two NOPs in case of branch execution. When "inserting" zero NOPs, all instructions are executed (i.e., the succeeding two instructions and then the branched instruction). If one NOP is inserted, then the subsequent instruction is executed, and one NOP is added, so the next instruction executed is the branched one. When two NOPs are inserted, then after the branch instruction, the next instruction to be executed is the branched one. See the examples as follows.

Care should be exercised when several jump instructions are sequenced, as the first jump will happen, but immediately after executing one instruction from the branched place, the next jump will occur, and the next instruction to be executed will come from the second branched address. This is demonstrated in the following example.

Example: Successive jump instructions

```
L1: Jmp L5;
L2: jmp L7;
L3: ADD ALU, REGB, 4;
L4: MOV REGC, 0x8000;
NEXT_COMMAND:
  . . .
  . . .

L5: SUB ALU, REGD, REGC;
L6: XOR REGB, REGB, REGC;
  . . .

L7: MOV REGA, 5;
L8: ADD ALU, REGC, REGB;
  . . .
```

Description: The sequence of instructions that will be executed is L1, L2, L3, L5, L7, L8, … Although jump L5 was clearly the first instruction, the program finds itself in another branch of the program (dictated by the second jump to L7).

Control hazards may be used in a productive way, to save machine cycles, as described in the following example.

Example: Suppose you want to jump to label NEWLABEL if bit 3 of UREG[0] is zero; otherwise, continue. Suppose further that before this jump you want to perform two actions, say GETting 4 bytes of data into UREG[6] and MOVing 0x8000 to UREG[7].

```
if (!UREG[0].BIT[3]) Jmp NEWLABEL |_NOP0;
Get UREG[6], 0(RD_PTR), 4;
Mov UREG[7], 0x8000, 4;
NEXT_COMMAND:
    . . .
    . . .
NEWLABEL: . . .
```

Description: When no NOPs are inserted into the pipeline, the actual order of execution will be: Get → Mov → NEXT_COMMAND or NEWLABEL.

The next example illustrates which commands are in the pipeline when zero, one or two NOPs are used, with the instructions as follows:

Example:
```
L1: mov UREG[1], 2, 2;
L2: jmp DISCARD | _NOPx;
L3: mov UREG[2], UREG[3], 4;
L4: sub ALU, UREG[3], UREG[1], 4;
    . . .
DISCARD: halt;
```

In case of jmp DISCARD | _NOP0 (in preceding L2), the pipeline, as a function of the clock cycles, will contain the instructions as follows:

	Clock Cycles						
	CC1	**CC2**	**CC3**	**CC4**	**CC5**	**CC6**	**CC7**
Fetch	L2 (jump)	L3 (mov)	L4 (sub)	Discard: Halt			
Decode	L1	L2	L3	L4	Discard: Halt		
E1		L1	L2	L3	L4	Discard: Halt	
E2			L1	L2	L3	L4	Discard: Halt

In case of jmp DISCARD | _NOP1, the pipeline, as a function of the clock cycles, will contain the instructions as follows:

	Clock Cycles						
	CC1	**CC2**	**CC3**	**CC4**	**CC5**	**CC6**	**CC7**
Fetch	L2 (jump)	L3 (mov)	L4 (sub)	Discard: Halt			
Decode	L1	L2	L3	NOP	Discard: Halt		
E1		L1	L2	L3	NOP	Discard: Halt	
E2			L1	L2	L3	NOP	Discard: Halt

In case of jmp DISCARD | _NOP2, the pipeline, as a function of the clock cycles, will contain the instructions as follows:

	Clock Cycles						
	CC1	**CC2**	**CC3**	**CC4**	**CC5**	**CC6**	**CC7**
Fetch	L2 (jump)	L3 (mov)	L4 (sub)	Discard: Halt			
Decode	L1	L2	L3	NOP	Discard: Halt		
E1		L1	L2	NOP	NOP	Discard: Halt	
E2			L1	L2	NOP	NOP	Discard: Halt

11.2 WRITING NP MICROCODE

In this section we describe the assembly language of the NP and how to use it.

11.2.1 Assembly Language Overview

This section presents an overview of the assembly language used to write microcode applications for the four distinct types of TOPs, which constitute the programmability of the NP network processor. TOPparse, TOPresolve, and TOPmodify

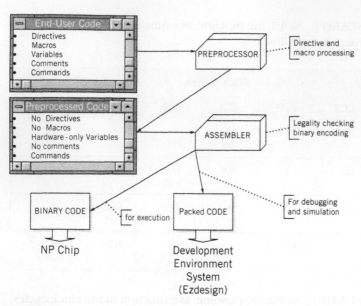

FIGURE 11.2

NP assembly language processing

have similar instruction sets (described in a following section), while TOPsearch has a distinct database-oriented instruction set. Nonetheless, all of the instruction sets share a common syntax and style, and can be collectively described as a single language.

The language consists of two layers, transparent to the user: the *preprocessor* (or macro-assembler) layer and the *assembler* layer. Program compilation of the program is performed in two stages (see Figure 11.2).

This subsection describes the assembler, the lower part of the two stages. As a rule, one command of the language described here corresponds to a single line in the instruction memory. Macros, variables and directives are described later in Section 11.3.

11.2.2 Command Notation Convention

The general format for an NP command is:

```
<LABEL:> <if (CONDITION)> COMMAND <OPERANDS>;
```

Pointed brackets (< >) indicate optional features. For example, a command may or may not be preceded by a label and/or a conditional statement. Not every command can be conditional, and the number and type of operands are command-dependent.

Commands are case-insensitive and newline insensitive (newlines are counted as white space) to increase program readability; a semicolon rather than a newline serves as a delimiter between commands. Comments are C and C++ styled, as demonstrated in the following example:

```
/*comment
and comment*/
//comment until a line end
```

The following example illustrates the syntax of the NP assembly language:

```
        .
        .
        .

BREAK_POINT_LAB:
    mov CNT, 2, 2;                  //initialize counter
    mov SREG[5], 0, 2;             //initialize SREG[5] register
WRITE_ALL_UREGS_LAB:
    if (!HOST_RDY)
            jmp WRITE_ALL_UREGS_LAB | _NOP2; //insert 2 NOPs in
                                             //pipeline
    loop WRITE_ALL_UREGS_LAB;       //no NOPs inserted
                                    //in pipeline

        .
        .
        .
```

11.2.3 Special Symbols

Table 11.1 describes the special symbols used by the NP compiler. Several character combinations are used as special symbols, as outlined in Table 11.2.

11.2.4 Resource ID, Indirect Access, and Address References

To access different devices (e.g., registers, functional units), the NP assembly language uses a pointer mechanism similar to high-level languages such as C. With the pointer mechanism, the programmer need not know the actual address of the memory area referred to by the pointer.

Table 11.1 Special Symbols

Symbol	Description	Example
$	The current PC (program counter) value.	mov UREG[0], $, 2; jz $ + 10;
&	Create immediate data with indirect access format (see the following section).	mov UREG[1], &UREG[0], 2;
+ following ")"	Auto-increment base register or index register after a read/write operation.	get UREG[1], 20(RD_PTR)+, 2;

Table 11.2 Character Combinations Used as Special Symbols

Combination	Description
\r	The "carriage return" character, ASCII 10.
\n	The "new line" character, ASCII 13.
\t	The tab character.
\\	The backslash character.
\"	The quotation mark character.
\xx, where xx is a two-digit hexadecimal number	An ASCII character with the code xx.

When a resource name (for example, UREG[2].BYTE[3]) is prefixed by an ampersand character (&UREG[2].BYTE[3]), it is interpreted as the actual hardware address of it, byte 3 in user register 2 in this example (which is 0x0218; the addresses are all known to the EZchip Compiler, but not necessarily to the programmer).

There are two indirect registers that can be used for various purposes. For example, an indirect register can be used to load an element of a vector that is loaded into UREG registers, as follows (there are potentially 64 byte values).

```
mov IND_REG0, &UREG[2].BYTE[3], 2;     //Now IND_REG0 contains a
    .                                  //reference to UREG[2].BYTE[3],
    .                                  //the first element

mov UREG[0], UREG[IND_REG0], 1;        //move 1 byte from UREG[2].BYTE[3],
    .                                  //which is the first element

Add IND_REG0, IND_REG0, 4, 2;          //point to next required
    .                                  //element, which happens to
    .                                  //be the next index in UREG
    .                                  //register block (4 bytes
    .                                  //difference)

mov UREG[0], UREG[IND_REG0], 1;        //move 1 byte from UREG[3].BYTE[3],
                                       //the next required element
```

11.2.5 Addressing Modes

There are many addressing modes supported by NP language, including immediate, register (and even specific bytes or bits in the registers), indirect, memory direct and base-indexed, TCAM groups, structures, and others. These are detailed in the chapters that describe each of the TOP engines, according to its memories and specific register interfaces.

11.2.6 Little Endian and Big Endian Notation

Networking generally operates in big endian[1] notation (i.e., the most significant byte of the field comes first). EZchip's NP uses little endian notation.

In either notation, there is very often a need to transfer data from memory (which, in some cases, contains network data) to registers for processing, and vice versa. This may require the conversion from big endian to little endian notation, or vice versa, on the fly. For example, an IP address that is received from the network in network order (big endian), and that has to be processed by the NP, which reads little endian, must be converted before processing.

11.3 PREPROCESSOR OVERVIEW

The preprocessor handles tokens, macros, and file inclusions. The commands supported are `macro` (and `endmacro`), `define` (and `undefined`), `include`, and several conditional commands: `if`, `else`, `endif`, `ifdef`, and `ifndef`. The meaning and usage of these preprocessor commands are very similar to those of C and similar languages, and are described briefly in Appendix A of this chapter.

11.4 DEVELOPING AND RUNNING NP APPLICATIONS

There are two main software components that are used for developing, and later enabling, the implementation and usage of the NP in customized equipment. The first software component is EZdesign package, which is used for the development phase and includes several development tools, such as the Microcode Development Environment, which is a common GUI wrapper to all tools. The second software component is EZdriver, a software API package that enables the user to design embedded NP systems, based on any host processor or any operating system.

---

[1] There are mainly two byte-ordering representations of any data, called *little endian* and *big endian*. Little endian is byte ordering (either as positioned in memory or as transmitted), according to which the least significant byte comes first. Big endian is the opposite ordering; that is, the most significant byte comes first. In networks, byte ordering is called "network order," and is generally big endian due to historical reasons, where communication equipment (switches, routers, etc.) starts to switch or to route based on the most significant bytes that flow into the equipment, representing numbers, prefixes, or more meaningful addresses than the bytes that follow.

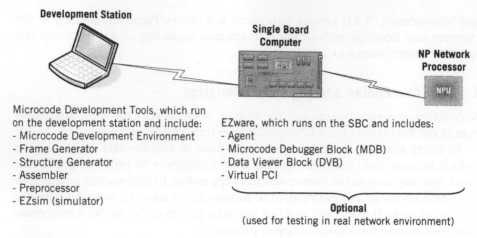

Development Station

Single Board Computer

NP Network Processor

Microcode Development Tools, which run on the development station and include:
- Microcode Development Environment
- Frame Generator
- Structure Generator
- Assembler
- Preprocessor
- EZsim (simulator)

EZware, which runs on the SBC and includes:
- Agent
- Microcode Debugger Block (MDB)
- Data Viewer Block (DVB)
- Virtual PCI

Optional
(used for testing in real network environment)

FIGURE 11.3

Sample EZdesign system architecture

11.4.1 EZdesign Software Toolset and Libraries

EZdesign is a set of design and testing software tools for developers. It allows designers to create, verify, and implement NP applications to meet specific functionality and performance targets. Its general setup is depicted in Figure 11.3.

EZdesign can be used for developing NP applications on a stand-alone PC, with no NP attached. It can even be used to simulate and debug the NP applications, when network traffic can be artificially generated. For testing in real network environment, a development system is required (a single board computer (SBC) that contains EZware, a software package that interfaces between the NP and the developing system). EZdesign components include:

Microcode Development Environment (MDE): A unified GUI for editing and debugging code, including setting breakpoints, single-stepping program execution and access to internal resources. MDE enables code editing, view of memory and register contents, performance charting, macro recording and script execution. MDE is used in development and debugging of code on both the simulator and the actual network processor.

  The MDE and a short description of how to install and use it can be downloaded according to the instructions in Chapter 1. As described before, this package enables programming, debugging and simulation of the NP, without having to attach the MDE to a development system (however, no real network environment can be tested).

Simulator: The simulator provides cycle-accurate simulation of the EZchip network processor for code functionality testing and performance optimization.

Assembler and Preprocessor: These components generate code for execution on EZchip's network processors.

Frame Generator: A GUI guiding the programmer through the process of creating frames, layer by layer. The frame generator allows for the generation of frames of different types, protocols and user-defined fields. The resulting frames can be used for testing and simulating the NP code.

Structure Generator: A GUI enabling the definition of data structures used by the EZchip network processor for forwarding and policy table lookups (e.g., hash, trees), their keys, and associated result information.

To support debugging, interrupts both from the host to a TOP as well as from a TOP to the host are supported. Interrupts sent from the host may halt a TOP operation in order to read it for debugging. The host runs a service routine that reads the TOP's memory and adds breakpoints by using a set-bit at branches to indicate the location in the code. This mechanism is used by the EZdesign Microcode Development Environment to allow programmers to develop and debug the microcode.

11.4.2 EZdriver Control Processor API Layer

EZdriver is a toolset that facilitates the development of the control path software for EZchip network processor-based systems. It enables applications that run on the control CPU to communicate with the EZchip network processor. EZdriver consists of routines that are executed on the control CPU and provide an API for interfacing with the network processor. It includes the chip configuration, microcode loading, creation and maintenance of lookup structures, sending and receiving frames to and from the network processor, as well as configuration and access to the statistics block.

EZware, the interface to the EZdesign toolset, can be utilized optionally to allow debugging and simulation of the NP application on the target system, the customized network equipment.

11.5 TOP COMMON COMMANDS

TOPparse, TOPresolve, and TOPmodify share common language conventions and commands. Each of these TOPs also requires some specific commands that are unique to their function. In the following chapters that describe each of the TOPs, only the specific commands are described. It should be noted that although the commands' syntax in each of the TOPs is almost identical, and the operand usage is also very similar, there are some differences in the operands' usage, and these are detailed in the appendices of each of the following chapters.

In this section, we provide a list of the common commands, divided into five categories: resource initialization commands, move commands, branch commands, arithmetic (ALU) commands, and the halt command.

11.5.1 Resource Initialization

Initialization commands are "executed" by the host rather than the TOP engines; they indicate the initial values given to various registers prior to executing the program in the instruction memory. There are several resource initialization commands in each of the TOP engines, but there is one common command for all TOP engines:

```
LDREG MREG[n], value;
```

Load the Mask register MREG[n] (used for ALU operations) with a value.

Example: LDREG MREG[0], 0x0F;

Description: Move 0x0F value to the first mask register, MREG[0].

11.5.2 Move Commands

Move commands copy data (bits, bytes, and words) from various sources to various destinations. The following commands are used in almost identical ways by the TOP engines. There are some differences in naming, usage, and optional operands that are supported by some TOP engines and not others, but it does not change the general meaning of these commands. The complete and accurate use of these commands is listed in the chapters that detail each of the TOP engines.

It should be noted that the last four move instructions in this section are not supported by the TOPmodify, mostly due to the fact that the TOPmodify is the last engine in the NP pipeline. Hence, there is no need for it to output keys or messages, nor is there need for headers for these keys and messages.

It should also be noted that the TOPmodify has in all move commands an optional operand, jpe, that instructs the TOPmodify to jump after the move operation according to the FAST_REG, a mechanism for allowing TOPmodify to execute predetermined segments of require code. Move semantics depend on the type of the source and the destination, as depicted in Figure 11.4.

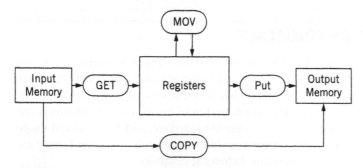

FIGURE 11.4

Data movement semantics

11.5.2.1 *Mov DST, SRC, SIZE*

Move SIZE bytes from source (SRC) to destination (DST).

Example: `Mov UREG[0], UREG[1], 4;`

Description: Move 4 bytes from UREG[1] to UREG[0].

Note: TOPmodify has one more optional operand—`jpe`, which is used to optionally jump according to the FAST_REG, after the move operation.

11.5.2.2 *MovBits DST, SRC, SIZE*

Move SIZE bits from source (SRC) to destination (DST).

Example: `MovBits UREG[0].BIT[0], UREG[1].BIT[2], 5;`

Description: Move 5 bits from UREG[1] from bit 2 to UREG[0] starting with bit 0.

Note: TOPmodify has one more optional operand—`jpe`, which is used to optionally jump according to the FAST_REG, after the move operation.

11.5.2.3 *Get DST, SRC, SIZE*

Move SIZE bytes from the source memory[2] (SRC) to a destination (DST) register.
   Note: TOPparse and TOPmodify have several other optional operands. One of these, `EOFjump` (in TOPparse), determines whether a jump will occur at the end of a frame. `jpe` operand (in TOPmodify) is used to optionally jump according to the FAST_REG, after the move operation.

Example: `Get UREG[0], 23(0)+, 2;`

Description: Load two bytes from input memory address 23 to UREG[0] register. Auto-increment the base pointer of the input memory to 23 + 2 = 25.

11.5.2.4 *Put DST, SRC, SIZE*

Move[3] up to four bytes (SIZE) from source register (SRC) to destination memory (DST).

Example (in the case of TOPparse): `PutKey OFF_KMEM(KBS2)+, UREG[0], 4;`

Description: Put 4 bytes from UREG[0] to the key memory, in address KBS2+OFF_MEM. Auto-increment KBS2 to KBS2+ OFF_KMEM+4.

Note: TOPmodify has one more optional operand—`jpe`, which is used to optionally jump according to the FAST_REG, after the move operation.

---

[2]Usually, the move is from input memory; however, in TOPmodify, SRC points to the frame memory (FMEM), and for reading the input memory, it uses a similar command, that is, `GetRes DST, SCR, SIZE`.
[3]In TOPparse and TOPresolve, only keys are output, so this command is called `Putkey DST, SCR, SIZE`.

11.5.2.5 Copy DST, SRC, SIZE

Move SIZE bytes from the source memory (SRC) to the destination memory (DST).

Note: TOPparse and TOPmodify have several other optional operands. One of these, EOFjump (in TOPparse), determines whether a jump will occur at the end of a frame. jpe operand is used to optionally jump according to the FAST_REG, after the move operation.

Example (in the case of TOPparse): Copy 10(KBS0)+, 12(RD_PTR)+, 4;

Description: Copy four bytes from FMEM address RD_PTR+12, to KMEM address KBS0+10. Auto-increment RD_PTR to RD_PTR+12+4. Auto-increment also KBS0 to KBS0+10+4.

11.5.2.6 Mov4Bits DST, SRC [, MASK]

Move four bits from source (SRC) to destination (DST). Use MASK to indicate which bits should be inverted during the move. This command is not supported by TOPmodify.

Example: Mov4Bits UREG[2].BITS[0,3,5,7], SREG[4].BITS[3,5,7,9];

Description: Move bits 3, 5, 7, 9 of SREG[4] to UREG[2] bits 0, 3, 5, 7.

Example: Mov4Bits FBLK[4].BITS[1,3,5,7], SREG[13].BITS[3,5,7,11], 0xE;

Description: The inverse mask is 0xE = 1110. The lowest mask bit is used for the inverse of the source register's highest bit. So the result will be:

inverse(SREG[13].bit3) -> FBLK[4].bit1 // as inverse bit 3 (value = 1) is used
inverse(SREG[13].bit5) -> FBLK[4].bit3 // as inverse bit 2 (value = 1) is used
inverse(SREG[13].bit7) -> FBLK[4].bit5 // as inverse bit 1 (value = 1) is used
SREG[13].bit11   -> FBLK[4].bit7 // as inverse bit 0 (value = 0) is used
E.g., if SREG[13].BITS[3,5,7,11] = 1111, then FBLK[4].BITS[1,3,5,7] = 0001

11.5.2.7 PutHdr DST, SRC, SIZE

Move SIZE bytes from source (SRC) to DST, the header registers (i.e., HREG). This command is not supported in TOPmodify (since it does not write keys or messages).

Example: PutHdr HREG[HBS0].BYTE[1], UREG[1], 2;

Description: Put 2 bytes from UREG[1] to header with number from HBS0, offset in header—1.

11.5.2.8 PutHdrBits DST, SRC, SIZE

Move SIZE bits from source (SRC) to DST, the header registers (i.e., HREG). This command is not supported in TOPmodify (since it doesn't write keys or messages).

Example: `PutHdrBits HREG[0].BIT[3], UREG[1].BIT[2], 5;`

Description: Put 5 bits from UREG[1] bit 2 to header 0, bit offset 3 in header.

11.5.2.9 *PutHdr4Bits DST, SRC [, MASK]*

Move four bits from source (SRC) to DST, the header registers (i.e., HREG). Use MASK to indicate which bits to invert while moving. This command is not supported in TOPmodify (since it doesn't write keys or messages).

Example: `Mov UREG[1], 0x7, 1;`
`        nop;`
`        PutHdr4Bits HREG[HBS1].BITS[3,5,7,9], UREG [1], 5;`

Description: Put 4 bits from UREG[1] to header with number from HBS1, bits 3, 5, 7, 9:

HREG[HBS1].BITS[3,5,7,9] should be 1, 0, 1, 1. This is because the inverse mask 0x5 = 0101, it will be used as 1010 as follows:

0111   SRC or UREG[1]
1010   inverse mask
. . .
1101   Result of HREG, i.e., bit 9 = 1, bit 7 = 1, bit 5 = 0, and bit 3 = 1

11.5.3 Halt Command

Halt is used to finish packet processing and to release it to the next stage in the pipeline, or to multicast the frame several times, or to abort any further processing of the packet and to discard it. These options are defined by the MODE operand of the command.

TOPmodify is the last TOP engine in the NP pipeline; hence, it is responsible to decide how the frame is to be handled (i.e., to which output it should be sent, if any, and whether the frame should be rewritten to the frame memory or not). This is determined by an additional operand that the Halt command has in the TOPmodify engine.

Example: `HALT HALT_UNIC;`

Description: Halt execution of the TOP engine, and release the frame, the keys and messages to the next stage.

11.5.4 ALU Commands

Arithmetic Logic Unit commands are used for arithmetic and logic calculations. There are two formats of ALU commands, one with two source-operands and the other with one source-operand:

`    COMMAND  DST,SRC1,SRC2,SIZE,[,MREG[,MODE[,JPE]]];`
                              // See commands in Table 11.3.

or

`    COMMAND  DST,SRC,SIZE,[,MREG[,MODE[,JPE]]];`
                              // See commands in Table 11.3.

Definition: Calculate the *COMMAND* on the sources (SRC, or SRC1 and SCR2) of SIZE bytes, and put the result in the destination (DST). Use masks in MREG, according to the required MODE, to mask the source operands if required. jpe is used only by TOPmodify to optionally jump according to the FAST_REG, after the ALU operation.

Table 11.3 lists the TOP ALU operations and flag dependencies. The inverse can be performed on the logical functions "And" and "Xor", by entering a negation mark (!) before the operand.

Example: Add ALU, UREG[3], 0x20, 4, MREG[0], _ALU_FRST;

Description: ALU = (UREG[3] & MREG[0]) + 0x20;

Example: XOR UREG[2], ! UREG[3], ! UREG[4], 4;

Description: This performs inverse UREG[3] Xor inverse UREG[4] and writes to UREG[2].

Example: NumOnes UREG[2], UREG[4].BYTE[1], 1;

Description: This counts the number of "1" bits in UREG[4].BYTE [1] and writes the count into UREG[2].

Table 11.3 TOP ALU Operations

Name	Description	Source Operands	Flag Dependencies
And	DST = SRC1 & SRC2	2	OV, ZR, SN
Xor	DST = SRC1 ^ SRC2	2	OV, ZR, SN
Or	DST = SRC1 I SRC2 (Performed by macro)	2	OV, ZR, SN
Not	DST = !SRC1 (Performed by macro)	1	OV, ZR, SN
Add	DST = SRC1 + SRC2	2	OV, ZR, SN
Sub	DST = SRC1 - SRC2	2	OV, ZR, CY, SN
Addc	DST = SRC1 + SRC2 + Carry	2	OV, ZR, SN
Subb	DST = SRC1 - SRC2 - Carry	2	OV, ZR, CY, SN
Decode	Set bit k in the DST, when k = SRC	1	ZR
Encode	DST = k when bit k is the most significant set bit in SRC	1	ZR
NumOnes	DST = the number of bits equal to 1 in SRC. 1 byte only	1	ZR

Example: `AND ! UREG[1], UREG[3], UREG[4], 4;`

Description: This performs UREG[3] And UREG[4] and writes the inverse to UREG[1].

11.5.5 Jump Commands

Jump commands instruct the NP to jump to a given label in the microcode. Some of the commands are conditional jumps, and some jump with pushing and popping program addresses. At the end of the jump, optional `NOP_NUM` (0, 1 or 2) NOPs may be inserted into the pipeline following the jump command (to disable execution of the commands that immediately follow the jump instruction). The NOPs are effective only if the jump command is executed.

The standard Jump command format is:

```
if (CONDITION)command LABEL [| NOP_NUM];
```
                              // See commands in Table 11.4.

or

```
command LABEL [| NOP_NUM];
```
                              // See commands in Tables 11.4 and 11.5.

Example:

```
if (FLAGS.BIT [F_ZR_TOP])
    Jmp LAB | _NOP2;
```

or

```
JZ LAB | _NOP2;
```

Description: If flag Zero is set, jump to label LAB. Two NOPs are inserted. The two commands are exactly the same, as the second format is merely a macro that deploys the first format command, which is a single command (a single machine instruction).

Examples:

```
Jstack _NOP1; //Jump unconditionally to address taken from PC_STACK,
             //and insert one NOP.

if ( FLAGS.BIT [F_ZR_TOP] )

Jstack _NOP1; //Jump to address taken from PC_STACK only if flag
             //ZERO is set and insert one NOP.

Jmp  LAB1;    //Jump always

if ( FLAGS.BIT [F_ZR_TOP] )  // Jump to LAB1 only if flag ZERO is set
Jmp  LAB1;
JZ   LAB1;    //Jump to LAB1 only if flag ZERO is set
```

Table 11.4 TOP Jump Commands

Name	Description	Conditions for jump
Jmp	Jump unconditionally.	Any (using condition)
Jcam	Indirect jump—address jump from CAMO register (not available in TOPresolve).	MH
Jstack	Indirect jump—address from PC_STACK register.	Any (using condition)
Return	Indirect jump—address from PC_STACK register. Equivalent to Jstack.	Any (using condition)
Loop	Jump with loop counter decrement.	LP
Call	Call and save current address in PC stack.	Any (using condition)
CallCam	JCam and save current address in PC stack (not available in TOPresolve).	MH
CallStack	JStack and save current address in PC stack.	Any (using condition)

Table 11.5 TOP Jump Macro Commands

Name	Description	Name	Description
JB	Jump if below (CY)	JNS	Jump if not sign (!SN)
JC	Jump if carry (CY)	JA	Jump if above (!CY & !ZR)
JNAE	Jump if not above or equal (CY)	JNBE	Jump if not below or equal (!CY & !ZR)
JAE	Jump if above or equal (!CY)	JBE	Jump if below or equal (CY I ZR)
JNB	Jump if not below (!CY)	JNA	Jump if not above (CY I ZR)
JNC	Jump if not carry (!CY)	JGE	Jump if greater or equal (SN == OV)
JE	Jump if equal (ZR)	JNL	Jump if not less (SN == OV)
JZ	Jump if zero (ZR)	JL	Jump if less (SN != OV)
JNE	Jump if not equal (!ZR)	JNGE	Jump if not greater or equal (SN != OV)
JNZ	Jump if not zero (!ZR)	JNLE	Jump if not less or equal (!ZR & S == OV)
JO	Jump if overflow (OV)	JG	Jump if greater (!ZR & SN == OV)
JNO	Jump if not overflow (!OV)	JLE	Jump if less or equal (ZR I SN != OV)
JS	Jump if sign (SN)	JNG	Jump if not greater (ZR I SN != OV)

The standard assembly notation for the jump commands is provided by macros predefined in a library header file, and listed in Table 11.5.

The format of the Call, CallCam, and CallStack commands is a bit different from the jump command, and also includes a number of NOPs (RET_NOP_NUM) that should be inserted before returning, as follows:

```
Call function [ | NOP_NUM [ , RET_NOP_NUM ] ];
```

or

```
CallCam function [ | NOP_NUM [ , RET_NOP_NUM ] ];
                                      //This command is not available in
                                      //TOPresolve
```

or

```
CallStack function [ | NOP_NUM [ , RET_NOP_NUM ] ];
```

Definition: Call a function, and return from it to the next address upon completion (return command in the function). Insert NOP_NUM NOPs before jumping and RET_NOP_NUM NOPs before returning.

11.6 SUMMARY

NP programming is described in this chapter from its very basic programming model through the programming style. The chapter also explains the usage of software packages that enable the NP to be developed and run in target, customized network equipment. In the following chapters, a detailed programming of each of the TOPs is given, along with an example of how the TOP is used.

APPENDIX A
PREPROCESSOR COMMANDS

The preprocessor commands are listed in this appendix by their groups; that is, macros, definitions, includes, and conditional commands.

Macro Commands

The command format is as follows:

```
macro <NAME> <PARAM_LIST>;              // may be also written #macro
```

where `<PARAM_LIST>` takes the form `<TOKEN>, [...], <TOKEN>=<DEFAULT>, [...]`

A macro preprocessor command defines a macro with the name `NAME`. All code between this command and the command `"endmacro"` will be placed in the assembly code each time the preprocessor encounters a line in the source such as:

```
<NAME> <parameters>;
```

The comma-separated items of `PARAM_LIST` will be equated (as per #define) with the corresponding items of `"parameters"`. If `"parameters"` is superfluous, extra parameters are ignored and a warning is issued. If there are insufficient parameters, the default values (if they exist) of the missing parameters are used.

Note that any preprocessor command inside a macro will be executed at every expansion of the macro. A #define inside a macro has global scope for the whole program unless it is mentioned as a parameter in `PARAM_LIST`, in which case it is local. If an external token has a value (e.g., per #define) and the same name is used for a macro parameter, the parameter value will be used inside the macro locally; a warning will be issued to ensure that this is what the programmer meant.

Macros can use other macros inside them, called nested macros; there is even the possibility of defining a macro within a macro, although nothing may be gained from it.

Terminating a macro definition is done by the following command:

```
endmacro;   // may be also written #endmacro, end macro or #end macro.
```

Define Commands

The command format is as follows:

```
#define <TOKEN> <VALUE>;
```

where `<VALUE>` is taken as all the text between `"#define"` and the semicolon except for the first token (whitespace-separated).

Define preprocessor command sets a meaning to a token. Any time the token occurs in the text, it is replaced by a full evaluation of VALUE as it is currently (unless the token occurs as a parameter of certain preprocessor commands, where this obviously has no meaning). That is to say, white space-separated tokens inside VALUE that currently are #defined (or, within a macro, passed as parameters of this macro), are expanded to their values. Moreover, if the resulting text is a calculable arithmetical/logical expression, the preprocessor evaluates it numerically. Note that if A is defined as B+C, redefinitions of B will affect the value of A.

A subsequent #define of the same token replaces the previous one and a warning is issued. #undef removes it. Terminating a definition is done by the following command:

```
#undef <TOKEN>;
```

Include Command

The command format is as follows:

```
#include "<FILENAME>";
```
    // Angular brackets may be used instead of the square quotes, or neither.

<FILENAME> may be either a simple name or a complete path.

Include preprocessor command inserts the whole text of the file FILENAME into the program, and causes execution of all the preprocessor commands in it as well. If FILENAME is a simple name (without a path), it is looked for in the current directory and in directories mentioned in the INCLUDE environment variable of the user; if directory name include spaces, the whole list should be quoted in the command line.

Conditional Commands

Conditional preprocessor commands identify a block of code that is to be expanded in the source, or not, according to a condition. The command format is as follows:

```
#if <LOGICAL_CONDITION>;
```
or
```
#ifdef <TOKEN>;
```
or
```
#ifndef <TOKEN>;
```

The "if" preprocessor command evaluates the logical_condition (including tokens in it), and decides whether it is true or false. The "ifdef" will result true if TOKEN is defined, and the ifndef will result true if TOKEN is undefined. The following lines (up to the else or the endif preprocessor commands) will be included in the program text if the condition is true; otherwise, they will be ignored. Preprocessor commands in this block will be treated as well. #if-s may be nested as many times as memory allows.

Additional conditional commands are:

`#endif;`

Marks the end of an `IF` **block.**

`#else;`

Acts as a normal ELSE in an IF statement.

`#elif <LOGICAL_CONDITION>;`   **// may be also written** `#elsif`

Equivalent to "`#else; #if <LOGICAL_CONDITION>;`**"—except that an extra** `#endif` **is not needed.**

`#elifdef <TOKEN>;`   **// may be also written** `#elsifdef`

Equivalent to "`#else; #ifdef <TOKEN>;`**"—except that an extra** `#endif` **is not needed.**

`#elifndef <TOKEN>;`   **// may be also written** `#elsifndef`

Equivalent to "`#else; #ifndef <TOKEN>;`**"—except that an extra** `#endif` **is not needed.**

Parsing

The first stop of the packet, as well as of our description of packet processing in EZchip, is parsing: the stage in which the structure of the packet is analyzed and keywords are parsed for further lookups.

TOPparse decodes, analyzes, and extracts frame headers and significant fields corresponding to all seven layers of the packet. This includes Medium Access Control (MAC) addresses, Virtual Local Area Network (VLAN) tags, Ethernet frame types, Multi-Protocol Label Switching (MPLS) tags, Internet Protocol (IP) options, IP addresses, ports, HyperText Transfer Protocol (HTTP) information, Real Time Protocol (RTP) header, and so on. The headers and fields extracted by TOPparse are passed as keys to TOPsearch.

This chapter outlines the TOPparse architecture, its functional blocks (called devices), and its instruction set, in order to provide a general understanding of the functionality of this Task Optimized Processor (TOP) engine. A simple example is also given to demonstrate the use of the TOPparse. For those readers who would like more complex examples, or who intend to write programs themselves, detailed descriptions as well as lists of the most important registers and instructions can be found in the appendices to this chapter.

12.1 INTERNAL ENGINE DIAGRAM

The block diagram (Figure 12.1) illustrates the internal blocks of a single TOPparse engine.

12.1.1 Instruction Memory

The instruction memory contains microcode instructions for the TOPparse engines.

12.1.2 Pipeline Control

The pipeline control receives a command from the instruction memory, decodes the command, and distributes the control signals to each block.

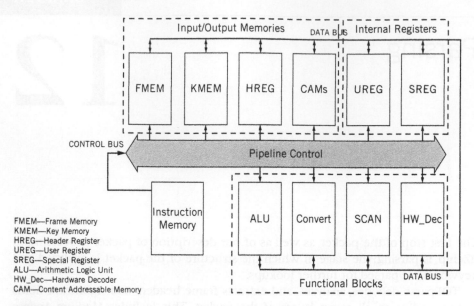

FIGURE 12.1

TOPparse engine internal block diagram

12.1.3 Data Bus

The data bus supports a data flow between the TOPparse blocks.

12.1.4 Frame Memory

Frame Memory (FMEM) is a cache memory that contains a portion of the frame that resides in the frame memory of the Network Processor (NP). The loading of segments into the FMEM is completely transparent to the programmer who "sees" the entire frame.

12.1.5 Key Memory

The Key Memory (KMEM) is a memory that stores extracted frame fields as keys to be forwarded to the TOPsearch I stage. The memory also stores messages that are being passed along to TOPresolve. All the internal TOPparse blocks can write to the KMEM indirectly via the Key Base (KBS) address pointer register.

12.1.6 Register Blocks (User-Defined Register, Specific Register, and Header Register)

The User-Defined Register (UREG) contains general-purpose registers for data processing. Specific Registers (SREG) are dedicated to specific TOPparse functions.

The Header registers (HREG) contain descriptions of the KMEM contents, which include information such the size of the key and what to do with the key. The registers are described in Section 12.2, and some are detailed in Appendix A of this chapter.

12.1.7 Arithmetic Logic Unit

The 32-bit general purpose Arithmetic Logic Unit (ALU) performs computational functions. ALU source operands are by definition registers in any TOPparse device. To use an operand from the memory, the programmer must first copy it to a register and then use it as an ALU source. ALU feedback registers are used for consecutive operations in order to avoid data hazards.

12.1.8 Content Addressable Memories

Content Addressable Memories (CAMs) compare keys to entries to find a match and corresponding results. There are a wide variety of CAMs for rapid matching of well-known frame identifiers, for example, Bridge Protocol Data Unit (BPDU), IP protocol, IP options, IP addresses, and IP ports. ASCII identifiers can also be matched quickly, for example, HTTP "GET," HTTP "POST," or Real Time Streaming Protocol (RTSP) "Cseq."

In TOPparse, there are four CAM tables depending on the width of the entry: 8, 16, 32, and 64 bits. In the 8/16/32-bit binary CAM tables, the size of the key must match the size of the entry exactly, whereas in the 8-character (64-bit) ternary CAM table, "don't care" bits can be used to define characters in the keys that are irrelevant. For instance, if a 64-bit key has 8 characters and 5 of the characters are set to "don't care," then a 3-character key can be used. The ternary CAM also supports case insensitivity as defined by one bit per character.

All CAM results are 16 bits and may be used to indicate a jump to a specified address in the code memory. A CAM table may be divided into sections for layer 2 (L2), layer 3 (L3) protocols, and so on. Table entries may be divided into groups, up to eight in number, to support keys with multiple results. Three group bits are used to specify to which group a CAM entry corresponds. Keys can be downloaded to the CAMs during resource initialization (see Section 12.4.1).

12.1.9 Convert Block

The Convert Block converts an ASCII string to a binary number. The ASCII string is either a decimal number (maximum of five characters) or a hexadecimal number (maximum of four characters), depending upon the command parameters. The Convert Block is useful in applications such as Network Address Translation (NAT).

12.1.10 Hardware Decoder Block

The hardware decoder performs preprocessing of frames, up to the IP Options, before microcode execution begins. It is designed to support IP Version4 (IPv4) and IP Version6 (IPv6) with/without MPLS using Ethernet or Point-to-Point Protocol

(PPP) encapsulation. The hardware decoder can traverse up to four labels in the MPLS label stack.

The hardware decoder can also be configured to search for up to 64 bits of user-defined data. At initialization, the host must set the hardware decoder to process either Ethernet or PPP frames. The hardware decoder block writes into the Hardware registers (HWARE), which is described in Section 12.2.1, and detailed in Appendix A of this chapter.

12.1.11 Text Scanning Block

The SCAN block is a text-scanning engine for identifying delimiters and tokens in text strings. SCAN searches the contents of FMEM for tokens, which are character strings located between defined delimiters (such as a space or tab). Any 8-bit character may be defined as a delimiter. Programmers may define up to 16 delimiters and use masking to determine which of the delimiters are used in the current SCAN operation. SCAN operations are performed either forwards or backwards along the frame.

The SCAN engine can extract the string between two delimiters. The strings can be a URL or significant key words. URLs can be passed as keys for TOPsearch, while keywords can be used for searching in the TOPparse TCAM. The SCAN block performs four operations:

- *Find delimiter*—locates the next defined delimiter in the bit stream.
- *Find nondelimiter*—locates the next character that is not defined as a delimiter.
- *Find token*—performs "find delimiter" followed by "find nondelimiter," which leads to the beginning of the next token.
- *Get token*—performs "find nondelimiter" followed by "find delimiter." The data scanned can be written to a specified destination, such as KMEM or a register.

The SCAN contains four output registers that are updated following each scan operation. The registers specify the start (SCAN_START), the end (SCAN_STOP), the size of the scanned data (SCAN_SIZE), and the termination cause (SCAN_TC).

When the SCAN operation reaches the end of the frame, it can generate an end-of-frame (EOF) interrupt to a subroutine. When the SCAN operation terminates due to the limit or due to an EOF, the reason is indicated in the termination cause register.

The SCAN block also changes the value of two registers during its execution: IND_REG[0] when writing to a register, and KBS0 when writing to KMEM.

At the end of SCAN operations, the character that results in the termination (either delimiter found or nondelimiter found) is written to the SCAN_TC. If, as mentioned before, the reason for the termination is that EOF occurred, this is also indicated in the SCAN_TC (see instructions descriptions, Section 12.4.5 and Appendix A of this chapter).

12.2   TOPPARSE REGISTERS

There are two sets of registers in the EZchip TOPparse; one that is accessible directly through the TOPparse microcode, and the other, which must first be initialized by the host (through the host interface) prior to the TOPparse execution, and only then can be read by the TOPparse microcode.

12.2.1   Microcode Registers

The tables that follow list the TOPparse registers and structures that are accessible from the microcode. Detailed descriptions of the registers are given in Appendix A of this chapter. All registers can be written to in bits or bytes. All registers are accessed by the register name and the index in square brackets, example, UREG[3], or by their defined name, example, LIM_REG (provided, of course, that they were defined as such in the source program or in the libraries). The registers are intended for specific purposes and, consequently, have dedicated input sizes. For instance, attempting to write 8 bits into a 5-bit register will result in only the first five bits being written. The R/W column indicates whether the register is read, write or both.

The UREG registers (Table 12.1) are composed of 16 registers, each of 32 bits, and are used for general purpose, though some may have specific purposes depending on the use of the other blocks of the TOPparse. The UDB register (which is UREG[0]) is zeroed for each new frame.

Index	Register name	Size (bits)	Bytes (bits)	Description	R/W
0	UDB	32 bits	4 (32)	User-defined bits. Each bit may serve as a condition in an IF statement.	R/W
1	OFFS_FMEM0	32 bits	4 (32)	Possible index to frame memory address.	R/W
2	OFFS_FMEM1	32 bits	4 (32)	Possible index to frame memory address.	R/W
3	LIM_REG	32 bits	4 (32)	Register for limit parameter in SCAN operations.	R/W
4	OFFS_KMEM	32 bits	4 (32)	Possible index to KMEM address.	R/W
5	DELIM_MASK	32 bits	4 (32)	Indirect delimiter mask for scan operation (16 lsb).	R/W
. . .					
15		32 bits	4 (32)	General-purpose register.	R/W

Table 12.1 TOPparse UREG Registers

The Functional Block (FBLK) registers (Table 12.2) are composed of 14 variable length registers, and contain registers for the SCAN, ALU, CAM, and convert operations.

The Specific Registers (SREG) are composed of 16 variable length registers (see Table 12.3). The RD_PTR register (which is SREG[0]) is zeroed for each new frame. The key Header register (HREG) file (Table 12.4) contains 16 registers, each of 24 bits. The Hardware (HWARE) registers contain 10 variable bits registers, which are initialized by the hardware decoder block according both to the arriving frame and to a set-up register that tells the hardware decoder how to decode the incoming frame. (See Table 12.5.)

Table 12.2 TOPparse FBLK Registers

Index	Register name	Size (bits)	Bytes (bits)	Description	R/W
0	SCAN_START	14 bits	4 (14)	Start address of found token. See Section 12.4.5.	R
1	SCAN_STOP	14 bits	4 (14)	End address of found token.	R
2	SCAN_SIZE	14 bits	4 (14)	Size of found token.	R
3	SCAN_TC	22 bits	4 (22)	Scan result bitmap and termination cause.	R
4	ALU	32 bits	4 (32)	ALU register. See ALU Commands in Appendix C of this chapter.	R/W
5	CAMI	32 bits	4 (32)	CAM in register. See CAM Operations in Appendix C of this chapter.	R/W
6	CAMIH	32 bits	4 (32)	CAM in high register.	R/W
7	CAMO	16 bits	4 (16)	CAM out register.	R
8	CNVI	32 bits	4 (32)	Convert in register. See Section 12.4.4.	R/W
9	CNVIH	8 bits	4 (8)	Convert in high register.	R/W
10	CNVO	16 bits	4 (16)	Convert result.	R
13	SCAN_REG_HI	16 bits	4 (16)	Interface to statistics block.	R/W

Table 12.3 TOPparse SREG Block

Index	Register name	Size (bits)	Bytes (bits)	Description	R/W
0	RD_PTR	14 bits	4 (14)	FMEM base address.	R/W
1	CNT	8 bits	4 (8)	Counter for loops.	R/W

Table 12.3 (*continued*)

Index	Register name	Size (bits)	Bytes (bits)	Description	R/W
2	PC_STACK	16 bits	4 (16)	Stack for PC used for subroutines	R/W
3	FLAGS	16 bits	4 (16)	Flag register. See Appendix A of this chapter.	R
4	SIZE_REG	5 bits	4 (5)	Dedicated size register.	R/W
5	IND_REG0	16 bits	2 (3 + 2 + 4)	Dedicated offset register for indirect access to all devices.	R/W
5	IND_REG1	16 bits	2 (3 + 2 + 4)	Dedicated offset register for indirect access to all devices.	R/W
6	STAT_REG	32 bits	4 (32)	Interface to statistics block.	R/W
7	HOST_REG	32 bits	4 (32)	Interface to host.	R/W
8	RFD_REG0	10 bits	4 (10)	Interface to RFD (received frame descriptor) register.	R/W
9	RFD_REG1	26 bits	4 (26)	Interface to RFD register.	R/W
10	EOF_ADDR	10 bits	4 (10)	Jump address in case of EOF. See Section 12.4.5.	R/W
11	HBS0 HBS1 HBS2 HBS3	4 × 4 bits	4 (4 + 4 + 4 + 4)	Indirect headers pointer.	R/W
12	KBS0 KBS1	2 × 9 bits	4 (9 + 9)	Base addresses for key memory.	R/W
13	KBS2 KBS3	2 × 9 bits	4 (9 + 9)	Base addresses for key memory.	R/W
15	NULL_REG	NA	NA	Dump register.	R

Table 12.4 TOPparse HREG Block

Index	Register name	Size (bits)	Bytes (bits)	Description	R/W
0		24 bits	4 (24)	HREG registers, containing headers of the keys and messages, as described in Chapter 13.	R/W
. . .					
15		24 bits	4 (24)		R/W

Table 12.5 TOPparse HWARE Block

Index	Register name	Size (bits)	Bytes (bits)	Description	R/W
0	HD_REG0	32 bits	4 (32)	Hardware decoding; see Appendix A.	R
1	HD_REG1	32 bits	4 (32)	Hardware decoding.	R
2	HD_REG2	32 bits	4 (32)	Hardware decoding.	R
3	HD_REG3	32 bits	4 (32)	User defined data per in port. See CREG description in Appendix A.	R
4	HD_REG4	32 bits	4 (32)	User defined data per in port. See CREG description in Appendix A.	R
6	HISTORY0 HISTORY1	2 × 10 bits	4 (10 + 10)	Instructions in pipeline.	R
7	HISTORY2 HISTORY3	2 × 10 bits	4 (10 + 10)	Instructions in pipeline.	R
8	UNIT_NUM	4 bits	1 (4)	TOP engine number.	R
8	HOST_BITS	8 bits	3 (8)	User defined bits for NP chip.	R
9	SOURCE_PORT_ REG	8 bits	1 (8)	Source port from hardware decoder.	R

It should be noted that HD_REG3 and HD_REG4 (HWARE[3] and HWARE[4]) contain values according to the ingress port of the frame and the content of the CREG registers, for example, for a frame that entered the network processor in port 0, HD_REG3 receives the contents of CREG0. Thus, these can be used to share data between the host and the microcode, according to the ingress port. Further description is provided in Appendix A of this chapter.

12.2.2 Host Registers

All host registers are initialized by the host; however, the initial values for some are inserted into the NP-1 microcode for loading. These initialization commands are "executed" by the host prior to executing the program in the instruction memory.

Table 12.6 lists all of TOPparse's host registers and indicates which registers have their initial values in the NP microcode. A detailed description of some of the registers is provided in Appendix A of this chapter.

As described previously, CREG registers are shared with the HD_REG3 and HD_REG4 (HWARE[3] and HWARE[4]), according to the ingress port of the frame;

Table 12.6 TOPparse Host Registers

Address	Name	Description	Init
0x00	INT_REG	Interrupt register.	Host only
0x01 − 0x05	WIDE_LOAD[5:0]	Load SRAM instruction register and CAMs.	Host only
0x08	HOST_CONF	Host configuration register for decoding.	Host only
0x09	UNIT_MASK	Enables TOPparse engines.	Host only
0x40 − 0x4F	MREG[15:0]	ALU mask register.	Microcode
0x50	Reserved	Reserved.	
0x51 − 0x54	DEL_VECTOR[3:0]	Load SCAN delimiters register, contains 16 characters.	Host only
0x56 − 0x65	CREG[15:0]	User-defined port data register.	Host only
0x67	BR_ADDR	Service routine address.	Host only
0x80 − 0x85	HOST_REG[5:0]	Host debug register.	
0x8C	MCODE_BR_INT	Microcode execution or break point command.	
0x8D	STATUS_REG	Status register.	

for example, a frame that entered the network processor in port 0, HDREG3 receives the contents of CREG0.

12.3 TOPPARSE STRUCTURES

TOPparse maintains various types of memories, as described in Table 12.7. Key Memory is accessed indirectly via the KBS register.

12.4 TOPPARSE INSTRUCTION SET

This section lists the specific instructions of the TOPparse microcode that are not common NP instructions described in Chapter 11.

The microcode execution begins on the first instruction for each new frame. After the frame parsing, programmers are responsible for writing the keys for TOPsearch in KMEM (PutKey instruction) and the key headers in HREG (PutHdr

Table 12.7 TOPparse Memories and CAMs

Name	Description
FMEM	Frame memory
KMEM	Key memory
BCAM8	Content Addressable Memory (CAM, a table for exact searches)
BCAM16	CAM
BCAM32	CAM
TCAM64	Ternary CAM (TCAM, a table for pattern searches)

instruction). A header must accompany each single key, since each may be processed by a separate TOPsearch engine. Each key header contains the structure number (and therefore the type of the structure), according to which a search will be done. Compound keys (complex searches that involve several lookups) are defined by a single header. The format of the header is located in Section 13.4 in the next chapter.

Programmers are also responsible for writing a TOPparse messages destined for TOPresolve in KMEM (PutKey instruction). The messages are actually keys that, instead of being processed by TOPsearch, are passed as is to TOPresolve. A message may include fields from the hardware decoder, notably the frame pointer, frame length, and time stamp, as well as other data obtained from the frame and needed down the line. The format is located in Section 13.4 in the next chapter.

12.4.1 Resource Initialization

Initialization commands are "executed" by the host rather than by the TOPs; they indicate the initial values given to the mask registers and to various CAMs prior to executing the program in the instruction memory. Table 12.8 lists the initialization commands.

12.4.2 Move Commands

Move commands copy data (bits, bytes, and words) from various sources to various destinations, as described in the previous chapter (EZchip Programming). Move command names (the opcode mnemonic) indicate the type of move, as defined in Table 12.9 and Figure 12.2.

Table 12.8 Initialization Commands

Syntax	Description
LDREG CREG [*n*], value;	Load value to the CREG register (see Appendix A) *n* – 0–15 value—number to load to this register.
LDCAM CAM [CAMgp], KEY, value;	Load KEY to a specific CAM. CAM—specific CAM (BCAM8, BCAM16, BCAM32) CAMgp—group number (0, 1, 2, ... 7) KEY—number to load to this CAM value—either a 16-bit numeral or a label. Possibly the number of NOPs to insert, written as Label_name I # NOPs, where #NOPs is _NOP0, _NOP1 or _NOP2).

LDTCAM CAMgp, str1, str2, value;	Load string to the Text CAM (TCAM64).
	CAMgp—group number (0, 1, 2, ... 7)
	str1—"string" or you can use "\t","\r","\n", "\xx", or "\?"
	  where xx is hexadecimal number and ? is a wildcard.
	
	Limitation: The backslash may be specified with "\\". However, the sequence "\\?" results in the character "?", and "\\\?" results in the two characters "\" and "?".
	
	str2 = 0 means case sensitive;
	    = 1 means case insensitive.
	value—the result of the search in the CAM (LookCam command) for the specific key.

For example, LDTCAM TCAM64 [0], "ABCD??34", "00110000", VALUE;	This example searches for ABCD??34, ABCd??34, ABcD??34 or ABcd??34, where ?? are any two characters.
LDDV string	Load string of the delimiters to the DEL_VECTOR register (see Appendix A). Up to 16 delimiters characters) are supported.
For example, LDDV ";:-/"	This example loads four delimiters: semicolon (bit 0), colon (bit 1), dash (bit 2) and slash (bit 3). These may be masked in scan operations.

12.4.3 LookCam DST, SRC, TYPE, SIZE

Definition: Search in a CAM indicated by TYPE. Use source (SRC) for search key and put result in destination (DST), which is the CAMO. If the type is TCAM, search for SIZE bytes.

Example: `LookCam CAMO, CAMI, BCAM8 [1], 1;`

Description: Search in BCAM8, group 1 key from CAMI and put result in CAMO if match.

Table 12.9 List of All Move Commands

Name	Description	Max. Size*
Mov	Move bytes from immediate value or register to register	4 bytes
MovBits	Move bits from immediate value or register to register	16 bits
Mov4Bits	Move 4 separate bits	4 bits
Get	Move bytes from memory (FMEM) to register	8 bytes
PutKey	Move bytes from register or immediate value to memory (KMEM)	4 bytes
Copy	Move bytes from memory to memory	8 bytes
PutHdr	Write bytes from immediate value or register to header	3 bytes
PutHdrBits	Write bits group from immediate value or register to header	16 bits
PutHdr4Bits	Write 4 bits with immediate values or from register to header	4 bits

Maximum size that can be moved; also limited by the size of the register.
NOTE: *Immediate offsets are limited to 9 bits, which indicate a size of 255 bytes.*

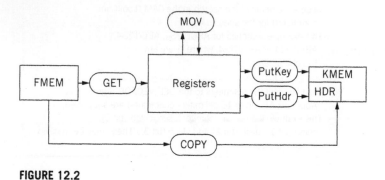

FIGURE 12.2

TOPparse move commands name prefix mnemonic

12.4.4 **Convert SRC, SIZE [, BASE]**

Definition: Convert numbers from ASCII to binary format. The ASCII string can be either a decimal number (maximum of five characters) or a hexadecimal number (maximum of four characters) depending upon the BASE. This operation is useful in applications such as NAT. The source (SRC) is of SIZE bytes, and the result is written into the CNVO.

Example: `Convert UREG[7], 3, _DEC_PRS;`

Description: Convert 3 character string from UREG[7] to binary number. If UREG[7] contains an incorrect decimal number in ASCII format, the CV flag is set.

12.4.5 **Scan Operations**

Scan operations enable the programmer to use one instruction to scan the frame rapidly to search for multiple tokens or multiple specific characters. Scan operations are composed of the four commands listed in Table 12.10. The overall format of scan operations is:

> *Command* `DST,SRC,DELIM,LIMIT[,DIR[,EOF]];`
>
> // See commands in Table 12.10.

Definition: The scan `command` starts scanning at the source point (`SRC`) up to a `LIMIT`, and copies the result to the destination (`DST`). It uses a mask (`DELIM`) on the delimiter vector it maintains, in order to choose the desired delimiters that will be used for the search. `DIR` is used to indicate a forward or backward search, and `EOF` indicates whether an end-of-frame (EOF) jump will be executed in case of an EOF. For a detailed use of the scan operations, please refer to Appendix C of this chapter.

12.4.6 **Conditional Commands**

Some of the instructions can be preceded by conditional statements that function as part of the instructions, that is, they are coded in one machine instruction, and are executed at the same time as the unconditional statements. These TOPparse commands are listed in Table 12.11.

Table 12.10 Scanning Operations

Name	Description
FindDel	Find delimiter in frame, that is, scan the frame for the first enabled CHAR in the vector of delimiters
FindNonDel	Find nondelimiter in frame, that is, scan the frame for the first character that is not a delimiter
FindToken	Find token in frame, that is, scan for a delimiter and then scan for a nondelimiter character
GetToken	Get token in frame, that is, scan for a nondelimiter character and then for a delimiter character

Table 12.11 TOPparse Conditional Commands

Mov	Call
MovBits	CallStack
Get	PutHdr
Copy	PutHdrBits
PutKey	Convert
Jmp	Return
Jstack	

12.5 EXAMPLE

In this example, the hardware decoder inspects each frame to determine whether or not it has a VLAN tag. The resulting information is passed as a message to TOPmodify, where it is used to determine the proper treatment for the frame. If the frame is found to have a VLAN tag already, a CSIX header is added and the frame is transmitted to the Virtual Output Queue; otherwise, a VLAN tag is inserted to the frame header prior to its transmission.

In the following TOPparse example, the NP processes the frames on the ingress path—that is, the frames that arrive from the network and that are transmitted to the switching fabric. The NP examines the arriving frames for a VLAN tag. The results of this inspection are put into a message, and the frame and the message are sent to the next TOP in the pipeline for further processing.

Definitions and macro files for this sample application are located after the sample codes and explanations.

12.5.1 Program Flow

Prior to execution of TOPparse microcode, the NP hardware decoder automatically inspects each incoming frame and provides useful information about the frame for microcode use, which is found in the HWARE register. The TOPparse program's steps are as follows.

1. A TOPparse-to-TOPresolve message is prepared containing information extracted by the hardware decoder and destined for subsequent TOPs in the pipeline. The message is passed through TOPsearch I using the same mechanism as that used to pass search keys and key headers.

   TOPparse passes the message (as well as the keys) to TOPsearch I engines, via the KMEM (NP-1 internal key memory). Each message and each key is accompanied by a header that is passed via HREG (NP-1 internal header

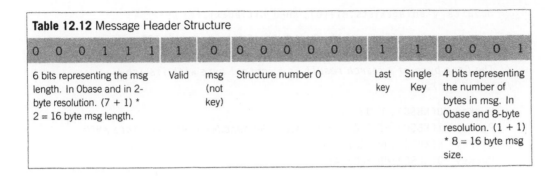

Table 12.12 Message Header Structure

0 0 0 1 1 1	1	0	0 0 0 0 0 0	1	1	0 0 0 1
6 bits representing the msg length. In 0base and in 2-byte resolution. (7 + 1) * 2 = 16 byte msg length.	Valid	msg (not key)	Structure number 0	Last key	Single Key	4 bits representing the number of bytes in msg. In 0base and 8-byte resolution. (1 + 1) * 8 = 16 byte msg size.

structure), providing details regarding the message or the key. The message header structure in this example, according to Table 12.12, is as follows:

```
HW_HEADER = 0x1E031 = 0001 1110 0000 0011 0001
```

2. The hardware decoder inspects whether an arriving frame contains a VLAN TAG, and sets a bit in HD_REG0 (bit 17), accordingly. TOPparse passes this information in the same message used to pass the hardware registers.
3. TOPparse finishes working on the frame and Halt HALT_UNIC passes the current frame to TOPsearch I.

12.5.2 TOPparse Sample Microcode

The following code implements the TOPparse stage of the VLAN Example:

```
Eztop Parse;
#include "mcglobal.h"    // Global definition file, contains predefined
                         // constants for NOPs, flags, etc.

#include "TOPParse.h"    // TOPparse definition file, provides
                         // recognizable names for registers and flags.

#include "hdreg.h"       // Global definition file supplied to provide
                         // recognizable names to the HW decoded registers.

// Constants:
#define  HW_HEADER          0x1E031;
#define  vlanTagExists      UDB.BYTE[0];

START:                   //Beginning of microcode sequence
////////// Step 1: Create TOPparse-TOPresolve message //////////
   Mov KBS0, 0, 2;  //Initialize KBS0 to point to the base of KMEM

// Step 2: Initialize vlanTagExists with the value HD_REG0 bit 17
// (0 = TAG does not exist, 1 = TAG exist).
```

```
    MovBits vlanTagExists.BIT[0], sHR0_bitTagExists, 1;
```

// *Always set the LSB bit of the first byte of the message. (TOPresolve*
// *must receive the message and the search results in the same format.*
// *The first byte of a search result is a control byte with its lsb a*
// *valid bit).*

```
    PutKey  0(KBS0), 1, 1;
    PutKey  1(KBS0),HD_REG0,4; //Insert HW decoder registers into KMEM
    PutKey  5(KBS0),HD_REG1,4;
    PutKey  9(KBS0),HD_REG2,4;
```

// *Send whether the arriving frame contains a VLAN TAG as a message*
// *byte to TOPresolve.*

```
    PutKey  13(KBS0),vlanTagExists,1;
    PutHdr  HREG[0],HW_HEADER,3; // Insert the message header to HREG
```

//*Step 3:* ///
```
    Halt    HALT_UNIC; //pass frame to the next stage in the pipeline
```

12.5.3 Definition Files

There are three header files used in this example, the "mcglobal.h," the "TOPParse.h," and the "hdreg.h" files.

The microcode global **Definition** file (mcglobal.h) is as follows:

```
#ifndef _mcglobal_h_        ;
#define _mcglobal_h_        ;
...
// HALT type defines
#define HALT_UNIC    0      ; // unicast
...
#endif /* _mcglobal_h_ */   ;
```

The hardware decoding **Definition** file (hdreg.h) is as follows:

```
#ifndef _hdreg_h            ;
#define _hdreg_h            ;

#ifndef sHR0                ;
#define sHR0      HD_REG0   ;
#endif                      ;
// length
...
#define sHR0_bitTagExists     sHR0.BIT [17] ;// 1
...
#endif                      ;
```

The TOPparse Definition file (TOPparse.h) is as follows:

```
#ifndef _TOPparse_h_                    ;
#define _TOPparse_h_                    ;
#define _TOPPARSE    ___TOPPARSE        ;
// UREG
#define UDB       UREG [0]              ;
...
// SREG
#define KBS0      SREG [12]             ;
...
// HWARE
#define HD_REG0   HWARE [0]             ;
#define HD_REG1   HWARE [1]             ;
#define HD_REG2   HWARE [2]             ;
...
#endif /* _TOPparse_h_ */               ;
```

12.5.4 **TOPsearch I Structures**

Table 12.13 describes the TOPsearch I data structures, which TOPparse assumes or builds while running the microcode. In our example, a message created in TOPparse for TOPresolve is placed in structure number 0 on the ingress path. A message header (or key header) must accompany each message (or key) passed to TOPsearch I, as described previously.

12.5.5 **TOPparse–TOPresolve Hardware Decoded Message**

Table 12.14 provides the format of the 14-byte message (padded to 16-byte) from TOPparse to TOPresolve via TOPsearch I on the ingress path. This message is placed in TOPsearch I structure #0. This message contains information extracted by the HW decoder destined for TOPs down the line.

Table 12.13 TOPsearch I Structure

Name	Structure type	Structure number	Path	Used for	Key size (bytes)	Result size (bytes)
TOPparse–TOPresolve message	(Message)	0	Ingress	L2 + L3, VLAN tag flag		

Table 12.14 TOPparse to TOPresolve Message

Field name	Byte offset	Size (bits)	Note
Valid	0	1	Always 1 for messages
To host	0	1	1—send to host
TTL_EXP	0	1	From HD_REG0 b28
Ctrl reserved bits	0	5	
HD_REG0	1–4	32	HW decoding
HD_REG1	5–8	32	HW decoding
HD_REG2	9–12	32	HW decoding
vlanTagExists	13	1	Indicates whether the arriving frame contained a VLAN tag field.

12.6 SUMMARY

The internal structure of TOPparse is described in this chapter, including its registers and data structures, as well as its blocks (functional units) and instruction set. The appendices of this chapter provide more details about the registers and the instruction set for those who need them either to understand a code or to write one. In the next chapter, we describe very briefly the TOPsearch I engine, which is the next processor in the EZchip network processor pipeline.

APPENDIX A
DETAILED REGISTER DESCRIPTION

A detailed description of some TOPparse devices and registers is given in this appendix. Though some of the information here duplicates that which can be found in the body of the chapter, the details and descriptions are fleshed out considerably.

MICROCODE REGISTERS

The microcode registers contain registers that the microcode access for normal operations during program execution. These registers include functional, specific, and hardware registers.

Functional Block Registers

The Functional Block (FBLK) registers contain registers for the SCAN, ALU, CAM, and convert operations, as described in Table A12.1.

Table A12.1 TOPparse Detailed FBLK Registers	
Register name	**Description**
SCAN_TC	22-bit register specifying the SCAN termination cause. Bits 0–15 refer to the 16 different delimiters and nondelimiters that can be user-defined. When the SCAN operation terminates on 'find delimiter' or 'find token', the corresponding delimiter vector in bits 0–15 is set. When SCAN terminates on 'find nondelimiter' or 'find token', then bits 0–7 contain the first nondelimiter that caused the termination of the operation.
	Bit 20 indicates that the limit of the SCAN has been reached, for example, if scanning was limited to 20 bytes and it finished scanning those bytes. (NOTE: Bits 19:16 are reserved.)
	Bit 21 indicates that SCAN reached the end of the frame.
CAMI/CAMIH	CAM input register—an 8-byte register divided into two 4-byte registers. A unique mechanism enables writing to it as if it were a single 8-byte register. When it reaches the end of the first register, data will automatically wrap into the second register. This enables writing from memory 8 bytes at a time.
CAMO	CAM output register—16-bit register containing the results of the CAM search. CAMO can either be used as an immediate data store for 16 bits, or a jump address for bits 0–13 and specifying the number of NOPs used for the branch in bits 14–15.
CNVI/CNVIH	Convert input register—a 5-byte register divided into two registers: 4 bytes and 1 byte. A unique mechanism enables writing into it as if it were a single 5-byte register. When it reaches the end of the first register, data will automatically wrap into the second register.
CNVO	16-bit Convert output register.
STAT_REG_HI	16-bit operand for the statistics block See the STAT_REG register for the address and command.

Specific Registers

The SREG contains specific registers for the counter block, flags, indirect access, and interfaces to external devices, as described in Table A12.2.

Table A12.2 TOPparse SREG Registers

Register name	Description
CNT	8-bit register to support loops. The program branches on the counter and checks its value. If the counter value is not zero, then the program jumps to the start of the loop. Each branch on the value of the counter automatically decrements the counter by one. That is, it counts down per each loop. Up to 256 repetitions of a particular sequence of code are supported. The depth of the Counter register is one. The counter can read/write to a register in order to implement nested loops.
PC_STACK	16-bit register for call commands (that is branch + set bit to push to stack). For each call command, the content (branch address +3 − #NOPs) is automatically written to PC_STACK. The PC_STACK has a depth of one. The PC_STACK can be read/write by the user in order to build nested call commands. The call command contains the number of NOPs that follow the relevant return command.
SIZE_REG	5-bit register for an indirect size. The size is specified in either bits or bytes, depending on the instruction.
HBS	Indirect address of the index field in the format: device + index + byte + bit.
KBS	Four 9-bit base addresses for KMEM. There are four KBS registers that enable TOPparse to write four keys simultaneously. KMEM is always accessed in the format: base address + index. Auto increments the base address (offset + base + size).
NULL_REG	Dump register for unwanted data. Writing to NULL_REG has no effect and reading from it implies zero.

EOF_ADDR	10-bit end of address register. Each time that FMEM is accessed and requested to read data beyond the end of the frame there are two possibilities:
	    1. Interrupts to a different routine in the code.
	    2. If the interrupt has been disabled, then garbage is read/written.
	
	In both situations, the EOF flag is set. The two most significant bits specify the number of NOPs.
	NOTE: To read data in the frame, but not in the current window, machine operation is stalled until the DMA loads the next window. As the window always moves forward along the frame, the frame should be parsed from beginning to end. Going back beyond the boundary of the current windows can result in having to read the frame from the beginning again, which is time consuming.

FLAGS	16-bit flag register. The defined bits and bit fields within the FLAGS register control specific operations and indicate the status of the network processor (see Table A12.3). ZR—zero flag CY—carry flag SN—sign flag OV—overflow flag (0 = no overflow; 1 = overflow)

Table A12.2 (*continued*)

Register name	Description
	LP—loop counter register (0 = CNT register is not zero; 1 = CNT register is zero) ST—statistics register (0 = interface ready; 1 = not ready) HT—host register (0 = interface ready; 1 = not ready) MH—CAM match flag (0 = no match; 1 = match) RD—RFD interface (0 = interface ready; 1 = not ready) CV—Convert command (0 = no error; 1 = ended with error) LM—SCAN command (0 = limit not reached; 1 = limit reached) ED—end of frame (0 = not EOF; 1 = reached EOF) CO—convert overflow (1 = overflow; 0 = normal)

NOTE: *In microcode, the flags are referred to as FLAGS.BIT [F_x_PRS], where x is the flag name from the table below; for example, FLAGS.BIT [F_ZR_PRS].*

Table A12.3 TOPparse Flags

bit#	15	14	13	12	11	10	9	8	7	6	5	4	3	2	1	0
			CO	ED	LM	CV	RD	MH	HT	ST	LP		OV	SN	CY	ZR

The following registers (STAT_REG, RFD_REG0 and RFD_REG1) act as output interface (OUT_IF) registers to access external blocks, even though they belong to the SREG device (see Table A12.4). Programmers must check the ready flag to ensure that these registers can be written to. Attempting to write to one of these blocks when it is not ready will result in data loss.

Table A12.4 SREG Output Interface Registers

Register name	Description
STAT_REG	32-bit interface to the statistics block: address (b31:8), reserved (b7:3), and command (b2:0). Valid commands are increment (000), decrement (101) and reset (001). Increment and decrement commands use the operand in the STAT_REG_HI register.
RFD_REG0	10-bit interface to the RFD table: (7:0) data and (9:8) command. Command bits: 1—Recycle, 2—Read, and 3—Write. The command bits are the trigger for setting the ready flag. Data cannot be written to RFD_REG0 or RFD_REG1 until the flag is cleared.
RFD_REG1	26-bit interface to the RFD table. The register format differs depending on the NP mode as configured. In NP-1c mode, for example, its format is:(15:0) pointer, (20:16) number of buffers, (24:21) port number, and (25) multicast.

Hardware Registers

The Hardware Registers (HD_REG) contain five 32-bit registers specifying the information that was hardware decoded from a frame (see Table A12.5). TOPparse automatically performs the initial classification of ingress frames (up to the IP options) without requiring microcode instructions. It is designed to support IPv4 and IPv6 with/without MPLS using Ethernet or PPP encapsulation. Up to four MPLS labels may be traversed. The label that comes after the L2 level is referred to as "top" and the label that comes before the L3 level is "last."

These registers are read only and may be read like any other register. Using the offsets, additional information may be obtained from the frame. The values of the fields for the most common protocols for Layers 2 and 3 may be accessed since their location in the frame is fixed related to either the beginning of the frame or the pointer to Layer 3 offset (L2 size).

Usually programmers will send information, such as the frame pointer, frame length, and time stamp as a message, to the TOPs down the line. The hardware decoder may be set by the host to support either Ethernet frames or PPP (see Table A12.5).

The hardware decoder can also be configured to search for up to 64 bits of user-defined data as defined by the CREG register and HD_REG 3/4, described in the following. A check mark in the Egress path column in Table A12.5 indicates that the field is set on the egress path as well.

Table A12.5 Contents of HD_REG Based on Hardware Decoding

Contents and description	Offset (bits)	Size (bits)	Egress path
HD_REG 0			
Layer 2 header size. Five LSB are valid.	0–4	5	–
Reserved, set to 0.	5–7	3	–
Direction frame was received from: 0—RX (ingress path); 1—TX (egress path).	8	1	√
Layer 3 type: 0—L3 undefined; 1—L3 is IP (Ethernet protocol type 0x0800 or 0x86DD); 2—L3 is MPLS unicast (0x8847); 3—reserved; 4—L3 is MPLS multicast (0x8848).	9–11	3	–
First buffer length control (required for TOPresolve and TOPmodify). It contains 7 if frame length > 448; otherwise, add one for each 64 bytes starting from one.	12–14	3	√
Layer 2 protocol type: 0—undefined; 1—"Old" Ethernet (EtherType 2); 2—PPP	15–16	2	–
Existence of TAG: 0—TAG does not exist; 1—TAG exists	17	1	–
CFI bit of VLAN tag	18	1	–
DA format. Only one of b21:19 will be set. Defined even if the L2 protocol type is undefined. DA = BROADCAST (FF FF FF FF FF FF) DA = MULTICAST (byte[0].bit[0] ==1) DA = BPDU (01 80 C2 00 00 xx)	19 20 21	1 1 1	– – –

Table A12.5 (*continued*)

Contents and description		Offset (bits)	Size (bits)	Egress path
Only one of b26:22 will be set. For MPLS (unicast or multicast), indicates the top label location.				
DIP = BROADCAST	One label in MPLS label stack	22	1	–
DIP = MULTICAST: 0xE******* (IPv4)/FF*... DIP.byte[1].bit[4] == 1 (IPv6)	Two labels in MPLS label stack	23	1	–
DIP = Well-known multicast: 0xE00000**/FF*... DIP.byte[1].bit[4] == 0 (IPv6)	Three labels in MPLS label stack			
IPv6 multicast – RFC 2373		24	1	–
IP version: 0—IPv4, 1—IPv6	Four labels in MPLS label stack	25	1	–
First fragment if IP fragmentation	More than four MPLS labels	26	1	–
Is fragmentation allowed? From IP header byte[6].bit[6].		27	1	–
Is TTL expired? TTL == 0 or TTL == 1. For IP frames, from IP header. For MPLS frames, from top label. Not valid for other frame types.		28	1	–
TTL expired value.		29	1	–
Indicates if RX chip = TX chip. This is when a frame received on the egress path originated from the ingress of the same NP-1 chip. The TX channel number equals the channel number in the frame's CSIX header.		30	1	√
IPv4 error detection. Set when the header length < 20, IP total length > frame length, or IP total length < IP header length.		31	1	–
HD_REG 1				
Time stamp for determining when the frame entered the NP-1c (should be included in a message from TOPparse to TOPmodify)		0–15	16	√
Frame_ptr—logical pointer to the first frame buffer (should be included in a message from TOPparse to TOPmodify)		16–31	16	√
HD_REG 2				
Frame_length		0–15	16	√
Number of buffers. Starting from one, add one for the first 448 bytes, then add one more for each additional 512 bytes.		16–20	5	√
Source port of NP-1c When set to operate in NP-1c mode, the source port is 4 bits and written to the SOURCE_PORT_REG register. HD_REG 2 contains the 3 lower bits. On the egress path, it contains the 3 LSB bits from the 7 priority bits. This information represents in which of the 8 ECFD queues the frame entered.		21–23	3	–
Channel number. Should be used in CSIX/XGMII header. Initialized by the host.		24–31	8	√

(*continued*)

Table A12.5 Contents of HD_REG Based on Hardware Decoding (*continued*)

Contents and description	Offset (bits)	Size (bits)	Egress path
HD_REG 3			
User-defined data per in port. Can be used for port-specific information such as the PVID configuration. Defined by the host using the driver API. See detailed explanation in CREG registers.		32	√
HD_REG 4			
User-defined data per in port. Defined by the host using the driver API. See HD_REG3 section.		32	√

HOST REGISTERS

Most registers are initialized with a high-level Application Program Interface (API) interface, which is not described here. The only registers that are described in this appendix are those that are mentioned in this chapter and in the sample programs in the book. For more details, see [118].

HOST_CONF Register

The ETH/PPP setting determines the type of frames processed by the hardware decoder. The data written to this register (Table A12.6) by the host is accessible from the microcode via the HOST_BITS register, which is HWARE[8].

MREG[15:0] Register

The LDREG command in the microcode instructs the loader to load the user-defined masks to each of these registers (see Section 12.4.1). One of the 16 masks (Table A12.7) may be used in an ALU operation (see Appendix C, pages 493–494).

DEL_VECTOR[3:0] Register

The LDDV command in the microcode instructs the loader to load the user-defined delimiters (Table A12.8) to the delimiter vector (see Section 12.4.1).

Table A12.6 HOST_CONF Register

Name	Bits#	Description	Init value
ETH/PPP	b0	0—perform hardware decoding for PPP frames.	0
	W	1—perform hardware decoding for Ethernet frames.	

Table A12.7 MREG[15:0] Register

Name	Bits#	Description	Init value
ALU_MASK	b31:0	Used for ALU mask.	

Table A12.8 SCAN Delimiter Register

Name	Bits#	Description	Init value
DEL_VECTOR	b31:0	Used for loading SCAN delimiters.	

Table A12.9 CREG Register[15:0]

Name	Bits#	Description	Init value
CREG	b31:0	Used for defining HD_REG3/4.	

CREG Register[15:0]

The CREG register is used for defining the 64 user-defined bits in HD_REG3 and HD_REG4. HD_REG3 and HD_REG4 receive the values of the pair of CREG registers (Table A12.9) according to the port from which the frame entered; for example, for a frame that enters in port 0, the value of HDREG3 is taken from CREG[0]. There are 16 registers for 8 ports: CREG[0] & CREG[1] –> port 0, CREG[2] & CREG[3] –> port 1, and so on. When using a 16 port/channel configuration, CREG[0] –> port 0, CREG[2] –> port 1, CREG[4] –> port 2 ... CREG[1] –> port 8, CREG[3] –> port 9 ... CREG[15] –> port 15. Configuration is per port.

APPENDIX B
TOPPARSE ADDRESSING MODES

Table B12.1 provides the numbers, names, and syntaxes of the addressing modes that are relevant to the TOPparse. Devices in this table refer to registers or structures: for example, UREG, FBLK. The numbers in the first column are used in this chapter to indicate the addressing modes supported by operands. **Bold** typeface indicates required text. *Italic* typeface indicates text that must be replaced with the appropriate value.

Table B12.1 Addressing Modes of TOPparse

No.	Name	Syntax	Description
1	Immediate	123 or 0x12 or "abcd" or $ or label	123—decimal number 0x12—hexadecimal number "abcd"—4 bytes = 0x61626364 $—program counter (PC) label—program label.
2	Register	*device* [*number*]	For the device, see registers listed in 12.2. *number*—index of the register in the device array.
3	Register (byte-specific)	*device* [*number1*]**.byte** [*number2*]	*number1*—index of the register in the device array *number2*—byte number in this register
4	Register (bit-specific)	*device* [*number1*]**.bit** [*number2*]	*number2*—bit number in this register
5	Register (four-bit)	*device* [*number1*]**.bits** [*n1,n2,n3,n4*]	*n1...n4*—specific bits in this register
6	Indirect	*device* [*IND_REG*]	
7	Base-index	offset (*base*) or offset (*base*)**+**	*base*—base registers. *offset*—immediate or some register. + is used for auto-increment.
8	Direct	*number* (*0*) or (*number*)**+**	Can be used for Get or Copy commands. For src in TOPparse copy and get instructions, use number (**0**)**+**.
9	Memory (byte-specific)	**hreg** [*HBS*]**.byte**[*num2*]	
10	Memory (bit-specific)	**hreg** [*HBS*]**.bit**[*num2*]	
11	Memory (four-bit)	**hreg** [*HBS*]**.bits** [*n1,n2,n3,n4*]	
12	Device group	**bcam8** [*num1*] or **bcam8 [ureg[0]]** **bcam16**[*num1*] or **bcam16[ureg[0]]** **tcam** [*num1*] or **tcam [ureg[0]]**	*num1*—immediate 3 bits

APPENDIX C
TOPPARSE DETAILED INSTRUCTION SET

Detailed description of TOPparse commands' operands are listed next. For even more detailed explanations, see [118].

MOVE COMMANDS

Move commands copy data (bits, bytes, and words) from various sources to various destinations.

Mov DST, SRC, SIZE

Definition: Move SIZE bytes from source (SRC) to destination (DST).

Operands: DST, the destination, can be a register (or a byte in it) or an indirect addressing mode (type 2, 3, or 6 in Table B12.1). Valid registers are UREG, ALU, CAMI, CAMHI,[1] EOF_ADDR, CNVI, CNVIH,[2] RD_PTR, CNT, PC_STACK, SIZE_REG, IND_REG0..1, STAT_REG, RFD_REG0..1, HBS0..3, KBS0..3, or NULL_REG (see Section 12.2.1).

SRC, the source, can be an immediate 32-bit value, a register (or a byte in it) or an indirect addressing mode (type 1, 2, 3, or 6 in Table B12.1). Register or indirect addressing modes can be any of the registers except STAT_REG and STAT_REG_HI (see Section 12.2.1).

Size can be either an immediate value (1 to 4) or a register content, that is, the SIZE_REG (which is SREG[4]) value (0–4). This corresponds to addressing mode type 1 or 2 in Table B12.1.

MovBits DST, SRC, SIZE

Definition: Move SIZE bits from source (SRC) to destination (DST).

Operands: DST, the destination, can be bit-specific position in a register (e.g., UREG[0].bit[5]) or an indirect addressing mode (type 4 or 6 in Table B12.1). Valid registers are UREG, ALU, CAMI, CAMHI,[3] EOF_ADDR, CNVI, CNVIH,[4] RD_PTR, CNT, PC_STACK, SIZE_REG, IND_REG0..1, STAT_REG, RFD_REG0..1, HBS0..3, KBS0..3, or NULL_REG (see Section 12.2.1).

---

[1] `mov cami.byte[2], src, 4;` changes CAMIH
[2] `mov cnvi.byte[1], src, 4;` changes CNVIH.
[3] `movbits cami.bit[24], src, 15;` does not change CAMIH.
[4] `movbits cnvi.bit[24], src, 15;` does not change CNVIH.

SRC, the source, can be an immediate 16-bit value, a register (or a specific bit in it) or an indirect addressing mode (type 1, 4, or 6 in Table B12.1). Register or indirect addressing modes can be any of the registers except STAT_REG and STAT_REG_HI (see Section 12.2.1).

Size can be either an immediate value (1 to 16) or a register content, that is, the SIZE_REG (which is SREG[4]) value (0–16). This corresponds to addressing mode type 1 or 2 in Table B12.1.

Mov4Bits DST, SRC [, MASK]

Definition: Move four bits from source (SRC) to destination (DST). Use MASK to indicate which bits should be inverted during the copy. Bits do not need to be adjacent. When copying different bits to the same destination, the value written is logically 'ORed' between the bits.

Operands: DST, the destination, can be a four-bit specification in a register (e.g., UREG[0].bits[1, 5, 6, 8]) (addressing mode type 5 in Table B12.1). Valid registers are UREG, ALU, CAMI, CAMHI, EOF_ADDR, CNVI, CNVIH, RD_PTR, CNT, PC_STACK, SIZE_REG, IND_REG0..1, STAT_REG, RFD_REG0..1, HBS0..3, KBS0..3, or NULL_REG (see Section 12.2.1).

SRC, the source, can be an immediate four-bits, or a four-bit specification in a register (e.g., UREG[0].bits[1, 5, 6, 8]) (addressing mode type 1 or 5 in Table B12.1). The bit offsets must be specified when the source is not immediate. The register can be any of the registers except STAT_REG and STAT_REG_HI (see Section 12.2.1).

MASK, is a bitmap, mask of bit inversion: move bit (0) and move inverse of bit (1). By default MASK is 0x0 (bit string "0000"). If the source is immediate, a mask is not valid. The mask is an immediate value 0 to 15.

Get DST, SRC, SIZE [, EOFjump]

Definition: Move SIZE bytes from FMEM memory (SRC) to a destination register (DST). Use EOFjump to jump to EOF_ADDR if end of frame occurred while getting data.

Get can be preceded by conditional statements (and it will still function as one instruction in machine code and in execution time). A conditional command is impossible if the source is in the frame memory and not in TOPparse's internal FMEM.

Operands: DST, the destination, can be a register (or a byte in it) or an indirect addressing mode (type 2, 3, or 6 in Table B12.1). Valid registers are

UREG, ALU, CAMI, CAMHI,[5] EOF_ADDR, CNVI, CNVIH,[6] RD_PTR, CNT, PC_STACK, SIZE_REG, IND_REG0..1, STAT_REG, RFD_REG0..1, HBS0..3, KBS0..3, or NULL_REG (see Section 12.2.1).

SRC, the source, can be a direct or a base-index addressing mode, referring to FMEM (type 7 or 8 in Table B12.1), as can be seen in the examples that follow.

Size can be either an immediate value (1 to 4) or a register content, that is, the SIZE_REG (which is SREG[4]) value (0 to 4). This corresponds to addressing mode type 1 or 2 in Table B12.1.

EOFjump indicates that a jump to the address in EOF_ADDR should be executed if "end of frame" happens while reading from the FMEM (_JEOF_PRS), or do not jump (_NJEF_PRS). The default is _NJEF_PRS.

+ indicates that auto-increment of the base address register (e.g., RD_PTR) should be done. When the auto-increment operation is used, the RD_PTR will be set to base + index + size at the end of the operation.

PutKey DST, SRC, SIZE

Definition: Move SIZE bytes from SRC to DST, the key memory (KMEM). KMEM may only be accessed indirectly via the KBS register.

Operands: DST, the destination, is a base indexed reference to the KMEM (addressing mode type 7 in Table B12.1).

SRC, the source, can be an immediate 32-bit value, a register (or a byte in it) or an indirect addressing mode (type 1, 2, 3, or 6 in Table B12.1). Register or indirect addressing modes can be any of the registers except STAT_REG and STAT_REG_HI (see Section 12.2.1).

Size can be either an immediate value (1 to 4) or a register content, that is, the SIZE_REG (which is SREG[4]) value (0 to 4). This corresponds to the type 1 or 2 addressing mode in Table B12.1.

+ indicates that auto-increment of the base address register (e.g., KBS0) should be done. When the auto-increment operation is used, the KBS will be set to base + index + size at the end of the operation.

---

[5] mov cami.byte[2], src, 4; changes CAMIH
[6] mov cnvi.byte[1], src, 4; changes CNVIH.

Copy DST, SRC, SIZE [, EOFjump]

Definition: Move SIZE bytes from SRC, the frame memory (FMEM) to DST, the key memory (KMEM). Use EOFjump to jump to EOF_ADDR if end of frame occurred while getting data.

Copy can be preceded by conditional statements (and it will still function as one instruction in machine code and in execution time). A conditional command is impossible if the source is in the frame memory and not in TOPparse's internal FMEM.

Operands: DST, the destination, can be a base-index addressing mode, referring to KMEM (addressing mode type 7 in Table B12.1), as can be seen in the examples that follow.

SRC, the source, can be a direct or base-index addressing mode, referring to FMEM (type 7 or 8 in Table B12.1), as can be seen in the examples that follow.

Size can be either an immediate value (1 to 8) or a register content, that is, the SIZE_REG (which is SREG[4]) value (0 to 8). This corresponds to addressing mode type 1 or 2 in Table B12.1.

EOFjump indicates that a jump to the address in EOF_ADDR should be executed if "end of frame" happens while reading from the FMEM (_JEOF_PRS), or do not jump (_NJEF_PRS). The default is _NJEF_PRS.

+ indicates auto-increments of the base address registers (e.g., RD_PTR, KBS0) for each interface (FMEM, KMEM, and so on). When the auto-increment option is used, the RD_PTR will be set to base + index + size and each of the four base registers KBS*n* will be set to base register + index + size at the end of the operation. Auto-increment can be used for both the source and destination.

PutHdr DST, SRC, SIZE

Definition: Move SIZE bytes from source (SRC) to DST, the header registers, that is, HREG.

Operands: DST, the destination, can be a register or a byte specific memory with respect to HREG (addressing mode type 2 or 9 in Table B12.1) as can be seen in the examples that follow. In other words, DST refers to HREG through an immediate index or through HBS register-like index (indirect access to HREG); for example, HREG[0] and HREG[HBS1].

SRC, the source, can be an immediate 24-bit value, a register (or a byte in it) or an indirect addressing mode (type 1, 2, 3, or 6 in Table B12.1). Register or indirect addressing modes can be any of the registers except STAT_REG or STAT_REGHI (see Section 12.2.1).

Size can be either an immediate value (1 to 3) or a register content, that is, the SIZE_REG (which is SREG[4]) value (0 to 3). This corresponds to addressing mode type 1 or 2 in Table B12.1.

PutHdrBits DST, SRC, SIZE

Definition: Move SIZE bits from source (SRC) to DTS, the header registers, that is, HREG.

Operands: DST, the destination, can be a bit-specific position in a register or a bit-specific position in memory with respect to HREG (addressing mode type 4 or 10 in Table B12.1) as can be seen in the examples that follow. In other words, DST refers to HREG through an immediate index or through HBS register-like index (indirect access to HREG); for example, HREG[0] and HREG[HBS1].

SRC, the source, can be an immediate 16-bit value, a register (or a bit position in it) or an indirect addressing mode (type 1, 2, 4, or 6 in Table B12.1). Register or indirect addressing modes can be any of the registers except STAT_REG or STAT_REGHI (see Section 12.2.1).

Size can be either an immediate value (1 to 16) or a register content, that is, the SIZE_REG (which is SREG[4]) value (0 to 16). This corresponds to addressing mode type 1 or 2 in Table B12.1.

PutHdr4Bits DST, SRC [, MASK]

Definition: Move four bits from source (SRC) to DST, the header registers, that is, HREG. Use MASK to indicate what bits to invert while moving.

Operands: DST, the destination, can be four specific bits in a register or four specific bits in memory with respect to HREG (addressing mode type 5 or 11 in Table B12.1) as can be seen in the examples that follow. In other words, DST refers to HREG through specific bits index or through HBS specific bits of register index (indirect access to HREG); for example, HREG[0] and HREG[HBS1].

SRC, the source, can be an immediate four-bits, or a four-bit specification in a register (e.g., UREG[0].bits[1, 5, 6, 8]) (addressing mode type 1 or 5 in Table B12.1). The bit offsets must be specified when the source is not immediate. Register can be any of the registers except STAT_REG or STAT_REGHI (see Section 12.2.1).

MASK, is a bitmap, mask of bit inversion: move bit (0) or move inverse of bit (1). By default MASK is 0x0 (bit string "0000"). If the source is immediate, a mask is not valid. The mask is an immediate value (0 to 15).

JUMP COMMANDS

Jump commands instruct the NP to jump to a given label in the microcode. Some of the commands are conditional jumps, and some jump with pushing and popping program addresses. At the end of the jump, optional NOP_NUM (0, 1 or 2) NOPs may be inserted into the pipeline following the jump command (to disable execution of the commands that immediately follow the jump instruction). The NOPs are effective only if the jump command is executed. The standard Jump command format is:

```
if ( CONDITION )
command LABELS [|NOP_NUM];
                        // See commands in Table 11.4, Chapter 11.
```

or

```
command LABELS [|NOP_NUM];
                        // See commands in Tables 11.4 and 11.5, Chapter 11.
```

The format of the Call, CallCam, and CallStack commands is a bit different from the jump command, and also includes a number of NOPs (RET_NOP_NUM) that should be inserted before returning, as follows:

```
Call function [|NOP_NUM [, RET_NOP_NUM]];
```

or

```
CallCam function [|NOP_NUM [, RET_NOP_NUM]];
```

or

```
CallStack function [|NOP_NUM [, RET_NOP_NUM]];
```

Definition: Call a function, and return from it to the next address on completion (return command in the function). Insert NOP_NUM NOPs before jumping and RET_NOP_NUM NOPs before returning.

Operands: CONDITION is any valid logical operation on a flag bit or any of the UDB bits. The negation mark (!) may precede any condition.

LABELS is the place in the program where the jump is to take place.

function is the name of the function to be called.

NOP_NUM shows the number of NOPs to be inserted after the branch, to prevent the following command from entering the pipeline.

RET_NOP_NUM shows the number of NOPs to be inserted on return from the called function.

ARITHMETIC LOGIC UNIT COMMANDS

Arithmetic Logic Unit commands are used for arithmetic and logic calculations. There are two formats of ALU commands, one with two source-operands and the other with one source-operand:

COMMAND DST,SRC1,SRC2,SIZE,[,MREG[,MODE]];

// See commands in Table 11.3, Chapter 11.

or

COMMAND DST,SRC,SIZE,[,MREG[,MODE]];

// See commands in Table 11.3, Chapter 11.

Definition: Calculate the *COMMAND* on the sources (SRC, or SRC1 and SCR2) of SIZE bytes, and put the result in the destination (DST). Use masks in MREG, according to the required MODE, to mask the source operands if required.

Operands: DST, the destination, receives the result of the ALU block operation, together with the ALU register, that is, the result is entered into both the destination and the ALU register. DST can be a register (or a byte in it) or an indirect addressing mode (type 2, 3, or 6 in Table B12.1), and may use any of the following registers: UREG, ALU, CAMI, CAMHI,[7] EOF_ADDR, CNVI, CNVIH,[8] RD_PTR, CNT, PC_STACK, SIZE_REG, IND_REG0..1, STAT_REG, RFD_REG0..1, HBS0..3, and KBS0..3 (see Section 12.2.1).

SRC or SRC1, the first source operand or the only source operand can be a register (or a byte in it) or an indirect addressing mode (type 2, 3, or 6 in Table B12.1). Register or indirect addressing modes can be any of the registers except STAT_REG or STAT_REGHI (see Section 12.2.1).

SRC2, the second source, can be an immediate 24-bit value, a register (or a byte in it) or an indirect addressing mode (type 1, 2, 3, or 6 in Table B12.1). Register or indirect addressing modes can be any of the registers except STAT_REG or STAT_REGHI (see Section 12.2.1).

SIZE is the size of the ALU operation in bytes. Zero (0) is not a valid size. Size can be either an immediate value (1 to 4) or a register content, that is, or the SIZE_REG (which is SREG[4]) value (1 to 4). This corresponds to addressing mode type 1 or 2 in Table B12.1.

MREG indicates one of the 16 ALU mask registers (see Table A12.7).

---

[7] mov cami.byte[2], src, 4; **changes CAMIH.**
[8] mov cnvi.byte[1], src, 4; **changes CNVIH.**

MODE indicates how to use the mask:

_ALU_NONE—no masking done
_ALU_FRST—masking SRC1; that is, SRC1 = SRC1 & MASKREG
_ALU_SCND—masking SRC2; that is, SRC2 = SRC2 & MASKREG
_ALU_BOTH—masking both SRC1 and SRC2; that is,
        SRC1 = SRC1 & MASKREG, SRC2 = SRC2 & MASK_REG

When the explicit destination is not the ALU register, all four bytes of the ALU register are updated, even if the instruction SIZE was less than 4. For example, "Add UREG[7], UREG[5], UREG[6], 1;" adds the contents of UREG[5] and UREG[6] and writes 1-byte to UREG[7] and 4-bytes to ALU. The ALU register cannot be used with an offset other than 0 (zero) when it is used for destination, source1 or source2. Examples of improper usage of the ALU register:

```
Add ALU.BYTE [1], UREG[0].BYTE[2], 2, 1 ;
Add ALU, UREG[0].BYTE[2], ALU.BYTE [2], 1;
Add ALU, ALU.BYTE [3], UREG[0].BYTE[2], 1;
```

CONTENT ADDRESSABLE MEMORY OPERATIONS

There is one CAM instruction in the TOPparse CAM operation. See Section 12.1.8 for a description of the TOPparse CAMs.

LookCam DST, SRC, TYPE, SIZE;

Definition: Search in a CAM indicated by TYPE. Use source (SRC) for search key and put result in destination (DST), and in the CAMO register. If the type is TCAM, search for SIZE bytes. All CAM operations also write to the CAMO register. If a match is found, the MH bit is set.

              CAMs are divided into user-defined groups according by increasing their key size by three bits. The first three bits represented indicate the CAM group. Each CAM table may be divided into up to eight groups.

Operands: DST, the destination, can be the CAMO register (see Section 12.2.1).

              SRC, the source, can be a register (or a byte in it) or an indirect addressing mode (type 2, 3, or 6 in Table B12.1). Register or indirect addressing modes can be any of the registers except STAT_REG or STAT_REGHI (see Section 12.2.1).

              TYPE indicates the CAM type (BCAM8, BCAM16, BCAM32 or TCAM64) and its group number, for example BCAM8[0]. Alternatively, the 3 Least Significant Bits (LSB) of the UDB may be used to indicate the group number, example, BCAM16[UREG0] (addressing mode type 12 in Table B12.1).

SIZE is only valid when searching the text CAM (TCAM64) because CAMI may be 1–8, and is either an immediate value (1 to 8) or a register content, that is, the SIZE_REG (which is SREG[4]) value (0 to 8). This corresponds to type 1 or 2 of the addressing modes in Table B12.1.

In order to reduce data hazards, it is advisable to Get data from the FMEM into CAMI and then perform a lookup in the next instruction.

CONVERT OPERATIONS

Convert Operation is used to convert ASCII formatted numbers to their binary format. This operation is useful in applications such as NAT.

Convert can be preceded by conditional statements (and it will still function as one instruction in machine code and in execution time).

```
Convert SRC, SIZE [, BASE];
```

Definition: Convert numbers from ASCII to binary format. The ASCII string can be either a decimal number (maximum of five characters) or a hexadecimal number (maximum of four characters) depending on the BASE. This operation is useful in applications such as NAT. The source (SRC) is of SIZE bytes, and the result is written into the CNVO.

Operands: SRC, the source, can be a register (or a byte in it) or an indirect addressing mode (type 2, 3, or 6 in Table B12.1). Register or indirect addressing modes can be any of the registers except STAT_REG or STAT_REGHI (see Section 12.2.1).

       SIZE is either an immediate value (1 to 5) or a register content, that is, the SIZE_REG (which is SREG[4]) value (0 to 5). This corresponds to addressing mode type 1 or 2 in Table B12.1.

       BASE indicates whether the source is a decimal number (_DEC_PRS) or a hexadecimal number (_HEX_PRS).

It should be noted that by putting the value in the CNVI register and then converting it, programmers can avoid a data hazard.

SCAN OPERATIONS

Scan is a very useful operation that allows rapid searches in the frame for particular strings, using delimiters to define the searches. The vector that contains these delimiters (a host register, DEL_VECTOR, see Appendix A) should be initialized before performing scan operations (see Section 12.4.1). The LDDV initialization instruction can store up to 16 characters in the delimiter vector, for use as delimiters. These delimiters may then be masked for each scan operation. Scan instructions are executed by the SCAN block of the TOPparse, and they use four SCAN registers in the Functional Block (FBLK) registers, as well as the delimiter vector, which is a host register (DEL_VECTOR) mentioned before. Scan operations are composed of four commands, listed in Table 12.10. The overall format of scan operations is:

Command DST,SRC,DELIM,LIMIT[,DIR[,EOF]];

// See commands in Table 12.10.

Definition: The scan *command* scans a memory, using a mask to choose the desired delimiters.

Operands: DST, the destination, can be used for a copy of the scanned information. During the SCAN operation, data from the starting point of the scan toward the scan stop may be copied in the same clock cycle to a user-defined destination. The number of characters copied is limited by the destination size. The destination can be a register (or a byte in it), an indirect or a based-indexed addressing mode (type 2, 3, 6, or 7 in Table B12.1). Register-, indirect- or base-indexed addressing modes can use one of these registers: KMEM, NULL_REG, CAMI, CAMIH, CNVI, or CNVIH (see Section 12.2.1).

If the destination is a register, up to the first 4 bytes (from the starting point) may be copied. If the destination is CAMI, up to 8 bytes may be copied. If the destination is KMEM, up to the key length may be copied. If the programmer does not wish any data to be copied, NULL_REG may be used as the destination. NOTE: *When the destination is KMEM, the KBS0 register is used for the scan operation and will contain undefined data at the end of the operation. Similarly, when the destination is CAMI, CONVI or NULL_REG, the IND_REG0 register is used for the scan operation.*

SRC, the source, indicates where to begin the scan, and can be given in either direct or base-index addressing mode, referring to FMEM (type 7 or 8 in Table B12.1), as can be seen in the examples that follow.

DELIM is the delimiter bitmap, which can be either immediate, indirect (DELIM_MASK register, which is UREG[5]), or the delimiter that terminated the previous scan (SCAN_TC register). A mask is used to indicate which of the delimiters to use in the specific command. For example, let's assume that LDDV ";:-/" was used to load four delimiters: semicolon (0), colon (1), dash (2), and slash (3). If the programmer would like to scan using the colon and the slash only (e.g., for a URL analysis), then bits 1 (for colon) and 3 (for slash) should be masked as 1, while the other bits should be masked as 0. This creates a mask of 1010 or decimal 10 (0xA). Then, in order to use these two delimiters, one should simply code use 10 for DELIM:

```
FINDDEL CAMI, (RD_PTR)+, 10
```

Defines may be created (and used) to facilitate the use of masks, example,

```
#define COLON_AND_SLASH 10;
#define DASH_ONLY 4;
```

LIMIT indicates the depth of the SCAN (tested bytes number). LIMIT may be LIM_REG (which is UREG[3]) register or immediate value; in both cases, the value should be in the range of 1 to 0x3FF. If the SCAN operation is terminated by LIMIT, the LM flag is set.

DIR indicates the SCAN direction, that is, it is either forwards (_FRWD_PRS) or backwards (_BCKW_PRS). The default is _FRWD_PRS.

EOF is used as an exception handler, that is, to jump to the address in EOF_ADDR if the SCAN operation terminates due to end of frame (_JEOF_PRS), or not to jump (_NJEF_PRS). The default is _NJEF_PRS. If EOF is reached, the ED flag is set (and jumps to an address in EOF_ADDR, if the EOF parameter is set).

+ is the indication for auto-increment, which automatically increments the base address registers (e.g., RD_PTR) for each memory (FMEM, KMEM, and so on) depending on the destination.

The SCAN group of registers (in the FBLK registers) contains scanning information including the start and end of scanning and the scanning result, and is associated with the scan operations as defined in each of the scan commands. If the auto-increment option is used, then the RD_PTR may be affected as well. Table C12.1 below summarizes the results that are written to each of the three SCAN registers upon completion of the SCAN operation (the forth SCAN register, SCAN_TC, containing the termination cause, is described in Appendix A, and briefly in each of the operations that follow). In Table C12.1 "*del*" means the place where a delimiter was found in the scan and "*non del*" means the position of any character that is not delimiter that was found during the scan.

FindDel DST, SRC, DELIM, LIMIT [, DIR [, EOF]]

Find Delimiter (FindDel) scans for the first enabled CHAR in the delimiter vector from a specified location. If the auto-increment option is set at source, the RD_PTR will point to the found character at the end of the operation. The auto-increment option can be set in the destination, when the destination is KMEM. The data written to the destination consists of all the characters from the starting point

Table C12.1 Summary of SCAN Operation Results

SCAN Operation	SCAN_START	SCAN_STOP	RD_PTR If AI option is used	Data to destination
FindDel	start point	del + 0	del + 0	SCAN_START... SCAN_STOP - 1
FindNonDel	start point	non del + 0	non del + 0	SCAN_START... SCAN_STOP - 1
FindToken	del + 0	non del + 0	non del + 0	SCAN_START... SCAN_STOP - 1
GetToken	non del + 0	del + 0	del + 0	SCAN_START... SCAN_STOP - 1

NOTE: *If a scan operation with auto-increment is followed by a second scan operation, then one NOP must be inserted for the update.*

to the delimiter without the delimiter itself. The returned size is the number of data characters. SCAN_TC contains a bit vector set according to the scan results: bit 21 is set if the scan reached end of frame before finding a match, bit 20 indicates if the limit of the scan was reached before finding a match, or bits 0-15 of SCAN_TC indicate which of the delimiters was found in the scan.

Examples: `FindDel CAMI, (RD_PTR), 1, 100, _FRWD_PRS, _JEOF_PRS;`
                `FindDel CAMI, (RD_PTR)+, 1, 100, _FRWD_PRS, _JEOF_PRS;`

Description: Find the delimiter that is defined in the bitmap position 0 of the delimiter vector, by scanning forward the frame memory from RD_PTR address, with limit depth of 100 bytes. Jump if end of frame; write result in CAMI. Set RD_PTR to the delimiter position (if found), and the SCAN_STOP to the place of the delimiter in the frame memory (if found). Set SCAN_TC to the appropriate scan termination cause. If, as in this example, the DELIM is 1 and the delimiter vector was loaded by `LDDV` ";:-/", then a character in the frame memory that equals "/" will stop the scan.

FindNonDel DST, SRC, DELIM, LIMIT [, DIR [, EOF]]

Find Nondelimiter (FindNonDel) scans from a specified location in the frame memory (SRC) for the first character that is not in the delimiter vector (enabled according to the DELIM). If the auto-increment option is set at source, the RD_PTR will point to the found character at the end of the operation. The auto-increment option can be set in the destination, when the destination is KMEM. The data written to the destination consists of all the characters from the starting point to the first nondelimiter without the last character. The returned size is the number of data characters. Either the low significant byte of SCAN_TC (bits 0–7) contains the character found, or bit 20 of SCAN_TC is set to indicate that the limit of the scan was reached, or bit 21 of SCAN_TC is set to indicate that end of frame was reached before the nondelimiter character was found.

Example: `FindNonDel CAMI, (RD_PTR), 3, 100, _FRWD_PRS, _JEOF_PRS;`

Description: Find nondelimiter (first symbol not equal to delimiters) from bitmap 0 or 1 of the delimiter vector in the frame memory by scanning forward from RD_PTR address with limit depth of 100. Jump if end of frame; write result in CAMI. If, as in this example, the DELIM is 3 and the delimiter vector was loaded by `LDDV` ";:-/", then any character that is not a "-" or a "/" will stop the scan.

FindToken DST, SRC, DELIM, LIMIT [, DIR [, EOF]]

Find Token (FindToken) performs Find Delimiter and then Find Nondelimiter in a single instruction. If the auto-increment option is set at source, the RD_PTR will point to the found nondelimiter at the end of the operation. The auto-increment option can be set in the destination, when the destination is KMEM. The data written to the destination is all the characters between found delimiter to found nondelimiter. The low significant bytes of SCAN_TC contain the character (nondelimiter) found (see FindDel explanation of SCAN_TC).

Example: `FindToken CAMI, (RD_PTR), 3, 100, _FRWD_PRS, _JEOF_PRS;`

Description: Find token: set two pointers by Find Delimiter and Find Nondelimiter; pointers are stored in SCAN_START and SCAN_STOP registers; the size is in SCAN_SIZE register, jump if end of frame; write result in CAMI.

GetToken DST, SRC, DELIM, LIMIT [, DIR [, EOF]]

Get Token (GetToken) performs Find Nondelimiter and then Find Delimiter in a single instruction. If the auto-increment option is set in the source, the RD_PTR will point to the found delimiter at the end of the operation. The auto-increment option can be set in the destination, when the destination is KMEM. The data is only the data of the second operation (from the found nondelimiter to the found delimiter). SCAN_TC contains the termination cause.

Example: `GetToken CAMI, (RD_PTR), 3, 100, _FRWD_PRS, _JEOF_PRS;`

Description: Get token: set two pointers by Find Nondelimiter and Find Delimiter; pointers results in SCAN_START and SCAN_STOP registers; the size is in SCAN_SIZE, jump if end of frame; write result in CAMI.

HALT COMMAND

The halt command is used either to finish the packet processing and release it to the next stage in the pipeline, or to multicast the frame several times, or to abort any further processing of this packet and discard it. It is the most useful instruction and the only one that is mandatory, and it has the following format:

```
Halt Mode;
```

Mode is a two-bit immediate value, and it has these predefined constants:

HALT_UNIC Indicates that the current TOP has completed processing this frame and the next TOP can get the data.

HALT_DISC Instruction to discard this frame and begin processing the next frame.

HALT_MULC This is not a true halt and in fact may often be followed by other instructions. HALT HALT_MULC notifies TOPsearch I that it can take these keys and begin processing. TOPparse will resume parsing the same frame again to build new keys for it. Although multicast frames are usually duplicated by TOPresolve or TOPmodify, this could be used to duplicate a frame and send different keys with each copy. Once the last set of keys is ready for TOPsearch I, use HALT HALT_UNIC to terminate the processing of the current frame.

HALT_BRKP For debug mode. This freezes this specific TOPparse engine until the host releases it.

NOTE: *The two commands preceding a Halt command should not be jump commands.*

NOTE: *A hardware mechanism ensures that keys in KMEM are passed to TOPsearch I without programmers having to implement it.*

CHAPTER

Searching

13

Searching is an important function in every network processor, and EZchip architecture implements this utility in dedicated TOP processors that are integrated into the EZchip NP pipeline. The search engine is actually used twice along the pipeline. It is first used after the TOPparse in order to search keys that are produced during the analysis and parsing phase, and its results are handed to the TOPresolve for decision making. The second use of TOPsearch is after the TOPresolve has finished processing the results and making decisions, and further searches are required; for example, figuring out port numbers, IP addresses, and so on. The two TOPsearches, termed I and II, are quite similar, although there are some difference in the parameters each can receive, the number of them, search types (trees, hash-tables, etc.), and memories.

   In this chapter, we describe TOPsearch I very briefly. TOPsearch II is not described in this book, since the details that are given for TOPsearch I are sufficient for using TOPsearch II at the same level.

   TOPsearch engines are capable of handling complex searches, and even capable of running search procedures that are coded into its instruction memory, as in the other TOP engines. As the programming of TOPsearch is not essential for understanding how NP works, we shall not address it here. We will, however, describe the very frequent use of TOPsearch by showing how to program and carry out various simple lookup operations of the TOPsearch engine, such as, providing the TOPsearch with search keys and getting results that match these keys.

13.1 INTRODUCTION

TOPsearch I performs table look-ups by matching the keys received from TOPparse to the various search structures (data structures maintained by the search engine). Search structures are located in various memories that the NP uses, both internal memory and external memory. There are three types of data structures: trees, hash tables, and direct access tables.

   TOPparse creates and sends keys and messages to the TOPsearch I by transferring them to the Key Memory (KMEM). Messages are simply data structures that

TOPparse builds and sends via the TOPsearch to the TOPresolve, as a mean for providing raw data to the TOPresolve. TOPsearch ignores these messages other than to move them to the Output Memory (OMEM) for TOPresolve's input. There are two kinds of keys that TOPparse can create for TOPsearch I: single keys and compound keys.

A *single key*, as its name implies, is a simple key of variable length that contains one search key; for example, IP address, port number. A single key may contain an aggregate combination of several fields, extracted from the frame, in one key, that will be searched in a single operation; for example, <source port><protocol> for looking for the assigned priority of this specific port of that protocol.

A *compound key* is built from several keys that are combined together from various fields extracted from the frame. The keys are combined in such a way that the TOPsearch I has to execute a macrocode procedure for implementing this compound search. The compound key that TOPparse provides is not necessarily the full key, and information can be added by TOPsearch's macrocode procedure, as a function of a sequence of search results. The compound key can also include a long key structure, for TOPsearch to obtain keys longer than 38 bytes directly from the NP-1 frame memory.

For single keys, TOPsearch macrocode is not required and the structure number, defined in the structure definition table indicates the precise search structure (both specific repository and the specific search method, i.e., hash, tree, etc.) for the lookup. For compound keys, the structure number points to the beginning of the macrocode sequence. TOPsearch may also access the Statistics block to retrieve any counter value by using its counter ID as an index to a direct table.

13.2 INTERNAL ENGINE DIAGRAM

The following block diagram (Figure 13.1) illustrates the internal components of a single TOPsearch engine. These components, internal functional units, registers, and memories are described in this section.

13.2.1 Instruction Memory

Macrocode instructions are used for building compound searches. In these types of searches the results of one search can trigger another search in a separate structure.

13.2.2 Key Register and Header Register

The Key Register (KREG) stores the keys and messages, and the Header Register (HREG) stores the key and message headers received from TOPparse. Keys received

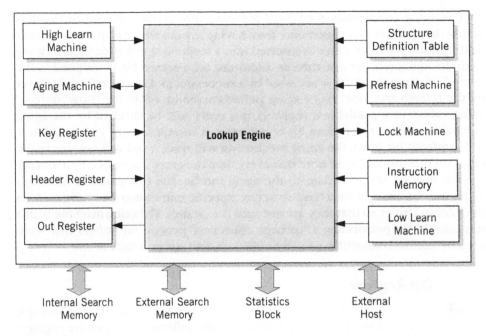

FIGURE 13.1

TOPsearch internal engine diagram

from TOPparse are up to 38 bytes in length; keys longer than 38 bytes are read directly from the frame memory according to the pointer provided by TOPparse. Messages are up to 64 bytes. A header accompanies each key, indicating whether it is a single key, a compound key or a message key, and its structure number.

13.2.3 Structure Definition Table

The structure definition table contains one row for each search structure defined. A structure number indicates the precise row number in the structure definition table. For single keys, the structure number is contained in the TOPparse key. For compound keys, the TOPsearch macro command has a STR_NUM parameter.

13.2.4 Learning, Refreshing, and Aging Machines

The host updates structures upon initialization as well as dynamically. At the same time, the dynamic updating of hash tables relies on the NP's hardware aging, refresh and learning machines. Layers 2 and 3 learning is based on traffic ("*low learn*"), whereas further learning is obtained from the TOPresolve or the host learning ("*high learn*").

The aging mechanism is engaged automatically by the NP's hardware, thus saving the application programmer from having to put a time stamp for the packet in the result. When an entry is inserted into a hash table that is defined to work with the aging mechanism, then an additional bit, a refresh bit, is assigned to the entry. Each time the entry is accessed by a successful lookup, the refresh bit is set by the refresh machine. Every aging period (based on either the "slow" or "fast" aging values in a predefined register), this entry will be checked by the aging mechanism and if the refresh bit of the entry is reset, the entry will be deleted. If the refresh bit is set, the aging mechanism will reset it and will not delete the entry on this round. Please note that every time the entry is accessed by a lookup, this bit may be set according to the aging mechanism configuration. Although aging may be enabled for a hash structure, specific entries may be configured via the EZchip Driver, so that they are not aged (i.e., static). The aging machine is also responsible for performing a "garbage collection" process on buffers that were used by the hash algorithm to solve collisions, and are no longer in use.

13.2.5 Out Register

Since the same frame is distributed among many TOPsearch engines for its lookups, an out register (OREG) for each TOPsearch engine collects the respective engine's results. All of the search results for the frame are then forwarded to a single result memory (RMEM) in TOPresolve. TOPresolve is notified when all the results of a frame have been forwarded to the respective RMEM.

13.2.6 Memory Cores and External Memory

TOPsearch has embedded memory cores, containing user-defined search structures. These include trees, hash tables, and direct tables. For systems requiring additional memory, structures may also be stored in external memory.

13.2.7 Statistics Counters

All the TOPsearch engines access the statistics counters stored in external memory. Arbitration and ordering mechanisms are in place. TOPsearch can read and write to any counter and the counters can be viewed as a direct lookup table where the counter ID is the index to the direct table. The statistical counters can be used in conjunction with the search engine for a variety of applications, such as stateful classification and packet processing.

13.2.8 Lock Machine

TOPsearch maintains frame ordering and proper learning by using a locking mechanism that the lock machine provides. This is required, for example, for stateful classification and packet processing.

This mechanism guarantees that hash table entries in the search memory are searched and updated according to the frame ordering. When a lookup is performed in a hash table and an entry is matched, the row in the search memory in which the entry is located is locked. This "ordinary" locking mechanism prevents the hash entry from being accessed by another engine (or the host processor) until it is unlocked. In ordinary lookups, this lock function is invisible to the programmer and the entry will be unlocked as soon as the lookup is completed. However, when the lock ordering mechanism is used, the entry is locked and will not be released when the lookup is completed; rather, it will only be released after the entry has been updated by the high learn feature. It is the programmer's responsibility in the TOPresolve microcode to fill-in the learn registers (LRN_REG) instructing the high learn (H_LRN) mechanism to unlock the entry following the hash entry's update. The LOCK_ORDER bit indicates to H_LRN whether it should unlock the hash table entry. Failure to do so may suspend the processing of subsequent frames. The lock ordering mechanism is activated for the NP by default.

13.3 TOPSEARCH I STRUCTURES

NP supports a variety of search data structures, and maintains a definition table that describes the actual data structures that the TOPsearch uses, as described briefly in this section.

13.3.1 Structure Definition Table

The structure definition table is built by the EZchip Driver software. The structure definition table is composed of one row for each search structure defined. The STRUC_NUMBER from TOPparse indicates the specific row in the table that defines the required searched structure. The row also determines in which type of structure to search (STRUCT_TYPE), the size of the key (KEY_SIZE), the size of the result (RESULT_SIZE), and other relevant information.

13.3.2 Tree Structures

Trees are convenient search structures for keys of varying lengths or keys that contain wildcards. In a typical binary tree, each bit in the key is examined and a branch is created based on the bit value. If the bit is 0, the tree branches to the left. If the key bit value is 1, the tree branches to the right. The procedure continues until the leaf. In a binary tree, the length of the longest path equals either the number of bits in the key or the number of keys supported by the tree, whichever is less. To reduce path length, Patricia trees (actually tries) can be used, as they examine the keys and skip all identical bits. This is explained further in Chapter 5.

13.3.3 Hash Table Structures

Programmers may create a variety of hash tables of different sizes for different searches, such as L2 MAC addresses, L3 routing, L4 session switching, and so on. Hash tables and their entries may be either stored in the core memories or in the external memory as defined in the structure definition table. Built-in memory provides fast access to the memory for efficient hashing. The host can add and delete hash table entries.

Dynamic updating of the hash tables is possible, and is done by the NP's hardware aging, refresh, and learning machines, as described previously. The low-learn machine updates the hash tables from Layers 2 and 3 traffic, while high-learn is done by the TOPresolve and/or the host, based on algorithms, statistics, counters, and so on. Hash table learning is described in more details in a later section.

13.3.4 Direct Table Structure

Direct tables offer the most efficient searching. In a direct access table, the key is the actual address of the row in the table structure. Direct tables can be stored in internal or external memory.

13.4   INTERFACE TO TOPPARSE (INPUT TO TOPSEARCH)

This section describes the structure of the key header and the key that TOPparse sends to TOPsearch I. TOPparse microcode is responsible for writing the keys for TOPsearch I (including a message for TOPresolve) in KMEM and the key headers in HREG. A header must accompany each single key, since each may be processed by a separate TOPsearch I engine. Compound keys are defined by a single key header.

13.4.1 Frame Parameters

Frame pointers and other information that is required to maintain frame ordering, or to use frame contents when required, is passed from TOPparse to TOPsearch via hardware. This is transparent to the programmer, and besides having some optional control (which is not described in this book), the programmer does not have to be concerned with this at all.

13.4.2 Key/Message Header Structure

The 3-byte key header contains key information. The key is either a key for a simple lookup, a compound lookup, or a message that is being passed directly to TOPresolve.

The header indicates the key size (KEY_SIZE), whether it is a single or a compound search, whether it is the frame's last key (KEY_LAST), the key's structure number (STRUCT_NUMBER), and whether it is a message to be forwarded to TOPresolve (MSG).

For a single search, the structure number indicates in which of the structures the TOPsearch should conduct its search. For a compound search, the structure number multiplied by two (x2) points to the macrocode instruction.

TOPparse uses the compound field validation bits (COMPND_FIELD_V) to indicate if a specific field exists within the frame. (Certain fields, e.g., VLAN tag, may not exist in every frame.) TOPsearch checks the validity of each key before using the key in a compound search. Table 13.1 describes the key/message header to be written to the HREG register by TOPparse.

Table 13.1 Key/Message Header Structure

Name	Bits #	Byte #	Description
KEY_SIZE(3:0)	b3:0	0	Up to 38 bytes for keys or 16/32/64 bytes for messages in 8-byte resolution. *Note: Keep in mind that this field is zero based.*
SINGLE/COMPOUND	b4	0	1—Single lookup key or MSG key. 0—Compound key.
KEY_LAST	b5	0	Indicate last key or message for the frame. Set to 1 for the last element regardless of whether it is a key or a message. Should be set once per frame.
STRUCT_NUMBER(5:0)	b11:6	1–0	If single key, indicates a specific data structure from the up to 64 structures. If compound key, indicates the address in the macrocode divided by two (/2).
LKP/MSG/COMPND_FIELD_V(0)	b12	1	If single key, 1—Lookup key (LKP). 0—Key is MSG from TOPparse (write result). If compound key, validation of key bit 0.
SINGLE_KEY_VALID/ COMPND_FIELD_V(1)	b13	1	If single key, 1—Key Valid 0—Key Not Valid. If compound key, validation of key bit 1.
KEY_LENGTH(5:0)/ COMPND_FIELD_V(7:2)	b19:14	2–1	Lookup key length in byte resolution, up to 38 bytes. If message, length is in 2-byte resolution, 16/32/64 bytes. *Note: Keep in mind that this field is zero based.* If compound key, validation of key bits (7:2).
COMPND_FIELD_V(10:8)	b22:20	2	If compound key, validation of key bits (10:8).
WRITE_TO_RESOLVE	b23	2	Must be set to 0. In NP-1c, the key is always passed to TOPresolve.

Table 13.2 Single Key Format

Name	Bits #	Byte #	Description
KEY		up to 38	Includes several frame fields used as one key (e.g., session key).

Table 13.3 Compound Key Format

Name	Bits #	Byte #	Description
KEY		up to 38	Includes several frame fields, used as several keys.

13.4.3 Key Formats

The key follows the key header. There are three main types of key formats corresponding to single keys, compound keys (which may include long keys), and messages.

13.4.3.1 Single Key Format

A single key may contains various fields extracted from the frame by TOPparse and combined into a single key for a single lookup. Up to 38 arbitrary bytes may used as a single TOPsearch key. No prefix is required (see Table 13.2).

13.4.3.2 Compound Key Format

A compound key may contains various fields extracted from the frame by TOPparse to build different keys. This is not necessarily the full key and information can be added from a previous search. The use of compound keys (and TOPsearch programming) is not described in this book. Detailed instructions are provided in [118] (see Table 13.3).

13.4.3.3 Long Key Format

Since keys from TOPparse are limited to 38 bytes, the contents of longer keys, such as a long URL, are fetched directly from the frame memory. The long key format is used for such searches; it includes two pointers to the frame memory, and is a special case of the compound key. Long keys are handled similarly to compound keys; they call TOPsearch macrocode procedures, and are not detailed in this book.

13.4.3.4 Message Key Format

The message key format refers to TOPparse messages destined for TOPresolve (see Table 13.4). TOPsearch combines these messages with all other frames' search results sent to TOPresolve.

The message is actually a key that, instead of being processed by TOPsearch, is passed as is to TOPresolve. The message may include fields from the TOPparse

Table 13.4 Message Key Format

Name	Bits #	Byte #	Description
ONE	b0	0	Reserved for VALID bit. Set to 1.
CONTROLS(15:1)	b15:b1	1–0	Different control bits related to the entry (e.g., MATCH, forward to host, discard, etc.). Content and bit ordering are user defined.
OPTIONAL_ADDITIONAL_ STRUCTURES	b512:b16	64–2	Additional structure such as: Frame PTR, Port #, Timestamp, Frame length, Host_Cause_Vec, Session_Key for high learning, and so on.

hardware decoder, notably the time stamp and the frame_ptr fields (see in the TOPparsing chapter, the description of the Hardware register), as well as other data obtained from the frame and needed down the line.

13.5 INTERFACE TO TOPRESOLVE (OUTPUT OF TOPSEARCH)

The Search Results Structure is the structure of the results that are forwarded to TOPresolve. Up to 16 TOPsearch results and messages from TOPparse for an entire frame can be sent to the same TOPresolve RMEM. The result written to TOPresolve RMEM includes the structure number and the result pointer.

It is important to note that in TOPresolve, the results of the search can be read simply by referring to the RMEM and providing the structure number from which the result should be read, as an index (e.g., Get ALU,10(12),2; for loading two bytes from the RMEM, structure number 12, offset 10, to the ALU register). For more details, see the next chapter on TOPresolve programming.

13.5.1 Frame Parameters

There are several parameters that are passed to TOPresolve from TOPsearch by hardware circuitry, and that are transparent to the programmer. One important parameter is the RMEM number, which indicates to which RMEM all the results were written.

13.5.2 General Result Format

In general, the results from all different types of structures are comprised of control bits and the result (data) itself. Figure 13.2 and Table 13.5 show the general formate of a result.

In all results, the first three bytes are composed of control bits that may be user-defined. The first two bytes are referred to as user-defined control bits and can

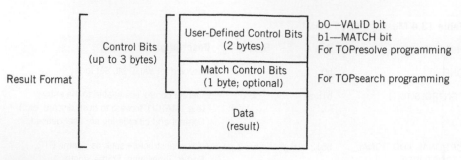

FIGURE 13.2

General result format

Table 13.5 Result Format			
Name	**Bits #**	**Byte #**	**Description**
CONTROL BITS			
ONE	b0	0	1—VALID bit
ONE	b1	0	1—MATCH bit
CONTROLS(14:2)	b14:2	1–0	0
ZERO	b15	1	0
MATCH_COND(3:0)	b19:16	2	0
MATCH_ALL	b20	2	0
NEW_MATCH_BITS	b23:21	2	0
DATA (RESULT)			Size is determined by the structure definition table. Sizes may differ depending on whether the search structure is stored in internal or external memory.

be easily accessed by TOPresolve instructions, such as GetRnd4Bits, which may compare the control bits from two results. Byte 0 bits 1:0 are best reserved for the VALID and MATCH bits, respectively, and should be set to one.

The third byte of control bits is generally referred to as match bits and may be used in the TOPsearch programming. These match bits also offer compatibility with the NP result format. The result is transferred to the TOPresolve RMEM.

It should be noted that when TOPresolve reads the result, it starts from offset 0 of the result's entry in the structure; that is, it sees the control bits, starting from the valid and match bits. Also note that in creating the results in the structure entries, either in high-learn, low-learn, from the host or manually with the EZdesign structure generator's tool, this format has to be kept; that is, the first three bytes should be left for controls, especially the first two bits (valid and match).

13.6 HASH TABLE LEARNING

The aging, refresh, and learning machines can dynamically update hash tables. Layers 2 and 3 learning is obtained from the macrocode (low learn), whereas further learning is obtained from TOPresolve or host learning (high learn). Learning requested from TOPresolve microcode has higher priority than that from the macrocode.

An example of how the NP creates a new session with the TOPresolve high learn mechanism is as follows: A frame initiating a new session (e.g., SYN frame of TCP protocol) enters the network processor. TOPparse creates a key with the session information and TOPsearch performs a lookup in a hash table containing the known sessions. Since it is a new session, no match will be found. TOPresolve gets the TOPsearch results (with the "not found" indication), and places them into a new hash entry. TOPresolve microcode then instructs the H_LRN block to update the session hash table in the search memory (internal or external) with the new hash entry, effectively creating a known session. Before updating an entry, the H_LRN block checks to see if the entry already exists in the table and if it should be overwritten (CREATE/UPDATE COMMAND).

Thus, when subsequent frames from this session arrive, the relevant information will be found in the session hash table. When the session is closed, TOPresolve microcode sends a delete entry message to the H_LRN block.

13.6.1 Low-Learning Interface

Hardware Layers 2 and 3 learning (low learning) is done as frames fly through the NP ("from the wire" so to speak). This learning is conducted by TOPsearch through two mechanisms: MAC hash table learning (Layer 2) and IP hash table learning (Layer 3). The learning mechanisms are driven by the TOPsearch macrocode whose data is interpreted according to one of two templates: MAC or IP. The template selection is based upon the structure number that must be provided in the Learn command. A transparent mechanism looks up the structure number in the Structure Definition Table to determine the structure type (according to the key size—4, 8, or 10). If the structure type indicates a hash table, then it is a MAC entry. If it indicates a tree + hash, then it is an IP entry.

TOPparse microcode or TOPsearch compound search must prepare a key in the data format according to the MAC or IP learning templates, to be used in the learn instruction, before a TOPsearch Learn command is executed. Since we are not describing the compound searches and TOPsearch programming, we will not elaborate more on low-learning; the interested reader is referred to [118].

13.6.2 TOPresolve High-Learning Interface

TOPresolve can add or delete an entry from any hash table using the high learn (H_LRN) feature, by using its learn registers (LRN_REG, described in Chapter 14). TOPsearch receives these learn-entries from TOPresolve in an H_LRN interface format, which varies according to the NP type (i.e., NP-1, NP-1c, NP-2, etc.). The general format, however, is shown in Figure 13.3.

Programmers are responsible for writing to the TOPresolve learn registers (LRN_ REG) in this format. No macrocode programming is required in TOPsearch I for carrying the learning itself. The example in the next section will clarify this. The detailed format of the basic message header of the H_LRN interface is shown in Table 13.6.

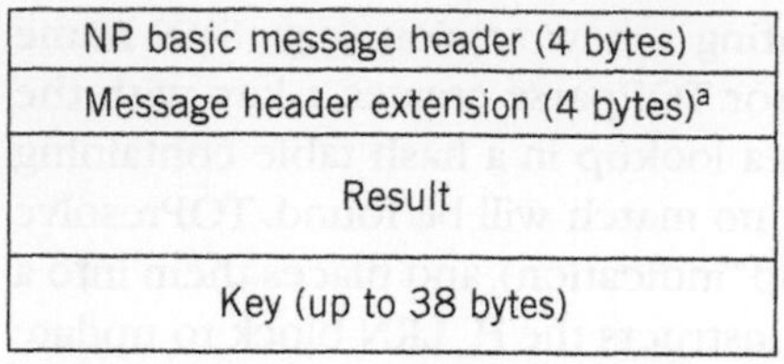

| NP basic message header (4 bytes) |
| Message header extension (4 bytes)[a] |
| Result |
| Key (up to 38 bytes) |

[a]NP-1 does not support the message header extension

FIGURE 13.3

General format of TOPresolve H_LRN interface

Table 13.6 TOPresolve H_LRN Interface

Name	Bits #	Byte #	Description
LRN_INFO_ LENGTH(4:0)	b4:0	0	Learn data length in 4-byte resolution. Does not include message header and header extension. In add operation = size of key + result. In delete operation = size of key. In partial update[a] = size of key + 4. Minimum length is 8 bytes (or 12, if TOPresolve ordering mechanism is activated). If less than minimum, must be padded.
ZEROS	b7:5	0	Reserved, set to 0.
STRUCT_ NUMBER(5:0)	b13:8	1	High learn structure ID. (ID of hash structure to be updated by H_LRN)
CREATE/UPDATE COMMAND[b]	b14	1	Indicates what H_LRN should do if it finds an existing entry in the hash table. If Add Entry (b16 = 1) and entry exists: 1-Create mode = do no overwrite existing entry; 0-Update mode = overwrite existing entry. Otherwise if does not exist, create new entry.
ZERO	b15	1	Reserved, set to 0.
COMMAND	b16	2	Indicates the purpose of the message: 0-delete entry 1-add entry.
ONE	b17	2	Reserved for Refresh bit.

Table 13.6 (*continued*)

Name	Bits #	Byte #	Description
STATIC	b18	2	Static bit: 0-entry may be aged; 1-aging will not affect the entry.
LOCAL	b19	2	Local bit: 0-entry learned from remote TOPresolve (via host message); 1-entry learned from TOPresolve in this chip.
SEND_MSG	b20	2	1-send message to host.
INCR_COUNTER	b21	2	0-do not increment counter 1-increment counter (bits 59:24 in ENTRY)
ZERO	b31:22	2–3	Reserved, set to 0.
Result[c]		4–	
Key			

[a] *NP-1 does not support partial update.*
[b] *NP-1 does not support this bit, and must be 0.*
[c] *Result and Key are not provided for the delete command.*

13.7 EXAMPLE

We provide two examples to show the usage of the TOPsearch I engine. The first example is of conducting a simple search, and the other is of activating the high-learn mechanism.

13.7.1 Searching

The following TOPparse code describes how to initiate a simple search for an IP address that will be conducted by TOPsearch in a hash table (structure #7) that contains IP addresses. In this example, the Source IP address (SIP) is the last search key that TOPparse creates, and it is the second key that is sent to TOPsearch:

```
//Some definitions
#define IP_SIP_OFFSET    12;   // the Source IP address offset in the
                               // frame
//
#define KEY_SIZE         4;    // Source IP is IP address, i.e., 4
                               // bytes long
#define Simple_KEY       1;    // indicate that the searching is single
#define Last_key         1;    // in this example it is the last key
```

```
#define STR                 7;  // the structure number to look in
#define Lookup_Key          1;  // indicates that it is not a message
#define Single_VALID        1;  // validate it as a single key, not a
                                // message
//
// create the header from the definitions, shifted to their bit
// positions
#define HDR ( ((KEY_SIZE - 1)>>3) | (Simple_KEY<<4) | (Last_KEY<<5) |
(STR<<6) | (Lookup_KEY<<12) | (Single_VALID<<13) | ((KEY_SIZE-1)<<14));
//
// Now the actual code—two instructions: put data in KMEM,
//                                  and create Header register (HREG)
//
Copy 0(KBSO), IP_SIP_OFFSET(RD_PTR), 4; // Copy 4 bytes: KMEM <-FMEM
PutHdr HREG[2], HDR, 3; // This is the second key, for searching SIP
```

13.7.2 Learning

The following code shows how to use the TOPresolve high-learn register to activate the high-learn mechanism. Please note that we did not discuss the TOPresolve yet; nevertheless, it is important to have it here, in the context of the search structures, and the interested reader may refer to the next chapter, the TOPresolve programming, in case of difficulties.

```
// Some definitions
/* basic message header fields (4 bytes) */
//
// prepare the header information length
#define KEY_SIZE            4; // Source IP is the key, i.e., 4 bytes
                               // long
#define RES_SIZE            32; // result is 32 bytes long
// prepare the header fields
#define LRN_INFO_LEN (((((KEY_SIZE+3)>>2)<<2)+RES_SIZE-1)>>2);
#define STR                 7; // the structure number
#define CREATE_UPDATE_MODE  1; // HL_CREATE—don't change entry if exists
#define CMD_MODE            1; // HL_ADD_CMD - this is to add
#define STATIC_MODE         0; // HL_NO_STATIC - entry may be aged
#define LOCAL_MODE          1; // HL_LOCAL - TOPresolve teaches the
                               // entry
#define SEND_MSG_MODE       0; // HL_NO_SEND_MSG - do not send to host
#define INCR_CNT_MODE       0; // HL_NO_INCR_CNT - do not increment
                               // counter
//
// create the High-learn header from the definitions above,
//                          shifted to their correct bit positions
#define NP1_MSG_HDR ( (LRN_INFO_LEN<<0) | (STR_NUM<<8) |
(CREATE_UPDATE_MODE<<14) | (CMD_MODE<<16) | (STATIC_MODE<<18) |
```

```
(LOCAL_MODE<<19) | (SEND_MSG_MODE<<20) | (INCR_CNT_MODE<<21) );
//

// prepare the first result byte (control bits of the search structure),
// in particular the valid bit and the match bits.
#define Valid_bit_offset          0; //HL_VALID_BIT_OFF
#define Match_bit_offset          1; //HL_MATCH_BIT_OFF
//

#define RES_B0_B3 ( (1<<Valid_bit_offset) | (1<<Match_bit_offset) );
//
#define WRITE_ALL_LRN_REGS 3; // number of LRN_REGs to use in learning
#define WRITE_LRN_REG0_1    1; // -"-
// Here is the actual code:
WAIT_NO_FLAG F_LN_RSV; //macro that makes sure the High Learn is free
//

// Now start by writing the header to LRN_REG0,
// then the result to LRN_REG1 (first 4 bytes), and the next 4 bytes
// to LRN_REG2 and the next 4 bytes to LRN_REG3.
Mov  LRN_REG0, NP1_MSG_HDR,  4; // LRN_REG0 contains the header
Mov  LRN_REG1, RES_B0_B3,    4; // LRN_REG1 contains result bytes 0-3
Mov  LRN_REG2, something..., 4; // LRN_REG2 contains result bytes 4-7
Mov  LRN_REG3, something..., 4; // LRN_REG3 contains result bytes 8-11
//

Mov  LRN_SIZE, WRITE_ALL_LRN_REGS, 1; // this ignites the learning
                                      // with 4 learn registers

Nop;
//

HL_2ND_WAIT L_DISCARD, 0; //a macro that makes sure that the second
// learning phase is possible
// Continue the learning with bytes 12-27 in the four learn register
Mov  LRN_REG0, something..., 4; // LRN_REG0 contains result bytes 12-15
Mov  LRN_REG1, something..., 4; // LRN_REG1 contains result bytes 16-19
Mov  LRN_REG2, something..., 4; // LRN_REG2 contains result bytes 20-23
Mov  LRN_REG3, something..., 4; // LRN_REG3 contains result bytes 24-27
//

Mov  LRN_SIZE, WRITE_ALL_LRN_REGS, 1;  // this ignites the learning
                                       // with 4 learn registers

Nop;
//

WAIT_NO_FLAG F_LN_RSV; // macro that makes sure the High Learn is
                       // free
//

Mov  LRN_REG0, something...,       4; // LRN_REG0 has the last result
                                      // bytes 28-31
Mov  LRN_REG1, KEY_to_be_used..., 4; // now copy the key
Mov  LRN_SIZE, WRITE_LRN_REG0_1, 1; // this ignites the learning
                                    // with 2 learn registers
```

13.8 **SUMMARY**

The TOPsearch engine is an important ingredient of the NP processor; although we describe it only briefly here, it is one of the most significant factors in achieving good performance. EZchip technology supports a search engine that uses an instruction set similar to the rest of the processors in its pipeline, and integrates this engine with the others; it is not a peripheral or attached processor, as other architectures usually make it. In the next chapter, we return to the regular processing unit of the NP pipeline by looking at the TOPresolve.

Resolving

The next stop of the packet, as well as of our description of packet processing in EZchip, is resolving. Based on the results of the searched items and the messages coming from the parsing phase, the resolving stage determines the actions to be taken on the frame in the next stage—modifying. In other words, TOPresolve is the decision point for handling frames. It receives messages and search results from the TOPparse and TOPsearch I engines, respectively, and processes them to make decisions for TOPmodify about how to manipulate the frame.

This chapter outlines the TOPresolve architecture, its blocks (called devices), and its instruction set. A simple example is also given to demonstrate TOPresolve's use. As with other chapters in this part, a detailed description of blocks, registers, and instructions is given in the appendices for readers who would like more complex examples or who intend to write programs themselves.

14.1 INTERNAL ENGINE DIAGRAM

The block diagram (Figure 14.1) illustrates the internal blocks of a single TOPresolve engine.

14.1.1 Instruction Memory Block

Instruction memory contains microcode instructions for TOPresolve engines.

14.1.2 Pipeline Control Block

The pipeline control receives a command from the instruction memory, decodes it, and distributes the control signals to each block.

14.1.3 Data Bus

The data bus supports a data flow between the TOPresolve blocks.

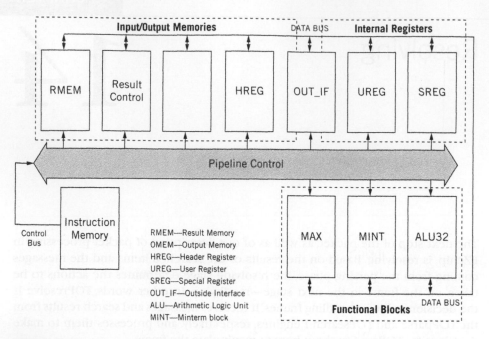

FIGURE 14.1

TOPresolve internal engine diagram

14.1.4 **Result Memory and Result Control**

The Result Memory (RMEM) is a buffer that retains search results from TOPsearch I and messages from TOPparse. The first 16 bits of each result are control bits that are also loaded into the Result Control block.

14.1.5 **Register Blocks (User-Defined Register, Specific Register, and Header Register)**

The User-Defined Register (UREG) contains general-purpose registers for data processing. The Specific Register (SREG) contains special registers of differing lengths, dedicated to specific TOPresolve functions. The Header Register (HREG) block contains descriptions of the Output Memory (OMEM) contents, including information such as which key to use, the size of the key, and what to do with the key.

The registers are described in Section 14.2, and some are detailed in Appendix A of this chapter.

14.1.6 **Arithmetic Logic Unit**

The Arithmetic Logic Unit (ALU32) is a 32-bit general purpose ALU used for performing computational functions. ALU source operands are any of the TOPresolve registers. To use an operand from the memory, the programmer must first copy the operand to a register before using it as an ALU source. An ALU feedback register is used for consecutive operations in order to avoid data hazards.

14.1.7 Minterm Block

The Minterm block (MINT) solves logical nested IF_ELSE_IF equations in a minimum number of clocks without branches. Users can select a complex logical operation from up to 12 predefined operations. The Minterm block receives several predefined bit wide inputs from any block, and outputs several predefined bit wide results, depending upon the logical operation chosen.

For example, with a 32-bit wide input, Minterm operation #2 returns 4 logical values ("0" or "1"). Each result is an OR of four logical values, which are themselves functions of the inputs to Minterm. Specifically, each of these four logical values is the result of comparing two-input-bits with a predefined two-bit-pattern (an outcome of an XORNOT operation of the two-input-bits with the corresponding two-bit-pattern values), where the matching results ("1" for identical, "0" otherwise) are then masked by an AND operation with two adjacent mask bit inputs. And there are also more complicated operations. All operations are explained in a clearer way in Appendix C.

14.1.8 Max Block

The MAX block calculates maximum/minimum and index values based on the input. The 8-bit input can originate from any other block. Output, calculated within one clock, includes the max/min value derived from the input and its index. The index is the number of the result that contains the max/min value. The result of the max/min calculation is written to the MAX_OUT register.

For example, if the input contains QoS setting information from three registers (results #1-3), the Max block determines the frame's maximum QoS setting. A frame may have a Layer 2 QoS setting, a Layer 3 QoS setting, and a Layer 4 QoS setting, with each setting being the result of a search in a different structure. MAX will take these three results (see Table 14.1) and determine the setting that takes precedence (i.e., the max value). The Max block also indicates what result the max had (i.e., in this example, it is result #2, which contains the Layer 3 QoS setting).

Table 14.1 Example of Max Block Calculation

Input		Output
Result #1	Layer 2 QoS setting	Max = Layer 3 QoS setting
Result #2	Layer 3 QoS setting	Index = 2
Result #3	Layer 4 QoS setting	

14.1.9 Output Memory

The OMEM is a memory that stores search keys for TOPsearch II. The memory also stores messages that are being passed along to TOPmodify. All the internal TOPresolve blocks can write to the OMEM indirectly via the OBS register.

14.1.10 Result Control Block

The result control block holds control bits written by TOPsearch I. These bits are used for sophisticated retrieval of information from the results, for various control operations.

14.1.11 Outside Interface Block

The Outside Interface Block (OUT_IF) provides access to external blocks such as the learn machine, Receive Frame Descriptor (RFD) table, statistics block, and host.

TOPresolve learning and tracking functions send a key and its result back to TOPsearch I. TOPsearch I either adds or updates the search structures based on the key and result that it receives from TOPresolve. The key and result contain information such as the state of a session.

14.1.12 TOPresolve High Learn Ordering Mechanism

In addition to the ordering mechanisms in TOPsearch, TOPresolve has an ordering mechanism that guarantees the ordering of messages sent to the high learn block (H_LRN) of the TOPsearch I. When activated (using the HOST_CONF register described in the following section), messages are sent to H_LRN according to frame ordering. This ensures that the updates to the TOPsearch core memories (both internal and external) will be performed in the same order that the frames arrived to the network processor (NP). In other words, the first frame updates the memory before the second frame, which comes before the update of the third frame, and so on. This ensures that high learn messages are queued in TOPresolve engines and unable to access the H_LRN block until the previous update is completed. This feature is useful for the stateful classification of flows when using the high learn state or the software counter feature.

For example, identifying a specific sequence of frames—A, B, C—without the use of the lock ordering feature is outlined as follows:

1. The TOPresolve ordering mechanism should be enabled from the host.
2. When receiving frame A, use high learn to add an entry with state A using the conditional state update feature of high learn.
3. When receiving frame B, use high learn to update the entry with state B (only if the current state is A).
4. The same for frame C, updating the state to C only if the current state is B.
5. When TOPsearch I reads this entry and passes it to TOPresolve, it will be possible to know whether there was frame sequence A, B, C, if the state field has C in it.

When activated, the programmers may determine which operation indicates the end of the high learning processes (again, by using the HOST_CONF register).

In the default setting, all the high learn messages are processed by the H_LRN block at the end of the microcode sequence (i.e., at HALT). The programmer may, alternatively, "signal" the end of the high learn messages for optimization purposes. Once all the high learn messages have been written to the learn registers (LRN_REGs, as detailed in the following section), the programmer is then responsible for writing something to the END_H_LRN register (it is irrelevant what is written). Writing to this register instructs the H_LRN arbiter to finish processing this engine and to move to the next engine. If ordering is not enabled, then the H_LRN arbiter uses round robin to select between messages from the TOPresolve engines. When activating this mechanism, high learn messages must be larger than 16 bytes.

14.2 TOPRESOLVE REGISTERS

There are two sets of registers in the EZchip TOPresolve: one that is accessible directly through the TOPresolve microcode, and one that can be accessed by the TOPresolve microcode only after it is first initialized by the host (through the host interface).

14.2.1 Microcode Registers

The following tables list the TOPresolve registers and structures accessible from the microcode. Detailed descriptions of the registers follow in Appendix A of this chapter. The R/W column indicates whether the register is read, write or both. The Init column indicates the register's initial value. All registers can be written to in bits or bytes. All registers are accessed by the register name and the index in square brackets (e.g., SREG[4]), or by their defined name (e.g., CNT) provided, of course, that they were defined as such in the source program, or in the libraries. The registers are intended for various and specific purposes and, consequently, have dedicated input sizes. For instance, attempting to write 8 bits into a 5-bit register will result in only the first five bits being written.

The UREG register is composed of 16 general purpose registers, 32 bits each, as described in Table 14.2. The UDB register (which is UREG[0]) is zeroed for each new frame.

The ALU32 register block is composed of just one register of 32 bits, as described in Table 14.3. The MAX register block is composed of two registers, as described in Table 14.4. The Minterm (MINT) register block is composed of two registers, as described in Table 14.5. The SREG are composed of 16 variable length registers, as described in Table 14.6. The Outside Interface (OUT_IF) register block is composed of 15 registers of variable length, as described in Table 14.7. The header key (HREG) register block is composed of

Table 14.2 UREG Register Block

Index	Register name	Size (bits)	Bytes (bits)	Description	R/W	Init
0	UDB	32 bits	4 (32)	User-defined bits. Each bit may serve as a condition in an IF statement.	R/W	0
...						
15		32 bits	4 (32)	General-purpose register.	R/W	

Table 14.3 ALU32 Register Block

Index	Register name	Size (bits)	Bytes (bits)	Description	R/W	Init
0	ALU	32 bits	4 (32)	ALU feedback register. See Section 14.1.6.	R/W	0

Table 14.4 MAX Register Block

Index	Register name	Size (bits)	Bytes (bits)	Description	R/W	Init
0	MAX_O	12 bits	4 (12)	Out register. See Sections 14.1.8 and 14.4.3.	R	0
1	MAX_I	8 bits	4 (8)	In register.	R/W	0

Table 14.5 MINT Register Block

Index	Register name	Size (bits)	Bytes (bits)	Description	R/W	Init
0	MINT_O	16 bits	4 (16)	Out register. See Sections 14.1.7 and 14.4.2.	R	
1	MINT_I	32 bits	4 (32)	In register.	R/W	

Table 14.6 SREG Register Block

Index	Register name	Size (bits)	Bytes (bits)	Description	R/W	Init
0	RBS0, RBS1, RBS2, RBS3	4 × 6 bits	4 (6 + 6 + 6 + 6)	Four RMEM base address registers.	R/W	
1	OFF_RMEM0 OFF_RMEM1 OFF_RMEM2 OFF_RMEM3	4 × 7 bits	4 (7 + 7 + 7 + 7)	Four RMEM offset registers.	R/W	
2	OBS0, OBS1, OBS2, OBS3	4 × 7 bits	4 (7 + 7 + 7 + 7)	Four OMEM base address registers.	R/W	
3	OFF_OMEM0 OFF_OMEM1 OFF_OMEM2 OFF_OMEM3	4 × 7 bits	4 (7 + 7 + 7 + 7)	Four OMEM offset registers.	R/W	
4	CNT	8 bits	4 (8)	See Loop commands.	R/W	0
5	PC_STACK	16 bits	4 (16)	See Call commands.	R/W	0
6	SIZE_REG	5 bits	4 (5)	Indirect size register for selected commands (Mov, ALU).	R/W	0
7	IND_REG0 IND_REG1	2 × (3 + 2 + 4 bits)	4 (2 × (3 + 2 + 4))	Dedicated offset register for indirect access to all devices (see Figure A14.1).	R/W	0
8	ST_GRP0 ST_GRP1	2 × 3 bits	4 (3 + 3)	Two structure group registers.	R/W	0
9	HBS0, HBS1, HBS2, HBS3	4 × 3 bits	4 (3 + 3 + 3 + 3)	Four indirect access registers to HREG.	R/W	0
10	FLAGS	12 bits	4 (12)	Flags. See Figure A14.3 in Appendix A.		0
11	HISTORY0 HISTORY1	2 × 10 bits	4 (10 + 10)	Two instructions in pipeline.	R	0
12	HISTORY2 HISTORY3	2 × 10 bits	4 (10 + 10)	Two more instructions in pipeline.	R	0

(*continued*)

Table 14.6 SREG Register Block (*continued*)

Index	Register name	Size (bits)	Bytes (bits)	Description	R/W	Init
13	UNIT_NUM	4 bits	1 (4)	Number of the TOP engine.	R	
14	RX_CNT_HI	24 bits	4 (24)	Status of ingress queues (high).	R	
15	TX_CNT_HI	24 bits	4 (24)	Status of egress queues (high).	R	

Table 14.7 OUT_IF Register Block

Index	Register name	Size (bits)	Bytes (bits)	Description	R/W	Init
0	STAT_REG	32 bits	4 (32)	Interface to statistics block.	R/W	0
1	HOST_REG	32 bits	4 (32)	Host register.	R/W	0
2	LRN_REG0	32 bits	4 (32)	Learn register 0.	R/W	0
3	LRN_REG1	32 bits	4 (32)	Learn register 1.	R/W	0
4	LRN_REG2	32 bits	4 (32)	Learn register 2.	R/W	0
5	LRN_REG3	32 bits	4 (32)	Learn register 3.	R/W	0
6	LRN_SIZE	2 bits	4 (2)	Learn size register.	R/W	0
7	CTRL_REG	3 bits	1 (3)	TOPmodify frame length control.	R/W	0
	RESERVED	8 bits	1 (8)	Reserved byte set to 0.		
	FR_PTR	16 bits	3 (16)	TOPmodify frame pointer register.	R/W	0
8	RFD_REG0	10 bits	4 (10)	Interface to RFD register.	R/W	0
9	RFD_REG1	26 bits	4 (26)	Interface to RFD register.	R/W	0
10	RX_CNT	32 bits	4 (32)	Status of ingress queues.	R	

Table 14.7 (*continued*)

Index	Register name	Size (bits)	Bytes (bits)	Description	R/W	Init
11	TX_CNT	32 bits	4 (32)	Status of egress queues.	R	
13	END_H_LRN	32 bits	4 (32)	End of H_LRN for ordering.	W	
14	STAT_REG_HI	16 bits	4 (16)	Interface to statistics block.	R/W	

Table 14.8 HREG Register Block

Index	Register name	Size (bits)	Bytes (bits)	Description	R/W	Init
0	HREG0	24 bits	4 (24)	HREG registers, containing headers of the keys and messages, as described in the TOPsearch chapter.	R/W	
...						
7	HREG7	24 bits	4 (24)		R/W	

eight registers (one for each key or message created for next stages in the NP pipeline), each of 24 bits, as described in Table 14.8.

14.2.2 Host Registers

All host registers are initialized by the host; however, the initial values for some are inserted into the NP microcode for loading. These initialization commands are "executed" by the host prior to executing the program in the instruction memory. Table 14.9 lists all of TOPresolve's host registers and indicates which registers have their initial values in the NP microcode.

Table 14.9 TOPresolve Host Registers

Address	Name	Description	Initialization
0x00	INT_REG	Interrupt register.	Host only
0x00	INT_REG	Interrupt register.	Host only
0x01 – 0x04	WIDE_LOAD[3:0]	SRAM instruction register.	Host only
0x08	BR_ADDR	Service routine address.	Host only
0x09 – 0x18	MREG[15:0]	ALU mask register.	Microcode
0x19	HOST_CONF	Learn ordering and channel support register.	Host only
0x20 – 0x25	HOST_REG[5:0]	Host debug register.	Host only
0x2C	MCODE_BR_INT	Microcode execution or break point command.	Host only
0x2D	STATUS_REG	Status register.	Host only

Table 14.10 TOPresolve Memories

Name	Description
RMEM	Result memory (from TOPsearch I)
Result control block	Result memory (from TOPsearch I)
OMEM	Output memory (to TOPsearch II)

14.3 TOPRESOLVE STRUCTURES

TOPresolve's microcode instructions access three "external" memories: the result control block and the RMEM (both of which contain input data for TOPresolve), and the OMEM (which is used by TOPresolve for output). The result control block and RMEM—the inputs—contain the results of the previous pipelined engine (TOPsearch I), while OMEM will be used by the next pipelined engine (TOPsearch II). These three kinds of memories are described in Table 14.10.

Output Memory is accessed indirectly via the OBS register. It is important to note that RMEM, the input memory, cannot be accessed as a continuous memory, like regular direct accessed memory, but can only be accessed using the structure number of the search operation (or the message). In other words, if a key or message were written to the input memory of TOPsearch I using some structure (as indicated in the key header in the HREG), the RMEM could be read by TOPresolve only by indicating this structure number as a base.

14.4 TOPRESOLVE INSTRUCTION SET

This section examines each of the instructions in the TOPresolve microcode that are unique to this TOP. The common NP instructions that are shared by all of the TOPs are described in the EZchip programming chapter.

The microcode execution begins on the first instruction for each new frame. After examining and processing the input TOPsearch I results and the TOPparse messages and then making decisions, programmers are responsible for writing the output keys for TOPsearch II in the OMEM (PutKey instruction) and the key headers in HREG (PutHdr instruction). A header must accompany each key. The format of the header is located in Section 13.4.

Programmers are also responsible for writing TOPresolve messages destined for TOPmodify in OMEM (PutKey instruction). The messages are actually keys that, instead of being processed by TOPsearch II, are passed as is to TOPmodify. The messages may include fields from the TOPparse hardware decoder, notably the frame pointer, frame length, first frame buffer, and number of frame buffers. A message may also contain other data obtained in TOPsearch I and TOPresolve; for example a 32-bit map indicating up to 32 modifications for the TOPmodify Accelerator, needed in TOPmodify. The format is located in Section 13.4.

Programmers are also responsible for filling in the CTRL_REG and FR_PTR registers. These registers are automatically read by TOPmodify, enabling it to pre-fetch the first frame buffer if it requires modification. These registers have a separate interface and do not need to be passed through a message, unless the programmer wishes to use them in the TOPmodify microcode.

Most of the commands in the following subsections are described briefly, with examples. In most cases, this should be enough for understanding and writing complex code; however, the reader who needs deeper utilization of the commands is referred to Appendix C of this chapter.

14.4.1 Move Commands

Move commands copy data (bits, bytes, and words) from various sources to various destinations, as described in the EZchip Programming chapter. There are four prefixes for move commands, depending on where the data is being moved (see Figure 14.2 and Table 14.11):

- Use Get for moving data from RMEM to TOPresolve registers.
- Use Mov for moving data within the TOPresolve registers.
- Use Put for moving data from TOPresolve registers to OMEM.
- Use Copy for moving data from RMEM to OMEM.

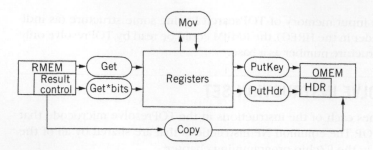

FIGURE 14.2

TOPresolve move commands name prefix mnemonic

Table 14.11 List of Move Commands

Name	Description
Mov	Move bytes from immediate value or register to register
MovBits	Move bits from immediate value or register to register
Mov4Bits	Move 4 separate bits
Get	Move bytes from memory to register
PutKey	Move bytes from register or immediate value to memory
Copy	Move bytes from memory to memory
PutHdr	Write bytes from immediate value or register to header
PutHdrBits	Write bits group from immediate value or register to header
PutHdr4Bits	Write 4 bits with immediate values or from register to header
GetRndBits	Move bits from result control
GetFixBits	Move bits from result control
GetRnd4Bits	Move bits from result control

The additional move commands in TOPresolve that are not described in the common NP instructions (described in Chapter 11) are those that use the result control block bits, as follows.

14.4.1.1 *GetRndBits DST, SRC0, SRC1, SRC2, SRC3, SRC4, SRC5, SRC6, SRC7*

Definition: Move eight random bits from the result control block (SRC0..7) to destination (DST). The term random refers to bits that are arbitrarily addressed and not necessarily adjacent (like in Random Access Memory); actually, they are programmable-selected. This instruction enables random bits from any two structure groups (each structure group contains up to eight structures) and writes them to the destination.

This is the only command that uses the structure groups (see ST_GRP in the Specific Registers section of Appendix A in this chapter). If a structure number does not exist, "0" is written.

Example:
```
GetRndBits UREG[1], STR_GRP0.INDEX[5].BIT[4],
                    STR_GRP0.INDEX[1].BIT[4],
                    STR_GRP0.INDEX[2].BIT[0],
                    STR_GRP0.INDEX[6].BIT[3],
                    STR_GRP1.INDEX[5].BIT[12],
                    STR_GRP1.INDEX[7].BIT[4],
                    STR_GRP1.INDEX[0].BIT[15],
                    STR_GRP1.INDEX[5].BIT[4];
```

Description: STR_GRP0 and STR_GRP1 are aliases for SREG[8].BYTE[0] and SREG[8].BYTE[1], respectively. (They are defined in TOPresolve.h.) These are not really two byte size resources, but rather two 3-bit size resources. Each one indicates a structure group number (0–7), giving the programmer random access to the 16 structures within these two groups. INDEX is the three least significant bits (lsb) specifying the row (0–7) within the group, using all the bits to build a byte. This example loads the eight bits from the various structures to UREG[1].BIT[0], and STR_GRP0.INDEX[5].BIT[4], for example, indicates ST_GRP0, structure 5 within this group, and bit 4.

14.4.1.2 GetFixBits DST, SRC0, SRC1, SRC2, SRC3, SRC4, SRC5, SRC6, SRC7

Definition: Move eight bits from the result control block (SRC0..7) to destination (DST). This instruction builds a byte from eight bits within any two offsets. If structure number does not exist, '0' will be written.

Example:
```
GetFixBits UREG[1], STRNUM[55].BIT[15],
                    STRNUM[0].BIT[15],
                    STRNUM[55].BIT[15],
                    STRNUM[15].BIT[15],
                    STRNUM[13].BIT[4],
                    STRNUM[51].BIT[4],
                    STRNUM[55].BIT[4],
                    STRNUM[7].BIT[4];
```

Description: Move eight bits from the structures indicated with corresponding offsets to UREG[1].BIT[0]. Each of the offset bits may only have one of two values (e.g., 15 or 4), hence, the mnemonic FixBits.

14.4.1.3 GetRnd4Bits DST, SRCO [,SRC1 [,SRC2 [,SRC3 [,SIZE]]]]

Definition: Move up to four bits (SIZE) from the result control block (SRC1..4) to destination (DST). All structure numbers and offsets (0..15) can be used. If a structure number does not exist, "0" will be written.

Example:
```
GetRnd4Bits UREG[1], STRNUM[0].BIT[0],
                     STRNUM[5].BIT[2],
                     STRNUM[19].BIT[1],
                     STRNUM[19].BIT[0];
```

Description: Move bit 0 from structure 19 to UREG[1].BIT[0].
Move bit 1 from structure 19 to UREG[1].BIT[1].
Move bit 2 from structure 5 to UREG[1].BIT[2].
Move bit 0 from structure 0 to UREG[1].BIT[3].

14.4.2 Minterm Operations

Minterm operations are a toolkit consisting of multiparameter bit-wise operations, all being ORs of groups of input bits ANDed together. The various operations are dictated by an "operation" operand. These operations are used to perform complex logical decisions in a single clock. The mnemonic *Minterm* originates from the name given to elements of Conjunctive Normal Form in switching theory. The format of the Minterm operation is:

```
Minterm DST, OPER, INVMASK, MASK4, CTRL, SIZE;
```

Definition: Calculate logical formulas, according to OPER. Results of the Minterm operation can be up to two bytes, sent to the destination (DST). The input is from the 32-bit MINT_I register.

14.4.3 SetMaxMin Operations and Max Block

The Max functional block executes the SetMaxMin operations to find the maximum or minimum value of a series of arbitrary length. The format of the SetMaxMin operation is:

```
SetMaxMin SRC, MODE;
```

Definition: Find max/min value in a sequence (according to MODE) and write it to MAX_O.BYTE[0].

Example:
```
SetMaxMin UREG [13], _MAX_RST;
Mov MAX_IN. UREG [14], 1;
Mov MAX_IN. UREG [15], 1;
```

Description: These three commands calculate the maximum of values from UREG[13], UREG[14], and UREG[15]. The one-byte result will be written to MAX_OUT.BYTE[0]. The index of the maximum value (in a sequence of up to 16 values) will be written in MAX_OUT. BYTE[1].

Example: If a series is 1, 9, 0, 8, 5, 10, 3,
            then its index is 0, 1, 2, 3, 4, 5, 6.
            When 1 and 9 are compared, 9 is the max value and its index is 1. After
            comparing the entire series, the max value is 10, and the index is 5.

14.4.4 Conditional Commands

Some of the instructions can be preceded by conditional statements that function
as part of the instructions, that is, they are coded in one machine-instruction, and
are executed at the same time as the unconditional statements. These TOPresolve
commands are listed in Table 14.12.

Table 14.12 TOPresolve Conditional Commands	
Mov	Jmp
MovBits	Call
Get	PutHdr
Copy	PutHdrBits
PutKey	ALU operations

14.5 EXAMPLE

This sample application continues the example of the TOPparse chapter, for transmitting the ingress frames with a VLAN TAG to the Virtual Output Queue (VOQ)
with the application code written for each NP TOP processor.

This example adds a field (VLAN tag) to the frame header, if required, based
on inspection by the hardware decoder as to whether the incoming frame has a
VLAN TAG. This information is passed through a message to TOPmodify, where the
proper treatment of the frame is decided.

14.5.1 Frame Handling Overview

The NP operates on both ingress and egress paths. As described in the Parsing
phase in Chapter 12, the NP processes frames on the ingress path; that is, frames
that arrive from the network and are transmitted to the switching fabric. The
TOPparse receives ingress frames, and the NP's hardware decoder automatically
inspects each incoming frame and provides useful information about the frame for
microcode use prior to execution of the NP microcode.

The decoded information is passed by a message from TOPparse to TOPresolve,
accompanied with a decision byte. TOPresolve forwards the message to TOPmodify
and initializes the internal buffer of TOPmodify as the first frame buffer. According
to the message (decision byte), TOPmodify decides whether to add a Common

0	0	0	1	1	1	1	0	0	0	0	0	0	0	1	1	0	0	0	1

6 bits representing the msg length. In Obase and in 2-byte resolution. (7 + 1) * 2 = 16 byte msg length.	valid	msg (not key)	structure number 0	last key	single key	4 bits representing the number of bytes in msg. In Obase and 8-byte resolution. (1 + 1) * 8 = 16 byte msg size.

FIGURE 14.3

Message header structure

Switch Interface Specification (CSIX) header or a VLAN tag and a CSIX header to the current frame header, prior to its transmission to the VOQ in Mode 1.

14.5.2 Data Flow and TOP Microcode

The subsections that follow describe the data flow for TOPresolve, including the relevant microcode. Definition and macro files for this sample application are located in Section 14.5.4.

TOPresolve does not do much, actually; it just passes the message received from TOPparse along to TOPmodify (via TOPsearch II). It also initializes the internal buffer to be used in TOPmodify as the first frame buffer. The entire code is eight instructions, organized in two steps:

1. Pass the hardware decoded data to TOPmodify. This data was originally sent by TOPparse as a message in the first 16 bytes of KMEM. The "copy" command copies data from the RMEM (input memory of TOPresolve) to OMEM (output memory of TOPresolve). The message header structure in this example, according to Figure 14.3, is as follows:

```
HW_MESSAGE_HEADER = 0x1E031 = 0001 1110 0000 0011 0001
```

2. TOPresolve can define whether the internal buffer of TOPmodify will be initialized. By initiating the value of CTRL_REG with the frame's first buffer length, TOPresolve defines that the internal TOPmodify buffer will be initialized with the contents of the first frame buffer.

14.5.3 TOPresolve Microcode

The following code implements the TOPresolve stage.

```
EZtop Resolve;

#include "mcglobal.h"      // Global definition file supplied with  EZdesign,
                           // with predefined constants for NOPs, flags, etc.
#include "TOPResolve.h"    // TOPresolve definition file supplied, provides
                           // recognizable names for registers and flags.
#include "hdreg.h"         // Global definition file supplied to provide
                           // recognizable names to the HW decoded registers.
```

```
// Constants:
#define HW_HEADER          0x1E031;
#define HW_HD_REG0_OFF     1;
#define HW_FR_PTR_OFF      7;
#define HW_OBS             OBS0;
#define HW_STRUCT          0;
#define HD_REG0            UREG[1];

START:                     // Beginning of microcode sequence
Mov HW_OBS, 0, 2;          // Initialize the base pointer HW_OBS

// Initialize FR_PTR with the logical pointer to the first frame buffer.
// The pointer is extracted from HD_REG1 b31:16 (passed via a message
// from TOPparse).
Get FR_PTR, HW_FR_PTR_OFF(0), 2;

// Initialize the local register (HD_REG0) with the value of the HW
// decoder register HD_REG0 (passed via a message from TOPparse).
Get HD_REG0, HW_HD_REG0_OFF(0), 4;

///////////////////////////////////////
// Step 1: Copy message to TOPsearch II
        PutHdr  HREG[0], HW_HEADER, 3;
        Copy    0(HW_OBS)+, 0(HW_STRUCT), 8;
        Copy    0(HW_OBS)+, 8(HW_STRUCT), 8;

// Step 2: Initialize the internal buffer of TOPmodify with the
// contents of the first frame buffer.
    MovBits CTRL_REG, sHR0_bits1stBufLenCtrl, 3;

    Halt HALT_UNIC; //pass frame to the next TOP stage in the pipeline
```

14.5.4 Definition Files

There are three header files used in this example: the microcode global definition file (mcglobal.h), the hardware decoding definition file (hdreg.h), and the TOPresolve definition file (TOPresolve.h).

The microcode global definition file (mcglobal.h) is described in Chapter 12. The hardware decoding definition file (hdreg.h) is also described in Chapter 12 but is using here another definition:

```
#ifndef _hdreg_h                              ;
#define _hdreg_h                              ;

#ifndef sHR0                                  ;
#define sHR0                        HD_REG0   ;
```

```
#endif                                              ;
...

#define sHRO_bits1stBufLenCtrl      sHRO.BIT [12];// 3
...

#endif;
```

The TOPresolve definition file (TOPresolve.h) is as follows:

```
#ifndef _TOPresolve_h_                              ;
#define _TOPresolve_h_                              ;
...

// SREG
...

#define OBSO               SREG [2].BYTE [0]    ;
...

// Outside interfaces
...

#define CTRL_REG           OUT_IF [7]           ;
#define FR_PTR             OUT_IF [7].BYTE [2]   ;
...

#endif /* _TOPresolve_h_ */                         ;
```

14.5.5 Structures and Message Formats

Several data structures and formats are used to pass the information, keys, and results between the previous pipeline engines and the TOPresolve, and between TOPresolve and the following pipeline engine—the TOPmodify. These structures are described in the subsections.

14.5.5.1 TOPsearch I Structures

Table 14.13 describes the TOPsearch I data structures (also described in the Parsing chapter). The message from TOPparse to TOPresolve is placed in structure number 0 on the ingress path.

14.5.5.2 TOPsearch II Structures

As described in Table 14.14, TOPsearch II data structures in our example are precisely like the TOPsearch I data structures described earlier. The message from TOPresolve to TOPmodify is placed in structure number 0 on the ingress path. A message header must accompany the message passed to TOPsearch II.

14.5.5.3 TOPparse–TOPresolve Hardware Decoded Message

Table 14.15 presents the format of the 14-byte message (padded to 16 bytes) from TOPparse to TOPresolve via TOPsearch I on the ingress path (as described in Chapter 12). This message is placed in TOPsearch I structure #0, and contains information extracted by the HW decoder destined for TOPs down the line.

Table 14.13 TOPsearch I Strucures

Name	Structure type	Structure number	Path	Used for	Key size (bytes)	Result size (bytes)
TOPparse-TOPresolve message	(Message)	0	Ingress	L2 + L3, VLAN tag flag	–	–

Table 14.14 TOPsearch II Data Structures

Name	Structure type	Structure number	Path	Used for	Key size (bytes)	Result size (bytes)
TOPresolve-TOPmodify message	(Message)	0	Ingress	L2 + L3, VLAN tag flag	–	–

Table 14.15 TOPparse–TOPresolve Message

Field name	Byte offset	Size (bits)	Note
Valid	0	1	Always 1 for messages
To host	0	1	1—send to host
TTL_EXP	0	1	From HD_REG0 b28
Ctrl reserved bits	0	5	
HD_REG0	1–4	32	HW decoding
HD_REG1	5–8	32	HW decoding
HD_REG2	9–12	32	HW decoding
vlanTagExists	13	1	Indicates whether the arriving frame contained a VLAN tag field

14.5.5.4 TOPresolve–TOPmodify Message

The message from TOPresolve to TOPmodify, via TOPsearch II, in our example is identical to the TOPparse to TOPresolve message described above, and outlined in Table 14.16. This 14-byte message (padded to 16 bytes) is placed in TOPsearch II structure #0. This message contains the HW decoded information, including whether the original frame contains a VLAN tag field.

Table 14.16 TOPresolve–TOPmodify Message

Field name	Byte offset	Size (bits)	Note
Valid	0	1	Always 1 for messages
To host	0	1	1—send to host
TTL_EXP	0	1	From HD_REG0 b28
Ctrl reserved bits	0	5	
HD_REG0	1–4	32	HW decoding
HD_REG1	5–8	32	HW decoding
HD_REG2	9–12	32	HW decoding
vlanTagExists	13	1	Indicates whether the arriving frame contained a VLAN tag field

14.6 SUMMARY

The TOPresolve is a middle engine whose job is to make decisions based on the analysis and findings of the TOP engines that precede it have done, and to pass these decisions on to the TOP engines that follow it, TOPmodify (and potentially also through the extra step of lookup). The TOPresolve is also responsible for updating the search database that the NP maintains for its TOPsearch operations, for allowing stateful analysis, as well as enabling dynamic and adoptive program algorithms.

The internal structure of TOPresolve is described in this chapter, including its registers and data structures, as well as its blocks (functional units) and instruction set. The appendices of this chapter provide more details about the registers and the instruction sets for those who need them either to understand a code or to write one.

In the next chapter, we very briefly describe the TOPmodify engine, which is the next processor in the EZchip network processor pipeline.

APPENDIX A
DETAILED REGISTER DESCRIPTION

A detailed description of all TOPresolve devices and registers is given in this appendix. Though some of the information here duplicates what can be found in the body of the chapter, the details and descriptions are fleshed out considerably.

MICROCODE REGISTERS

The microcode registers are accessed by the microcode for normal operations during program execution, and include user-defined, functional, and specific registers.

User-Defined Register

The UREG contains 16 general-purpose registers for data processing; UREG[0] is the UDB register.

Register name	Description
UDB	User-defined bits. Programmers can impose a condition on each bit, or its inverse, and act accordingly.

Arithmetic Logic Unit Register

Register name	Description
ALU	32-bit ALU feedback register with the last result of the ALU calculation. Used to perform adjacent ALU calculations and avoid a data hazard.

MAX Register

Register name	Description
MAX_O	12-bit output from the MAX block: 8 bits for the min/max and 4 bits for the index value (0–15). MAX_O can be used as a source register for a move instruction.
MAX_I	8-bit input for the MAX block. Input is read for debugging only.

Minterm Register

Register name	Description
MINT_O	1–16 bits of output from the Minterm block. The number of bits depends on the Minterm equation used (see Section 14.1.7). Output can also be written automatically to another destination as either one or two bytes (with all unused bits set to 0). To write to another destination in bits, the register must wait for one instruction (to prevent a data hazard) and then use the MovBit command.
MINT_I	32-bit input for the Minterm block for calculating logical formulas (see Section 14.1.7).

Specific Registers (SREG Table)

Register name	Description
RBS	6-bit RMEM base address based on the TOPsearch structure number (0–63). This provides TOPresolve with indirect access to the results in RMEM. Four RBS registers enable TOPresolve to work on four results simultaneously.
OFF_RMEM	7-bit RMEM offset specifying the result (up to 128 bytes). Four OFF_RMEM registers enable TOPresolve to work on four results simultaneously. Autoincrement (offset + size) updates this register to read additional bytes from the same result in the RMEM.
OBS	7-bit OMEM base address for writing keys for TOPsearch II/TOPmodify. Four OBS registers enable TOPresolve to write four keys simultaneously. Auto increments the base address (offset + base + size).
OFF_OMEM	7-bit OMEM offset for writing keys for TOPsearch II/TOPmodify. Four OFF_OMEM registers enable TOPresolve to write four keys simultaneously.
CNT	8-bit register for loop support. The program branches on the counter and checks the counter's value; if the value is not zero, the program jumps to the start of the loop. Each counter value branch automatically decrements the counter by one, that is, counting down at each loop taken. Up to 256 repetitions of a particular sequence of code are supported. The Counter register depth is one; thus, the counter can write/read to a register to implement nested loops.
PC_STACK	16-bit register for call commands (i.e., branch + set bit to push to stack). At each call command, the content (branch address +3-#NOPs) is automatically written to PC_STACK. The PC_STACK has a depth of one; thus, the PC_STACK can be written/read by the user to build nested call commands. The call command contains the number of NOPs that follow the relevant return command.
SIZE_REG	5-bit register for an indirect size. The size may be specified in either bits or bytes, depending on the instruction.
IND_REG	9-bits contains the indirect address for a Dword, byte, and bit. Source. Destination addresses are written in device#, Dword#, byte#, and bit# formats. The device number is always specified as immediate, whereas the other three destinations (Dword#, byte#, and bit#) can be addressed either immediately or indirectly using the IND_REG register. See Figures A14.1 and A14.2 for IND_REG format.
ST_GRP0 ST_GRP1	Two 3-bit registers specifying the structure group. Structures are divided into 8 groups. Another 3 bits from the GetRndBits command specify the particular structure within that group (relative to the group). (See figure on facing page.)
HBS	Four 3-bit registers for indirect access to header keys in HREG. Supports up to eight keys per frame.
FLAGS	12-bit flag register. The defined bits and bit fields within the FLAGS register control selected operations and indicate the status of the network processor (see Figure A14.3). ZR—zero flag CY—carry flag SN—sign flag OV—overflow flag (0=no overflow; 1=overflow) LP—loop counter register (0=CNT register is not zero; 1=CNT register is zero) ST—statistics register (0=interface ready; 1=not ready) HT—host register (0=interface ready; 1=not ready)

Register name	Description
	LN—learn ready flag (0=interface ready; 1=not ready) RD—RFD interface (0=interface ready; 1=not ready) PN—H_LRN pending flag interface (0 = TOPresolve engine not pending; 1=TOPresolve engine is pending for H_LRN machine to begin receiving its learn message) *Note:* In microcode, the flags are referred to as FLAGS.BIT (F_x_RSV), where x is the flag name from Table A14.3. (e.g. FLAGS.BIT (F_ZR_RSV)).

History	Four 10-bit registers specifying the four instructions currently in the pipeline:
	  HISTORY0 Byte 0–1: Fetch stage
	  HISTORY1 Byte 2–3: Decode stage
	  HISTORY2 Byte 0–1: Execution 1 stage
	  HISTORY3 Byte 2–3: Execution 2 stage

UNIT_NUM	4-bit register that numbers each of the TOPresolve engines. This enables a specific TOPresolve engine to be referenced directly in the code.
RX_CNT_HI	In NP-1 and NP-1c: 24-bit register containing the status of the ingress upper queues. See RX_CNT register for overall status and status of the lower ports. b23:0 RFD upper ports (3*8) − 3 most significant bit (msb) (bits 10:8) of Rx per port counter bits 10:0. Indicates the number of RFD pointers being used in the Rx portion of the RFD table (see RX_CNT).
TX_CNT_HI	In NP-1 and NP-1c: 24-bit register containing the status of the egress upper queues. See TX_CNT register for overall status and status of the lower ports. b23:0 RFD upper port (3*8) − 3 msb of Tx per port counter (11 bit) Indicates the number of RFD pointers being used in the Tx portion of the RFD table (see TX_CNT).

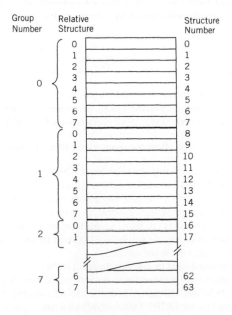

SREG ST_GRP

Diagram of 64 structures divided into 8 groups with 8 structures each

byte 1								byte 0							
7	6	5	4	3	2	1	0	7	6	5	4	3	2	1	0
				Dword#							byte#			bit#	

FIGURE A14.1

IND_REG0 format

byte 3								byte 2							
7	6	5	4	3	2	1	0	7	6	5	4	3	2	1	0
				Dword#							byte#			bit#	

FIGURE A14.2

IND_REG1 format

bit#	15	14	13	12	11	10	9	8	7	6	5	4	3	2	1	0
					PN		RD	LN	HT	ST	LP		OV	SN	CY	ZR

FIGURE A14.3

TOPresolve flags

Outside Interface Registers

The OUT_IF registers access external blocks. Programmers must check the ready flag to ensure that these registers can be written to. Attempting to write to one of these blocks when it is not ready will result in a loss of the data.

Register name	Description
STAT_REG	32-bit interface to the statistics block: address (b31:8), reserved (b7:3), and command (b2:0). Valid commands are increment (000), decrement (101) and reset (001). Increment and decrement commands use the operand in the STAT_REG_HI register.
HOST_REG	32-bit interface to the host. Microcode programmers should not write to this register.
LRN_REG	Four 32-bit registers containing up to 128 bits of TOPresolve learn information for the TOPresolve learning feature. The minimum learn size is 16 bytes (12 bytes if operating in "NP-1 mode"). The data may occupy from one to four of these registers depending on the size of the learn information. Learn information can be written into the four LRN_REG registers (16 bytes) several times over; however, the first time the four control bytes (four lsbs of LRN_REG b31:0) must be written and sent first. For the structure format, refer to Section 13.6.2. Programmers are responsible for checking the status of the LN flag; if it is set, then the LRN_REG registers must not be written to because the learn interface is busy.Based on the STRUCT_NUMBER, TOPsearch knows the size of the key and the size of the result and takes these from the ENTRY. It then calculates the hash signatures.

Register name	Description

LRN_SIZE	2-bit register specifying which LRN_REG register should be written.
	    0—write LRN_REG0 only.
	    1—write LRN_REG0 and LRN_REG1 only.
	    2—LRN_REG0, LRN_REG1, and LRN_REG2 only.
	    3—write all four LRN_REG registers
	Once the LRN_SIZE_REG is written into, it triggers the ready flag. None of the LRN_REG registers can be written into until the ready flag is cleared.

CTRL_REG	3 bits of hardwired information for the TOPmodify block containing the number of times that the TOPmodify needs to access the frame memory (64 bytes at a time) to prefetch the first frame buffer. This enables optimization of TOPmodify's prefetch mechanism. These are b2:0 in byte 0. If set to zero, then TOPmodify will not fetch the frame buffer.
FR_PTR	16-bit frame pointer for TOPmodify. These are b31:16 in bytes 2–3. In addition to being placed in this register, the frame pointer may be passed to TOPmodify in a message for use in the TOPmodify microcode.
RFD_REG0	10 or 11-bit interface to the RFD table: (7:0) data, (9:8) command. The command bits are the trigger for the ready flag to be set. Data cannot be written to RFD_REG0 or RFD_REG1 again until the flag is cleared. **Command bits:** 1—Recycle (based on the multicast bit in RFD_REG1, the recycle operation will either check (1) or ignore (0) the multicast counter field in RFD table); 3—Write to the multicast counter field in the RFD table (the multicast bit in RFD_REG1 should be set). The number of RFD entries that are updated with this multicast counter field is based on the value of the RFD_REG1 number of buffers field: 1 RFD entry if the number of buffers is 1, or 2 entries if the number of buffers is greater than 1. In NP-2 Bit 10 is used for reading one of the RFD budget counter according to the port field in RFD_REG1 (all "1" indicate reading ALL_PORT counter) and bit 15 in RFD_REG1 used for selecting RX counter or TX counters. Writing the command "100" to bits 10:8 will trigger a reading from the RFD counter. Data cannot be written to RFD_REG0 or RFD_REG1 again until the flag is cleared, and then the counter value will be in bits 24:16.

RFD_REG1	26 to 28-bit interface to the RFD table. The register format differs depending on whether the chip is operating in NP-1, NP-1c, or NP-2 mode, as configured in the HOST_CONF register. When in different modes, the number of bits used for the port number differs.
	In NP-1 mode:
	    (15:0) pointer, (20:16) number of buffers, (23:21) port number, (24) multicast, and (25) set to zero.
	In NP-1c mode:
	    (15:0) pointer, (20:16) number of buffers, (24:21) port number, and (25) multicast.
	In NP-2 mode:
	    (15:0) pointer, (20:16) number of buffers, (26:21) port number, and (27) multicast.

RX_CNT	32-bit register containing the status of the ingress queues. This register is a counter showing the number of RFD pointers in use, which indicates the number of active buffers in the system. The counters are increased as pointers are taken, and decreased as pointers are recycled. These states may be used by TOPresolve as watermarks.

Register name	Description
RX_CNT (*cont'd*)	In NP-1 and NP-1c the RX_CNT register fields are separate counters indicating the number of RFD pointers being used in the Rx portion of the RFD table: 2:0 VOQ total—bits 11:9 out of VOQ CNT counter bits 11:0. Host—number of Rx RFD pointers that are used as the host sends frames to the VOQ and receives frame from the HTFD queue. This provides an overall indication of congestion at the host interface and could be used in restricting the host's bandwidth consumption. VOQ total—number of counters presently in the VOQ. 7:3 RFD total—bits 10:6 out of Rx RFD counter bits 10:0. RFD total—total number of Rx RFD pointers being used by the IFDMA, the host and the TOPmodify block. This provides an overall indication of congestion on the ingress path. 31:8 RFD port (3*8)—3 msb (bits 10:8) of Rx per port counter bits 10:0. These are the lower 8 ports/channels; see RX_CNT_HI register for the status of the upper 8 ports. RFD port—number of Rx RFD pointers being used by each IFDMA. This provides a indication of congestion at each NP-1 input port from the network.
TX_CNT	32-bit register containing the status of the egress queues. This register is a counter showing the number of RFD pointers in use, which indicates the number of active buffers in the system. The counters are increased as pointers are taken, and decreased as pointers are recycled. These states may be used by TOPresolve as watermarks. In NP-1 and NP-1c the TX_CNT register fields are separate counters indicating the number of RFD pointers being used in the Tx portion of the RFD table: 2:0 host—(9..7) bits of RxTx Host counter (11 bit) Host—number of Tx RFD pointers used for sending frames to the host on the egress path, although this is not a standard path implemented by the NP-1. 7:3 RFD total—(10..6) of Tx RFD counter (11 bit) RFD total—total number of Tx RFD pointers being used by the EBDMA and the TOPmodify block. This provides an overall indication of congestion on the egress path. 31:8 RFD port (3*8)—3 msb of Tx per port counter (11 bit). These are the lower 8 ports/channels; see TX_CNT_HI register for the status of the upper 8 ports. RFD port—number of Tx RFD pointers being used by each ETFD. This provides an indication of congestion on the queues to the network.
END_H_LRN	32-bit register that when written to indicates that the TOPresolve engine has completed its messages to the high learn block (H_LRN). This will release the H_LRN arbiter from waiting so it may process messages from the next TOPresolve engine. This register is only used when TOPresolve ordering is enabled (H_LRN_ORDERING) and the H_LRN_END_MODE bit is set.
STAT_REG_HI	16-bit operand for the statistics block. See the STAT_REG register for the address and command.

Header Register

The HREG contains the headers for the keys in the OMEM—eight registers for eight headers (24 bits each). These are for the eight searches that may be performed in TOPsearch II. HREG can be accessed indirectly from the HBS register.

HOST REGISTERS

Most registers are initialized with a high-level Application Program Interface (API) interface, which is not described here. The only registers that are described in this Appendix are those that are mentioned in this chapter and in the sample programs in the book. For more details, see [118].

MREG[15:0] Register

The LDREG command in the microcode instructs the loader to load the user-defined masks to each of these mask-registers (Figure A14.4).

NAME	BITS#	DESCRIPTION	INIT VALUE
ALU_MASK	b31:0	Used for ALU mask.	0

FIGURE A14.4

ALU mask[15:0] register

APPENDIX B
TOPRESOLVE ADDRESSING MODES

Table B14.1 provides the numbers, names, and syntaxes of the addressing modes that are relevant to the TOPresolve. Devices in this table refer to registers or structures; for example UREG, SREG. The numbers in the first column are used in this chapter to indicate the addressing modes supported by operands. Bold typeface indicates required text. *Italic* typeface indicates text that must be replaced with the appropriate value.

Table B14.1 TOPresolve Addressing Modes

No.	Name	Syntax	Description
1	Immediate	123 or 0x12 or "abcd" or $ or label	123 – decimal number 0x12 – hexadecimal number "abcd" – 4 bytes = 0x61626364 $ - program counter (PC) label – program label
2	Register	*device* [*number*]	For the device, see the registers listed in Section 14.2. *Number* – index of the register in the device array
3	Register (byte-specific)	*device* [*number1*].**byte** [*number2*]	*number1* – index of the register in the device array *number2* – byte number in this register
4	Register (bit-specific)	*device* [*number1*].**bit** [*number2*]	*Number2* – bit number in this register
5	Register (four-bit)	*device* [*number1*].**bits** [*n1,n2,n3,n4*]	*n1..n4* – specific bits in this register
6	Indirect	*device* [*IND_REG*]	
7	Base-index	*offset* (*base*) or *offset* (*base*) +	*Base*–base registers. *Offset*—immediate or some register
9	Memory (byte-specific)	**hreg** [*HBS*].**byte**[*num2*]	
10	Memory (bit-specific)	**hreg** [*HBS*].**bit**[*num2*]	
11	Memory (four-bit)	**hreg** [*HBS*].**bits** [*n1,n2,n3,n4*]	

Table B14.1 (*continued*)

No.	Name	Syntax	Description
12	Control registers in result memory (bit specific access with group registers)	**sreg[8].byte[**n1**].index[**n2**].bit[**n3**]**	n1 − (0, 1) − group number register n2 − (0, 1,..7) − structure index n3 − (0, 1,...31) − bit offset
13	Control registers in result memory (bit specific access without group registers)	**strnum [**num1**].bit[**num2**]**	num1 − (0,1...63) structure number num2 − (0,1,...31) − bit offset ...
14	Structure number-offset	offset **(**str**)** or offset **(**str**)** +	str − register or immediate. Access to result memory. Offset − immediate or some register

APPENDIX C
TOPRESOLVE DETAILED INSTRUCTION SET

Detailed description of TOPresolve commands' operands are listed in this Appendix. For even more detailed explanations, see [118].

MOVE COMMANDS

Move commands copy data (bits, bytes, and words) from various sources to various destinations.

Mov DST, SRC, SIZE

Definition: Move SIZE bytes from source (SRC) to destination (DST).

Operands: DST, the destination, can be a register (or a byte in it) or an indirect address mode (type 2, 3, or 6 in Table B14.1). Valid registers are UREG, ALU, MAX_I, MINT_I, SREG registers (except FLAGS, HISTORY0...3, UNIT_NUM, RX_CNT_HI and TX_CNT_HI), and OUT_IF registers (except RX_CNT and TX_CNT) (see Section 14.2.1).

SRC, the source, can be an immediate 32-bits value, a register (or a byte in it) or an indirect addressing mode (type 1, 2, 3, or 6 in Table B14.1). Register or indirect addressing modes can be any of the registers (see Section 14.2.1).

Size can be either an immediate value (1–4) or a register content; that is, the SIZE_REG (which is SREG[6]) value (0–4). This corresponds to type 1 or 2 in Table B14.1.

MovBits DST, SRC, SIZE

Definition: Move SIZE bits from source (SRC) to destination (DST).

Operands: DST, the destination, can be a register, a specific bit in a register (e.g., UREG[0].bit[5]) or an indirect addressing mode (type 2, 4, or 6 in Table B14.1). Valid registers are UREG, ALU, MAX_I, MINT_I, SREG registers (except FLAGS, HISTORY0...3, UNIT_NUM, RX_CNT_HI, and TX_CNT_HI), and OUT_IF registers (except RX_CNT and TX_CNT) (see Section 14.2.1).

SRC, the source, can be an immediate 16-bits value, a register (or a byte in it) or an indirect addressing mode (type 1, 2, 4, or 6 in Table B14.1). Register or indirect addressing modes can be any of the registers (see Section 14.2.1).

Size can be either an immediate value (1 to 16) or a register content; that is, the SIZE_REG (which is SREG[6]) value (0–16). This corresponds to type 1 or 2 in Table B14.1.

Mov4Bits DST, SRC [,MODE]

Definition: Move four bits from source (SRC) to destination (DST). MODE indicates how to move the four bits. Moving several bits with the Mov4Bits command will OR them.

Operands: DST, the destination, can be four specific bits in a register (e.g., UREG[0].bits[1,3,5,9]) (type 5 in Table B14.1). Valid registers are UREG, ALU, MAX_I, MINT_I, SREG registers (except FLAGS, HISTORY0...3, UNIT_NUM, RX_CNT_HI, and TX_CNT_HI), and OUT_IF registers (except RX_CNT and TX_CNT) (see Section 14.2.1).

SRC, the source, can be an immediate 4-bit value, or a specific 4 bits in a register (type 1 or 5 in Table B14.1). Register can be any of the registers (see Section 14.2.1). An offset must be specified when the source is not immediate.

MODE indicates how to copy the bits from the source to the destination. It is immediate value, 1 byte of 0–255 in value. Each 2 bits indicate the moving mode for each bit from SRC: move bit (0), move inverse of bit (1), move "1" (2), or move "0" (3). The default MODE is 0x0 (bit string "00000000"). When SRC is immediate, MODE cannot be used.

Get DST, SRC, SIZE

Definition: Move SIZE bytes from RMEM (according to the SRC structure) to a destination (DST) register.

Operands: DST, the destination, can be a register, a specific byte in a register (e.g., UREG[0].byte[2]) or an indirect addressing mode (type 2, 3, or 6 in Table B14.1). Valid registers are UREG, ALU, MAX_I, MINT_I, SREG registers (except FLAGS, HISTORY0...3, UNIT_NUM, RX_CNT_HI, and TX_CNT_HI), and OUT_IF registers (except RX_CNT and TX_CNT) (see Section 14.2.1).

SRC, the source, can be an immediate 8-bit value, or a register, indicating the structure number in RMEM (type 15 in Table B14.1). Auto-increment is on the index only, not the base. Thus, additional bytes of the same result (structure) can be read from RMEM.

Size can be either an immediate value (1 to 4) or a register content; that is, the SIZE_REG (which is SREG[6]) value (0–4). This corresponds to type 1 or 2 in Table B14.1.

PutKey DST, SRC, SIZE

Definition: Move SIZE bytes from source (SRC) to OMEM (DST').

Operands: DST, the destination, can be a base-indexed addressing mode (type 7 in Table B14.1), referring to OMEM.

SRC, the source, can be an immediate 32-bit value, a register (or a byte in it) or an indirect addressing mode (type 1, 2, 3, or 6 in Table B14.1). Register or indirect addressing modes can be any of the registers (see Section 14.2.1).

Size can be either an immediate value (1 to 4) or a register content; that is, the SIZE_REG (which is SREG[6]) value (0–4). This corresponds to type 1 or 2 in Table B14.1.

Copy DST, SRC, SIZE

Definition: Move SIZE bytes from source (SRC, which is RMEM) to destination (DST, which is OMEM).

Operands: DST, the destination, can be a base-indexed addressing mode (type 7 in Table B14.1), referring to OMEM.

SRC, the source, can be an immediate 8-bit value, or a register, indicating the structure number in RMEM (type 15 in Table B14.1).

Size can be either an immediate value (1 to 8) or a register content; that is, the SIZE_REG (which is SREG[6]) value (0–8). This corresponds to type 1 or 2 in Table B14.1.

PutHdr DST, SRC, SIZE

Definition: Move SIZE bytes from source (SRC) to destination (DST, which is the header register, HREG). Move bytes from immediate value or register to header.

Operands: DST, the destination, can be the HREG register (or a byte in it) or an indirect address mode (type 2, 3, or 9 in Table B14.1). Access to HREG is either through an immediate index or indirectly via an HBSn register-like index; for example HREG[0] and HREG[HBS1].

SRC, the source, can be an immediate 24-bit value, a register (or a byte in it) or an indirect addressing mode (type 1, 2, 3, or 6 in Table B14.1). Register or indirect addressing modes can be any of the registers (see Section 14.2.1).

Size can be either an immediate value (1–3) or a register content; that is, the SIZE_REG (which is SREG[6]) value (0–3). This corresponds to type 1 or 2 in Table B14.1.

PutHdrBits DST, SRC, SIZE

Definition: Move SIZE bits from source (SRC) to destination (DST, which is the header register, HREG).

Operands: DST, the destination, can be the HREG register (or a specific bit in it) or a bit specific memory (HREG) addressing mode (type 2, 4, or 10 in Table B14.1). Access to HREG is either through an immediate index or indirectly via an HBSn register-like index; for example, HREG[0] and HREG[HBS1].

SRC, the source, can be an immediate 16-bit value, a register (or a specific bit in it) or an indirect addressing mode (type 1, 2, 4, or 6 in Table B14.1). Register or indirect addressing modes can be any of the registers (see Section 14.2.1).

Size can be either an immediate value (1 to 16) or a register content; that is, the SIZE_REG (which is SREG[6]) value (0–16). This corresponds to type 1 or 2 in Table B14.1.

PutHdr4Bits DST, SRC [,MODE]

Definition: Move four bits from source (SRC) to destination (DST). MODE defines the way these four bits are moved.

Operands: DST, the destination, can be four specific bits in memory HREG (type 5 or 11 in Table B14.1).

SRC, the source, can be an immediate 4-bit value, or a specific 4 bits in a register (type 1 or 5 in Table B14.1). The register can be any of the registers (see Section 14.2.1). An offset must be specified when the source is not immediate.

MODE indicates how to copy the bits from the source to the destination. It is an immediate value, 1 byte of 0–255 in value. Each 2 bits indicate the moving mode for each bit from SRC: move bit (0), move inverse of bit (1), move "1" (2), or move "0" (3). The default MODE is 0x0 (bit string "00000000"). When SRC is immediate, MODE cannot be used.

GetRndBits DST, SRC0, SRC1, SRC2, SRC3, SRC4, SRC5, SRC6, SRC7

Definition: Move eight random bits from the result control block (SRC0..7) to destination (DST). The term random refers to bits that are arbitrarily addressed and not necessarily adjacent (like in Random Access Memory); actually, they are programmable-selected. This instruction enables random bits from any two structure groups (each structure group contains up to eight structures) and writes them to the destination.

Operands: DST, the destination, can be either a register or a byte-specific register (type 2 or 3 in Table B14.1). Valid registers are UREG[0..7], and MINT_I.

SRC0..7, the sources, are result control bits, pointed by the structure group stored in ST_GRP0..1, which is SREG[8] and the index of the structure within the group (type 13 in Table B14.1).

GetFixBits DST, SRC0, SRC1, SRC2, SRC3, SRC4, SRC5, SRC6, SRC7

Definition: Move eight bits from the result control block (SRC0..7) to destination (DST). This instruction builds a byte from eight bits within any two offsets. If structure number does not exist, "0" will be written.

Operands: DST, the destination, can be either a register or a byte-specific register (type 2 or 3 in Table B14.1). Valid registers are UREG[0..7], and MINT_I.

SRC0..7, the sources, are result control bits, pointed by the structure number (type 14 in Table B14.1). In all SRCs there can be only two offsets.

GetRnd4Bits DST, SRC0 [,SRC1 [,SRC2 [,SRC3 [, SIZE]]]]

Definition: Move up to four bits (SIZE) from the result control block (SRC1..4) to destination (DST). All structure numbers and offsets (0..15) can be used. If a structure number does not exist, "0" will be written.

Operands: DST can be either a register or a bit-specific register (type 2 or 4 in Table B14.1). Valid registers are UREG, ALU, MAX_I, MINT_I, SREG registers (except FLAGS, HISTORY0..3, UNIT_NUM, RX_CNT_HI, and TX_CNT_HI), and OUT_IF registers (except RX_CNT and TX_CNT) (see Section 14.2.1).

SRC0..7, the sources, are result control bits, pointed by the structure number (type 14 in Table B14.1). In all SRCs there can be only two offsets. SIZE indicates how many bits to move, and contains immediate value (1–4).

JUMP COMMANDS

Jump commands instruct the NP to jump to a given label in the microcode. Some of the commands are conditional jumps, and some jump with pushing and popping program addresses. At the end of the jump, optional NOP_NUM (0, 1, or 2) NOPs may be inserted into the pipeline following the jump command (to disable execution of the commands that immediately follow the jump instruction). The NOPs are effective only if the jump command is executed. The standard Jump command format is:

```
if (CONDITION)
command LABEL [| NOP_NUM];
                        // See commands in Table 11.4, Chapter 11.
```

or

```
command LABEL [| NOP_NUM];
                        // See commands in Tables 11.4 and 11.5, Chapter 11.
```

The format of the Call and CallStack commands is a bit different from the jump command, and also includes a number of NOPs (RET_NOP_NUM) that should be inserted before returning, as follows:

```
Call function [|NOP_NUM [,RET_NOP_NUM]];
```

or
```
CallStack function [| NOP_NUM [, RET_NOP_NUM]];
```

Definition: Call a function, and return from it to the next address upon completion (return command in the function). Insert NOP_NUM NOPs before jumping and RET_NOP_NUM NOPs before returning.

Operands: CONDITION is any flag bit or UDB register bit. The negation mark (!) may precede any condition. For flags, see Figure A14.3.

LABELS is the place in the program where the jump is to take place.

function is the name of the function to be called.

NOP_NUM shows the number of NOPs to be inserted after the branch, to prevent the following command from entering the pipeline.

RET_NOP_NUM shows the number of NOPs to be inserted upon return from the called function.

ARITHMETIC LOGIC UNIT OPERATIONS

Arithmetic Logic Unit commands are used for arithmetic and logic calculations. There are two formats of ALU commands, one with two source-operands and the other with one source operand:

```
COMMAND DST,SRC1,SRC2,SIZE,[,MREG[,MODE]];
                        // See commands in Table 11.3, Chapter 11.
```

or
```
COMMAND DST,SRC,SIZE,[,MREG[,MODE]];
                        // See commands in Table 11.3, Chapter 11.
```

Definition: Calculate the COMMAND on the sources (SRC, or SRC1 and SCR2) of SIZE bytes, and put the result in the destination (DST). Use masks in MREG, according to the required MODE, to mask the source operands if required.

Operands: DST, the destination, receives the result of the ALU block operation, together with the ALU register itself; that is, the result is entered into both the destination and the ALU register. DST can be a register (or a byte in it) or an indirect addressing mode (type 2, 3, or 6 in Table B14.1), and may use any of the following registers: UREG, ALU, MINT_I, MAX_I, OUT_IF registers (except RX_CNT and TX_CNT) and SREG registers (except FLAGS, HISTORY0..3 UNIT_NUM, RX_CNT_HI, and TX_CNT_HI), see Section 14.2.1.

SRC or SRC1, the first source operand or the only source operand can be a register (or a byte in it) or an indirect addressing mode (type 2, 3, or 6 in Table B14.1). Register or indirect addressing modes can be any of the registers (see Section 14.2.1).

SRC2, the second source, can be an immediate 8-bit value, a register (or a byte in it) or an indirect addressing mode (type 1, 2, 3, or 6 in Table B14.1). Register or indirect addressing modes can be UREG, ALU, and SREG registers (except HISTORY0..3 and UNIT_NUM), see Section 14.2.1.

SIZE is the size of the ALU operation in bytes. Zero (0) is not a valid size. Size can be either an immediate value (1 to 4) or a register content; that is, or the SIZE_REG (which is SREG[6]) value (1 to 4). This corresponds to type 1 or 2 in Table B14.1.

MREG indicates one of the 16 ALU mask registers (see MREG[15:0] Register section in Appendix A).

MODE indicates how to use the mask:

_ALU_NONE—no masking done
_ALU_*FRST*—masking SRC1; that is, *SRC1 = SRC1 & MASKREG*
_ALU_*SCND*—masking SRC2; that is, SRC2 = *SRC2 & MASKREG*
_ALU_*BOTH*—masking both SRC1 and SRC2; that is, *SRC1 = SRC1 & MASKREG*, SRC2 = SRC2 & MASK_REG

The inverse can be performed on the logical functions "AND" and "XOR," by entering a negation mark (!) before the operand.

When the explicit destination is not the ALU register, all four bytes of the ALU register are updated, even if the instruction SIZE was less than 4. For example, "Add UREG[7], UREG[5], UREG[6], 1;" adds the contents of UREG[5] and UREG[6] and writes 1-byte to UREG[7] and 4-bytes to ALU. The ALU register cannot be used with an offset other than 0 (zero) when it is used for destination, source1 or source2. Examples of *improper* usage of the ALU register follow:

```
Add ALU.BYTE [1], UREG[0].BYTE[2], 2, 1 ;
Add ALU, UREG[0].BYTE[2], ALU.BYTE [2], 1;
Add ALU, ALU.BYTE [3], UREG[0].BYTE[2], 1;
```

MINTERM OPERATIONS

Minterm operations are a toolkit consisting of multiparameter bit-wise operations, all being ORs of groups of input bits ANDed together. The various operations are dictated by a "operation" operand. These operations are used to perform complex logical decisions in a single clock. The mnemonic *Minterm* originates from the name given to elements of Conjunctive Normal Form in switching theory. The format of the Minterm operation is:

```
Minterm DST, OPER, INVMASK, MASK4, CTRL, SIZE;
```

Definition: Calculate logical formulas, according to OPER. Results of the Minterm operation can be up to two bytes, sent to the destination (DST). The input is from the 32-bit MINT_I register.

Operands: DST, the destination, receives the result of the ALU operation, together with the ALU itself; that is, the result is entered into both the destination and the ALU. DST can be a register (or a byte in it) or an indirect addressing mode (type 2, 3, or 6 in Table B14.1), and may use any of the following registers: UREG, ALU, MINT_I, MAX_I, OUT_IF registers (except RX_CNT and TX_CNT) and SREG registers (except FLAGS, HISTORY0..3 UNIT_NUM, RX_CNT_HI, and TX_CNT_HI), see Section 14.2.1.

OPER indicates the number of the Minterm operation and is an immediate value (0–11).

INVMASK indicates whether the value itself or its inverse is taken. Logically this is XORNOT between the INVMASK and the source (MINT_I). The operand is an immediate 32-bit mask.

MASK4 is used for masking the operands, as shown in the operation tables. The operand is an immediate 4-bit mask.

CTRL indicates which byte of MINT_O, and is an immediate value (0 or 1); if size is 1, is output to DST.

SIZE indicates the size of the result, and can be either an immediate value (1 or 2) or a register content; that is, the SIZE_REG (which is SREG[6]) value (0–2). This corresponds to type 1 or 2 in Table B14.1.

Minterm operations are presented in Table C14.1 and Figures C14.1 and C14.2. Conventions used: AND equations are represented by AB; OR equations are represented by $A + B$. M is a bit representing the result of the comparison between 2 adjacent bits from MINT_I to the same bits in INVMASK. Mi corresponds to bits $2i$-1 and $2i$ in both MINT_I and INVMASK. $MASK_n$ corresponds to the nth bit of MASK4.

In Figures C14.1 and C14.2, M is a bit representing the result of the comparison between two adjacent bits from MINT_I to the same bits in INVMASK; that is, Mi corresponds to bits $2i$-1 and $2i$ in both MINT_I and INVMINT (see Table C14.1).

Table C14.1 Minterm Operations

Operation #0

OUT0 = MASK0 M1   M1 = MINT_I[1:0] compared to INVMASK[1:0]
OUT1 = MASK0 M2   M2 = MINT_I[3:2] compared to INVMASK[3:2]
OUT2 = MASK0 M3   M3 = MINT_I[5:4] compared to INVMASK[5:4]
OUT3 = MASK0 M4
OUT4 = MASK1 M5
OUT5 = MASK1 M6
. . .
OUT7 = MASK1 M8
OUT8 = MASK2 M9
. . .
OUT11 = MASK2 M12
OUT12 = MASK3 M13
. . .
OUT15 = MASK3 M16   M16 = MINT_I[31:30] compared to INVMASK[31:30]

Operation #1

OUT0 = MASK0 M1 + MASK0 M2
OUT1 = MASK0 M3 + MASK0 M4
OUT2 = MASK1 M5 + MASK1 M6
OUT3 = MASK1 M7 + MASK1 M8
OUT4 = MASK2 M9 + MASK2 M10
OUT5 = MASK2 M11 + MASK2 M12
OUT6 = MASK3 M13 + MASK3 M14
OUT7 = MASK3 M15 + MASK3 M16

Operation #2

OUT0 = MASK0 M1 + MASK0 M2 + MASK0 M3 + MASK0 M4
OUT1 = MASK1 M5 + MASK1 M6 + MASK1 M7 + MASK1 M8
OUT2 = MASK2 M9 + MASK2 M10 + MASK2 M11 + MASK2 M12
OUT3 = MASK3 M13 + MASK3 M14 + MASK3 M15 + MASK3 M16

Operation #3

OUT0 = (MASK0 M1 + MASK0 M2 + MASK0 M3 + MASK0 M4) + (MASK1 M5 + MASK1 M6 + MASK1 M7 + MASK1 M8)
OUT1 = (MASK2 M9 + MASK2 M10 + MASK2 M11 + MASK2 M12) + (MASK3 M13 + MASK3 M14 + MASK3 M15 + MASK3 M16)

Table C14.1 (*continued*)

Operation #4

OUT0 = MASK0 M1 + MASK0 M2 + MASK0 M3 + MASK0 M4 + MASK1 M5 + MASK1 M6 + MASK1 M7 + MASK1 M8 + MASK2 M9 + MASK2 M10 + MASK2 M11 + MASK2 M12 + MASK3 M13 + MASK3 M14 + MASK3 M15 + MASK3 M16

Operation #5

OUT0 = MASK0 M1 M2
OUT1 = MASK0 M3 M4
OUT2 = MASK1 M5 M6
OUT3 = MASK1 M7 M8
OUT4 = MASK2 M9 M10
OUT5 = MASK2 M11 M12
  . . .
OUT8 = MASK3 M15 M16

Operation #6

OUT0 = MASK0 M1 M2 + MASK0 M3 M4
OUT1 = MASK1 M5 M6 + MASK1 M7 M8
OUT2 = MASK2 M9 M10 + MASK2 M11 M12
OUT3 = MASK3 M13 M14 + MASK3 M15 M16

Operation #7

OUT0 = MASK0 M1 M2 + MASK0 M3 M4 + MASK1 M5 M6 + MASK1 M7 M8
OUT1 = MASK2 M9 M10 + MASK2 M11 M12 + MASK3 M13 M14 + MASK3 M15 M16

Operation #8

OUT0 = MASK0 M1 M2 + MASK0 M3 M4 + MASK1 M5 M6 + MASK1 M7 M8 + MASK2 M9 M10 + MASK2 M11 M12 + MASK3 M13 M14 + MASK3 M15 M16

Operation #9

OUT0 = MASK0 M1 M2 M3 M4
OUT1 = MASK1 M5 M6 M7 M8
OUT2 = MASK2 M9 M10 M11 M12
OUT3 = MASK3 M13 M14 M15 M16

Operation #10

OUT0 = MASK0 M1 M2 M3 M4 + MASK1 M5 M6 M7 M8
OUT1 = MASK2 M9 M10 M11 M12 + MASK3 M13 M14 M15 M16

Operation #11

OUT0 = MASK0 M1 M2 M3 M4 + MASK1 M5 M6 M7 M8 + MASK2 M9 M10 M11 M12 + MASK3 M13 M14 M15 M16

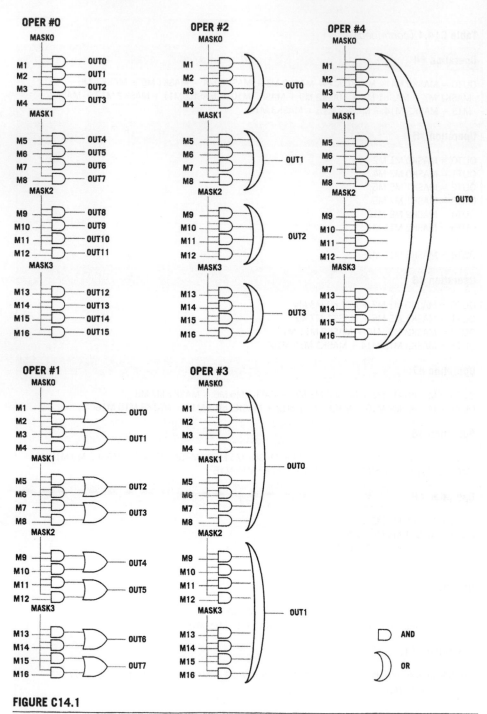

FIGURE C14.1

Chart illustrating minterm operations 0–4

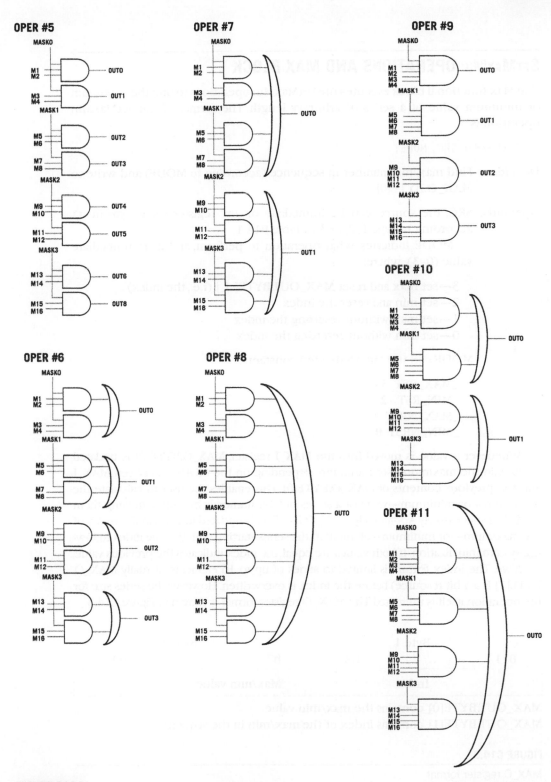

FIGURE C14.2

Chart illustrating minterm operations 5–11

SETMAXMIN OPERATIONS AND MAX BLOCK

The Max functional block executes the SetMaxMin operations to find the maximum or minimum value of a series of arbitrary length. The format of the SetMaxMin operation is:

```
SetMaxMin SRC, MODE;
```

Definition: Find max/min number in sequence (according to MODE) and write to MAX_O.BYTE[0].

Operands: SRC, the source, can be immediate or any register (or a byte in it), according to type 1, 2, or 3 in Table B14.1.

MODE indicates what operation to perform, and is an immediate value (0-3), where:

3—set max and reset MAX_OUT.BYTE[1] (i.e., the index)
2—set min and reset the index
1—set max without resetting the index
0—set min without resetting the index

MODE can use the predefined constants:

_MAX_RST 3
_MIN_RST 2
_MAX_NRST 1
_MIN_NRST 0

Whenever a value is moved into the MAX_I register, MAX_O.BYTE[0] is updated to contain the maximum—or minimum, depending on the MODE—between MAX_I and the previous contents of MAX_O.BYTE[0]. This enables the user to calculate the maximum—or minimum—value of a series of 8-bit numerals by simply moving them to MAX_I successively. Additionally, MAX_O.BYTE[1] is updated to contain the *index* of the maximum—or minimum—element in the series starting with 0 as the index following system initialization. If both values are equal, the index indicates the previous value.

Note: The index feature is limited to series of up to 16 elements (actually MAX_O. BYTE[1] is a 4-bit resource) before the index is overwritten. However, the series size for the operation itself is unlimited. The MAX_O register format is shown in Figure C14.3.

	Byte 1			Byte 0	
B11	...	b8	b7	...	b0
Index			Max/min value		

MAX_OUT.BYTE[0] contains the max/min value.
MAX_OUT.BYTE[1] contains index of the max/min in the sequence.

FIGURE C14.3

MAX_O register format

HALT COMMAND

The halt command is used either to finish the packet processing and release it to the next stage in the pipeline, or to multicast the frame several times, or to abort any further processing of this packet and discard it. It is the most useful command and the only one that is mandatory, and it has the following format:

```
HALT MODE;
```

Mode is a two-bit immediate value. Mode has the following predefined constants:

HALT_UNIC Indicates that the TOPresolve has completed processing this frame and the next TOP can get the data.

HALT_DISC Instruction to discard this frame and begin processing the next.

HALT_MULC This is not a true halt and in fact may often be followed by other instructions. HALT HALT_MULC notifies TOPsearch II that it can take these keys and wait for additional keys that TOPresolve is building for this frame. This could be used to duplicate a frame and send different keys with each copy. Once the last set of keys is ready for TOPsearch II, use HALT HALT_UNIC to terminate the processing of the current frame.

HALT_BRKP For debug mode. This freezes this specific TOPresolve engine until the host releases it.

Modifying

The last step in packet processing, as well as of our description of EZchip processing, is modifying: the phase in which actual modifications are made to the packets, determining the forwarding, queueing, and routing parameters, and releasing the processed packets to their destinations.

TOPmodify decodes modify instructions, as well as messages from previous stages and TOPsearch II results, and modifies the frame contents accordingly. An unlimited number of bytes anywhere in the frame and in any order can be overwritten or modified. Modified frames and added headers are written to the frame memory (if required; e.g., for backplane switching).

Three possible modification types are available according to TOPmodify microcode programming. Data within the frame may be:

- *Inserted*—'new data' is inserted to the frame; for example, VLAN or MPLS tags.
- *Cut*—'old data' is removed from the frame; for example VLAN or MPLS tags.
- *Overwritten*—current frame contents are overwritten; for example, MAC or IP address, TOS bits, or TTL (of course, information can also be altered for illegal or military purposes as well).

15.1 INTRODUCTION

In order to write microcode for TOPmodify effectively, programmers should understand how the frames are linked in the frame memory and how the first frame buffer is pre-fetched. On the ingress path, frames from the network are stored in the frame memory buffers by the Ingress Front DMA (the IFDMA, the internal block for frames arriving from external interface). Since most modifications to frames are done only in their headers (i.e., at their beginning), EZchip's NP uses a mechanism that allows such modifications easily, without the need to move and adjust the entire frame's contents. When storing in the first frame buffer, the IFDMA renders the first 64 bytes free and stores only the first 448 bytes of the frame. These first 64 bytes are reserved for modification information, and an additional frame header

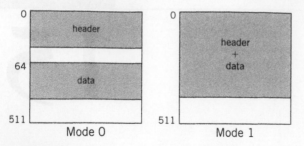

FIGURE 15.1

Two modes in which TOPmodify writes to the first memory buffer

can be added at the TOPmodify stage. The remainder of the frame is stored in a sequence of linked 512-byte buffers, using their entire space (i.e., without reserving 64 bytes).

TOPmodify interfaces with the frame memory and loads the 512-byte frame buffers into its FMEM memory. This process may be optimized to read only 64-byte chunks in which insert, cut or overwrite is necessary. The programmer may opt to calculate this optimal number at the TOPresolve stage and put it to the TOPresolve OUT_IF registers (i.e., CTRL_REG and FR_PTR). A zero value indicates that no changes are to be made to the first frame buffer, which is then not read. If changes are to be made in the second or third frame buffer (bytes 449 and on), the programmer must fetch these buffers through the Receive Frame Descriptor (RFD) interface (the RFD points to the frame buffers in the frame memory).

After modifying the frame, TOPmodify writes the first memory buffer of a frame to the frame memory in one of two modes (see Figure 15.1). In Mode 0, the frame header[1] is stored in the first 64 bytes (or less, as required), and data in bytes 64–511 (or less). This mode is used when TOPmodify is not required to read the frame from the frame memory. It is the faster method and is implemented in order to ease the load on the frame memory arbiter (which is the module that arbitrates between the many "customers" of I/O resources and bandwidth of the memory).

In Mode 1, TOPmodify writes the frame header and data contiguously. This mode can be used, for example, in a configuration in which two NP processors are interconnected via a crossbar and modifications are being performed on the egress path. In this case, TOPmodify does not read the frame on the ingress path. Buffer mode 1 must be used to add a protocol header (not a switch fabric one) to the frame when sending it to the Egress Transmit Frame Descriptor (ETFD, the internal block for a frame exiting from the NP)—that, is when sending it to the network.

---

[1]"Header" refers to the switch fabric header and EZheader that are appended to the first frame buffer.

As mentioned before, the frame buffers, other than the first buffer, contain 512 bytes. If data is to be inserted, then the frame buffer must be broken down into two frame buffers. The RFD table is updated to add the new buffer to the linked list.

Modification of frame buffers differs slightly for the first frame buffer. The first frame buffer actually has bytes 0–63 free (that is, it only contains 448 of the possible 512 bytes). Since most modifications are to a frame's header, this leaves enough space for insertion of the header's information. Existing data is shifted (by read/write operations) in 8-byte portions to make room for new data, which is also written to the frame in 8-byte portions.

15.2 INTERNAL ENGINE DIAGRAM

The block diagram in Figure 15.2 illustrates the internal blocks of a single TOPmodify processor; these blocks are described in the following subsections.

15.2.1 Instruction Memory

Instruction memory contains microcode instructions for TOPmodify engines.

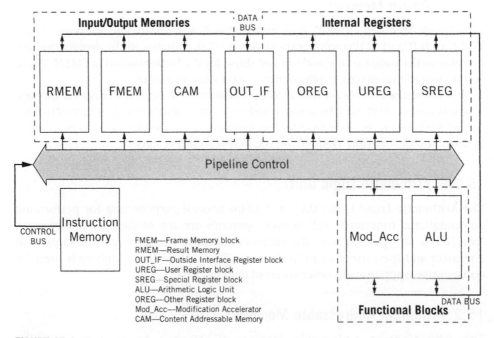

FIGURE 15.2

TOPmodify internal engine diagram

15.2.2 Pipeline Control Block

The pipeline control receives a command for decoding from the instruction memory and distributes control signals to each block.

15.2.3 Data Bus

The data bus supports a data flow between the TOPmodify blocks.

15.2.4 Frame Memory Block

Before starting work on a new frame, its first frame buffer (with 448 bytes) is pre-fetched from the frame memory, and loaded in the Frame Memory Block (FMEM). Modifications may then be made in any of the 512-byte portions of the frame. Although most frame modifications are performed on the header in the first buffer memory, the programmer can load an additional buffer to modify the contents of the payload. In order to fetch the next buffer, the next pointer is accessed from the linked list in the RFD table via the OUT_IF register. FMEM then uses this pointer to access the buffer via the frame memory arbiter. The next buffer is then loaded to FMEM via the OUT_IF register.

15.2.5 Result Memory

The Result Memory (RMEM) is a buffer that holds results and messages from TOPsearch II and TOPresolve respectively. RMEM contains all the results that are relevant to the entire frame, and not just those for the buffer loaded in FMEM. These results include messages, configuration, URL, VLAN, and so on.

For instance, a multicast that is to be transmitted to several VLANs via the destination output port may be transmitted only once across the system's switching fabric. Each frame is then modified individually at the egress side to match the appropriate destination VLAN (from the TOPsearch II result).

15.2.6 Arithmetic Logic Unit

The Arithmetic Logic Unit (ALU) is a 32-bit general-purpose unit for performing computational functions. ALU source operands are any of the TOPmodify registers. To use an operand from the memory, the programmer must first copy it to a register and then use it as an ALU source. The ALU feedback register is used for consecutive operations in order to avoid data hazards.

15.2.7 Content Addressable Memory

The 8-bit Content Addressable Memory (CAM) can be used for complex branches, verification of constants or comparison of configuration bits. Keys can be downloaded to the CAM during resource initialization (see `LDCAM BCAM8 [CAMgroup]`, `KEY`, `value` instruction in Section 15.5.1).

15.2.8 Register Blocks (User-Defined Register, Specific Register and Other Register)

The User-defined Register (UREG) is a block of user-defined, general purpose registers. The Specific Register (SREG) block contains special registers for pointers, indexes, and the loop counter, that are used by the TOPmodify. The Other Register (OREG) block of registers contains other registers for the ALU and CAM. The registers are described in Section 15.3, and some are detailed in Appendix A of this chapter.

15.2.9 Modification Accelerator Block

The Modification Accelerator block (MOD_ACC) is used to perform rapid multiple branches, in order to jump to various procedures that execute specific modifications to the frame. Up to 32 such modification procedures can be addressed and jumped to, according to a decision made in a previous TOP engine, such as TOPresolve.

A 32-bit map message from TOPresolve, indicating up to 32 possible modifications to be performed on the frame, is set and transferred to TOPmodify within one of the results from TOPsearch II. After writing this 32-bit map to the FAST_REG register, the MOD_ACC block enables rapid branching only to those modification procedures that need to be performed, according to the 32-bit map. Without the MOD_ACC block, each of the 32 bits would require an individual check. The 32 labels/addresses that point to the procedures in the microcode program, which correspond to the 32-bit map message and to the operations performed via the Modification Accelerator, are downloaded from the host (see `LDREG LREG[n]`,`label`) instruction, during resource initialization (see Section 15.5.1).

Essentially, the block performs an ELSE-IF branch in a single command, and this happens at any of the TOPmodify instructions that have the `jpe` operand (see below). The MOD_ACC examines the modify bits in the FAST_REG register, from the least-significant-bit (lsb) to the most significant bit (msb), until it reaches the first bit that is set, indicating the corresponding LREG register that contains the address of the procedure for this modification. The code then jumps to this corresponding location. After this modification is done, the bit is reset and the MOD_ACC block jumps according to the next set modify bit to perform modification, when and if the block is called again.

The MOD_ACC is triggered by the code; a `jpe` operand (which is a bit in the instruction op-code) indicates that there should be a jump to the procedure pointed to by the MOD_ACC block. This jump is done immediately after the execution of the instruction itself, and subject to the control hazard mechanism (i.e., the next two successive instructions begin execution before the first instruction of the procedure that is called for).

15.2.10 Outside Interface Block

The Outside Interface Block (OUT_IF) enables access to the RFD table, frame memory arbiter, host debug, and statistics. It also transfers data to the next stage.

Outside Interface Block interfaces with the frame memory arbiter to load additional buffers to FMEM. The programmer performs four functions on the RFD table: read, write, fetch free buffers, and recycle. If frames are discarded in TOPmodify, the RFD table is accessed via the OUT_IF register to recycle the pointers.

A function is sometimes halted while waiting for data from external blocks such as the frame memory arbiter or RFD table.

15.3 TOPMODIFY REGISTERS

As with other TOP engines, TOPmodify has two sets of registers: one that is accessible directly through the TOPmodify microcode, and the other, which must first be initialized by the host (through the host interface) prior to the TOPmodify execution, and which can only then be accessed by the TOPmodify microcode.

15.3.1 Microcode Registers

The tables below list the TOPmodify registers and structures accessible from the microcode. All registers can also be read by the host for debugging. Detailed descriptions of the registers follow in Appendix A of this chapter. The R/W column indicates whether the register is read, write or both. The Init column indicates the register's initial value. All registers can be written to in bits or bytes. All registers are accessed by the register name and the index in square brackets, for example, OREG[1], or by their defined name; for example, ALU (provided, of course, that they were defined as such in the source program, on in the libraries). The registers are intended for various and specific purposes and, consequently, have dedicated input sizes. For instance, attempting to write 8 bits into a 5-bit register will result in only the first five bits being written.

The UREG block is composed of 14 general-purpose registers, each of 32 bits, as shown in Table 15.1. Only the 32 bits of the UDB (UREG[0]) are zeroed with each new frame.

The OREG block is composed of 10 variable-bit registers, as shown in Table 15.2. The SREG block is composed of 14 variable-bits registers, as shown in Table 15.3. The Output Interface (OUT_IF) register block is composed of 14 variable-bits registers, as shown in Table 15.4.

Table 15.1 TOPmodify UREG Registers

Index	Register name	Size (bits)	Bytes (bits)	Description	R/W	Init
0	UDB	32 bits	4 (32)	User-defined bits. Each bit may serve as a condition in an IF statement.	R/W	zeroed for each frame
. . .						
13		32 bits	4 (32)	General-purpose register.	R/W	

Table 15.2 TOPmodify OREG Registers

Index	Register name	Size (bits)	Bytes (bits)	Description	R/W
0	FAST_REG	32 bits	4 (32)	Modification accelerator input.	R/W
1	ALU	32 bits	4 (32)	ALU feedback register. See Section 15.2.6.	R/W
2	CAMI	8 bits	2 (8)	CAM in register. See Section 15.2.7.	R/W
3	CAMO	16 bits	2 (16)	CAM out register. See Section 15.2.7.	R
4	HISTORY0 HISTORY1	2×10 bits	4 (10 + 10)	Instructions in pipeline.	R
5	HISTORY2 HISTORY3	2×10 bits	4 (10 + 10)	Instructions in pipeline.	R
6	BUF_STATUS0	4 + 16 bits	4 (4 + 16)	Buffer and queue status register.	R
7	PORT_SEL	5 bits	1 (4)	Port status.	R/W
8	UNIT_NUM	4 bits	1 (4)	Number of the TOP engine.	R
9	BUF_STATUS1	16 bits	2 (16)	Buffer and queue status register.	R

Table 15.3 TOPmodify SREG Registers

Index	Register name	Size (bits)	Bytes (bits)	Description	R/W
0	CNT	8 bits	1 (8)	See Loop commands.	R/W
0	RESERVED	8 bits	1 (8)	Reserved bits, set to 0.	

(continued)

Table 15.3 TOPmodify SREG Registers (*continued*)

Index	Register name	Size (bits)	Bytes (bits)	Description	R/W
0	PC_STACK	16 bits	2 (16)	For call and return in Jump commands.	R/W
1	FLAGS	18 bits	3 (18)	Flags. See Figure A15.1 in Appendix A.	R
2	SIZE_REG	5 bits	1 (5)	Indirect size register for some commands (Mov, ALU).	R/W
2	RESERVED	8 bits	1 (8)	Reserved bits, set to 0.	
2	DISP_REG	9 bits	1 (9)	Displacement register. Static offset in FMEM.	R/W
3	IND_REG0 IND_REG1	$2 \times (3 + 2 + 4)$	$4 (2 \times (3 + 2 + 4))$	For indirect access to all devices.	R/W
4	FR_PTR_IND0 FR_PTR_IND1	2×10 bits	$4 (10 + 10)$	Indirect index for reading/writing to FMEM.	R/W
5	FR_PTR_IND2 FR_PTR_IND3	2×10 bits	$4 (10 + 10)$	Indirect index for reading/writing to FMEM.	R/W
6	FR_PTR_IND4 FR_PTR_IND5	2×10 bits	$4 (10 + 10)$	Indirect index for reading/writing to FMEM.	R/W
7	FR_PTR_IND6 FR_PTR_IND7	2×10 bits	$4 (10 + 10)$	Indirect index for reading/writing to FMEM.	R/W
8	FR_PTR0 FR_PTR1	2×9 bits	$4 (9 + 9)$	Indirect base for reading/writing to FMEM.	R/W
9	FR_PTR2 FR_PTR3	2×9 bits	$4 (9 + 9)$	Indirect base for reading/writing to FMEM.	R/W
10	FR_PTR4 FR_PTR5	2×9 bits	$4 (9 + 9)$	Indirect base for reading/writing to FMEM.	R/W
11	FR_PTR6 FR_PTR7	2×9 bits	$4 (9 + 9)$	Indirect base for reading/writing to FMEM.	R/W
12	RD_PTR_RES0 RD_PTR_RES1 RD_PTR_RES2 RD_PTR_RES3	4×7 bits	$4 (7 + 7 + 7 + 7)$	Indirect index for reading from RMEM.	R/W
13	STR_NUM0 STR_NUM1 STR_NUM2 STR_NUM3	4×4 bits	$4 (4 + 4 + 4 + 4)$	Indirect base (structure number) for reading from RMEM.	R/W

Table 15.4 TOPmodify OUT_IF Registers

Index	Register name	Size (bits)	Bytes (bits)	Description	R/W
0	SEND_REG0	32 bits	4 (32)	Frame descriptors.	R/W
1	SEND_REG1	26 bits	4 (26)	Frame descriptors.	R/W
2	SEND_REG2	18 bits	4 (18)	Frame descriptors.	R/W
3	QUEUE_NUM	11 bits	4 (11)	Interface data.	R/W
4	FR_STRT	9 bits	2 (9)	Start offset of data in first frame buffer.	R/W
5	NXT_BUF_PTR	16 bits	4 (16)	Pointer for next buffer.	R/W
5	NXT_BUF_SZ	10 bits	4 (10)	Size of next buffer.	R/W
6	RFD_CMD	17 bits	4 (17)	RFD command.	R/W
7	RFD_RD	25 bits	4 (25)	Read RFD entry.	R
8	RFD_WR	25 bits	4 (25)	Write RFD entry.	R/W
9	RFD_RD_CNT	8 bits	1 (8)	RFD read data.	R/W
9	RFD_WR_CNT	8 bits	1 (8)	RFD write data.	R/W
9	RFD_PTR	16 bits	2 (16)	RFD pointer.	R/W
10	RX_PRE_PTR	16 bits	2 (16)	Rx pre-fetch pointer to free buffer	R
10	TX_PRE_PTR	16 bits	2 (16)	Tx pre-fetch pointer to free buffer	R
11	STAT_REG	32 bits	4 (32)	Interface to statistics block.	R/W
12	HOST_REG	32 bits	4 (32)	Interface with host register.	R/W
13	STAT_REG_HI	16 bits	2 (16)	Interface to statistics block.	R/W

15.3.2 Host Registers

All host registers are initialized by the host; however, the initial values for some are inserted into the NP microcode for loading. These initialization commands (see Section 15.5.1) are "executed" by the host prior to executing the program in the instruction memory.

Table 15.5 lists all of TOPmodify's host registers and indicates which registers have their initial values in the NP microcode.

Table 15.5 TOPmodify Host Registers

Address	Name	Description	Initialization
0x00	INT_REG	Interrupt register.	Host only
0x01–0x03	WIDE_LOAD	SRAM instruction register.	Host only
0x09–0x18	MREG[15:0]	ALU mask register.	Host only
0x04–0x5F	LREG[31:0]	Modification acceleration register.	Host only
0x60–0x62	DISCARD_CNT_NUM[2:0]	Counter number configuration register.	Host only
0x64	BR_ADDR	Service routine address register.	Host only
0x65	FLOW_CFG	Flow control configuration register.	Host only
0x66	Reserved	Reserved.	
0x67	HOST_CONF	Chip's mode of operation register.	Host only
0x80–0x85	HOST_REG[5:0]	Host debug register.	Microcode
0x88	MCODE_BR_INT	Microcode execution or break point command register.	Host only
0x89	STATUS_REG	TOPmodify status register.	Host only

Table 15.6 TOPmodify Memories and CAMs

Name	Description
RMEM	Result memory
BCAM8	Content access memory (a table for search)
FMEM	Frame memory

15.4 TOPMODIFY STRUCTURES

TOPresolve's microcode instructions access three memories, as described in Table 15.6; the RMEM (which contains data from TOPresolve and TOPsearch II), CAM memory for internal, fast searches, and the FMEM.

It is important to note that RMEM, the input memory, cannot be accessed as a continuous memory, like regular direct accessed memory, but can only be accessed using the structure number of the search operation (or the message). In other words, if a key or message is written to the input memory of TOPsearch II using some structure (as indicated in the key header in the HREG), the RMEM can be read by TOPmodify only by indicating this structure number as a base.

15.5 TOPMODIFY INSTRUCTION SET

This section examines each of the specific instructions in the TOPmodify microcode that are not common NP instructions described in Chapter 11. The microcode execution begins on the first instruction for each new frame.

Following the decisions made by TOPresolve regarding a frame, and the results provided by TOPsearch II, programmers can use the TOPmodify microcode to modify the frame as required, to transmit it, and even to implement multiple instances of the frame. Registers are used to update the RFD table (a linked list of pointers to the buffers in the frame memory), and to indicate the number of copies of the frame that will be sent. By using the Halt command, multiple copies of the frame are forwarded to different queues. The RFD pointers are recycled only after all copies of the frame have been sent.

Before sending a frame to a queue, programmers are responsible for filling in the SEND_REG register with the relevant frame information. Often this is obtained from TOPparse and TOPresolve via messages.

Programmers are also responsible for writing the modified frame buffers to the frame memory before sending the frame to a queue (see Section 15.5.4). Several commands in the TOPmodify instruction set have a Jpe operand for use with the MOD_ACC (see Section 15.2.9).

Most of the commands in the following subsections are described briefly, with examples. In most cases, this should be enough for understanding and writing complex code; however, the reader who needs deeper utilization of the commands is referred to Appendix C of this chapter.

15.5.1 Resource Initialization

Initialization commands are "executed" by the host rather than the TOPs (see Table 15.7); they indicate the initial values given to the mask registers and CAM prior to executing the program in the instruction memory. These are the commands inserted into the NP microcode for loading.

Table 15.7 Initialization Commands

Syntax	Description
LDREG LREG [n], label;	Load a value to the LREG register n – 0–31 label—label to load to this register
LDCAM BCAM8 [CAMgroup], KEY, value;	Load a KEY to the CAM. CAMgroup—group number (0, 1, 2, …7) KEY—number to load to this CAM value—16 bit value (may be address + NOPs)

15.5.2 **Move Commands**

Move commands copy data (bits, bytes, and words) from various sources to various destinations, as described in Chapter 11. Please note that all move commands have the `jpe` operand, to indicate an optional jump determined by the Modification accelerator (and the FAST_REG contents). The following are the five prefixes for move commands, depending on where the data is being moved (see Figure 15.3 and Table 15.8).

- Use Get or GetRes for moving data to registers from FMEM or RMEM.
- Use Mov for moving data within the registers.
- Use Put for moving data from registers to FMEM.
- Use Copy for moving data from RMEM to FMEM.
- Use Write for moving data within the FMEM.

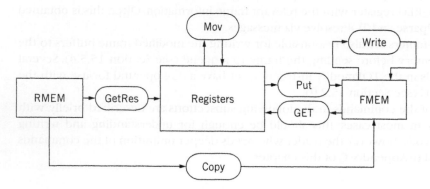

FIGURE 15.3

TOPmodify move commands name prefix mnemonic

Table 15.8 List of TOPmodify Move Commands

Name	Description
Mov	Move bytes from immediate value or register to register.
MovBits	Move bits from immediate value or register to register.
Get	Move bytes from frame memory to register.
GetRes	Move bytes from result memory to register.
Put	Move bytes from source register to frame memory.
Copy	Move bytes from memory to memory.
Write	Move bytes from frame memory to frame memory.

Please note that the Get command in TOPmodify reads data from the FMEM, and GetRes command is used for reading the RMEM. Both have similar format, as described in Chapter 11. The only additional command is Write:

```
Write DST, SRC, SIZE [,jpe]
```

Definition: Move SIZE data from the source (SRC, the frame memory FMEM) to the destination (DST, also the frame memory, FMEM). jpe is used to optionally jump according to the FAST_REG, after the move (see Section 15.2.9).

Examples:
```
Write 10(FR_PTR0), (FR_PTR1), 4;
Write FR_PTR_IND2(FR_PTR0), FR_PTR_IND0(FR_PTR2), 4;
```

Description: Write 4 bytes from frame memory (address from FR_PTR1) to the frame memory (address from FR_PTR0+10).

Write 4 bytes from frame memory (address from FR_PTR2 + value of FR_PTR_IND0) to the frame memory (address from FR_PTR0 + value of FR_PTR_IND2).

15.5.3 **LookCam DST, SRC, TYPE [, jpe]**

Definition: Search in BCAM8. Use source (SRC) for search key and put result in destination (DST), as well as in CAMO, which is OREG[3] register. jpe is used to optionally jump according to the FAST_REG, after the move (see Section 15.2.9).

Example: `LookCam UREG[0], UREG[1], BCAM8[1], _NJP_MDF;`

Description: Look in BCAM8, group 1 and search least significant byte from UREG[1]. If found, write result in UREG[0] and CAMO, and set MH flag (CAM match).

15.5.4 **Halt Command**

Frames can be sent to any combination of destinations by writing the descriptors to the SEND_REG register one at a time, and using the halt command to send each in turn. The halt command indicates type (e.g., unicast, multicast) and specific destination (e.g., Transmit Frame Descriptor [TFD] for network interface, Virtual Output Queue [VOQ] for switch interface, Host TFD [HTFD] for host interface, or none).

The pointer to the frame buffer is taken from the TX frame pointer in the SEND_REG register. The format of the Halt instruction is:

```
HALT DATA, MODE;
```

Definition: Halt is used to finish packet processing and to transmit it or to discard it. MODE determines whether this frame should be unicast, multicast, or discard, and DATA determines the frame's destination (FMEM and output queue).

Example: HALT _HALT_VQ_MDF, HALT_UNIC;

Description: Send the entire frame (all buffers) to the VOQ without writing the frame to the frame memory. The frame pointer must point to the first frame buffer.

Appendix C provides more details on the operands and the way halt works, as well as several sample scenarios.

15.5.5 **TOPmodify Conditional Commands**

Some of the instructions can be preceded by conditional statements that function as part of the instructions; that is, they are coded in one machine-instruction, and are executed at the same time as the unconditional statements. These TOPmodify commands are listed in Table 15.9.

Table 15.9 TOPmodify Conditional Commands	
Mov (only if source is not immediate)	Call
MovBits (only if source is not immediate)	Jmp
Get (only if source is not FMEM)	

15.6 **EXAMPLE**

This sample application continues the example of the TOPparse and TOPresolve chapters, for transmitting the ingress frames with a VLAN tag to the VOQ with the application code written for each NP TOP processor.

This sample focuses on the use of the MOD_ACC and the required steps to add fields (VLAN tag in this example) to the frame header. It is based on inspection by the hardware decoder as to whether each frame has a VLAN tag. This information is passed through a message to TOPmodify, where the VLAN tag is added if it did not already exist.

15.6.1 **Frame Handling Overview**

The NP operates on both ingress and egress paths. In this example, however, the NP processes frames on the ingress path; that is frames that arrive from the network and are transmitted to the switching fabric (see Chapter 12). The TOPparse receives ingress frames, and the NP's hardware decoder automatically inspects each incoming frame and provides useful information about it for microcode use prior to execution of the NP microcode.

The decoded information is passed by a message from TOPparse to TOPresolve, accompanied with a decision byte. TOPresolve forwards the message to TOPmodify and initializes the internal buffer of TOPmodify to be the first frame buffer. According to the message (decision byte), TOPmodify decides whether to add either a CSIX header alone or CSIX header plus a VLAN tag to the current frame header, prior its transmission to the VOQ in Mode 1.

15.6.2 Data Flow and TOP Microcode

The section that follows describes the data flow for TOPmodify, including the relevant microcode. Definitions and macro files for this sample application are located in Section 15.6.4.

TOPmodify is responsible for transmitting the ingress frames to the VOQ with a VLAN tag. TOPmodify checks whether the arriving frame contains a VLAN tag, in which case it does not add a VLAN tag field to the frame header.

The message from TOPresolve (via TOPsearch II) contains HD_REG0, HD_REG1, HD_REG2, and a decision byte. The HW decoder registers contain information about the frame that is important for its transmission. The decision byte acts as a flag to identify whether the current frame contains a VLAN tag field or not. If the original frame does not contain a VLAN tag, TOPmodify adds this field to the frame header. In order to forward the frames, TOPmodify updates the send registers (SEND_REG0, SEND_REG1, and SEND_REG2) and verifies that the FMEM (frame memory) is updated correctly. The code contains the following steps:

1. The FAST_REG register is reset to clear its contents and the new decision byte is extracted from the message and put into the FAST_REG.

2. The number of frame buffers and the source port are inserted into SEND_REG1. This data is found in bits 16–23 of HD_REG2 (sNumBuf). The operation is performed in two steps:
   - First, bits 16–31 of HD_REG2 are masked with M1_0x000000FF (due to the flag_ALU_FRST, where the first 8 bits of the mask are set) in order to take the first 8 bits (out of 16) as the first operand for the additional operation.
   - A value of 0 is added to the first operand and the result is inserted to SEND_REG1 bits 16–23.

3. The data interface register QUEUE_NUM is updated with the port number and the Quality of Service (QoS) setting (for each port, there are 8 different QoS settings). The value inserted to QUEUE_NUM is predefined in RX_OUT_DATA. An explanation of the RX_OUT_DATA structure follows:
   RX_OUT_DATA = 0x0201 = 0000 0010 0000 0001, where bits 0–6 define the port number (number 1) and bits 7–9 define the QoS (QoS = priority = 4).

4. The frames are stored in the FMEM (frame memory) in 512-byte buffers. Nevertheless, the first frame buffer contains only 448 bytes. The reserved 64 bytes of this buffer are used to add headers to the frame.

5. According to the _JPE_MDF flag that is appended to the command, the MOD_ACC examines the FAST_REG bits until it reaches the first bit that is set. The program jumps to the corresponding label (ADD_CSIX_HEADER) and the FAST_REG bit is reset. Please note that the two instructions that follow the jump command are also performed (this is an efficient use of the control hazards).

6. If the original frame did not contain a VLAN tag field, the structure of the internal frame buffer before modification is:

Data	Not defined	DA	SA	Other fields (may be referred to as the data field)
Byte offset	0–63	64–69	70–75	76–511
Field size (bytes)	64	6	6	Max Size 436 (448–12)

To send the frame to the VOQ with a VLAN tag, an 8-byte CSIX header must be added as the prefix of the frame, and a 4-byte VLAN tag field must be inserted between the SA field and the data field. This is implemented by:

- Pushing the DA field 4 bytes forwards.
- Pushing the SA field 4 bytes forwards. Thus, the structure of the internal frame buffer is:

Data	Not defined	DA	SA	Not defined	Other fields (may be referred to as the data field)
Byte offset	0–59	60–65	66–71	72–75	76–511
Field size (bytes)	60	6	6	4	Max 436 (448–12)

- Insert the VLAN tag field into the "hole" between the SA and data fields. The VLAN tag field is comprised of two fields; each with two bytes. The first field is Tag Type and its value is 0x8100. The second field is Tag Control and its value is 0xAAAA. Thus, the structure of the internal frame buffer is:

Data	Not defined	DA	SA	TAG type	TAG control	Other fields (may be referred to as the data field)
Byte offset	0–59	60–65	66–71	72–73	74–75	76–511
Field size (bytes)	60	6	6	2	2	Max 436 (448–12)
Notes				Always: 0x8100	0xAAAA	

7. The 8-byte CSIX header is appended to the beginning of the frame header. Thus, the structure of the internal frame buffer is:

Data	Not defined	CSIX header	DA	SA	TAG type	TAG control	Other fields (may be referred to as the data field)
Byte offset	0–51	52–59	60–65	66–71	72–73	74–75	76–511
Field size (bytes)	52	8	6	6	2	2	Max 436 (448–12)
Notes					Always: 0x8100	0xAAAA	

15.6.3 **TOPmodify Microcode**

The following code implements the TOPmodify stage (please note that it contains just 36 instructions; all the rest are comments):

```
EZtop Modify;

#include "mcglobal.h"      // Global definition file supplied with EZdesign,
                           // with predefined constants for NOPs, flags, etc.
#include "TOPModify.h"     // TOPmodify definition file supplied, provides
                           // recognizable names for registers and flags.
#include "hdreg.h"         // Global definition file supplied to provide
                           // recognizable names to the HW decoding registers

// Constants:
#define CSIX_HEADER_LEN 8;
#define CSIX_HEADER_LEN_MINUS1 CSIX_HEADER_LEN -1;
// for Obase usage
#define VLAN_TAG_LEN 4;
#define CSIX_BEGIN_NO_VLAN 64 - (CSIX_HEADER_LEN +VLAN_TAG_LEN);
                              // 52
#define CSIX_BEGIN_VLAN 64 - CSIX_HEADER_LEN;                    // 56
#define FRAME_BEGIN_NO_VLAN 64 - VLAN_TAG_LEN;                   // 60
#define MAX_1BUF_SIZE 448; // CSIX + 4bytes, moved from frame beginning
#define FIRST_DATA CSIX_HEADER_LEN + VLAN_TAG_LEN;
#define FIRST_DATA_MINUS1 FIRST_DATA - 1;
#define MASK_FIRST8BITS MREG[0];                    // after LDREG

#define RX_OUT_DATA            0x0201;
#define CSIX_REG1_BEGIN        0x0084;   // bits 21 - 23 = Qos = 2
#define CSIX_REG1_END          0x0040;   // bits 21 - 23 = Qos = 2
#define CSIX_REG2_BEGIN        0x0;
#define CSIX_REG2_END          0x0;
#define HW_STRUCT              0;
```

```
// Offsets in message
#define  HD_REG1_OFF              5;
#define  FR_PRT_OFF               7;
#define  HD_REG2_OFF              9;
#define  HW_SRC_CHANNEL_OFF      12;
#define  VLAN_TAG_EXISTS_OFF     13;


// Registers
#define  HD_REG2                 UREG[2];
#define  sFrameLen               HD_REG2.BIT[0];
#define  sNumBuf                 HD_REG2.BIT[16];
#define  uqMax1BufSize           UREG[0];


LDREG MREG[0], 0x00FF; // mask for first 8 bits - 1111 1111 = 0x00FF
// Set the label ADD_CSIX_HEADER to be the corresponding label for the
// first bit of the FAST_REG register.
LDREG LREG[0], ADD_CSIX_HEADER;


START:  // Beginning of microcode sequence
////////////////////////Step 1: Set FAST_REG////////////////////////////////
// Zero all four bytes of FAST_REG. Perform Xor of the ALU
// with itself and insert the result to the FAST_REG register.
   Xor       FAST_REG, ALU, ALU, 4;
// Extract byte 13 of the message received from TOPresolve
// and insert into FAST_REG.
   GetRes   FAST_REG, VLAN_TAG_EXISTS_OFF(HW_STRUCT), 1;

//////////////////////// Fill SEND_REGs ///////////////////////////////////////
// Init local register (HD_REG2) with the value of the HW decoder
// register HD_REG2 (passed via a message from TOPresolve).
   GetRes   HD_REG2, HD_REG2_OFF(HW_STRUCT), 4;
// Insert HD_REG1 (offset 5 in message structure) to
// SEND_REG0 (time stamp + frame pointer).
   GetRes   sSend_uxTime stamp, HD_REG1_OFF(HW_STRUCT), 4;

// Step 2: Insert NumBuf (5 bits) + SrcPort (3 bits) from HD_ REG2 b23:16
// (sNumBuff) to SEND_REG1 b25:16. (Note, the SrcPort field in the source
// (HD_REG2) is 3 bits and in the destination (SEND_REG1) it is 5 bits.)
   Add     sSend_bitBuffNum, sNumBuf, 0, 2, MASK_FIRST8BITS, _ALU_FRST;
   MovBits QUEUE_NUM, RX_OUT_DATA, 11;  // Step 3: port out + QOS

// Initialize DISP_REG. Fixed offset for writing to the internal buffer.
   MovBits DISP_REG, 0, 9;

// Step 4: Initialize register (uqMax1BufSize) with the value 448, the maximum
// first frame buffer length.
   MovBits uqMax1BufSize, MAX_1BUF_SIZE, 16;

// Initialize the first 8 bits of SEND_REG2: header length = 8 (7 LSB);
// buffer type = 0 (bit 8)
   MovBits sSend_bitHdrLen, CSIX_HEADER_LEN, 8;
```

```
// Insert HD_REG2 b13:0 to SEND_REG1 b13:0 for 14 bits of frame length.
   MovBits  sSend_uxFrameLen, sFrameLen, 14;
```

```
// Subtract 448 from the frame length (first 2 bytes of HD_REG2) to
// calculate whether the frame occupies more than one buffer. Insert the
// result into the ALU (and check the carry of the operation in the next line).

   Sub    ALU, sHR2_uxFrameLen, uqMax1BufSize, 2;

  Nop; // One NOP needed between the ALU operation and the carry
       // flag check.
```

```
// If (!FLAGS.BIT [F_CY_MDF]) // Frame occupies more than one buffer.
// SEND_REG2 b16:8 , 9 bits of header length with value 448 + 1
// (UqMax1BufSize in 0base is 449)
      MovBits  sSend_bit1stBufLen, uqMax1BufSize, 10;
```

```
// Entire frame is in the first frame buffer (i.e., frame less than 448 bytes).

   If     (FLAGS.BIT [ F_CY_MDF])
// Initialize SEND_REG2 b16:8 with 9 bits of header length extracted
// from the first 2 bytes of HD_REG2 (frame length field). The tenth
// bit is inserted to SEND_REG2 bit 17, multicast = 0. (If the frame
// length is shorter than 512 bytes, then bits 9-15 in the HD_REG2
// frame length field are reset and bit 10 can be used to reset the
// SEND_REG2 multicast bit.
     MovBits  sSend_bit1stBufLen, sHR2_uxFrameLen, 10;
     Nop;      // Avoid data hazard accessing sSend_bit1stBufLen.
```

```
// Add the length of the CSIX header to the first buffer length in 0base
// (7 bytes).
    Add sSend_bit1stBufLen, sSend_bit1stBufLen,
    CSIX_HEADER_ LEN_MINUS1, 2;
    MovBits sSend_bitBufType, 1, 1;  // SEND_REG2 bit 7, buffer type 1.
```

```
// Step 5: If VLAN TAG already exists, jump to ADD_CSIX_HEADER
// Initialize FR_PTR2 with a value of 56, and jump according to the
// first bit set in the FAST_REG.
   MovBits  FR_PTR2, CSIX_BEGIN_VLAN, 16, _JPE_MDF;
```

```
// The next two instructions are always executed.
// Initialize FR_PTR3 with the value of 52.
   MovBits  FR_PTR3, FRAME_BEGIN_NO_VLAN, 16;
// Initialize FR_STRT with the value of 56.
   MovBits  FR_STRT, CSIX_BEGIN_VLAN, 9;
```

```
///////////////////////////////////////////////////////////
// Step 6: The structure of an old Ethernet frame without tag:
//   DA(6 bytes) SA(6 bytes) ProtocolType(2 bytes) Data...
// The structure of an old Ethernet frame with tag:
//   DA(6 bytes) SA(6 bytes) TPID(2 bytes) TCI(2 bytes)
//                      ProtocolType(2 bytes) Data...
// Now insert tag - 4 bytes
```

```
    Write 0(FR_PTR3)+, 4(FR_PTR3), 6; // Move 6 bytes forward 4 bytes.
    Write 0(FR_PTR3)+, 4(FR_PTR3), 6; // Move next 6 bytes forward 4 bytes

// Change frame data - add Tag Type field (2 bytes).
    Put    0(FR_PTR3)+, 0x0081, 2;
// Change frame data - add Tag Control field (2 bytes).
    Put    0(FR_PTR3), 0xAAAA, 2;
    MovBits FR_PTR2, CSIX_BEGIN_NO_VLAN, 16;  // offset 52.
    MovBits FR_STRT, CSIX_BEGIN_NO_VLAN, 16;  // offset 52.

// Update the first buffer length because of the
// addition of the four bytes of the VLAN TAG.
    Add  sSend_bit1stBufLen, sSend_bit1stBufLen, VLAN_TAG_LEN, 2;
    Add  sSend_uxFrameLen, sHR2_uxFrameLen, VLAN_TAG_LEN, 2; //Update
// frame length

ADD_CSIX_HEADER:
/////////////////////////////////////////////////////////////////
// Step 7: Append 8-byte CSIX header to beginning of frame header:
    Put    0(FR_PTR2)+, CSIX_REG1_BEGIN, 2;
    Put    0(FR_PTR2)+, CSIX_REG1_END, 2;
    Put    0(FR_PTR2)+, CSIX_REG2_BEGIN, 2;
    Put    0(FR_PTR2)+, CSIX_REG2_END, 1;
    Copy   0(FR_PTR2)+, HW_SRC_CHANNEL_OFF(HW_STRUCT), 1;

/////////////////////////////////////////////////////////////////
VOQ_HALT:
// Write data from the internal buffer to FMEM and send the
// frame to the VOQ according to the parameters in the SEND_REGs and QUEUE_NUM.
    Halt  _WHALT_VQ_MDF, HALT_UNIC;
```

15.6.4 Definition Files

There are three header files used in this example: the microcode global definition file (mcglobal.h), the hardware decoding definition file (hdreg.h), and the TOPmodify definition file (TOPmodify.h).

The microcode global definition file (mcglobal.h) is described in Chapter 12, but is used here with another definition:

```
#ifndef _mcglobal_h_                              ;
#define _mcglobal_h_                              ;
...
// ALU commands
...
#define _ALU_FRST      1                          ;
...
// HALT type defines
#define HALT_UNIC      0                          ; // unicast
#endif /* _mcglobal_h_ */                         ;
```

The hardware decoding definition file (hdreg.h) is also described in Chapter 12, but is also used here with another definition:

```
#ifndef _hdreg_h                                    ;
#define _hdreg_h                                    ;
...
#ifndef sHR2                                        ;
#define sHR2                    HD_REG2             ;
#endif                                              ;
#define sHR2_uxFrameLen sHR2.BIT [0];// 16
...
#endif                                              ;
```

Last, the TOPmodify definition file (TOPmodify.h) is as follows:

```
#ifndef _TOPmodify_h_                               ;
#define _TOPmodify_h_                               ;
...
// OREG
#define FAST_REG                OREG [0]           ;
#define ALU                     OREG [1]           ;
...
// SREG
...
#define FLAGS                   SREG [1]           ;
...
#define DISP_REG                SREG [2].BYTE [2]  ;
...
#define FR_PTR2                 SREG [9]           ;
#define FR_PTR3                 SREG [9].BYTE [2]  ;
...
// OUT_IF
#define SEND_REG0               OUT_IF [0]         ;
#define SEND_REG1               OUT_IF [1]         ;
...
#define QUEUE_NUM               OUT_IF [3]         ;
#define FR_STRT                 OUT_IF [4]         ;
...
// Flag defines
...
#define F_CY_MDF                1                  ;
...
// Go to next procession
#define _JPE_MDF                1                  ;
...
// HALT data define
...
#define _WHALT_VQ_MDF           3
...
```

```
/*****************************************************/
#define sSend_uxTimestamp       SEND_REG0              ;
#define sSend_uxFramePtr        SEND_REG0.BYTE [2]     ;
#define sSend_uxFrameLen        SEND_REG1.BYTE [0]     ;
#define sSend_bitFrameDrp       SEND_REG1.BIT [15]     ;
#define sSend_bitBuffNum        SEND_REG1.BIT [16]     ;
#define sSend_bitSrcPort        SEND_REG1.BIT [21]     ;
#define sSend_bitHdrLen         SEND_REG2.BIT [0]      ;
#define sSend_bitBufType        SEND_REG2.BIT [7]      ;
#define sSend_bit1stBufLen      SEND_REG2.BIT [8]      ;
...
#endif /* _TOPmodify_h_ */                             ;
```

15.6.5 Structures and Message Formats

Several data structures and formats are used to pass the information, keys and results between the previous pipeline engines and the TOPmodify. These structures are described in this subsection.

15.6.5.1 TOPsearch II Structures

Table 15.10 describes the TOPsearch II data structures (also described in Chapter 14). The message from TOPresolve to TOPmodify is placed in structure number 0 on the ingress path.

A message header (or key header) must accompany each message (or key) passed to TOPsearch II.

15.6.5.2 TOPresolve–TOPmodify Message Format

In our example, the message from TOPresolve to TOPmodify, via TOPsearch II, is identical to the TOPparse to TOPresolve message described in Chapters 12 and 14, and outlined in Table 15.11. This 14-byte message (padded to 16 bytes) is placed in TOPsearch II structure #0. This message contains the hardware decoded information, including whether the original frame contains a VLAN tag field.

Table 15.10 TOPsearch II Structures

Name	Structure type	Structure number	Path	Used for	Key size (bytes)	Result size (bytes)
TOPresolve-TOPmodify message	(Message)	0	Ingress	L2 + L3 VLAN tag flag	–	–

Table 15.11 TOPresolve–TOPmodify Message

Field name	Byte offset	Size (bits)	Note
Valid	0	1	Always 1 for messages
To host	0	1	1—send to host
TTL_EXP	0	1	From HD_REG0 b28
Ctrl reserved bits	0	5	
HD_REG0	1–4	32	HW decoding
HD_REG1	5–8	32	HW decoding
HD_REG2	9–12	32	HW decoding
vlanTagExists	13	1	Indicates whether the arriving frame contained a VLAN tag field

15.7 SUMMARY

The TOPmodify engine is the last engine in the NP pipeline. It modifies the frame by either cutting it, adding to it, or altering its header fields or contents. The internal structure of TOPmodify is described in this chapter, including its registers and data structures, as well as its blocks (functional units) and instruction set. The appendices of this chapter provide more details about the registers and the instruction sets for those who need them either to understand a code or to write one.

In the next chapters we put everything together and show how to create the first program with network processors, using EZchip NP's coding, simulation and debugging tools.

APPENDIX A
DETAILED REGISTER DESCRIPTION

A detailed description of all TOPmodify devices and registers is given in this appendix. Though some of the information here duplicates that which can be found in the body of the chapter, the details and descriptions are fleshed out considerably.

The registers below, their bit functions and their overall roles are those of NP-1 and NP-1c. NP-2 and above NPs are slightly different, as more registers and bits are available to reflect more functions, ports, and functional blocks that are not described in this book.

MICROCODE REGISTERS

The microcode registers are accessed by the microcode for normal operations during program execution and include user-defined, functional, specific, and output interface registers.

User-Defined Registers

Register name	Description
UDB	32 user-defined bits that can be used for conditions (branches).

Other Registers

Register name	Description
FAST_REG	32-bit input for accelerating modifications, fixed addresses. The 32 labels/addresses corresponding to the 32 operations performed via the Modification Accelerator are downloaded from the host.
ALU	32-bit ALU feedback register with the last result of the ALU calculation. Used to perform adjacent ALU calculations and avoid a data hazard.
CAMO	16-bit CAM output register.
CAMI	8-bit CAM input register.
HISTORY	Four 10-bit registers specifying the four instructions currently in the pipeline. HISTORY0 Byte 0–1: Fetch stage HISTORY1 Byte 2–3: Decode stage HISTORY2 Byte 0–1: Execution 1 stage HISTORY3 Byte 2–3: Execution 2 stage

Register name	Description
BUF_STATUS0	20-bit register (4 bits + 16 bits) specifying buffer and queue status. bit 0—Tx_RFD_host_budget_expired bit 1—Rx_RFD_host_budget_expired bit 2—Tx_RFD_empty bit 3—Rx_RFD_empty bits 16:23—ETFD_modify_budget_expired[7:0] (per port). (The budget itself is specified in TX_PORT_BUDGET_LIMIT register in the ETFD block.) bits 24:31—Rx_RFD_modify_budget_expired[7:0] (per port) See BUF_STATUS1 below.
PORT_SEL	5-bit register. When inputting the register port number, flags are returned to indicate the Receive RFD port status. Msb 0 – Rx; 1 – Tx. When the chip is set to operate in NP-1 mode (in the HOST_CONF register, see Section 15.3.2), then the MSB is bit 3. When in NP-1c mode, the MSB is bit 4.
UNIT_NUM	4-bit register that numbers each of the TOPmodify engines. This enables a specific TOPmodify engine to be referenced directly in the code.
BUF_STATUS1	16-bit register specifying buffer and queue status. bits 7:0—ETFD_MODIFY_BUDGET_EXPIRED[15:8] (per port). (The budget itself is specified in the TX_PORT_BUDGET_LIMIT register in the ETFD block.) bits 15:8—RX_RFD_MODIFY_BUDGET_EXPIRED[15:8] (per port).

Specific Registers

Register name	Description
CNT	8-bit register to support loops. The program branches on the counter and checks its value. If the counter value is not zero, then it jumps to the start of the loop. Each branch on the counter value automatically decrements the counter by one. That is, the branch counts down each time that the loop is taken. Up to 256 repetitions of a particular sequence of code are supported. The depth of the Counter register is one, thus the counter can write/read to a register to implement nested loops.
PC_STACK	16-bit register for call commands (i.e., branch + set bit to push to stack). For each call command, the content (branch address +3 minus #NOPs) is automatically written to PC_STACK. The PC_STACK has a depth of one; thus, the PC_STACK can be written/read by the user in order to build nested call commands. The call command contains the number of NOPs that follow the relevant return command.
FLAGS	18-bit flags register. The defined bits and bit fields within the FLAGS register control specific operations and indicate the status of the network processor (see Figure A15.1). ZR—zero flag CY—carry flag SN—sign flag OV—overflow flag LP—counter register (0=CNT register is not zero; 1=CNT is zero) LP—counter register (0=CNT register is not zero; 1=CNT is zero) ST—statistics register (0=interface ready; 1=not ready) HT—host register (0=interface ready; 1=not ready) MH—last search in CAM flag (0=no match; 1=match)

(*continued*)

Register name	Description
	RAF—RFD address is ready flag (0=interface ready; 1=not ready) RDF—RFD data ready flag (0=interface ready; 1=not ready) RRP—RFD RX pre-fetch flag (0=interface ready; 1=not ready) RTP—RFD TX pre-fetch flag (0=interface ready; 1=not ready) RNB—RFD_EMPTY flag (0=free buffers available; 1=no free buffers) RBB—RFD_PORT_BUDGET_EXPIRED flag (0=not exceeded; 1=budget exceeded) RBE—RFD_EXPIRED_COMBINED flag (RNB I RBB) HBE—RFD_HOST_BUDGET_EXPIRED flag (0=not exceeded; 1=host budget exceeded) NBV—NEXT_BUFFER_VALID_FLAG (1=not ready; 0= next buffer has arrived) *NOTE: In microcode, the flags are referred to as FLAGS.BIT [F_x_MDF], where x is the flag name from Figure A15.1; for example, FLAGs.BIT [F_ZR_MDF].*
SIZE_REG	5-bit register for an indirect size. The size may be specified in either bits or bytes, depending on the instruction.
DISP_REG	9-bit register for a static offset in FMEM, used for read/write from/to the frame memory. Unlike the base, it is not updated by the auto-increment function. Real offsets = base + index + DISP_REG.
IND_REG0	9-bits containing the indirect address for a Dword, byte, and bit. Source and destination addresses are written in the format: device#, Dword#, byte#, and bit#. The device number is always specified as immediate, whereas the other three (Dword#, byte#, and bit#) can be addressed either immediately or indirectly using the IND_REG resister. Writing to this register is in the format shown in Figure A15.2.
IND_REG1	Identical to OFFS_REG0 (see Figure A15.3).
PRT_IND	Indirect indexes for reading/writing to FMEM. Each access to the FMEM is in the format: base address + index. The indexes are 10 bits in the signed format (2's-compliment).
PTR	Indirect bases for reading/writing to FMEM. Each access to the FMEM is in the format: base address + index. The base address for the FMEM is always indirect, that is, cannot be immediate. Base values are 9 bits (absolute FMEM address in unsigned format).
RD_PTR_RES	7-bit RMEM offset specifying the result (up to 128 bytes). Four RD_PTR_RES registers enable TOPmodify to work on four results simultaneously. Auto increment (offset + size) updates this register to read additional bytes from the same result in the RMEM.
STR_NUM	4-bit RMEM base address. Each access to RMEM is in the format: base address + index. Four STR_IND registers enable TOPmodify to work on four results simultaneously.

17	16	15	14	13	12	11	10	9	8	7	6	5	4	3	2	1	0
NBV	HBE	RBE	RBB	RNB	RTP	RRP	RDF	RAF	MH	HT	ST	LP		OV	SN	CY	ZR

FIGURE A15.1

TOPmodify flags

byte 1								byte 0							
7	6	5	4	3	2	1	0	7	6	5	4	3	2	1	0
				Dword#								byte#		bit#	

FIGURE A15.2

IND_REG0 format

byte 3								byte 2							
7	6	5	4	3	2	1	0	7	6	5	4	3	2	1	0
				Dword#								byte#		bit#	

FIGURE A15.3

IND_REG1 format

Outside Interfaces

The OUT_IF registers interface to external blocks.

Register name	Description
SEND_REG	Three registers (32 + 26 + 18 bits) that contain descriptors of the frame to be sent to the network (TFD), switching fabric (VOQ), and/or host (HTFD). Frames can be sent to any combination of these destinations by writing the descriptors to this register one at a time and using the halt command to send each in turn. Refer to the halt command for the specific destination. The halt command indicates the type (e.g. normal, multicast) and data (type of data being forwarded; that is, TFD, VOQ, HTFD, none).
	Programmers are responsible for filling in these fields according to the relevant semantics. Bits differ slightly depending on the destination of the data (see Table A15.1).
	Time stamp—received in a message originating from TOPparse.
	TX frame pointer—this is a pointer to the buffer that will be written or sent when using the halt command (see Section 15.5.4).
	Frame length—When sending the frame to a VOQ, the frame length includes the data length without the switch fabric header length.
	Number of buffers—identical to TOPparse HD_REG unless buffers were added by TOPmodify.
	First buffer length—Size of the data to be written by TOPmodify to memory.
	If buffer type =0, the first buffer length is equal to the header length. When sending short frames (less than 65 bytes) using buffer type =0 with no header (i.e., header length =0), set the first buffer length to 64 bytes.
	If buffer type =1, the first buffer length is equal to switch fabric header length plus length of the data in the first buffer. NOTE: Keep in mind that this field is zero-based. header length—Only valid when sending to the switch fabric. Includes the switch fabric header length.
	Buffer type—mode 0 (0) or mode 1 (1). See Section 15.1. *NOTE*: Buffer mode 1 must be used to add a header to the frame when sending it to the ETFD.
	Multiple—notifies the hardware to examine the MULTICAST_CNT field in the RFD table; otherwise, it is ignored and multiple copies of the frame will not be transmitted.

(continued)

Register name	Description
QUEUE_NUM	11 bits containing interface data such as priority settings or queue number for the frame descriptors that are sent to TFD, VOQ or HTFD. Programmers must write here each time that a frame is sent to a queue. TFD – For 8 ports/8 priorities (priority 8–10, port 0–2): 	10	9	8	7	6	5	4	3	2	1	0
---	---	---	---	---	---	---	---	---	---	---
priority			N/A					port		| For 16 ports/4 priorities (priority 9–10, port 0–3): | 10	9	8	7	6	5	4	3	2	1	0
---	---	---	---	---	---	---	---	---	---	---
priority		N/A						port		| VOQ – 10 bits (priority 8–10, ports 0–6): | 10	9	8	7	6	5	4	3	2	1	0
---	---	---	---	---	---	---	---	---	---	---
priority			N/A				port			| HTFD – 1 bit (priority 8, other bits are not valid): | 10	9	8	7	6	5	4	3	2	1	0
---	---	---	---	---	---	---	---	---	---	---
N/A		pri	port								
FR_STRT	Start of frame offset is 9 bits specifying the byte for the start of frame data. The data in the first frame buffer may no longer start at byte 64, depending upon the amount of data cut and inserted. This data is required for writing the buffer to the frame memory.
NXT_BUF_PTR	Pointer obtained from the RFD table to interface with another frame buffer in the frame memory. Writing here activates the next operation. Buffer size is in the NXT_BUF_SZ register. The NBV flag indicates if whether the next buffer has arrived. Bits 0:15.
NXT_BUF_SZ	10-bits specifying the size of the next buffer obtained from the RFD table; should be written with or before the NXT_BUF_PTR register as it activates the operation. Bits 16:25 (0—read 0 bytes; 1—read 1 byte)
RFD_CMD	2 bits specifying which RFD command: prefetch new pointer, read, write, and recycle. Writing here activates the interface. bits 0:1—command (00—prefetch; 01—recycle; 10—read; 11—write) bit 2—1 for multicast for recycle command; 0 for writing RFD_WR and RFD_WR_CNT for write command bits 8:12—number of buffers for recycle command bits 13:15—port for prefetch/recycle command bit 16—for 16-port configuration.
RFD_RD	25 bits specifying RFD read data. If the RFD_CMD is read, then the 33 bits of data are written to the RFD_RD and RFD_RD_CNT registers. Validation of data is performed via the appropriate flag. See Figure A15.1. bits 15:0—pointer bits 16:24—next buffer length

Register name	Description
RFD_WR	25 bits specifying RFD write data. If the RFD_CMD is write, then the 33 bits of data from the RFD_RD and RFD_RD_CNT registers are written to the RFD table. Validation of data is performed via the appropriate flag. See Figure A15.1. bits 15:0—pointer bits 16:24—next buffer length
RFD_RD_CNT	8 bits specifying RFD read data. If the RFD_CMD is read, then the 33 bits of data are written to the RFD_RD and RFD_RD_CNT registers.
RFD_WR_CNT	8 bits specifying RFD write data. If the RFD_CMD is write, then the 33 bits of data from the RFD_RD and RFD_RD_CNT registers are written to the RFD table.
RFD_PTR	16 bits specifying the RFD pointer. Programmers should first check the status of the RAF flag to see if the RFD address is ready.
RX_PRE_PTR	Pointer to new available frame buffers in the Rx frame memory, ready for use. The RRP flag indicates pointer presence.
TX_PRE_PTR	Pointer to new available frame buffers in the Tx frame memory ready for use. The RTP flag indicates pointer presence.
STAT_REG	32-bit interface to the statistics block: address (b31:8), reserved (b7:3), and command (b2:0). Valid commands are increment (000), decrement (101) and reset (001). Increment and decrement commands use the operand in the STAT_REG_HI register.
HOST_REG	Interface with the host. Microcode programmers should not write to this register.
STAT_REG_HI	16-bit operand for the statistics block. See the STAT_REG register for the address and command.

Table A15.1 SEND_REG Register Description

	TFD	OQ	HTFD
SEND_REG0 [31:0]	15:0—time stamp 31:16—Tx frame pointer	15:0—time stamp 31:16—Tx frame pointer	15:0_time stamp 31:16—Tx frame pointer
SEND_REG1 [57:32]	13:0—frame length 15:14—zero, reserved 20:16—number of buffers 25:21—source port number	13:0—frame length 15:14—zero, reserved 20:16—number of buffers 25:21—source port number	13:0—frame length 15:14—zero, reserved 20:16—number of buffers 25:21—source port number
SEND_REG2 [81:64]	6:0—zeros 7—buffer type 16:8—first buffer length 17—multiple	6:0—header length 7—buffer type 16:8—first buffer length 17—multiple	6:0—header length 7—buffer type 16:8—first buffer length 17—multiple

Table A15.2 ALU Mask Register

Name	Bits#	Description	Init value
MASK	b31:0	ALU mask.	

Table A15.3 LREG Register [31:0]

Name	Bits#	Description	Init value
LREG	b9:0	TOPmodify modification acceleration.	

Host Registers

Most registers are initialized with a high-level Application Program Interface (API), which is not described here. The only registers that are described in this appendix are those that are mentioned in this chapter and in the sample programs in the book. For more details, see [118].

MREG [15:0] Register

The LDREG command in the microcode instructs the loader to load the user-defined masks to each of these registers (see Section 15.5.1).

LREG Register [31:0]

These are 32 labels/addresses indicating the procedures in the microcode program corresponding to the 32 operations performed via the Modification Accelerator (see Section 15.2.9).

APPENDIX B
TOPMODIFY ADDRESSING MODES

Table B15.1 provides the numbers, names, and syntaxes of the addressing modes that are relevant to the TOPmodify. Devices in this table refer to registers or structures—for example, UREG, OREG. The numbers in the first column are used in this chapter to indicate the addressing modes supported by operands. **Bold** typeface indicates required text. *Italic* typeface indicates text that must be replaced with the appropriate value.

Table B15.1 TOPmodify Addressing Modes

No.	Name	Syntax	Description
1	Immediate	123 or 0x12 or "abcd" or $ or label	123—decimal number 0x12—hexadecimal number "abcd"—4 bytes = 0x61626364 $—program counter (PC) label—program label
2	Register	*device* [*number*]	For the device, see the devices listed for each TOP in its relevant part. *number*—index of the register in the device array
3	Register (byte-specific)	*device*[*num1*].byte[*num2*]	*num1*—index of the register in the device array *num2*—byte number in this register
4	Register (bit-specific)	*device*[*num1*].bit[*num2*]	*num2*—bit number in this register
6	Indirect	*device* [*IND_REG*]	
7	Base-index	*offset* (*base*) or *offset* (*base*)+	*base*—base registers. See the devices listed for each TOP in its relevant part. *offset*—immediate or some register.
8	Direct	*number* (0) or (*number*)+	Can be used for the Get command. For src in the Get commnad, use *number* (0)+.
12	Device group	bcam8 [*num1*] or bcam8 [ureg[0]]	*num1*—immediate 3 bits
15	Structure number-offset	*offset* (*str*) or *offset* (*str*)+	*str*—register or immediate. Access to result memory. *offset*—immediate or some register.

APPENDIX C
TOPMODIFY DETAILED INSTRUCTION SET

Detailed description of TOPmodify commands' operands are listed in this appendix. For even more detailed explanations, see [118].

MOVE COMMANDS

Move commands copy data (bits, bytes, and words) from various sources to various destinations.

Mov DST, SRC, SIZE [, jpe]

Definition: Move SIZE bytes from source (SRC) to destination (DST). jpe is used to optionally jump according to the FAST_REG, after the move (see Section 15.2.9).

Operands: DST, the destination, can be a register (or a byte in it) or an indirect addressing mode (type 1, 2, 3, or 6 in Table in B15.1). Valid registers are UREG, SREG registers (except FLAGS), OUT_IF registers (except HOST_REG, RFD_RD, RFD_RD_CNT, RX_PRE_PTR and TX_PRE_PTR), and OREG registers (except CAMO, HISTORY0..3, BUF_STATUS and UNIT_NUM) (see Section 15.3.1).

SRC, the source, can be an immediate 32-bit value, a register (or a byte in it) or an indirect addressing mode (type 1, 2, 3, or 6 in Table B15.1). Register or indirect addressing modes can be any of the registers, (see Section 15.3.1). When it is an immediate 32 bit, then destination is ALU only, and to move an immediate value other than 32 bits to a register (except ALU), use the MovBits command.

Size can be either an immediate value (1 to 4) or a register content; that is the SIZE_REG (which is SREG[2]) value (0–4). This corresponds to type 1 or 2 in Table B15.1. When an immediate address mode is used, the SIZE field is defaulted to 4 bytes; that is 32 bits long immediate value–irrelevant of the value written.

Jpe indicates whether to jump (_JPE_MDF) or not to jump (_NJP_ MDF), following examination of the modify bits in the FAST_REG. The examination starts at the least significant bit (LSB), and continues until it reaches the first bit that is set (which determines where to jump; see Section 15.2.9). The default is _NJP_MDF.

MovBits DST, SRC, SIZE [, jpe]

Definition: Move SIZE bits from source (SRC) to destination (DST). jpe is used to optionally jump according to the FAST_REG, after the move (see Section 15.2.9).

Operands: DST, the destination, can be a bit specific register or an indirect addressing mode (type 4 or 6 in Table B15.1). Valid registers are UREG, SREG registers (except FLAGS), OUT_IF registers (except HOST_REG, RFD_RD, RFD_RD_CNT, RX_PRE_PTR, and TX_PRE_PTR), and OREG registers (except CAMO, HISTORY0..3, BUF_STATUS and UNIT_NUM) (see Section 15.3.1).

SRC, the source, can be an immediate 16-bit value, a bit specific register or an indirect addressing mode (type 1, 4, or 6 in Table B15.1). Register or indirect addressing modes can be any of the registers (see Section 15.3.1).

Size can be either an immediate value (1–16) or a register content; that is, the SIZE_REG (which is SREG[2]) value (0–16). This corresponds to type 1 or 2 of the addressing modes in Table B15.1.

Jpe indicates whether to jump (_JPE_MDF) or not to jump (_NJP_MDF), following examination of the modify bits in the FAST_REG. The examination starts at the LSB, and continues until it reaches the first bit that is set (which determines where to jump; see Section 15.2.9). The default is _NJP_MDF.

Get DST, SRC, SIZE [, jpe]

Definition: Move SIZE bytes from the source (SRC, the frame memory, FMEM) to a destination (DST) register. jpe is used to optionally jump according to the FAST_REG, after the move (see Section 15.2.9).

Operands: DST, the destination, can be a register (or a byte in it) or an indirect addressing mode (type 2, 3, or 6 in Table B15.1). Valid registers are UREG, SREG registers (except FLAGS), OUT_IF registers (except HOST_REG, RFD_RD, RFD_RD_CNT, RX_PRE_PTR, and TX_PRE_PTR), and OREG registers (except CAMO, HISTORY0..3, BUF_STATUS and UNIT_NUM) (see Section 15.3.1).

SRC, the source, can be a based-indexed or a direct addressing mode (type 7 or 8 in Table B15.1), referring to FMEM.

Size can be either an immediate value (1 to 4) or a register content; that is the SIZE_REG (which is SREG[2]) value (0–4). This corresponds to type 1 or 2 in Table B15.1.

Jpe indicates whether to jump (_JPE_MDF) or not to jump (_NJP_MDF), following examination of the modify bits in the FAST_REG. The examination starts at the LSB, and continues until it reaches the first bit that is set (which determines where to jump; see Section 15.2.9). The default is _NJP_MDF.

GetRes DST, SRC, SIZE [, jpe]

Definition: Move SIZE bytes from the source (SRC, the result memory RMEM) to the destination (DST). jpe is used to optionally jump according to the FAST_REG, after the move (see Section 15.2.9).

Operands: DST, the destination, can be a register (or a byte in it) or an indirect addressing mode (type 2, 3, or 6 in Table B15.1). Valid registers are UREG, SREG registers (except FLAGS), OUT_IF registers (except HOST_REG, RFD_RD, RFD_RD_CNT, RX_PRE_PTR, and TX_PRE_PTR), and OREG registers (except CAMO, HISTORY0..3, BUF_STATUS and UNIT_NUM) (see Section 15.3.1).

SRC, the source, is the structure number of the required result or message in the result memory (RMEM) (type 15 in Table B15.1).

Size can be either an immediate value (1 to 4) or a register content; that is the SIZE_REG (which is SREG[2]) value (0-4). This corresponds to type 1 or 2 in Table B15.1.

Jpe indicates whether to jump (_JPE_MDF) or not to jump (_NJP_MDF), following examination of the modify bits in the FAST_REG. The examination starts at the LSB, and continues until it reaches the first bit that is set (which determines where to jump; see Section 15.2.9). The default is _NJP_MDF.

Put DST, SRC, SIZE [, jpe]

Definition: Move up to four bytes (SIZE) from source register (SRC) to destination (DST, which is the frame memory FMEM). jpe is used to optionally jump according to the FAST_REG, after the move (see Section 15.2.9).

Operands: DST, the destination, is a based indexed address mode (type 7 in Table B15.1), referring to FMEM.

SRC, the source, can be an immediate 16-bit value, a register (or a byte in it) or an indirect addressing mode (type 1, 2, 3, or 6 in Table B15.1). Register or indirect addressing modes can be any of the registers (see Section 15.3.1).

Size can be either an immediate value (1-4) or a register content; that is the SIZE_REG (which is SREG[2]) value (0-4). This corresponds to type 1 or 2 of the addressing modes, Table B15.1.

Jpe indicates whether to jump (_JPE_MDF) or not to jump (_NJP_MDF), following examination of the modify bits in the FAST_REG. The examination starts at the LSB, and continues until it reaches the first bit that is set (which determines where to jump; see Section 15.2.9). The default is _NJP_MDF.

Copy DST, SRC, SIZE [, jpe]

Definition: Move SIZE bytes from the source (SRC, which is the result memory RMEM) to the destination (DST, which is the frame memory, FMEM). Jpe is used to optionally jump according to the FAST_REG, after the move (see Section 15.2.9).

Operands: DST, the destination, is a based indexed addressing mode (type 7 in Table B15.1), referring to FMEM.

SRC, the source, is the structure number of the required result or message in the result memory (RMEM) (type 15 in Table B15.1).

Size can be either an immediate value (1 to 4) or a register content; that is the SIZE_REG (which is SREG[2]) value (0-4). This corresponds to type 1 or 2 of the addressing modes, Table B15.1.

Jpe indicates whether to jump (_JPE_MDF) or not to jump (_NJP_MDF), following examination of the modify bits in the FAST_REG. The examination starts at the LSB, and continues until it reaches the first bit that is set (which determines where to jump; see Section 15.2.9). The default is _NJP_MDF.

Write DST, SRC, SIZE [, jpe]

Definition: Move SIZE data from the source (SRC, the frame memory FMEM) to the destination (DST, also the frame memory, FMEM). jpe is used to optionally jump according to the FAST_REG, after the move (see Section 15.2.9).

Operands: DST, the destination, is a based indexed address mode (type 7 in Table B15.1), referring to FMEM.

SRC, the source, is a based indexed address mode (type 7 in Table B15.1), referring to FMEM.

Size can be either an immediate value (1-8) or a register content; that is the SIZE_REG (which is SREG[2]) value (0-8). This corresponds to type 1 or 2 in Table B15.1.

Jpe indicates whether to jump (_JPE_MDF) or not to jump (_NJP_MDF), following examination of the modify bits in the FAST_REG. The examination starts at the LSB, and continues until it reaches the first bit that is set (which determines where to jump; see Section 15.2.9). The default is _NJP_MDF.

JUMP COMMANDS

Jump commands instruct the NP to jump to a given label in the microcode. Some of the commands are conditional jumps, and some jump with pushing and popping program addresses. At the end of the jump, optional NOP_NUM (0, 1 or 2) NOPs may be inserted into the pipeline following the jump command (to disable execution of the commands that immediately follow the jump instruction). The NOPs are effective only if the jump command is executed. The standard Jump command format is:

```
if ( CONDITION )
    command LABEL [| NOP_NUM];
                            // See commands in Table 11.4, Chapter 11.
```

or

```
command LABEL [| NOP_NUM];
                            // See commands in Tables 11.4 and 11.5, Chapter 11.
```

The format of the Call, CallCam, and CallStack commands is a bit different from the jump command, and also includes a number of NOPs (RET_NOP_NUM) that should be inserted before returning, as follows:

```
Call function [ | NOP_NUM [ , RET_NOP_NUM ] ];
```

or

```
CallCam function [ | NOP_NUM [ , RET_NOP_NUM ] ];
```

or

```
CallStack function [ | NOP_NUM [ , RET_NOP_NUM ] ];
```

Definition: Call a function, and return from it to the next address on completion (return command in the function). Insert NOP_NUM NOPs before jumping and RET_NOP_NUM NOPs before returning.

Operands: CONDITION is any flag bit or UDB register bit. The negation mark (!) may precede any condition.

LABELS is the place in the program to where the jump is to take place.

function is the name of the function to be called.

NOP_NUM shows the number of NOPs to be inserted after the branch, to prevent the following command from entering the pipeline.

RET_NOP_NUM shows the number of NOPs to be inserted on return from the called function.

ARITHMETIC LOGIC UNIT OPERATIONS

Arithmetic Logic Unit commands are used for arithmetic and logic calculations. There are two formats of ALU commands, one with two source–operands and the other with one source–operand:

```
COMMAND DST,SRC1,SRC2,SIZE,[,MREG[,MODE[,JPE]]];
                            // See commands in Table 11.3, Chapter 11.
```

or

COMMAND DST,SRC,SIZE,[,MREG[,MODE[,JPE]]];
                                        // See commands in Table 11.3, Chapter 11.

Definition: Calculate the *COMMAND* on the sources (SRC, or SRC1 and SRC2) of SIZE bytes, and put the result in the destination (DST). Use masks in MREG, according to the required MODE, to mask the source operands if required. jpe is used to optionally jump according to the FAST_REG, after the move (see Section 15.2.9).

Operands: DST, the destination, receives the result of the ALU block operation, together with the ALU register itself; that is the result is entered into both the destination and the ALU. DST can be a register (or a byte in it) or an indirect addressing mode (type 2, 3, or 6 in Table B15.1), and may use any of the following registers: UREG, SREG registers (except FLAGS), OUT_IF registers (except HOST_REG, RFD_RD, RFD_RD_CNT, RX_PRE_PTR, and TX_PRE_PTR), and OREG registers (except CAMO, HISTORY0..3, BUF_STATUS and UNIT_NUM), see Section 15.3.1.

SRC or SRC1, the first source operand or the only source operand can be a register (or a byte in it) or an indirect addressing mode (type 2, 3, or 6 in Table B15.1). Register or indirect addressing modes can be any of the registers (see Section 15.3.1).

SRC2, the second source, can be an immediate 8-bit value, a register (or a byte in it) or an indirect addressing mode (type 1, 2, 3, or 6 in Table B15.1). Register or indirect addressing modes can be any of the registers (see Section 15.3.1).

SIZE is the size of the ALU operation in bytes. Zero (0) is not a valid size. Size can be either an immediate value (1–4) or a register content. Register content is taken from the SIZE_REG (which is SREG[4]), and valid values are 1–4. This corresponds to type 1 or 2 in Table B15.1.

MREG indicates one of the 16 ALU mask registers (see MREG[15:0] Register subsection in Appendix A).

MODE indicates how to use the mask:

_ALU_NONE—no masking done
_ALU_*FRST*—masking SRC1, that is *SRC1 = SRC1 and MASK*REG
_ALU_*SCND*—masking SRC2, that is *SRC2 = SRC2 and MASK*REG
_ALU_*BOTH*—masking both SRC1 and *SRC2*; that is *SRC1 = SRC1 and MASK*REG, SRC2 = SRC2, and MASK_REG

Jpe indicates whether to jump (_JPE_MDF) or not to jump (_NJP_MDF), following examination of the modify bits in the FAST_REG. The examination starts at the LSB, and continues until it reaches the first bit that is set (which determines where to jump; see Section 15.2.9). The default is _NJP_MDF.

When the explicit destination is not the ALU register, all four bytes of the ALU register are updated, even if the instruction SIZE was less than 4. For example, Add UREG[7],UREG[5],UREG[6],1; adds the contents of UREG[5] and UREG[6] and writes 1-byte to UREG[7] and 4 bytes to ALU. The ALU register cannot be used with an offset other than 0 (zero) when it is used for destination, source1 or source2. Examples of *improper* usage of the ALU register:

```
Add ALU.BYTE [1], UREG[0].BYTE[2], 2, 1;
Add ALU, UREG[0].BYTE[2], ALU.BYTE [2], 1;
Add ALU, ALU.BYTE [3], UREG[0].BYTE[2], 1;
```

CONTENT ADDRESSABLE MEMORY OPERATION

There is one CAM instruction in the TOPmodify CAM operation:

LookCam DST, SRC, TYPE [, jpe];

Definition: Search in a BCAM8. Use source (SRC) for a search key and put the result in destination (DST), and in the CAMO register (which is OREG[3]). If a match is found, the MH bit is set. jpe is used to optionally jump according to the FAST_REG, after the move (see Section 15.2.9).

The CAM table is divided into up to eight user-defined groups. BCAM8 is actually 11 bits, where the first three bits indicate the CAM group.

Operands: DST, the destination, can be a register (or a byte in it) or an indirect addressing mode (type 2, 3, or 6 in Table B15.1). Valid registers are: UREG, SREG registers (except FLAGS), OUT_IF registers (except HOST_REG, RFD_RD, RFD_RD_CNT, RX_PRE_PTR, and TX_PRE_PTR), and OREG registers (except CAMO, HISTORY0..3, BUF_STATUS and UNIT_NUM). In addition, as noted above, the CAMO register (which is OREG[3]), is always a destination (see Section 15.3.1).

SRC, the source, can be a register (or a byte in it) or an indirect addressing mode (type 2, 3, or 6 in Table B15.1). Register or indirect addressing modes can be any of the registers (see Section 15.3.1).

TYPE indicates the CAM group, for example, BCAM8[0]. Alternatively, the 3 LSBs of the UDB may also be used to indicate the group number; for example BCAM8[UREG0] (type 12 in Table B15.1).

In order to reduce data hazards, it is advisable to Get data from the FMEM into CAMI and then perform a lookup in the next instruction.

HALT COMMAND

The Halt command of the TOPmodify is somewhat different from other Halt commands, as no results are passed along the pipeline, but the actual frame is handled (e.g., transmitted, ignored, copied into the frame memory). Frames can be sent to any combination of destinations by writing the descriptors to the SEND_REG register one at a time, and using the halt command to send each in turn. The halt command indicates type (e.g., unicast, multicast) and specific destination (i.e., TFD for network interface, Virtual Output Queue [VOQ] for switch interface, HTFD for host interface, or none).

The pointer to the frame buffer is taken from the TX frame pointer in the SEND_REG register. The format of the Halt instruction is:

```
HALT DATA, MODE;
```

Definition: Halt is used to finish packet processing and to transmit the packet or to discard it. MODE determines if this frame should be unicast, multicast, or discarded, and DATA determines the frame's destination (FMEM and output queue).

Operands: DATA indicates the frame's destination, according to the predefined constants shown in Table C15.1.

Mode indicates how the TOPmodify terminates processing, according to the predefined constants shown in Table C15.2. The examples that follow the table show a series of instructions and when to use the halt command.

Table C15.1 Data Operands for Halt Instruction

Predefined DATA constants	Description
_WHALT_NO_MDF	Just write: do not send to queue, but write buffers. This writes the frame buffer (or the entire frame if the pointer is to the first frame buffer) to the frame memory. It is not sent to a queue.
_WHALT_TF_MDF	TFD + write: This sends the frame buffer (or the entire frame if the pointer is to the first frame buffer) to the TFD queue and writes it to the frame memory.
_WHALT_HT_MDF	HTFD + write: This sends the frame buffer (or the entire frame if the pointer is to the first frame buffer) to the HTFD queue and writes it to the frame memory.
_WHALT_VQ_MDF	VOQ + write: This sends the frame buffer (or the entire frame if the pointer is to the first frame buffer) to the VOQ and writes it to the frame memory.

(continued)

Table C15.1 Data Operands for Halt Instruction (*continued*)

Predefined DATA constants	Description
_HALT_NO_MDF	Nothing: do not send to queue and do not write to frame memory. This is not a true halt and in fact may often be followed by other instructions. HALT_HALT_NO_MDF temporarily halts and resumes operation on the same frame. Once all the frame buffers have been modified, use HALT _WHALT_xx_MDF to terminate the processing of the current frame and sent it to the specified queue.
_HALT_TF_MDF	TFD: This sends the frame buffer (or the entire frame if the pointer is to the first frame buffer) to the TFD queue. It is not written to the frame memory.
_HALT_HT_MDF	HTFD: This sends the frame buffer (or the entire frame if the pointer is to the first frame buffer) to the HTFD queue. It is not written to the frame memory.
_HALT_VQ_MDF	VOQ: This sends the frame buffer (or the entire frame if the pointer is to the first frame buffer) to the VOQ. It is not written to the frame memory.

Table C15.2 Mode Operands for Halt Instruction

Predefined MODE constants	Description
HALT_UNIC	Indicates that the current TOP has completed processing this frame and is ready to start processing the next frame.
HALT_DISC	Instruction to discard this frame and begin processing the next frame.
HALT_MULC	This is not a true halt and in fact may often be followed by other instructions. HALT HALT_MULC temporarily halts and resumes operation on the same frame. Once all the frame buffers have been modified, use HALT HALT_UNIC to terminate the processing of the current frame.
HALT_BRKP	For debug mode. This freezes this specific TOPmodify engine until the host releases it.

Example 1: A frame in which only the first frame buffer requires modification.

- Modify the first frame buffer.
- Use HALT _WHALT_VQ_MDF, HALT_UNIC to send the entire frame (all buffers) to the VOQ and write the frame to the frame memory.

Example 2: A frame in which two frame buffers require modification.

- Modify the first frame buffer.
- Use HALT _WHALT_NO_MDF, HALT_MULC to write the first frame buffer to the frame memory without sending it to a queue.
- Read second buffer and modify it.
- Use HALT _WHALT_NO_MDF, HALT_MULC to write the second frame buffer to the frame memory without sending it to a queue. After modifying buffer 2, the frame cannot be sent directly to a queue because the pointer is pointing to the second frame buffer.
- Use HALT _WHALT_VQ_MDF, HALT_UNIC to send the entire frame (all buffers) to the VOQ and write the frame to the frame memory. The frame pointer must be pointing to the first frame buffer.

Example 1: A frame in which only the first frame buffer requires modification.

- Modify the first frame buffer.
- Use HALT_VO_MOD..._HALT_QUE to send the entire frame (all but text) to the VOQ and write the frame to the frame memory.

Example 2: A frame in which two frame buffers require modification.

- Modify the first frame buffer.
- Use HALT..._HALT_NO_HBQ to write the first frame buffer to the frame memory without sending it to a queue.
- Read second buffer and modify it.
- Use HALT..._NO_MOD..._HALT_HBQ to write the second frame buffer to the frame memory without sending it to a queue. After modifying buffer 2 the frame cannot be sent directly to a queue because the pointer is pointing to the second frame buffer.
- Use HALT..._HALT_VO_MOD..._HALT_QUE to send the entire frame (all but text) to the VOQ and write the frame to the frame memory. The frame pointer must be pointing to the first frame buffer.

Running the Virtual Local Area Network Example

16

In the previous chapters, we used an example of Virtual Local Area Network (VLAN) processing to show how to program and use each of the Task Optimizet Processor (TOP) engines. In this chapter, we return to this example to describe how we actually use the EZchip development system; that is, how we load it, compile it, and debug it. Debugging here is very elementary, but once the reader will follow the procedure, he or she will be able to perform a complex debugging, which includes running to breakpoints, single step-through running, and watching internal registers, memories, frames, and so on.

In addition, this chapter provides a quick and basic review of how to define frames (that the simulator will use) and build the search-database (structures that also can be used during the debugging phase).

The purpose of this chapter is not to teach how to use the specific tool, but rather to show how to actually develop network processing software. For a comprehensive description of how to use the development environment, see [117].

16.1 INSTALLATION

First, we have to install the EZdesign kit, which includes EZmde (EZ Microcode Development Environment), frame, and structure generators. This kit is available at *http://www.cse.bgu.ac.il/npbook* (with its User's Guide). After the explanation of installation, we provide a detailed description of how to compile the VLAN code, and how to debug it.

Use Windows system (NT4.0, 2000, or XP) to install EZdesign according to the following steps (basically, all you have to do is to press the <Enter> key to accept all the defaults):

1. Run Setup.exe from the attached CD.

2. Click Next to start the wizard and to display the License Agreement window.

3. Approve License Agreement with Yes to display the Customer Information window.

4. Enter your user details:

   User Name: [System registration details or manual entry]
   Company Name: [System registration details or manual entry]
   Click Next to display the Choose Destination Location window.

5. Default location: C:\EZchip\EZdesign_demo.

   Browse if you want to define an alternative folder location.
   Click Next to display the Select Program Folder window.

6. Accept the default program folder name or enter preferred name and click Next to install the package.

7. The installation window monitors the progress of installation operations. You can Cancel the installation at any time. On completion, there may be a textbox asking about creating a new EZmcc workspace file association, if the 'wsc'-file association already exists. Answer Yes in this case (normally, choose the default, i.e., No).

8. The View Release Notes window replaces the current window. This window enables you to choose (optional) to open the Readme file after exiting the installation procedure. Click Finish to display the Finish Reboot window.

9. Before running the program, you must restart your computer. You can choose to restart your computer now or later. Click Finish to exit the installation process.

16.2 GETTING STARTED

Now we can run the VLAN example. First we'll launch the EZmde, and then we'll open the VLAN project.

16.2.1 Launching the Microcode Development Environment

Click Start>ALL Programs>EZdesign demo>EZmde, or double click 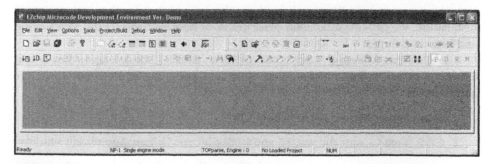 from your Windows desktop to launch EZmde. You'll get the main screen of EZdesign shown in Figure 16.1. You can now proceed to open the VLAN project for microcode management.

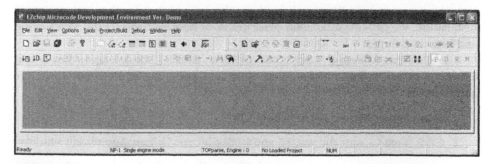

FIGURE 16.1

EZmde opening screen

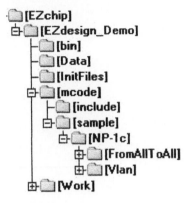

FIGURE 16.2

Microcode samples supplied

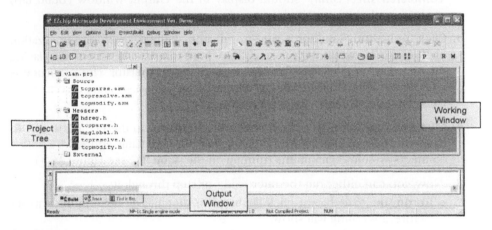

FIGURE 16.3

EZmde GUI—VLAN project loaded

16.2.2 Opening the Sample VLAN Project

To assist in GUI familiarization, the EZdesign package provides two sample microcode projects (*.prj) in the \mcode\sample folders shown in Figure 16.2.

We shall refer from this point on to the VLAN example. To open this sample project, proceed as follows:

1. Click Project/Build>Open Project.

2. Select the VLAN project and Open to load the project. Since this is the first time this project has been opened, there is no workspace, and you have to click OK at the prompt to create a new workspace for the project.

    When the project is loaded, three windowpanes (see Figure 16.3) are viewable in the GUI: Project Tree Window, Working Window and Output Window.

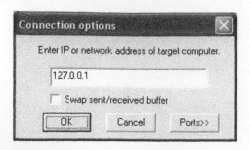

FIGURE 16.4

Connections options dialog window

3. Click Project/Build>Build All to compile the code. "*Compilation was completed successfully*" should display in the Output Window (Build tab) and "Compiled Project" in the status bar.

4. Click Tools>Run Simulator Scripts to run a pre-defined script for initialization of the system. As the script runs, DOS boxes appear on screen displaying the status of applications that are used by the simulator. The sequence is completed when "Initialization is finished" displays in a separate ezrunner.exe DOS window. Minimize all DOS windows.

5. Click Debug>Connect ... to connect the simulator to the PC, and then OK from the Connection Options dialog box. (The default address 127.0.0.1 connects system components running on the same PC; see Figure 16.4)

6. Now, you can either run the microcode, or step-through it:
   - To run the loaded code, click Debug>Go/Attach. If no breakpoint was set before, a warning will show.
   - To step-through the code, click Debug>Step Into.

   Step-through allows you to see the code's execution, instruction by instruction, and by using viewing options. For example, when choosing View>Local Memory, the Registers can be watched as the program steps through (see Figure 16.5).

After the code has been run, a message appears, letting you know that the simulation has ended (see Figure 16.6).

Now, the log files may be viewed in the ...\Debug\Output subfolder. For an explanation of the log files, see the EZdesign User's Guide and in [117].

Breakpoints can be inserted, code can be traced line-by-line (single step-through), register contents can be watched and memory contents can be analyzed.

All this is explained in detail in the EZdesign User's Guide and in [117].

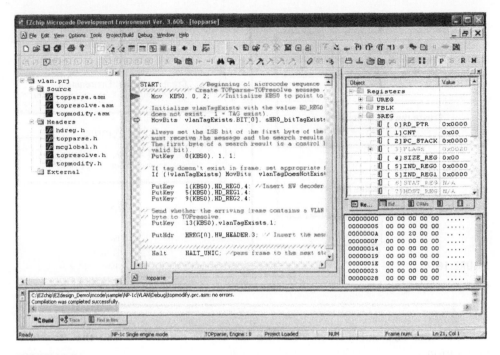

FIGURE 16.5

Debugging screen

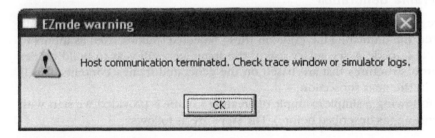

FIGURE 16.6

End of simulation message

16.3 MICROCODE DEVELOPMENT WORKFLOW

Although the VLAN example seems like a straightforward and simple task, usually a code requires a bit more than what we have done. Actually, even the VLAN example that we run through used a frame that was pre-prepared and a null structure that was also predefined. The flowchart shown in Figure 16.7 outlines the common EZdesign workflow.

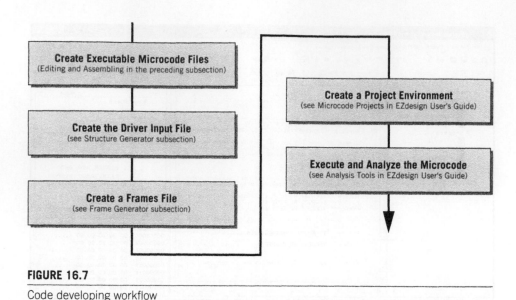

FIGURE 16.7

Code developing workflow

The next two subsections describe the frame generation and the structure generation.

16.3.1 Frame Generation

Frames are generated according to various random or predetermined methods, and are required for simulating the code on these "injected" frames, used as inputs to the executable code. A secondary use of the frame generation is to build entries in the search structures that are based on the generated frames' contents. This is described in the next subsection.

In the following, a simple example of creating a frame is provided. We start with EZmde running (as described before). The steps are as follows:

1. Click Tools>Frame / Structure Generator. We'll get the main frame generator screen shown in Figure 16.8.

2. Select the frame type you want to generate by picking the protocol that meets your requirements. At any rate, you have to define the frames by defining protocols according to their layers, that is, define layer 2 first, and then layer 3, and so on. For the sake of this example, we'll show how to create an Ethernet frame with no VLAN tag. We double click on Ethernet Without Tag (in the Layer 2 group of the Protocols), and we get the frame structure shown in Figure 16.9.

3. We can define how any field of the Ethernet header will be created by clicking the Edit button of this field. We can go on and choose one of the edit

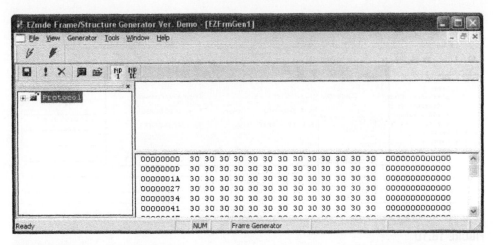

FIGURE 16.8

Main frame generator screen

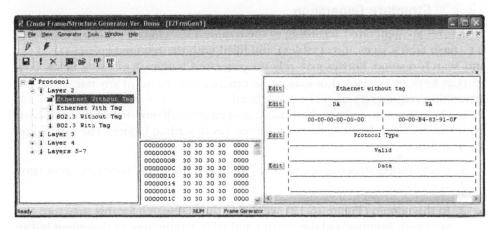

FIGURE 16.9

Frame generator screen

options (i.e., random, incremental, predefined, etc.) as described in detail in the EZdesign User Guide and in [117] (see Figure 16.10).

4. Eventually, after defining upper layers and all the required fields, we get the entire frame, and we can create the frame by clicking Generator>Run (Figure 16.11).

More options, such as saving, editing, creating, and manipulating the frames are described in the EZdesign User Guide and in [117].

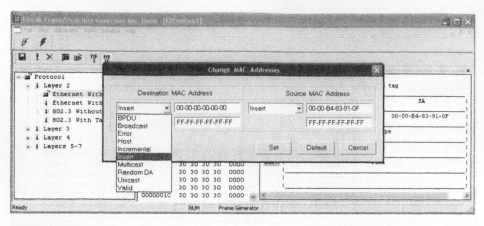

FIGURE 16.10

Frame field's options

16.3.2 Structure Generation

Structures are generated according to the required searching repository structures (i.e., hash tables, trees, direct tables), and filled with the required entries (i.e., keys and search results). Entries may rely on generated frames (e.g., using fields' contents as keys, or as results), or may be determined otherwise. At any rate, structure generation requires frames to be predefined.

In the following, a simple hash table generation is demonstrated. We start here from the previous frame generation screen (as described before). The next steps are as follows:

1. Click Tools>Structure Generator. We'll get the main structure generator screen.

2. In this screen (Figure 16.12), we define the TOPsearch for this structure, the structure number, and the entries and table parameters, as described in the detailed EZdesign User Guide.

3. We create a structure header by clicking the header H button ( ), and we get the fields that enable us to define the entries' parameters (Figure 16.13).

4. We define the number of key and result elements in each of the entries (as for filtering information, see the detailed Users' Guide [117]), and while doing so, we get a fourth column that allows us to define the keys and the results of the entries (Figure 16.14).

5. We can define the keys or the results by pressing the Edit button, and choosing the method by which it will be created (e.g., taken from a field or from an offset of a generated frame, fixed or random values). The example

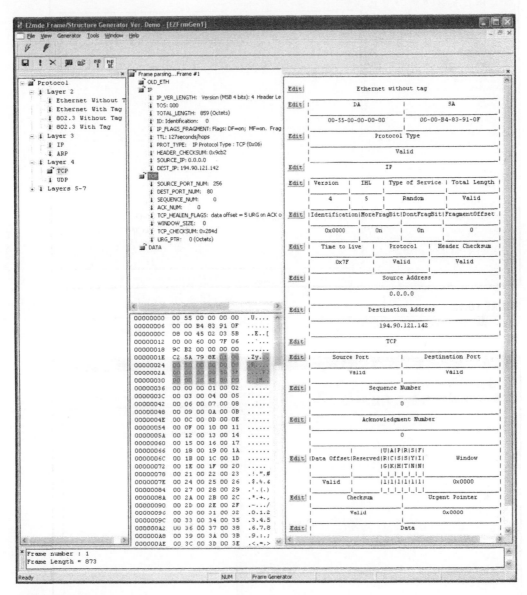

FIGURE 16.11

A defined frame

in Figure 16.15 shows how we determine the result according to the frame's field.

6. We continue by clicking the body B button (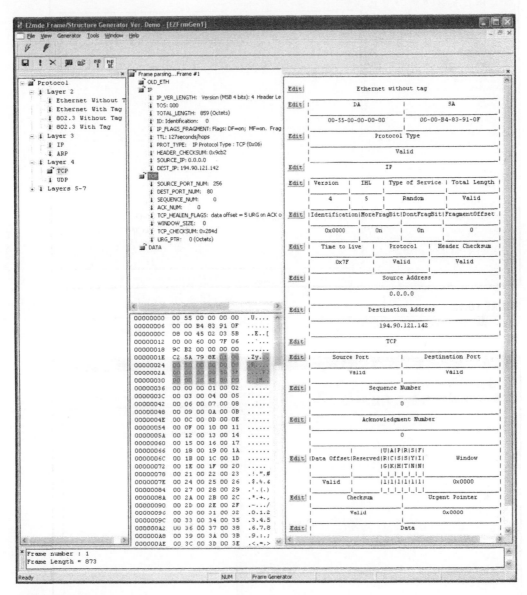), and we get the resulting structure (Figure 16.16).

7. Adding other structures is possible by going through steps 2–6.

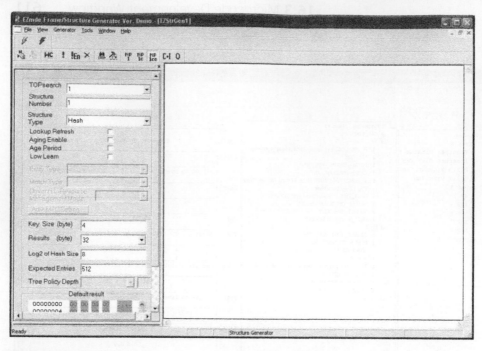

FIGURE 16.12

Main structure generator screen

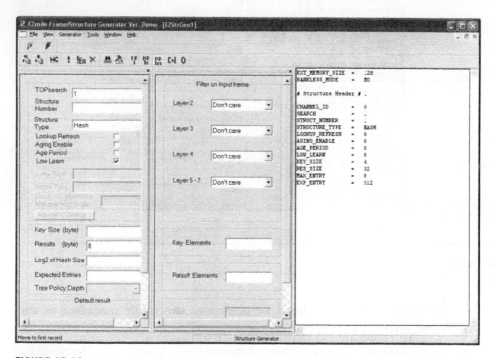

FIGURE 16.13

Structure definitions

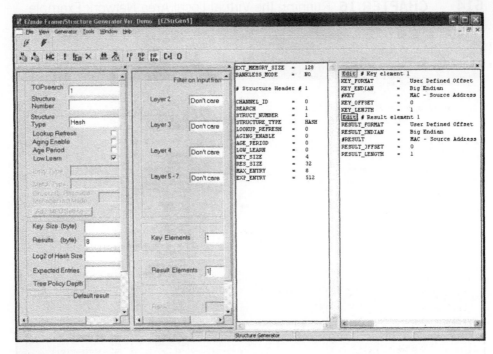

FIGURE 16.14

Keys and results definitions

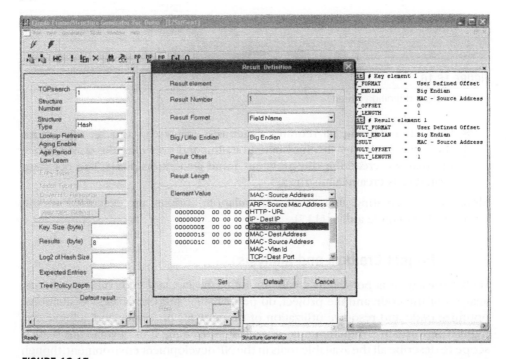

FIGURE 16.15

Result options screen

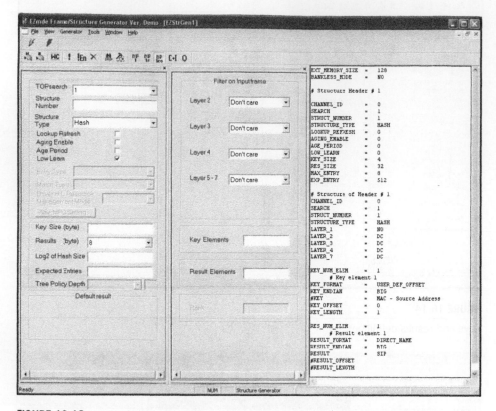

FIGURE 16.16

Final structure definition screen

8. Only after defining all required structures, can we go and actually create them. We do this by clicking Generator>Run and providing the frame binary file that we created in the previous frame generation phase. The resulting structure is created and stored.

More editing, creating, and other manipulation options are described in the EZdesign User Guide and in [117].

16.3.3 Project Creation and Analysis

With EZdesign, it is possible to create an entire project, and then run simulations, analysis of the code and the project, do performance and correctness evaluations, optimize code and resource utilization of the network processor, and achieve the highest performance requirements of packet processing. It is beyond the book's scope to describe all the available tools in the NP development environment, though some of them are described in the EZdesign User Guide. For a complete reference to EZdesign, the reader is referred to [117].

16.4 **SUMMARY**

This chapter provides an overview of how to create a project, compile it, and test (or debug) it. Obviously, the purpose was not to show how the EZmde works, but to provide a sense of a development environment for network processors. It is a bit different from a "regular" programming environment, as the process of developing a packet processing task requires the related definition of the inputs (frames and search repositories).

18.4 SUMMARY

This chapter provided an overview of how to create a project, compile it and test (or debug) it. Obviously, the purpose was not to show how the EZ-Kit works, but to provide a sense of a development environment for network processors. It is a bit different from a regular programming environment, as the process of developing a packet processing task requires the related definition of the inputs (frames and search repositories).

Writing Your First High-Speed Network Application

This chapter concludes the EZchip part of the book by demonstrating how to use the EZchip NP. We describe a high-speed network application, a multi-Gbps routing, and an "on the fly" screening filter for prescribed words that are to be identified and masked. This chapter shows how to design, write, run, debug, and simulate an application with the Microcode Development Environment.

17.1 INTRODUCTION

This example contains a demonstration application for inspecting, modifying, and re-routing HyperText Transfer Protocol (HTTP) frames. The example is written in a "teaching mode" in order to demonstrate the implementation of a network processor for high-speed application, and is not aimed to be efficient or even true (for example, check-summing is ignored); moreover, the example only uses NP instructions that were described in this book, and is therefore not comprehensive. As, in some cases, efficiency was sacrificed for readability, the reader should not look at this code as an example to use, but rather simply for using several methods of writing instructions, for explanation and learning purposes. Nevertheless, this example uses several of the important features of the NP processor, such as fast deterministic hash-table lookups, auto-learning of hash entries, scanning of frames for specific patterns in the payloads, and more.

The application inspects each frame for specific, predefined words, and according to one of three predefined operating modes, the application resolves what to do with the frame. If the packet isn't an IPv4 packet, it is discarded. If the packet isn't an HTTP packet, it is routed without further inspection. If the frame is an HTTP frame, the NP examines it and handles it according to the defined operating mode, as described in the following:

The first operating mode is the simplest—if a frame contains the "*hot string,*" which in our example is either "bomb" or "sex," this frame is discarded; otherwise, the frame is routed.

The second operating mode checks first if the incoming frame was received from a "*hot SIP*" (a Source IP Address that in the past had sent a frame containing the "hot string"). If it was, this frame is discarded. Otherwise, the frame is checked to see if it contains the "hot string." If it contains the "hot string," the frame is discarded, and the sender's SIP is stored for future blocking (i.e., it becomes a "hot SIP"). If the frame was clear, it is routed as usual.

The third operating mode is similar to the second operating mode, but in addition to what happens there, the frame that contained the "hot string" is now routed to a predefined destination IP address, where the three first letters in the "hot string" are replaced by "XXX."

The data structures that the program uses are three hash tables for look-ups, of which two are static (and have to be initially populated with data), and one is dynamically handled. The first hash data structure, the dynamic one, is used to keep the hot SIP addresses, and if the operating mode requires it, this table is updated with these SIPs at wire speed. Initially, this hash structure must be provided to the program, empty.

A second hash table contains DIP entries that must be provided. These Destination IP addresses are used for routing the frames, and only frames carrying DIPs that reside in this table will be routed. This table points to an output port of the NP through which the frame will be routed, as well as the destination MAC address of the next hop router that is attached to the output port of the NP (and thereby rendering the ARP[1] mechanism redundant).

The third hash table is used for describing the source MAC address for each of the output ports, as well as an alternative DIP address (for the diverted DIP in case of a "hot string"). This table is also static, that is, it should be populated prior to the execution and is not modified during the execution.

The first two tables described above (the one that contains SIP addresses and the one that contains DIP addresses) are used by the TOPsearch I, and keys are sent to the search engine from the TOPparse. The third table (the output ports) is used by TOPsearch II, and keys are sent to it from the TOPresolve after it determines the correct port to be used for routing. A detailed structure of these tables is provided in Section 17.3. Please note that in this example no code was written for the TOPsearch I or TOPsearch II, and these engines are supposed to perform just simple lookups in the hash tables that are in following pages.

17.2 DATA FLOW AND TOP MICROCODE

The general flow diagram of the entire application is shown in Figure 17.1. The following subsections describe the program for each TOP, including the relevant microcode. Each subsection includes an explanation of the program in the relevant TOP engine.

---

[1]Address Resolution Protocol, which is required to bind IP addresses (L3) to MAC addresses (L2).

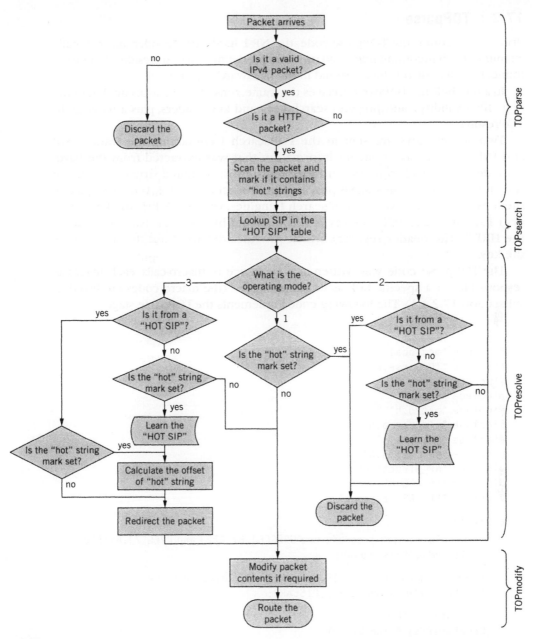

FIGURE 17.1

Application flowchart

17.2.1 **TOPparse**

Prior to executing the TOPparse code, the NP-1 hardware decoder automatically examines each incoming frame and provides predefined analysis results about each frame. The analysis results are stored in the hardware registers.

In a nutshell, the TOPparse receives the frame, runs several checks on that frame (e.g., IPv4 validity), and prepares search keys and key headers, plus a message to TOPresolve.

Two search keys are sent to the TOPsearch I containing the frame's SIP and DIP. The message contains information that was extracted from the hardware decoder (read from the hardware registers). A detailed structure of these search keys and the message is provided in subsection 17.3.2. TOPparse passes the message and the keys to TOPsearch I engines, via the KMEM (the key memory). Each message and each key is accompanied by a header that is passed via the HREG (the header register), which provides details about the message or the key.

The TOPparse code was written as a sequence of macro-calls; each macro is responsible for a specific task, as its name implies (the macro codes are listed in subsection 17.2.4.1). The following code implements the TOPparse stage.

```
EZtop Parse;
#include "mcglobal.h"
#include "TOPparse.h"
#include "hdreg.h"
#include "stat.h"
#include "macro.h"
#include "rfd_macro.h"
#include "network.h"
#include "NP_Common.h"
#include "NP_Structs.h"
#include "NP_PRS.h"
#include "NP_PRS_macros.h"

L_START:
    // Initialize the program, use UREG[6].BYTE[2] and UREG[7].BYTE[0]
    Inits byMsgCtrls, uqSIP;

    // Check if the frame is IPV4, and if not - discard frame
    IfL3NotIPv4DiscardFrame L_DISCARD;

    //Creating DIP Key
    CreateDIPKey ALIGN_0; //0

    //Creating SIP Key
    CreateSIPKey ALIGN_4; //4

    // Scan the frame, use parameters UREG[6].BYTE[2].BIT[4] and FBLK[1]
    //       for results of the scan (bit for found, and position)
    DetectSpecFrame byMsgCtrls.BIT[MSG_SPEC_FR], SCAN_STOP;
```

```
// Create a message, use UREG[6].BYTE[2], FBLK[1], and UREG[7].BYTE[0]
//          as inputs (bit for found, position, and SIP)
CreateMessage byMsgCtrls, SCAN_STOP, uqSIP;

Halt HALT_UNIC;
```

L_DISCARD:

```
    RFD_RECYC_PRS_RSV 0 /*mulc bit value*/, F_RD_PRS;
    Halt HALT_DISC;
```

17.2.2 **TOPresolve**

TOPresolve receives a message from TOPparse and, according to the defined operating mode, decides what to do with the frame. Then it provides this decision to TOPmodify (as a parameter in a message).

The defined operating mode is determined by using the GetRndBits instruction, which read bits from ST_GRP0. TOPresolve then jumps to the appropriate position in the program, according to the operating mode.

TOPresolve implements the required operations according to the operating mode, and its algorithm is quite simple. The main purpose of TOPresolve is either (a) to route the frame (using the TOPmodify, of course), if in operating mode 1, or (b) to activate the SIP learning mechanism (with the High Learn) and associated operations when applied (described in the Introduction of this chapter), or (c) to additionally calculate the offset of the "hot string" in the frame when required (in operating mode 3, for replacing the first three letters of the "hot string" by TOPmodify at a later stage).

Additionally, TOPresolve creates the appropriate message for TOPmodify (which is based on the message it received from TOPparse, appended with additional information). TOPresolve also creates an output port key for searching by the TOPsearch II, so that the correct port will be used by TOPmodify, along with the right Source MAC address of the NP.

At the end, TOPresolve either forwards the frame (at label L_FORWARD_MODE3 of the program) in case of a frame with a "hot string" in it, or routes the frame (at label L_L3_ROUT) as planned. The following code implements the TOPresolve stage.

```
EZtop Resolve;
#include "mcglobal.h"
#include "TOPresolve.h"
#include "cond.h"
#include "hdreg.h"
#include "stat.h"
#include "macro.h"
#include "rfd_macro.h"
#include "NP_Common.h"
#include "NP_Structs.h"
#include "NP_FastRegs.h"
#include "NP_RSV.h"
```

```
#include "NP_RSV_macros.h"

L_START:

    // If there are > 56<<5 (1792) occupied buffers, drop the frame,
    // since there is no point to continue, we have no room for
    // additional packets
    CriticalModeCheck RFD_CRITICAL_VAL, L_CRIT_MODE_DISCARD;

    // Perform all kinds of register initializations
    Inits;
    //Copy the result bits of operating mode to UDB.Byte[3] for decisions
    GetRndBits byFlags_UDB,
                ST_GRP0.INDEX [MSG_STR].BIT [MSG_MODE1],
                ST_GRP0.INDEX [MSG_STR].BIT [MSG_MODE2],
                ST_GRP0.INDEX [MSG_STR].BIT [MSG_MODE3],
         // One of mode bits is always set, use the rest for UDB flags
                ST_GRP0.INDEX [MSG_STR].BIT [VALID],
                ST_GRP0.INDEX [MSG_STR].BIT [VALID],
                ST_GRP0.INDEX [DIP_STR].BIT [MATCH],
      // DIP_STR is structure #1, bit1 is the match bit
                ST_GRP0.INDEX [SIP_STR].BIT [MATCH],
      // SIP_STR is structure #2, bit1 is the match bit
                ST_GRP0.INDEX [MSG_STR].BIT [MSG_SPEC_FR];

    //Check that we have a match in DIP_STR
    //In case that we don't have match, we can't route this frame, so discard it
    If (byFlags_UDB.BIT[2]) Jmp L_DISCARD | _NOP2; // UREG[0].BYTE[3].BIT[2]

    // Jump to the position in the program, according the operating mode

    If (byFlags_UDB.BIT[7]) Jmp L_MODE1 | _NOP2;
    If (byFlags_UDB.BIT[6]) Jmp L_MODE2 | _NOP2;
    If (byFlags_UDB.BIT[5]) Jmp L_MODE3 | _NOP2;
 // Otherwise—it is illegal (no operating mode), proceed to discard

L_DISCARD:
    // Discard frame
    Discardtreatment;
    Halt HALT_DISC;
//-----------------------------------------------------------------

L_MODE1:
// Check UDB.BIT[SPEC_FR] (found bit indicating the string is in frame)
    If (!byFlags_UDB.BIT[SPEC_FR]) Jmp L_L3_ROUT | _NOP2;
    Jmp L_MODE1_DISC | _NOP2;
//-----------------------------------------------------------------

L_MODE2:
    // Check UDB.Byte[3] for hot SIP
    If (byFlags_UDB.BIT[SIP_MATCH]) Jmp L_MODE2_DISC | _NOP2;
    // maybe 1st "SEX" or "BOMB" frame for that SIP
```

```
    If (!byFlags_UDB.BIT[SPEC_FR]) Jmp L_L3_ROUT | _NOP2;

    //Teach the Search1 database that there is one more hot SIP to
    // block in the future. . .
    // first forbidden frame for that SIP

    SipHighLearn;
      Jmp L_MODE2_DISC | _NOP2;
//-------------------------------------------------------------

L_MODE3:
    If (byFlags_UDB.BIT[SIP_MATCH]) Jmp L_MODE3_CHECK_DATA | _NOP2;
      If (!byFlags_UDB.BIT[SPEC_FR]) Jmp L_L3_ROUT | _NOP2;

      //Teach the Search1 database that there is one more hot SIP to block
      // in the future...
      SipHighLearn;

L_MODE3_CHECK_DATA:
    // SIP_MATCH case
    If (!byFlags_UDB.BIT[SPEC_FR]) Jmp L_L3_ROUTE | _NOP1;
    Mov byFastReg, L3_FAST_REG, 1; // 64
    //"Hot" word found; we want forward frame to special IP
    //Check offset of "hot" data

    CheckDataReplaceOffset
        byFastReg,                      //UREG[7].BYTE[3]
        uxWorkBufOff,                   //UREG[7]
        byWorkBufNum,                   //UREG[7].BYTE[2]
        byFlags_UDB.BIT[READ_MDF];      //UREG [0].BYTE[3].BIT[2]
//-------------------------------------------------------------
// Forward the frame

L_FORWARD_MODE3:
    //Create a message for TOPmodify
    CreateSearchIIMsg DIP_SPEC_PORT_OFF,    //3
                    SPEC_IP_FLAG_TRUE,      //1
                    byFastReg,              //UREG[7].BYTE[3]
                    uxWorkBufOff,           //UREG[7]
                    byWorkBufNum;           //UREG[7].BYTE[2]
    //Create a Search Key
    CreateOutPortCfgKey DIP_SPEC_PORT_OFF;  //3

    If (!byFlags_UDB.BIT[READ_MDF]) MovBits CTRL_REG, 0, 3;
    Halt HALT_UNIC;
//-------------------------------------------------------------
// Route the frame

L_L3_ROUT:
    // Create a message for TOPModify
```

```
    CreateSearchIIMsg DIP_OUT_PORT_OFF,        //1
                      SPEC_IP_FLAG_FALSE,       //0
                      L3_FAST_REG,              //1<<64
                      uxWorkBufOff,             // not relevant
                      byWorkBufNum;             // not relevant
    //Create a Search Key
    CreateOutPortCfgKey DIP_OUT_PORT_OFF;      //1
    Halt HALT_UNIC;

//-----------------------------------------------------------------

L_MODE1_DISC:
   Jmp L_DISCARD | _NOP2;
//-----------------------------------------------------------------

L_MODE2_DISC:
   Jmp L_DISCARD | _NOP2;
//-----------------------------------------------------------------

L_CRIT_MODE_DISCARD:
   Jmp L_DISCARD | _NOP2;
```

17.2.3 **TOPmodify**

TOPmodify is responsible for changing the frame according to the decision it receives from TOPresolve. Two modifications must be carried out: one is to replace the Source MAC address of the frame to that of the egress port of the NP; the other is to replace the Destination MAC address of the next hop device (e.g., router) according to the next hop address. An optional modification may be required, in operating mode 3; that is, to replace the first three letters of the "hot string" with "XXX," and to change the DIP address so that the frame will be diverted to the required place.

Replacing these letters, however, is not so simple, since the "hot string" may be located in two consecutive buffers (belonging to the same frame). TOPmodify is written in such a way that every possible case is handled, that is, the change operation might be required across frames. A few cases need to be taken into consideration: the strings "sex" or "bomb" might be contained in their entirety in the first memory buffer, but the strings can also be split over two buffers that are not consecutive in the memory structure. Therefore, each of the possible cases needs to be treated appropriately. After changing the frame (and taking care of the MAC and DIP addresses), the frame is sent away.

At the first stage, TOPModify initializes all necessary registers like SEND_REG's, FAST_REG and so on. At the second stage, TOPmodify jumps according to the FAST_REG to an appropriate program location, and modifies the "hot string" in the frame to "XXX." At the last stage, TOPmodify performs modifications of the frame at Layers 2–4 (at label L_L3 of the code), as discussed previously (to replace MAC addresses and possibly change the DIP if required). The following code implements the TOPmodify stage.

```
EZtop Modify;
#include "mcglobal.h"
#include "TOPmodify.h"
#include "stat.h"
#include "macro.h"
#include "rfd_macro.h"
#include "network.h"
#include "NP_FastRegs.h"
#include "NP_Common.h"
#include "NP_Structs.h"
#include "NP_MDF.h"
#include "NP_MDF_macros.h"
#include "Rfd.MDF.h"

LDREG LREG[F_FRST_BUF            ], L_FRST_BUF            ;
LDREG LREG[F_NOT_FRST            ], L_NOT_FRST            ;
LDREG LREG[F_FRST_SCND_BUF_x_xx ], L_FRST_SCND_BUF_x_xx ;
LDREG LREG[F_FRST_SCND_BUF_xx_x ], L_FRST_SCND_BUF_xx_x ;
LDREG LREG[F_TWO_BUFS_xx_x       ], L_TWO_BUFS_xx_x       ;
LDREG LREG[F_TWO_BUFS_x_xx       ], L_TWO_BUFS_x_xx       ;
LDREG LREG[F_L3                  ], L_L3                  ;


L_START:
    GetRes uxFramePtr, MSG_FR_PTR_OFF(MSG_STR), 2;  // UREG[1].BYTE[2]<-5(0)

    //Fill the FAST register for future jump
    GetRes FAST_REG, MSG_FAST_REG_OFF(MSG_STR), 1;   // OREG[0]<-19(0)

    // Get frame len, num bufs, src port
    GetRes SEND_REG1, MSG_HD_REG2_OFF(MSG_STR), 4;  // OUT_IF [1]<-7(0)
#ifdef EZ_NP_1c;
    GetRes tmpSrcPort, MSG_HD_REG9_OFF(MSG_STR), 1;  // UREG[6]<-10(0)
    Nop;
    // SRC port
    MovBits sSend_bitSrcPort,tmpSrcPort, 4;   // OUT_IF[1].BIT[21]<-UREG[6]
    Nop;
    // Zero msbits of src port
    MovBits SEND_REG1.BYTE[3].BIT[1], 0, 7;  // OUT_IF[1].BYTE[3].BIT[1]<-0
#else;
    // Zero msbits of src port
    MovBits SEND_REG1.BYTE[3], 0, 8;
#endif
    Nop;

    //Call InitRfdSrcPort macro
    InitRfdSrcPort sSend_bitSrcPort;

    //time stamp + frame pointer
```

```
        GetRes SEND_REG0, MSG_HD_REG1_OFF(MSG_STR), 4, _JPE_MDF;
            // Zero uxZeroFrPtr and uxL3FrPtr
        Mov uxZeroFrPtr, UDB, 4; // SREG[8]<-UREG[0]
            // Get Port + QoS
        GetRes QUEUE_NUM, MSG_OUT_PORT_OFF(MSG_STR), 2; // OUT_IF[3]<-27(0)
//------------------------------------------------------------------
//"Hot" data not is in the first buffer and also is not split between buffers
L_NOT_FRST:
#define uxCurPtr uqTemp;
    Mov uxCurPtr, uxFramePtr, 2; // UREG[5]<-UREG[1].BYTE[2]
    GetRes ALU, MSG_WORK_BUF_NUM_OFF(MSG_STR), 1; // OREG[1]<-26(0)
    Sub CNT, ALU, 1, 1;

FindCurBuffPtr uxCurPtr;
    Mov ALU, FULL_BUF_SIZE, 4; // OREG[1]<-512
    //Call ReadBufferToLocalMemory macro
    ReadBufferToLocalMemory uxCurPtr, ALU;

    //Call OneBufReplaceTreat macro
    OneBufReplaceTreat ZERO_OFF; // DISP_REG_VAL;

    Mov ALU, FULL_BUF_SIZE, 4;

    //Call WriteBufToGlobalMemory macro
    WriteBufToGlobalMemory uxCurPtr,
                           ALU,
                           ZERO_OFF;

    // Read first buffer
    Mov ALU, FULL_BUF_SIZE, 4;

    //Call ReadBufferToLocalMemory macro
    ReadBufferToLocalMemory uxFramePtr, ALU;

    Jmp L_L3 | _NOP2;

#undef uxCurPtr;

//------------------------------------------------------------------
//"Hot" data not is in the first buffer and is split between buffers:
// first letter is in buffer x and the rest of the "hot" word is in
// buffer x+1
L_TWO_BUFS_x_xx:
#define uxFrstPtr uqTemp;
#define uxScndPtr uqTemp.BYTE[2];

    //Calling the MACRO for replaces "hot" word
    TwoBufsTreatment uxFrstPtr,              //P_FRST_BUF_PRT,
                     1,                      //P_FRST_BUF_DATA_SIZE,
                     BUF_LAST_BYTE_OFF,      //P_FRST_BUF_DATA_OFF,
```

```
                              X_ASCII,                        //P_FRST_DATA,
                              uxScndPtr,                      //P_SCND_BUF_PTR,
                              2,                              //P_SCND_BUF_DATA_SIZE,
                              XX_ASCII;                       //P_SCND_DATA;

        Jmp L_L3 | _NOP2;

#undef uxFrstPtr;
#undef uxScndPtr;

//----------------------------------------------------------------
//"Hot" data is not in the first buffer, and is split between buffers:
// first two letters are in buffer x and the rest are is in buffer x+1
L_TWO_BUFS_xx_x:
#define uxFrstPtr uqTemp;
#define uxScndPtr uqTemp.BYTE[2];

        //Calling the MACRO for replaces "hot" word
        TwoBufsTreatment uxFrstPtr,                    //P_FRST_BUF_PRT,
                         2,                            //P_FRST_BUF_DATA_SIZE,
                         BUF_BEFORE_LAST_BYTE_OFF,     //P_FRST_BUF_DATA_OFF,
                         XX_ASCII,                     //P_FRST_DATA,
                         uxScndPtr,                    //P_SCND_BUF_PTR,
                         1,                            //P_SCND_BUF_DATA_SIZE,
                         X_ASCII;                      //P_SCND_DATA;
        Jmp L_L3 | _NOP2;

#undef uxFrstPtr;
#undef uxScndPtr;

//----------------------------------------------------------------
//"Hot" data is split between the first and the second buffer:
// first two letters of the "hot" word are in the first buffer,
// and the rest of the "hot" word are in the second buffer
L_FRST_SCND_BUF_xx_x:
#define uxScndBufPtr uqTemp;

    //Getting the pointer to second buffer
    GetRfdNextPtr uxFramePtr, uxScndBufPtr;

    //Calling the MACRO for replace "hot" word
    TwoBufsReplaceTreat uxFramePtr,                     //P_FRST_BUF_PRT,
                        2,                              //P_FRST_BUF_DATA_SIZE,
                        BUF_BEFORE_LAST_BYTE_OFF,       //P_FRST_BUF_DATA_OFF,
                        XX_ASCII,                       //P_FRST_DATA,
                        uxScndBufPtr,                   //P_SCND_BUF_PTR,
                        1,                              //P_SCND_BUF_DATA_SIZE,
                        X_ASCII;                        //P_SCND_DATA;
        Jmp L_L3 | _NOP2;

#undef uxScndBufPtr;
```

```
//-------------------------------------------------------------------
// "Hot" data is split between the first and the second buffer:
// the first letter of "hot" word is in first buffer,
// and the rest of the "hot" word is in the second buffer
L_FRST_SCND_BUF_x_xx:
#define uxScndBufPtr uqTemp;

    //Getting the pointer to second buffer
    GetRfdNextPtr sSend_uxFramePtr, uxScndBufPtr;

    //Calling the MACRO for replace "hot" word
    TwoBufsReplaceTreat uxFramePtr,          //P_FRST_BUF_PRT,
                        1,                    //P_FRST_BUF_DATA_SIZE,
                        BUF_LAST_BYTE_OFF,    //P_FRST_BUF_DATA_OFF,
                        X_ASCII,              //P_FRST_DATA,
                        uxScndBufPtr,         //P_SCND_BUF_PTR,
                        2,                    //P_SCND_BUF_DATA_SIZE,
                        XX_ASCII;             //P_SCND_DATA;
    Jmp L_L3 | _NOP2;

#undef uxScndBufPtr;

//-------------------------------------------------------------------
//The entire "hot" word is in the first buffer
L_FRST_BUF:

    OneBufReplaceTreat FRST_BUF_OFF; // DISP_REG_VAL;

//-------------------------------------------------------------------
//Modifying Layers 2-4
L_L3:
    MovBits DISP_REG, FRST_BUF_OFF, 9; // SREG[2].BYTE [2]<-64
    MovBits FR_STRT, FRST_BUF_OFF, 9; // OUT_IF [4]<-64

    GetRes uxL3FrPtr, MSG_L2_SIZE_OFF(MSG_STR), 1; // SREG[8].BYTE [2]<-1(0)
    MovBits uxFirstBufSize, FRST_BUF_SIZE, 16; // UREG[1]<-448

    GetRes byUDB_SpecIPFlag, MSG_SPEC_IP_FLAG_OFF(MSG_STR), 1;

    // L2 treatment
    Copy MAC_SA_OFF(uxZeroFrPtr), CFG_IF_MAC_OFF(CFG_STR), 4;
        //6(SREG[8])<-1(3)
    Copy MAC_SA_OFF_P4(uxZeroFrPtr), CFG_IF_MAC_OFF_P4(CFG_STR), 2;
    Copy MAC_DA_OFF(uxZeroFrPtr), MSG_NXT_HOP_MAC_OFF(MSG_STR), 4;
    Copy MAC_DA_OFF_P4(uxZeroFrPtr), MSG_NXT_HOP_MAC_OFF_P4(MSG_STR), 2;

    // TTL treatment
    Get ALU, IP_TTL_OFF(uxL3FrPtr), 1, _NJP_MDF;
    Sub ALU, ALU, 1, 1;
    Nop;
    Put IP_TTL_OFF(uxL3FrPtr), ALU, 1, _NJP_MDF;
```

```
                // lbuf len = min (adds start, 448) - 1
                Sub ALU, sSend_uxFrameLen, uxFirstBufSize, 2;
                    Mov sSend_bit1stBufLen, sSend_uxFrameLen, 2;
                If (A) Mov sSend_bit1stBufLen, uxFirstBufSize, 2;
                If (!byUDB_SpecIPFlag.BIT[0]) Jmp L_AFTER_SPEC_IP | _NOP0;
                    MovBits sSend_bitHdrLen, HDR_LEN0_BUF_TYPE1, 8;
                    Sub sSend_bit1stBufLen, sSend_bit1stBufLen, 1, 2;
                //Inserting Special IP address if needed
                Copy IP_DIP_OFF(uxL3FrPtr), CFG_SPEC_IP_OFF(CFG_STR), 4;

        L_AFTER_SPEC_IP:
            Mov sSend_uxFramePtr, uxFramePtr, 2;

            Halt _WHALT_TF_MDF, HALT_UNIC;
```

17.2.4 Definitions and Macro Files

This subsection contains the definition and macro files used in the sample code.

17.2.4.1 TOPparse Macros (TOP_PRS_macros.h)

```
/*================================================================
MACRO Inits: Initializing some necessary registers
Parameters: two registers, one for mode and the second for SIP
Output: RD_PTR       <- point to beginning of IP header
        (KBS0 = 0)   <- base address for message in KMEM
        (KBS1 = 32)  <- base address for key in KMEM
        P_uqSIP      <- holds Source IP
        P_byMsgCtrls <- holds valid bit, and is used later for control
================================================================*/
MACRO Inits P_byMsgCtrls, P_uqSIP;
    // FMEM base ptr <- HD_REG0 - L2 size
    Mov RD_PTR, sHR0_bitsLayer2Size, 1; // SREG[0]<-HWARE[0].BIT[0]

    //KBS0 & KBS1 - Base addr for key mem <- MSG_SIZE (=32) << 16
    Mov MSG_KBS, COM_KBS_AFTER_MSG_MSG_KBS_0_OFF, 4; // happens to be 2097152

    Mov P_byMsgCtrls, VALID_VAL, 1;

    // save Source IP for later treatment
    Get P_uqSIP, IP_SIP_OFF(RD_PTR), 4; // UREG<-FMEM

ENDMACRO;
/*================================================================
MACRO CreateMessage:Creating message
Parameters: P_byMsgCtrls - hold control bit
            P_DATA_OFF   - offset of the hot string
            P_SIP        - SIP

Output: KMEM holds the message
================================================================*/
```

```
MACRO CreateMessage P_byMsgCtrls, P_DATA_OFF, P_SIP;
    MovBits P_byMsgCtrls.BIT[MSG_MODE], CREGO_MODE, 3;

    PutKey MSG_HD_REGO_OFF(MSG_KBS), HD_REGO, 2;   // KMEM<-HD_REGO
    PutKey MSG_HD_REG1_OFF(MSG_KBS), HD_REG1, 4;
    PutKey MSG_HD_REG2_OFF(MSG_KBS), HD_REG2, 4;
#ifdef EZ_NP_1c //for NP1c mode
    PutKey MSG_HD_REG9_OFF(MSG_KBS), SOURCE_PORT_REG, 1;
#endif;

    // the SIP we saved at the beginning
    PutKey MSG_SIP_OFF(MSG_KBS), P_SIP, 4; // SREG[12]<-UREG[7].BYTE[0]

    // Take the SCAN_STOP (the pointer to the 's'/'b' found in
    // 'sex' /'bomb')
    PutKey MSG_DATA_OFF(MSG_KBS), P_DATA_OFF, 2; // 17(SREG[12])<-FBLK[1]

    PutKey 0(MSG_KBS), P_byMsgCtrls, 1;
    PutHdr HREG[0], MSG_HDR, 3; // HREG[0]<-0x3E013

ENDMACRO;

/*================================================================
MACRO CreateDIPKey: Creating DIP Key
Parameters: DIP offset relative to RD_PTR
Output: KMEM holds DIP on offset 32
        Filling HREG[1] with header for TOPSearchI
================================================================*/
MACRO CreateDIPKey P_ALIGN;
    Copy P_ALIGN(COM_KBS)+, IP_DIP_OFF(RD_PTR), 4;   // KMEM<-FMEM
    PutHdr HREG[1], DIP_HDR, 3; // HREG[1]<-0xF050

ENDMACRO;

/*================================================================
MACRO CreateSIPKey: Creating SIP Key
Parameters: SIP offset relative to RD_PTR
Output: KMEM holds SIP on offset 36
        Filling HREG[2] with header for TOPSearch
================================================================*/
MACRO CreateSIPKey P_ALIGN;

    Copy P_ALIGN(COM_KBS)+, IP_SIP_OFF(RD_PTR), 4;   // KMEM<-FMEM
    PutHdr HREG[2], SIP_HDR, 3; // HREG[2]<-0xF0B0

L_SIP_END:
ENDMACRO;

/*================================================================
MACRO IfL3NotIPv4DiscardFrame: Checking if frame is IPv4 or not
Parameters: Discard label
Output: YES => continue TOPParse
        NO => Jump to label P_L_DISCARD in TOPParse(Discard frame)
```

```
=============================================================*/
MACRO IfL3NotIPv4DiscardFrame P_L_DISCARD;
    Sub ALU, HD_REGO, HW_IPv4_BITS, 4, HW_IPv4_BITS_MASK, _ALU_FRST;
        Nop;

    // jump if not zero (i.e., not IPV4)
    If (!FLAGS.BIT[F_ZR_PRS]) Jmp P_L_DISCARD | _NOP2;
ENDMACRO;
/*=============================================================
MACRO DetectSpecFrame: This MACRO detecting if word "sex" or "bomb" are
exist in frame. Only TCP frames with HTTP port require the checking all
other frames routed.
Parameters: P_SPEC_FLAG—flag to update if "hot" word found
            P_SCAN_STOP—special register for saving offset of "hot" word
Output: Input registers updated

=============================================================*/
MACRO DetectSpecFrame P_SPEC_FLAG, P_SCAN_STOP;
    //Get IP protocol from the frame
    Get ALU, IP_PRT_OFF(RD_PTR), 1; // FBLK [4]<-9(SREG[0])
    Sub ALU, ALU, TCP_PROT, 1; // TCP_PORT is 6

    // take advantage of the data hazard in pipeline execution
    // Get Source and destination Ports (note that we read both ports)
    //                        and note also the little endian NP is using
    Get uxDPort, IP_TCP_SPRT_OFF(RD_PTR), 4; // UREG[5].BYTE[0]<-20(SREG[0])

    // now make the jump if it is not TCP
    If (!FLAGS.BIT[F_ZR_PRS]) Jmp L_FINISH_DETECT | _NOP2;

    // First compare SPORT to PORT# received from host
    Sub ALU, uxSPort, uxCREGO_HTTP_PORT, 2;

        // Second compare DPORT to PORT# received from host
        // Take advantage of the data hazard
        Sub ALU, uxDPort, uxCREGO_HTTP_PORT, 2;

    //Check if SPORT=HTTP - go scan if OK (first compare)
    If (FLAGS.BIT[F_ZR_PRS]) Jmp L_SCAN_HTTP_DATA | _NOP2;
        //if we get to this point it's mean that SPORT<>HTTP
        //Check if DPORT<>HTTP - go scan if OK (second compare)
        If (!FLAGS.BIT[F_ZR_PRS]) Jmp L_FINISH_DETECT | _NOP2;

L_SCAN_HTTP_DATA:
    //make sure scan will not stop in middle of longest possible frame
    Mov LIM_REG, MAX_SCAN_LIM, 2; //16000

    //Point to HTTP data, add to frame buffer pointer the TCP header-1
    //(for scan loop)
    Add RD_PTR, RD_PTR, IP_TCP_SIZE_M1, 4; //RD_PTR = start HTTP-1 (=39)
```

```
                    //put pointer to jump when frame scanning exhausts @ end of frame
                    Mov EOF_ADDR, L_FINISH_DETECT, 4;

L_DO_SCAN:
//scan and look for a 's','S','b' or 'B' from FMEM+1 (RD_PTR) and inc
                    FindDel NULL_REG, 1(RD_PTR)+, ALL_DELIMS_MASK, LIM_REG, _FRWD_PRS, _JEOF_PRS;

                    //if FinDel found a one of letters, we get 8 bytes from this point
                    Get CAMI, 0(RD_PTR), 8;

                    // checking in TCAM for "sex....." or "bomb...."
                    LookCam CAMO, CAMI, TCAM64[SPEC_GRP];
                        Nop;
                        Nop;
#ifdef EZ_STAT_NP_1c_MODE
                    Nop;
#endif

                    //tricky: jump at any rate, but in the pipe we have the
                    //conditional jump!
                    Jmp L_DO_SCAN | _NOPO;
                        //Now see if we had a match!
                        If (FLAGS.BIT[F_MH_PRS]) Jmp L_FINISH_DETECT | _NOP1;
                            // still in the pipe, before jumping mark SEX or BOMB
                            If (FLAGS.BIT[F_MH_PRS]) MovBits P_SPEC_FLAG, 1, 1;

L_FINISH_DETECT:
ENDMACRO;
```

17.2.4.2 TOPparse Definition File (TOP_PRS.h)

```
/*===================== constants ================================*/
#define COM_KBS_AFTER_MSG_MSG_KBS_0_OFFM (MSG_SIZE<<16);
#define VALID_VAL                        1;
#define ALIGN_O                          0;
#define ALIGN_4                          4;
#define TCP_PROT                         6;
#define sHRO_bitIpL3T_OFF                9;
#define HW_IPv4_BITS                     (1<<sHRO_bitIpL3T_OFF);

/*===================== frame offsets =========================*/
#define IP_TCP_SPRT_OFF (IP_BASE_SIZE+TCP_SPRT_OFF);
#define IP_TCP_SIZE_M1 (IP_BASE_SIZE+TCP_BASE_SIZE-1);

/*===================== key headers =========================*/
#define VALID_BITS_0             0;
#define SINGLE_VALID             1;

#define MSG_HDR (((MSG_SIZE - 1)>> 3) | (_SKEY_PRS<<4)    |
                 (_MKEY_PRS<<5)    | (MSG_STR << 6)       |
                 (_MSG_PRS<<12)    | (SINGLE_VALID<<13) |
                 (((MSG_SIZE - 1)>> 1)<<14));
```

```
#define MSG_HDR_MODE3 (((MSG_M3_SIZE - 1)>> 3) | (_SKEY_PRS<<4)      |
                       (_MKEY_PRS<<5)         | (MSG_STR << 6)       |
                       (_MSG_PRS<<12)         | (SINGLE_VALID<<13) |
                       (((MSG_M3_SIZE - 1)>> 1)<<14));

#define DIP_HDR (((DIP_KEY_SIZE - 1)>> 3) | (_SKEY_PRS<<4)      |
                 (_MKEY_PRS<<5)           | (DIP_STR << 6)      |
                 (_LKP_PRS<<12)           | (SINGLE_VALID<<13) |
                 ((DIP_KEY_SIZE-1)<<14));

#define SIP_HDR (((SIP_KEY_SIZE - 1)>> 3) | (_SKEY_PRS<<4)      |
                 (_LKEY_PRS<<5)           | (SIP_STR << 6)      |
                 (_LKP_PRS<<12)           | (SINGLE_VALID<<13) |
                 ((SIP_KEY_SIZE-1)<<14));

#define DIP_HDR_LAST (DIP_HDR | (_LKEY_PRS<<5));

/*===================== registers =============================*
#define MSG_KBS                    KBS0;
#define COM_KBS                    KBS1;

// do not use LIM_REG (UREG[3])
#define uxDPort                    UREG[5].BYTE[0];
#define uxSPort                    UREG[5].BYTE[2];
#define byMsgCtrls                 UREG[6].BYTE[2];
#define uqSIP                      UREG[7].BYTE[0];

/*===================== CREGS ============================*/
#define CREG0                      HD_REG3;
#define CREG1                      HD_REG4;

// CREG0_MODE:
// exactly one of bits 0-2 of CREG0 is set
// CREG0.BIT[0] – mcode works in mode 1
// CREG0.BIT[1] – mcode works in mode 2
// CREG0.BIT[2] – mcode works in mode 3
#define CREG0_MODE CREG0.BIT[0];

// CREG0_HTTP_PORT:
// contains real or proxy HTTP port number
#define uxCREG0_HTTP_PORT CREG0.BYTE[1];

LDREG CREG[0 ], 0x00005004;
LDREG CREG[2 ], 0x00005004;
LDREG CREG[4 ], 0x00005001;
LDREG CREG[6 ], 0x00005002;
LDREG CREG[8 ], 0x00005004;
LDREG CREG[10], 0x00005004;
LDREG CREG[12], 0x00005001;
LDREG CREG[14], 0x00005002;
```

```
/*===================== masks ====================*/
#define HW_IPv4_BITS_MASK MREG[0];

LDREG MREG[0], HW_IPv4_BITS;

/*===================== Scan defines ====================*/
LDDV "sSbB";

#define S_LOW_CASE_OFF              0;
#define S_UPPER_CASE_OFF            1;
#define B_LOW_CASE_OFF              2;
#define B_UPPER_CASE_OFF           3;

#define ALL_DELIMS_MASK        ((1<<S_LOW_CASE_OFF)     |
                                (1<<S_UPPER_CASE_OFF)   |
                                (1<<B_LOW_CASE_OFF)     |
                                (1<<B_UPPER_CASE_OFF)) ;

#define MAX_SCAN_LIM 16000;

/*===================== CAMs ====================*/
// TCAM64
#define SPEC_GRP                   0;

#define SPEC1                      "sex?????";
#define SPEC2                      "bomb????";

LDTCAM TCAM64[SPEC_GRP], SPEC1,    "11111111", 0;
LDTCAM TCAM64[SPEC_GRP], SPEC2,    "11111111", 0;
```

17.2.4.3 TOPresolve Macros (TOP_RSV_macros.h)

```
/*=============================================================
MACRO Inits: Initialization of TOPresolve registers
=============================================================*/
MACRO Inits;

#define byTemp_CtrlByte uqTemp5;

    // set Structure group register to 0
    Mov ST_GRP0, 0, 1;
    // initialize base to OMEM
    Mov HW_OBS, 0, 1;

    Get byTemp_CtrlByte, MSG_CTRL_REG_BYTE_OFF(MSG_STR), 2;

    //initialize the FR_PTR with the frame pointer, received from RMEM
    Get FR_PTR, MSG_FR_PTR_OFF(MSG_STR), 2;

    //Initialize control register with number of 64 bytes in 1st buffer
    MovBits CTRL_REG, byTemp_CtrlByte.BIT[CTRL_REG_OFF_IN_HW_BYTE],3;

#undef byTemp_CtrlByte;
```

```
ENDMACRO;
/*================================================================
MACRO Discardtreatment: Discard MACRO
================================================================*/
MACRO Discardtreatment;

#define sHR1              uqTemp5.BYTE[0];
#define sHR2              uqTemp6.BYTE[0];
#ifdef EZ_NP_1c;
   #define sHR9           uqTemp9.BYTE[0];
#endif;

   Get sHR1, MSG_HD_REG1_OFF(MSG_STR), 4; // UREG[5].BYTE[0]<-3(0)
   Get sHR2, MSG_HD_REG2_OFF(MSG_STR), 4; // UREG[6].BYTE[0]<-7(0)
#ifdef EZ_NP_1c;
   Get sHR9, MSG_HD_REG9_OFF(MSG_STR), 1; // UREG[9].BYTF[0]<-10(0)
#endif;

   RFD_RECYC_PRS_RSV 0 /*BIT_MULC_VAL*/, F_RD_RSV;

#undef sHR1;
#undef sHR2;
#undef sHR9;

ENDMACRO;
/*================================================================
MACRO CreateSearchIIMsg: Creating message for TOPmodify
Output: KMEM hold the message
        All needed parameters were transmitted as input of this MACRO
================================================================*/
MACRO CreateSearchIIMsg      P_OUT_PORT_OFF,
                             P_SPEC_IP_FLAG,
                             P_FAST_REG,
                             P_WORK_BUF_OFF,
                             P_WORK_BUF_NUM;

//next 3 instructions copy original message created by TOPparse
   Copy 0(HW_OBS),   0(MSG_STR), 8;
   Copy 8(HW_OBS),   8(MSG_STR), 8;
   Copy 16(HW_OBS), 16(MSG_STR), 8;

//next 3 instructions insert the result of DIP structure (#1) to message
   Copy MSG_OUT_PORT_OFF(HW_OBS), P_OUT_PORT_OFF(DIP_STR), 1;
   Copy MSG_OUT_QOS_OFF(HW_OBS), DIP_OUT_QOS_OFF(DIP_STR), 1;
   Copy MSG_NXT_HOP_MAC_OFF(HW_OBS), DIP_NXT_HOP_MAC_OFF(DIP_STR), 6;

   PutKey MSG_FAST_REG_OFF(HW_OBS), P_FAST_REG, 1;
   PutKey MSG_DATA_OFF(HW_OBS), P_WORK_BUF_OFF, 2;
   PutKey MSG_WORK_BUF_NUM_OFF(HW_OBS), P_WORK_BUF_NUM, 1;
   PutKey MSG_SPEC_IP_FLAG_OFF(HW_OBS), P_SPEC_IP_FLAG, 1;
   PutHdr HREG[0], RSV_MSG_HDR, 3;
```

```
ENDMACRO;
/*===============================================================
MACRO SipHighLearn: MACRO for High Learn
Output: After this MACRO we have "hot" SIP as a new entry in SIP data structure
===============================================================*/
MACRO SipHighLearn;

/* NP-1 message header fields (4 bytes) */
#define LRN_INFO_LEN
            (((((SIP_KEY_SIZE+3)>>2)<<2)+SIP_RES_SIZE-1)>>2);
#define STR_NUM                       SIP_STR;
#define CREATE_UPDATE_MODE            HL_CREATE;
#define CMD_MODE                      HL_ADD_CMD;
#define STATIC_MODE                   HL_NO_STATIC;
#define LOCAL_MODE                    HL_LOCAL;
#define SEND_MSG_MODE                 HL_NO_SEND_MSG;
#define INCR_CNT_MODE                 HL_NO_INCR_CNT;

#define RES_B0_B3     ((1<<HL_VALID_BIT_OFF)|
                       (1<<HL_MATCH_BIT_OFF));

/* prepare to write in 4 Dwords */
#define NP1_MSG_HDR   ((LRN_INFO_LEN<<HL_LRN_INFO_LEN_BIT_OFF)       |
                       (STR_NUM<<HL_STR_NUM_BIT_OFF)                 |
                 (CREATE_UPDATE_MODE<<HL_CREATE_UPDATE_MODE_BIT_OFF) |
                       (CMD_MODE<<HL_CMD_MODE_BIT_OFF)               |
                       (STATIC_MODE<<HL_STATIC_MODE_BIT_OFF)         |
                       (LOCAL_MODE<<HL_LOCAL_MODE_BIT_OFF)           |
                       (SEND_MSG_MODE<<HL_SEND_MSG_MODE_BIT_OFF)     |
                       (INCR_CNT_MODE<<HL_INCR_CNT_MODE_BIT_OFF));

    //wait for HL mechanism to be ready
    WAIT_NO_FLAG F_LN_RSV;   // if (SREG [10].BIT [8]) Jmp $ | _NOP2

    Mov LRN_REG0, NP1_MSG_HDR, 4;

#ifdef EZ_NP_1c;
#define LRN_RES_LEN ((SIP_RES_SIZE-1)>>2);
#define NP1C_MSG2_HDR (LRN_RES_LEN<<HL_RES_SIZE_BIT_OFF);

    Mov LRN_REG1, NP1C_MSG2_HDR, 4;
    Mov LRN_REG2, RES_B0_B3, 4; // 3
    // LRN_REG3 (contains result bytes 4-7) is not relevant
    Mov LRN_SIZE, WRITE_ALL_LRN_REGS, 1; // 3
     Nop;

HL_2ND_WAIT L_DISCARD, 0;

    // LRN_REG0 - LRN_REG3 (contains result bytes 8-23) is not relevant
    Mov LRN_SIZE, WRITE_ALL_LRN_REGS, 1;
```

```
    Nop;
    WAIT_NO_FLAG F_LN_RSV;

    // LRN_REG0 - LRN_REG1 (contains result bytes 24-31) is not relevant
    Get LRN_REG2, MSG_SIP_OFF(MSG_STR), 4; // copy key

    Mov LRN_SIZE, WRITE_LRN_REG0_1_2, 1;
#else;
    Mov LRN_REG1, RES_B0_B3, 4;
    // LRN_REG2 - LRN_REG3 (contains result bytes 4-11) is not relevant
    Mov LRN_SIZE, WRITE_ALL_LRN_REGS, 1;
    Nop;
    HL_2ND_WAIT L_DISCARD, 0;

    // LRN_REG0 - LRN_REG3 (contains result bytes 12-27) is not relevant
    Mov LRN_SIZE, WRITE_ALL_LRN_REGS, 1;
    Nop;
    WAIT_NO_FLAG F_LN_RSV;

    // LRN_REG0 (contains result bytes 28-31) is not relevant
    Get LRN_REG1, MSG_SIP_OFF(MSG_STR), 4; // copy key

    Mov LRN_SIZE, WRITE_LRN_REG0_1, 1;
#endif;

ENDMACRO;
/*=============================================================
MACRO CheckDataReplaceOffset: This MACRO check the offset of "hot"
word in frame and filling the FAST_REG for future modification
acceleration in TOPModify.
Output: Register P_FAST_REG updated for transmitting to TOPModify
=============================================================*/

MACRO CheckDataReplaceOffset  P_FAST_REG,
                              P_uxWorkBufOff,
                              P_byWorkBufNum,
                              P_READ_MDF_FLAG;
#define uxDataOff uqTemp5;

    Get uxDataOff, MSG_DATA_OFF(MSG_STR), 2; // UREG[5]<-17(0)
      Get P_uxWorkBufOff, MSG_DATA_OFF(MSG_STR), 2;
    // Check first buffer case
    Sub ALU, uxDataOff, MAX_FRST_BUF_DATA_OFF, 2; // 445
      Mov P_FAST_REG, FRST_BUF_FAST_REG, 1; // 65
    If (!A) Jmp L_END_CHECK | _NOP2;

    // Check first + second buffers case
    Sub ALU, uxDataOff, FRST_BUF_SIZE, 2; // 448
    MovBits P_READ_MDF_FLAG, 0, 1;
    If (!FLAGS.BIT[F_CY_RSV]) Jmp L_NOT_FRST_BUF | _NOP2;
```

```
        // Check "x"+"xx", "xx"+"x" cases
        Sub ALU, uxDataOff, FRST_BUF_SIZE_M1, 2; //447
        Jmp L_END_CHECK | _NOP0;

        If (FLAGS.BIT[F_ZR_RSV])
            Mov P_FAST_REG, FRST_SCND_BUF_x_xx_FAST_REG, 1; // 68
        If (!FLAGS.BIT[F_ZR_RSV])
            Mov P_FAST_REG, FRST_SCND_BUF_xx_x_FAST_REG, 1; // 72

L_NOT_FRST_BUF:

        // byWorkBufNum = (uxDataOff-448)/512 + 2;
        // uxWorkBufOff = (uxDataOff-448)% 512;
        Sub uxDataOff, uxDataOff, FRST_BUF_SIZE, 2; // 448
            Nop;
        Sub P_uxWorkBufOff, uxDataOff, 0, 2, M_0x000001FF, _ALU_FRST;
            MovBits P_byWorkBufNum, uxDataOff.BIT[9], 5;
        Mov P_FAST_REG, NOT_FRST_FAST_REG, 1; // 66

        // check two buffer replace case
        Sub ALU,P_uxWorkBufOff,MAX_FULL_BUF_DATA_OFF,2,M_0x000001FF,_ALU_FRST;
        Add P_byWorkBufNum, P_byWorkBufNum, 1, 1, M_0x0000001F, _ALU_FRST;
        // forward buf_num-1
        If (!A) Jmp L_END_CHECK | _NOP2;

        // Check "x"+"xx", "xx"+"x" cases
        Sub ALU, P_uxWorkBufOff, FULL_BUF_SIZE_M1, 2; // 511
            Mov P_FAST_REG, TWO_BUFS_xx_x_FAST_REG, 1;  // 80
        If (FLAGS.BIT[F_ZR_RSV])
            Mov P_FAST_REG, TWO_BUFS_x_xx_FAST_REG, 1; // 96

L_END_CHECK:
#undef uxDataOff;
ENDMACRO;

/*=======================================================================
MACRO CreateSearchIIMsg: Creating Search key for SearchII
Input: P_OUT_PORT_OFF - port for searching in structure#2
Output: KMEM hold the key (output port)
        HREG updated
========================================================================*/
MACRO CreateOutPortCfgKey P_OUT_PORT_OFF;
    Copy RSV_MSG_SIZE(HW_OBS), P_OUT_PORT_OFF(DIP_STR), 1;
    PutHdr HREG[1], CFG_HDR, 3;

ENDMACRO;

/*=======================================================================
MACRO CriticalModeCheck: this MACRO check if we have enough space in the RFD
========================================================================*/
```

```
MACRO CriticalModeCheck
                    PARAM_RFD_CRITICAL_VAL, PARAM_RFD_DISCARD_LABEL;
    Sub ALU, RX_CNT, PARAM_RFD_CRITICAL_VAL, 1, M_0x000000F8, _ALU_FRST;
    Nop;
    If (A) Jmp PARAM_RFD_DISCARD_LABEL | _NOP2;

ENDMACRO;
```

17.2.4.4 TOPresolve Definition File (TOP_RSV.h)

```
/*===================== constants ==========================*/
#define CTRL_REG_OFF_IN_HW_BYTE      4;
#define NO_TAG_ETHERNET_SIZE         14;
#define TAG_ETHERNET_SIZE            18;
#define WORD_SIZE                    3;
#define SPEC_IP_FLAG_FALSE           0;
#define SPEC_IP_FLAG_TRUE            1;

#define RFD_CRITICAL_VAL                ((0<<7) | (0<<6) | (1<<5) |
                                        (1<<4) | (1<<3));


#define MAX_FULL_BUF_DATA_OFF           (FULL_BUF_SIZE -WORD_SIZE);
#define MAX_FRST_BUF_DATA_OFF           (FRST_BUF_SIZE -WORD_SIZE);

#define FRST_BUF_SIZE_M1                (FRST_BUF_SIZE-1);
#define FULL_BUF_SIZE_M1                (FULL_BUF_SIZE-1);

#define RSV_CFG_SPEC_IP_OFF             (RSV_MSG_SIZE+CFG_SPEC_IP_OFF);

/*===================== high learn constants =====================*/
#define WRITE_LRN_REG0               0;
#define WRITE_LRN_REG0_1             1;
#define WRITE_LRN_REG0_1_2           2;
#define WRITE_ALL_LRN_REGS           3;

// fields values
// CREATE_UPDATE_MODE values
#define HL_CREATE                    1; // don't change entry if exists
#define HL_UPDATE                    0; // change entry if exists
// CMD_MODE values
#define HL_ADD_CMD                   1;
#define HL_RMV_CMD                   0;
// STATIC_MODE values
#define HL_STATIC                    1;
#define HL_NO_STATIC                 0;
//LOCAL_MODE values
#define HL_LOCAL                     1;
#define HL_NO_LOCAL                  0;
// SEND_MSG_MODE values
#define HL_SEND_MSG                  1;
```

```
#define HL_NO_SEND_MSG                  0;
// INCR_CNT_MODE values
#define HL_INCR_CNT                     1;
#define HL_NO_INCR_CNT                  0;

// fields offsets (in bits- from header start)
#define HL_LRN_INFO_LEN_BIT_OFF         0;
#define HL_STR_NUM_BIT_OFF              8;
#define HL_CREATE_UPDATE_MODE_BIT_OFF14;
#define HL_CMD_MODE_BIT_OFF             16;
#define HL_STATIC_MODE_BIT_OFF          18;
#define HL_LOCAL_MODE_BIT_OFF           19;
#define HL_SEND_MSG_MODE_BIT_OFF        20;
#define HL_INCR_CNT_MODE_BIT_OFF        21;

// bytes 0 - 3 fields offsets (in bits - from result start)
#define HL_VALID_BIT_OFF                0;
#define HL_MATCH_BIT_OFF                1;

#ifdef EZ_NP_1c;

#define HL_RES_SIZE_BIT_OFF             8;

#endif;
/*===================== key headers =========================*/
#define SINGLE_VALID                    1;
#define _SKEY_RSV                       1;
#define _LKEY_RSV                       1;
#define _MKEY_RSV                       0;
#define _LKP_RSV                        1;
#define _MSG_RSV                        0;

#define RSV_MSG_HDR    (((RSV_MSG_SIZE - 1)>> 3)      |
                        (_SKEY_RSV<<4)                |
                        (_MKEY_RSV<<5)                |
                        (MSG_STR << 6)                |
                        (_MSG_RSV<<12)                |
                        (SINGLE_VALID<<13)            |
                        (((RSV_MSG_SIZE - 1)>> 1)<<14));

#define CFG_HDR        (((CFG_KEY_SIZE - 1)>> 3)      |
                        (_SKEY_RSV<<4)                |
                        (_LKEY_RSV<<5)                |
                        (CFG_STR << 6)                |
                        (_LKP_RSV<<12)                |
                        (SINGLE_VALID<<13)            |
                        ((CFG_KEY_SIZE-1)<<14));
/*===================== masks =========================*/
#define M_0x000001FF MREG[0];
#define M_0x0000001F MREG[1];
#define M_0x000000F8 MREG[2];
```

```
LDREG MREG[0], 0x000001FF;
LDREG MREG[1], 0x0000001F;
LDREG MREG[2], 0x000000F8;

/*==================== registers ===========================*/
#define HW_OBSOBS0;

// UDB defines
#define byTemp_UDB                          UDB.BYTE[0];
#define byFlags_UDB                         UDB.BYTE[3];
#define READ_MDF                            2;
#define SIP_MATCH                           1;
#define SPEC_FR                             0;

#define uqTemp5                             UREG[5];
#define uqTemp6                             UREG[6];

#ifdef EZ_NP_1c;
  #define uqTemp9                           UREG[9];
#endif;

#define uxWorkBufOff                        UREG[7];
#define byWorkBufNum                        UREG[7].BYTE[2];
#define byFastReg                           UREG[7].BYTE[3];
```

17.2.4.5 TOPmodify Macros (TOP_MDF_macros.h)

```
/*===============================================================
MACRO TwoBufsReplaceTreat: replace "hot" word with "xxx" when "hot"
word is splitted between two buffers
===============================================================*/
MACRO TwoBufsReplaceTreat P_FRST_BUF_PRT,
                          P_FRST_BUF_DATA_SIZE,
                          P_FRST_BUF_DATA_OFF,
                          P_FRST_DATA,
                          P_SCND_BUF_PTR,
                          P_SCND_BUF_DATA_SIZE,
                          P_SCND_DATA;

    MovBits DISP_REG, 0, 9; // SREG[2].BYTE[2]<-0

    // second buffer treatment
    Mov ALU, P_SCND_BUF_DATA_SIZE, 4;
    ReadBufferToLocalMemory P_SCND_BUF_PTR, ALU;

    Get ALU, O(uxZeroFrPtr), P_SCND_BUF_DATA_SIZE, _NJP_MDF;
    Put O(uxZeroFrPtr), P_SCND_DATA, P_SCND_BUF_DATA_SIZE, _NJP_MDF;

    Mov ALU, P_SCND_BUF_DATA_SIZE, 4;
    WriteBufToGlobalMemory P_SCND_BUF_PTR,
                           ALU,
                           ZERO_OFF;
```

```
    // first buffer treatment
    Mov ALU, FULL_BUF_SIZE, 4;
    ReadBufferToLocalMemory P_FRST_BUF_PRT, ALU;

    Get ALU, P_FRST_BUF_DATA_OFF(uxZeroFrPtr),
                            P_FRST_BUF_DATA_SIZE, _NJP_MDF;
    Put P_FRST_BUF_DATA_OFF(uxZeroFrPtr), P_FRST_DATA,
                            P_FRST_BUF_DATA_SIZE, _NJP_MDF;

ENDMACRO;
/*==================================================================
MACRO FindWorkBuffsPtrs: find the buffers in which "hot" word is splitted
 Input:
        CNT - should be equal to iterations number (or work buff - 1)

 Output:
        P_uxFrstPtr - pointer to First replace buffer
        P_uxScndPtr - pointer to Second replace buffer
==================================================================*/
MACRO FindWorkBuffsPtrs P_uxFrstPtr, P_uxScndPtr;

READ_RFD_LOOP_LABEL:
    Mov P_uxFrstPtr, P_uxScndPtr, 2;

    // read uxBreakBufPtr Rfd entry
    Movbits RFD_PTR, P_uxScndPtr, 16;
    Movbits sRfd_bitCommand, READ_CMD, 3;     // READ_CMD is 2
    Nop;
    WAIT_NO_FLAG F_RDF_MDF;

    Loop READ_RFD_LOOP_LABEL | _NOP1;
        Mov P_uxScndPtr, sRfd_uxRdNxtPtr, 2;

ENDMACRO;
/*==================================================================
MACRO FindCurBuffPtr:
Input:
        CNT - should be equal to iterations number (or work buff - 1)
Output:
        P_uxCurPtr - pointer to replace buffer
==================================================================*/
MACRO FindCurBuffPtr P_uxCurPtr;

READ_RFD_LOOP_LABEL:

    // read uxBreakBufPtr Rfd entry
    Movbits RFD_PTR, P_uxCurPtr, 16;
    Movbits sRfd_bitCommand, READ_CMD, 3; // READ_CMD is 2
    Nop;
    WAIT_NO_FLAG F_RDF_MDF;

    Loop READ_RFD_LOOP_LABEL | _NOP1;
        Mov P_uxCurPtr, sRfd_uxRdNxtPtr, 2;
```

```
ENDMACRO;
/*================================================================
MACRO TwoBufsTreatment: this macro called is case that we now that our
"hot" word splitted between buffers but it not first and second buffer
================================================================*/
MACRO TwoBufsTreatment P_FRST_BUF_PRT,
                       P_FRST_BUF_DATA_SIZE,
                       P_FRST_BUF_DATA_OFF,
                       P_FRST_DATA,
                       P_SCND_BUF_PTR,
                       P_SCND_BUF_DATA_SIZE,
                       P_SCND_DATA;

   Mov P_SCND_BUF_PTR, uxFramePtr, 2;
   GetRes CNT, MSG_WORK_BUF_NUM_OFF(MSG_STR), 1; // buf_num-1

FindWorkBuffsPtrs P_FRST_BUF_PRT, P_SCND_BUF_PTR;

TwoBufsReplaceTreat P_FRST_BUF_PRT,
                    P_FRST_BUF_DATA_SIZE,
                    P_FRST_BUF_DATA_OFF,
                    P_FRST_DATA,
                    P_SCND_BUF_PTR,
                    P_SCND_BUF_DATA_SIZE,
                    P_SCND_DATA;

   Mov ALU, FULL_BUF_SIZE, 4; // 512
   WriteBufToGlobalMemory P_FRST_BUF_PRT,
                          ALU,
                          ZERO_OFF;

   // read first buffer
   Mov ALU, FULL_BUF_SIZE, 4;
   ReadBufferToLocalMemory uxFramePtr, ALU;

ENDMACRO;
/*================================================================
MACRO OneBufReplaceTreat:replace the "hot" word with "xxx" when "hot"
word is not splitted between buffers
================================================================*/
MACRO OneBufReplaceTreat DISP_REG_VAL;

#define uxDataPrt uxTempPrt;
   MovBits DISP_REG, DISP_REG_VAL, 9;
   GetRes uxDataPrt, MSG_DATA_OFF(MSG_STR), 2;
   Nop;
   Get ALU, 0(uxDataPrt), 3, _NJP_MDF;
   Put 0(uxDataPrt)+, XX_ASCII, 2, _NJP_MDF;
   Put 0(uxDataPrt), X_ASCII, 1, _NJP_MDF;

#undef uxDataPrt;

ENDMACRO;
```

17.2.4.6 TOPmodify Definition File (TOP_MDF.h)

```
/*====================== constants ======================*/
#define HDR_LENO_BUF_TYPE1          ((1<<7) | 0);
#define XX_ASCII                    0x5858;
#define X_ASCII                     0x58;

#define BUF_LAST_BYTE_OFF           (FULL_BUF_SIZE-1);
#define BUF_BEFORE_LAST_BYTE_OFF    (FULL_BUF_SIZE-2);

#define ZERO_OFF                    0;

/*===================== frame offsets =====================*/
#define MAC_SA_OFF_P4               (MAC_SA_OFF+4);
#define MAC_DA_OFF_P4               (MAC_DA_OFF+4);

/*===================== masks =====================*/
#define M_0x000003FF    MREG[0];
LDREG MREG[0], 0x000003FF;

/*===================== registers =====================*/
#define uxZeroFrPtr                 FR_PTR0;
#define uxL3FrPtr                   FR_PTR1;
#define uxTempPrt                   FR_PTR2;

#define byUDB_SpecIPFlag            UDB.BYTE[0];

#define uxFirstBufSize              UREG[1];
#define uxFramePtr                  UREG[1].BYTE[2];
#define uqTempUREG[5];

#ifdef EZ_NP_1c;
    #define tmpSrcPort              UREG[6];
#endif
```

17.2.4.7 Common Definition File (NP_common.h)

```
/*===================== constant =====================*/
#define FRST_BUF_SIZE       448;
#define FULL_BUF_SIZE       512;

#define FRST_BUF_OFF         64;

/*===================== frame offsets =====================*/
#define IP_TCP_CHK_OFF    (IP_BASE_SIZE+TCP_CHK_OFF);
```

17.2.4.8 Structures Definition File (NP_structs.h)

```
#define VALID                       0;
#define MATCH                       1;

//===================== Structures list =====================
#define MSG_STR                     0;
// searchI
#define DIP_STR                     1;
```

```
#define SIP_STR                              2;
// searchII
#define CFG_STR                              3;

//================= HW MESSAGE =================================
#define MSG_SIZE                             32;
#define RSV_MSG_SIZE                         32;

// control bits offsets
#define MSG_MODE                             1;
#define MSG_MODE1                            1;
#define MSG_MODE2                            2;
#define MSG_MODE3                            3;
#define MSG_SPEC_FR                          4;
#define MSG_TAG_EXISTS                       5;

// fields offsets
#define MSG_HD_REG0_OFF                      1;
#define MSG_L2_SIZE_OFF                      MSG_HD_REG0_OFF;
#define MSG_CTRL_REG_BYTE_OFF                (MSG_HD_REG0_OFF+1);
#define MSG_HD_REG1_OFF                      3;
#define MSG_FR_PTR_OFF                       (MSG_HD_REG1_OFF+2);
#define MSG_HD_REG2_OFF                      7;
#define MSG_FR_LEN_OFF                       MSG_HD_REG2_OFF;

//we will use last byte HD_REG2 for source port
//remember that
#ifdef EZ_NP_1c;
    #define MSG_HD_REG9_OFF                  (MSG_HD_REG2_OFF+3);
#endif;

#define MSG_SIP_OFF                          11;

#define MSG_DATA_OFF                         17; // or work buf
                                                 //offset for SearchII
// SearchII fields
#define MSG_FAST_REG_OFF                     19;
#define MSG_NXT_HOP_MAC_OFF                  20;
#define MSG_NXT_HOP_MAC_OFF_P4               (MSG_NXT_HOP_MAC_OFF+4);
#define MSG_WORK_BUF_NUM_OFF                 26; // mode 3 only
#define MSG_OUT_PORT_OFF                     27; // or special port
                                                 // for mode 3
#define MSG_OUT_QOS_OFF                      28;
#define MSG_SPEC_IP_FLAG_OFF                 29; // 1 byte

//===================== DIP structure =====================
#define DIP_KEY_SIZE                         4;
#define DIP_RES_SIZE                         32;

// fields offsets
#define DIP_OUT_PORT_OFF                     1;
```

```
#define DIP_OUT_QOS_OFF                          2;
#define DIP_SPEC_PORT_OFF                        3;
#define DIP_NXT_HOP_MAC_OFF                      4;

//======================= SIP structure =========================
#define SIP_KEY_SIZE                             4;
#define SIP_RES_SIZE                             32;

//======================= CFG structure =========================
#define CFG_KEY_SIZE                             1;
#define CFG_RES_SIZE                             16;

// fields offsets
#define CFG_IF_MAC_OFF                           1;
#define CFG_IF_MAC_OFF_P4               (CFG_IF_MAC_OFF+4);
#define CFG_SPEC_IP_OFF                          7;
```

17.2.4.9 FAST Register Definition File (NP_FastRegs.h)

```
// FAST_REG RX bits
#define F_FRST_BUF                  0;
#define F_NOT_FRST                  1;
#define F_FRST_SCND_BUF_x_xx        2;
#define F_FRST_SCND_BUF_xx_x        3;
#define F_TWO_BUFS_xx_x             4;
#define F_TWO_BUFS_x_xx             5;
#define F_L3                        6;

// FAST_REGs
#define L3_FAST_REG                    (1<<F_L3);
#define FRST_BUF_FAST_REG              ((1<<F_FRST_BUF)|(1<<F_L3));
#define NOT_FRST_FAST_REG              ((1<<F_NOT_FRST)|(1<<F_L3));
#define FRST_SCND_BUF_x_xx_FAST_REG
                       ((1<<F_FRST_SCND_BUF_x_xx)|(1<<F_L3));
#define FRST_SCND_BUF_xx_x_FAST_REG
                       ((1<<F_FRST_SCND_BUF_xx_x)|(1<<F_L3));
#define TWO_BUFS_xx_x_FAST_REG
                       ((1<<F_TWO_BUFS_xx_x)|(1<<F_L3));
#define TWO_BUFS_x_xx_FAST_REG
                       ((1<<F_TWO_BUFS_x_xx)|(1<<F_L3));
```

17.2.4.10 RFD Macros File (Rfd.MDF.h)

```
/*******************************************************************
MACRO InitRfdSrcPort:
*******************************************************************/
MACRO InitRfdSrcPort bitsSrcPort;
   WAIT_NO_FLAG F_RAF_MDF;

#ifdef EZ_RFD_NP_1c_MODE;
```

```
      MovBits sRfd_bitPort, bitsSrcPort, 4;
#else;
      MovBits sRfd_bitPort, bitsSrcPort, 3;
#endif;

ENDMACRO;

/*******************************************************************
MACRO ReadBufferToLocalMemory:
*******************************************************************/
MACRO ReadBufferToLocalMemory uxFramePtr, DATA_SIZE;
      Add NXT_BUF_SZ, DATA_SIZE, 0, 2, M_0x000003FF, _ALU_FRST;
      MovBits NXT_BUF_PTR, uxFramePtr, 16;
          Nop;
      If (FLAGS.BIT [F_NBV_MDF]) Jmp $ | _NOPO;
        Nop;
        Nop;
      Halt _HALT_NO_MDF, HALT_MULC;
          Nop;

ENDMACRO;

/*******************************************************************
MACRO PrefetchNewBuffer:
Assumptions: RFD src port must be initialized
*******************************************************************/
MACRO PrefetchNewBuffer uxNewPtr, PRE_PTR_REGISTER, PRE_FLAG;

      Mov uxNewPtr, PRE_PTR_REGISTER, 2;
      Nop;
      WAIT_NO_FLAG PRE_FLAG;

ENDMACRO;

/*******************************************************************
MACRO WriteBufToGlobalMemory:
Assumptions: RFD src port must be initialized
*******************************************************************/
MACRO WriteBufToGlobalMemory uxBufPtr,
                             uxDataSize,
                             uxDataOff;

#define CHUNK_SIZE 64;
#define CHUNK_SIZE_M1 (CHUNK_SIZE - 1);

      // align uxDataSize to fill last chunk:
      // ((x-1)/64 +1) * 64 = ... = (x+63) with 6 ls bits masked
      MovBits FR_STRT, uxDataOff, 9;

      Add ALU, uxDataSize, CHUNK_SIZE_M1, 2;
      MovBits ALU, 0, 6;
          MovBits sSend_uxFramePtr, uxBufPtr, 16;
      Sub sSend_bit1stBufLen, ALU, 1, 2;
```

```
        Halt _WHALT_NO_MDF, HALT_MULC;
            Nop;

    #undef CHUNK_SIZE;
    #undef CHUNK_SIZE_M1;

    ENDMACRO;

    /*******************************************************************
    MACRO GetRfdNextPtr:
    Assumptions: needs NOP before if is called after RFD command
    *******************************************************************/
    MACRO GetRfdNextPtr uxBufPtr, uxNextFragmPtr;
        WAIT_NO_FLAG F_RAF_MDF;
        Mov RFD_PTR, uxBufPtr, 2;
        MovBits sRfd_bitCommand, READ_CMD, 3; // READ_CMD is 2

            Nop;
        WAIT_NO_FLAG F_RDF_MDF;
        Mov uxNextFragmPtr, sRfd_uxRdNxtPtr, 2;

    ENDMACRO;
```

17.3 DATA STRUCTURES

This section describes the data structures used in the program. The functions of the data structures were described in the introduction of this chapter, and further detailed in the descriptions of each TOP engine.

Two data elements are described in this section: those that are created by the TOPparse and used in the TOPsearch I, and those that are created by TOPresolve and used by the TOPsearch II. One additional data element is provided through the CREG registers and serves for inputting parameters for the program (and can be used by TOPparse, by reading HD REG3 and HD_REG4). These input parameters are functions of the incoming port, that is, for each ingress port, the frames will be treated with different input parameters, as predefined by the corresponding CREG registers. The use of the data structures is described schematically in Figure 17.2.

17.3.1 CREGs

For each port from which frames are incoming, CREG predefines the mode in which we want to work. CREG also predefines the HTTP port numbers that are used for the TOPparse to decide that the frame is indeed an HTTP frame. Each port of the NP can define its own mode and the HTTP port numbers for all packets coming through it.

The operating mode is defined by CREG.BIT[0-2], which maps to HWARE[3]. BIT[0-2] (which is HD_REG3.BIT[0-2]). Only one mode bit should be set:

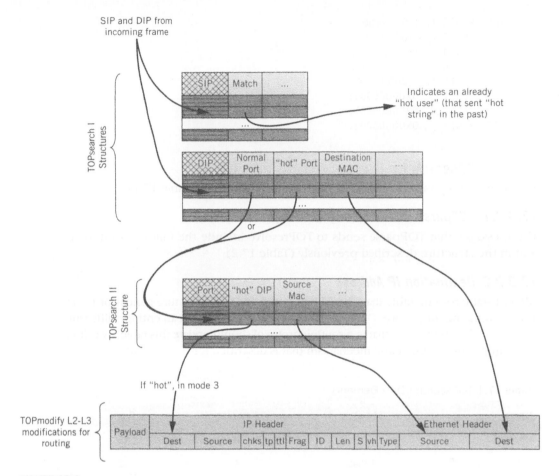

FIGURE 17.2

Data structures usage

CREG.BIT[0]—if set, the program works in operating mode 1
CREG.BIT[1]—if set, the program works in operating mode 2
CREG.BIT[2]—if set, the program works in operating mode 3

HTTP_PORT is defined for any real or proxy HTTP port number, in CREG. BYTE[1] and CREG.BYTE[2] (mapped into HWARE[3].BYTE[1-2], which are HD_REG3. BYTE[1-2]).

In this example, there are eight input ingress ports, and the definitions are such that all of them consider HTTP ports as 0x0050 (which is the known HTTP port 80). The operating mode for each of them varies (e.g., the first and second ports work in operating mode 3, the third is in mode 1, and the fourth port is in operating mode 2), according to the following sample code:

```
LDREG CREG[0 ], 0x00005004;
LDREG CREG[2 ], 0x00005004;
LDREG CREG[4 ], 0x00005001;
LDREG CREG[6 ], 0x00005002;
LDREG CREG[8 ], 0x00005004;
LDREG CREG[10], 0x00005004;
LDREG CREG[12], 0x00005001;
LDREG CREG[14], 0x00005002;
```

17.3.2 **TOPsearch I**

TOPsearch I receives the following three data elements (see Table 17.1).

17.3.2.1 *TOPparse -> TOPresolve Message*

The messages that TOPparse sends to TOPresolve include the following informa-tion in the structure described previously (Table 17.2).

17.3.2.2 *Destination IP Address*

DIP is a static routing table that should be created with the structure generator, prior to executing the microcode. Obviously, this should be handled in an entirely different way for "real" applications, that is, a routing protocol should update this table, by using, for example, the "High Learn" mechanism that is described for the "hot SIP" case.

Table 17.1 TOPsearch I Data Elements

Name	Structure #	Key size	Size
Message	0	32	32
DIP	1 (hash)	4	32
SIP	2 (hash)	4	32

Table 17.2 TOPparse Messages

Field name	Byte:bit offset	Size	Note
Valid bit	0:0	1 bit	1-valid bit
Mode	0:1	3 bits	b1—mode1 b2—mode2 b3—mode3 Only one bit should be set
Special flag	0:4	1 bit	1—"hot" word found
L2 size	1	1 byte	

Table 17.2 (*continued*)

Field name	Byte:bit offset	Size	Note
First buffer length control	2	1 byte	For TOPModify Initialization
HD_REG1	3	4 bytes	Hardware decoding HD_REG1, HD_REG2—for SEND_REG0, SEND_REG1, SEND_REG2 initialization in TOPmodify
HD_REG2	7	4 bytes	
SIP	11	4 bytes	Source IP for High Learn
Data offset	17	2 bytes	"sex" or "bomb" offset

Table 17.3 Search I DIP Search Result

Field name	Byte offset	Size	Note
Valid bit	0	1 bit	1-valid bit
Match bit	0	1 bit	1-match bit
Output Port	1	1 byte	This is the output port in the usual case
QoS	2	1 byte	
Special output port	3	1 byte	In case of a "hot string" and operating mode 3 this is output port
Destination MAC	4	6 bytes	Next Hop MAC

17.3.2.2.1 Key

Key is Destination IP—4 bytes.

17.3.2.2.2 Result

The result of a search by a destination IP address is two options for port output; the frame should be sent to one of them according to whether it contained a "hot string" or not (see Table 17.3 for the entire result).

17.3.2.3 Source IP Address

SIP is the data structure that enables the program to find the "hot SIP," that is, SIP addresses from which "hot strings" were sent. This structure uses the "High Learn" mechanism to learn the new hot IP's. It should be created before running the microcode, and shouldn't be populated, as the program will "learn" and populate the required entries on the fly.

Table 17.4 Search I SIP Search Result

Field name	Byte:bit offset	Size	Note
Valid bit	0:0	1 bit	1-valid bit
Match bit	0:1	1 bit	1-match bit

Table 17.5 TOPsearch II Data Elements

Name	Structure #	Key size	Result size
Message	0	32	32
Out port cfg	3	1	16

17.3.2.3.1 Key

Key is Source IP—4 bytes.

17.3.2.3.2 Result

The result is actually just an indication that such an IP was in the hash table, that is, the match bit is 1 (see Table 17.4).

17.3.3 TOPsearch II

TOPsearch II receives the following two data elements (see Table 17.5).

17.3.3.1 TOPresolve -> TOPmodify Message

The message that TOPresolve sends to TOPmodify includes the following information in the structure described previously (Table 17.6).

17.3.3.2 Output Port Configuration

"Output structure" is used for finding the source MAC address and the special IP address according to the output port. This structure has to be defined and populated before running the microcode. All the entries in this structure are static (i.e., they are not changed during the execution of the program).

17.3.3.2.1 Key

Key is Output Port—1 byte long.

17.3.3.2.2 Result

The result of the search is the source MAC address and the special IP address, to which "hot string" frames are diverted (see Table 17.7).

Table 17.6 TOPresolve Message

Field name	Byte:bit offset	Size	Note
Valid bit	0:0	1 bit	1-valid bit
Mode	0:1	3 bits	b1—mode1 b2—mode2 b3—mode3 Only one bit should be set
Special flag	0:4	1 bit	1—"hot" word found
L2 size	1	1 byte	
First buffer length control	2	1 byte	For TOPmodify Initialization
HD_REG1	3	4 bytes	Hardware decoding
HD_REG2	7	4 bytes	HD_REG1, HD_REG2—for SEND_REG0, SEND_REG1, SEND_REG2 initialization in TOPmodify
SIP	11	4 bytes	Source IP for High Learn
Data offset	17	2 bytes	"sex" or "bomb" offset in the frame
FAST_REG	19	1 byte	For modification acceleration
Destination MAC	20	6 bytes	Next hop MAC
Buffer number	26	1 byte	Operating mode 3 only
Output port	27	1 byte	Used for search II to put DIP
QoS	28	1 byte	
Special IP flag	29:0	1 bit	

Table 17.7 Search II Output Port Search Result

Field name	Byte:bit offset	Size	Note
Valid bit	0:0	1 bit	1- valid bit
Match bit	0:1	1 bit	1- match bit
Source MAC	1	6 bytes	Port Source MAC
Special IP	7	4 bytes	Destination IP address to be used instead of the original DIP if the frame contains a "hot string"

17.4 SUMMARY

This chapter provides a comprehensive example for writing ultra-high processing speed applications, for wire-speed packet analysis. Although it was not written for performance but rather for teaching purposes, it can do the job (with minor issues, e.g., checksum), and can demonstrate various ways to process packets by an NP.

This chapter also concludes the book, and the reader is encouraged to download (see instructions in Chapter 1) the simulator for writing and simulating such applications, as well as others.

List of Acronyms

AAA	Authentication, Authorization, and Accounting
AAL	ATM Adaptation Layer
AAS	Adaptive Antenna System
ABR	Available Bit Rate
AC	Access Connection/Attachment Circuit
ACL	Access Control List
ADM	Add/Drop Multiplexer
ADSL	Asymmetric Digital Subscriber Line
AF	Assured Forwarding (a Class of Service)
Agg Eth SW	Aggregation Ethernet Switch
Alloc-ID	Allocation Identifier of the upstream T-CONT
ALQD	Approximated LQD
ALU	Arithmetic Logic Unit
AMPS	Advanced Mobile Phone System
AP	Access Point
AP	Alternate Port (in Ethernet bridging)
API	Application Program Interface
APoM	Any Protocol over MPLS
APP layer	Application layer
AQM	Active Queue Management
AQT	Area-based QuadTtree
ARP	Address Resolution Protocol
ARQ	Automatic Request Response
AS	Autonomous System
ASIC	Application Specific Integrated Circuit
AS-ID	Autonomous System identifications
ASIP	Application Specific Instruction Set Processor
ASI-SIG	Advanced Switching Interconnect—Special Interest Group
ASN	Access Service Network
ASN	Autonomous System Number
ASN-GW	Access Service Network Gateway
ASON	Automatically Switched Optical Networks
ASP	Application Service Provider
ASSP	Application Specific Standard Product
ASTN	Automatic Switched Transport Network
ATCA	Advanced Telecom Computing Architecture
ATM	Asynchronous Transfer Mode
AToM	Any Transport over MPLS
AU	Administrative Unit (in SDH)
AUG	Administrative Unit Group (in SDH)
AUI	Attachment Unit Interface (in Ethernet)
AVQ	Adaptive Virtual Queue
BAN	Body Area Network
BCAM	Binary CAM
BCB	Backbone Core Bridge
B-DA	Backbone Destination MAC Address
BEB	Backbone Edge Bridge
BER	Bit Error Rate

BGP	Border Gateway Protocol
BIP	Bit Interleaved Parity
B-ISDN	Broadband Integrated Services Data Network
B-MAC	Backbone MAC
BMP	Best Matched Prefix
BNG	Broadband Network Gateway
BP	Backup Port (in Ethernet bridging)
BPDU	Bridge Protocol Data Unit
BPON	Broadband PON
bps	bits per second
BRAS	Broadband Remote Access Server
BRED	Balanced Random Early Detection
BS	Base Station
BSC	Base Station Controller (in 2G systems)
BSFQ	Bin Sort Fair Queuing
BSP	Bulk Synchronous Parallel
BSS	Base-Station Subsystem
BSSGP	Base Station System GPRS Protocol
BST	Binary Search Tree
B-Tag	Backbone VLAN tag
B-Tag TPID	Backbone Tag Identifier
BTS	Base Transceiver Station (in 2G systems)
B-VLAN	Backbone VLAN
BvN	Birkhoff von Neumann
C2C	Chip to Chip
CAC	Call Admission Control
CAM	Content Addressable Memory
CAS	Column Address Strobe
CATV	Cable TV
CBQ	Class-Based Queueing
CBR	Constant Bit Rate
CBS	Committed Burst Size
CBT	Core Based Tree
CC	Continuity Check (in Ethernet)
CCM	Continuity Check Message (in Ethernet)
CC-NUMA	Cache Coherent NUMA
CC-UMA	Cache Coherent UMA
CDG	CDMA Development Group
CDMA	Code Division Multiple Access
CDR	Committed Data Rate
CE	Congestion Experience
CE	Customer Edge or Equipment
CEN	Carrier Class Ethernet
CEOT	Circuit Emulation over Transport
CEP	Circuit Emulation over Packet
CEPT	European Conference of Postal and Telecommunications Administration
CES	Circuit Emulation Service
CESoETH	CES over Ethernet
CFD	Classify Frame Descriptor (in EZchip architecture)
CFI	Canonical Format Indicator (in Ethernet)
CFM	Connectivity Fault Management (in Ethernet)
CGMT	Coarse-Grained Multithreading

cHEC	Core HEC
CIDR	Classless Inter Domain Routing
CIFS	Common Internet File System
CIOQ	Combined Input Output Queued
CIR	Committed Information Rate
CISC	Complex Instruction Set Computing
CIST	Common and Internal Spanning Tree (in Ethernet bridging)
CLEC	Competitive Local Exchange Carrier
CLP	Cell Loss Priority
CLS	Controlled Load Service
CM	Cable Modem
CM	Configuration Message (in Ethernet bridging)
C-MAC	Customer MAC
CMP	Chip Multiprocessors
CMTS	Cable Modem Termination System
CN	Computer-Peripheral Network
CO	Central Office
COMA	Cache Only Memory Architecture
CoS	Class of Service
COW	Cluster of Workstations
CP	Common Part (in AAL type 5)
CPCS	Common Part Convergence Sublayer (in AAL type 5)
CPE	Customer Premises Equipment
CPI	Common Part Indicator (in AAL type 5)
CPI	Cycles per Instruction
CPIX	Common Programming Interface Forum
CPLD	Complex Programmable Logic Device
CPN	Customer Premises Network
CPU	Central Processing Unit
CRC	Cyclic Redundancy Check (or Code)
CREW	Concurrent Read Exclusive Write
CR-LDP	Constraint-Based LDP
CS	Circuit Switching
CS	Convergence Sub-layer
CSIX	Common Switch Interface Specification
CSMA/CD	Carrier Sense Multiple Access/Collision Detection
CSN	Connectivity Service Network
CST	Common Spanning Tree (in Ethernet bridging)
C-Tag	Customer VLAN tag
C-Tag TPID	Customer Tag Identifier
CTR	Committed Target Rate
CU	Control Unit
C-VLAN	Customer VLAN
CWDM	Coarse Wave Division Multiplexing
CX	Copper Physical (in Ethernet)
DA	Ethernet Destination MAC Address
DB	Designated Bridge (in Ethernet bridging)
DBR	Deterministic Bit Rate
DBRu	Dynamic Bandwidth Report upstream
DCE	Distributed Computing Environment
DCM	Distributed Call and Connection Management
DCOM	Distributed Component Object Model

DCS	Digital Cross Connect
DDR	Double Data Rate
DEI	Drop Eligible Indicator (in Ethernet)
DF	Default Forwarding (a Class of Service)
DiffServ	Differentiated Services
DIP	Destination IP Address
DLB	Dual Leaky Bucket
DLC	Digital Loop Carrier
DLP	Data Level Parallelism
DMA	Direct Memory Access
DOCSIS	Data over Cable Service Interface Specifications
DoD	Department of Defense
DP	Designated Port (in Ethernet bridging)
DPI	Deep Packet Inspection
DRAM	Dynamic RAM
DRDRAM	Direct Rambus DRAM
DRED	Dynamic Random Early Detection
DRR	Deficit Round-Robin
DRRM	Dual Round-Robin Matching
DS	Data Stream
DSAP	Designation Service Access Point
DSCP	Differentiated Services Code Point
DSL	Digital Subscriber Loop/Line
DSLAM	Digital Subscriber Line Access Multiplexer
DSM	Distributed Shared Memory
DST	Destination
DS-UWB	Direct Sequence—UWB
DTB	Dual Token Bucket
DVB	Digital Video Broadcast
DVI	Digital Visual Interface
DVMRP	Distance Vector Multicast Routing Protocol
DWDM	Dense WDM
EAP	Extensible Authentication Protocol
EBDMA	Egress Back DMA (in EZchip architecture)
EBRR	Eligibility-Based Round-Robin
EBS	Excess Burst Size
ECC	Error Correcting Code
ECFD	Egress Classify Frame Descriptor (in EZchip architecture)
ECMP	Equal Cost Multiple Path
ECN	Explicit Congestion Notification
ECT	ECN-Capable Transport
EDA	Electronic Design Automation
EDD	Earliest Due Date
EDD	Ethernet Demarcation Device
EDF	Earliest Deadline First
EDGE	Enhanced Data rates for GSM Evolution
eDSL	Ethernet DSL
EF	Expedited Forwarding (a Class of Service)
EFDMA	Egress Front DMA (in EZchip architecture)
EFM	Ethernet in the First Mile
EGP	Exterior Gateway Protocol
eHEC	Extension Header HEC

EIR	Excess Information Rate
EISA	Extended ISA
E-LAN	Ethernet LAN
E-Line	Ethernet Line
E-LMI	Ethernet Local Management Interface
eNB	evolved Node B
E-NNI	External Network to Network Interface
EO	End-Office
EOF	End of Frame
EPC	Evolved Packet Core
EPD	Early Packet Discard
EPIC	Explicitly Parallel Instruction Computing
EPL	Ethernet Private Line
EPON	Ethernet Passive Optical Network
EPROM	Erasable and Programmable Memory
EPS	Evolved Packet System
EREW	Exclusive Read Exclusive Write
ER-LSP	Explicitly Routed LSP
ES	Emulated Service
ESCON	Enterprise Systems Connection
ESF	Extended Super Frame
ESP	Ethernet Switched Path
ESPD	Early selective packet discard
ETFD	Egress Transmit Frame Descriptor (in EZchip architecture)
Eth	Ethernet
ETH layer	Ethernet layer
ETSI	European Telecommunications Standards Institute
ETTx	Ethernet-to-the-Home/Building/Office
E-UTRAN	Evolved Universal Terrestrial Radio Access Network
EVC	Ethernet Virtual Connection
EVC-ID	EVC Identification
EV-DO	Evolution-Data Optimized
EVPL	Ethernet Virtual Private Line
FA	Foreign Agent
FBLK	Functional Block Registers
FCRAM	Fast Cycle RAM
FCS	Frame Check Sequence
FDDI	Fiber Distributed Data Interface
FDM	Frequency Division Multiplexing
FEC	Forward Error Correction
FEC	Forwarding Equivalence Class (in MPLS)
FGMT	Fine-Grained Multithreading
FI	Fair Index
FICON	Fibre Connection
FID	Filtering Identifier (in Ethernet bridging)
FIFO	First In, First Out
FIS-tree	Fat Inverted Segment tree
FITL	Fiber In The Loop
FMEM	Frame Memory
FPGA	Field Programmable Gate Array
FPL	Functional Programming Language
FRED	Flow Random Early Detection

FTN	FEC-to-NHLFE
FTP	File Transfer Protocol
FTTB	Fiber to the Building
FTTC	Fiber to the Curb/Cabinet
FTTH	Fiber to the Home
FTTP	Fiber to the Premises
FTTZ	Fiber to the Zone
FW	Forwarder
GAN	Generic Access Network
GARP	Generic Attribute Registration Protocol
GbE	Gigabit Ethernet
GBIC	Gigabit Interface Converter
Gbps	Giga (10^9) bits per second
GCRA	Generic Cell Rate Algorithm
GEM	GPON Encapsulation Method
GERAN	GSM/EDGE Radio Access Network
GFC	Generic Flow Control
GFP	Generic Framing Procedure
GFP-F	Generic Frame Procedure—Frame mapped
GFP-T	Generic Frame Procedure—Transparent mapped
GFR	Guaranteed Frame Rate
GGSN	Gateway GPRS Support Node
GMII	Gigabit Media Independent Interface (in Ethernet)
GMPLS	Generalized Multi-Protocol Label Switching
GMRP	GARP Multicast Registration Protocol
GPON	Gigabit PON
GPP	General Purpose Processor
GPRS	General Packet Radio Service
GPS	Generalized Processor Sharing
GRE	Generic Routing Encapsulation
GRED	Gentle RED
GS	Guaranteed Service (a Class of Service)
GSM	Global System for Mobile—*Groupe Spécial Mobile*
GSMC	Gateway Mobile Switching Center
GSN	GPRS Support Node
GTC	GPON Transmission Convergence
GTP	GPRS Tunneling Protocol
GTP-U	GTP—User plane
GVRP	GARP VLAN Registration Protocol
HA	Home Agent
HD_REG	Hardware register (in EZchip architecture)
HDL	Hardware Description Language
HDLC	High-Level Data Link Control
HDMA	Host DMA (in EZchip architecture)
HDSL	High bit rate Digital Subscriber Line
HDTV	High Definition TV
HEC	Header Error Control
HFC	Hybrid Fiber and Coax
HiCuts	Hierarchical Intelligent Cutting
HLR	Home Location Register
H-NSP	Home Network Service Provider

HOL	Head of Line
HOLB	Head of Line Blocking
HPF	High Performance Fortran
H-PFQ	Hierarchical Packet Fair Queueing
HPNA	Home Phoneline Networking Alliance
HREG	Header Register (in EZchip architecture)
HSDPA	High Speed Downlink Packet Access
HSI	High-Speed Interconnect
HSPA	High Speed Packet Access
HSS	Home Subscriber Server
HSUPA	High Speed Uplink Packet Access
HTFD	Host Transmit Frame Descriptor (in EZchip architecture)
HTTP	Hyper Text Transfer Protocol
H-VPLS	Hierarchical VPLS
HW_DEC	Hardware Decoder Block (in EZchip architecture)
HWARE	Hardware registers (in EZchip architecture)
I/O	Inputs/Outputs
IAB	Internet Architecture Board
IANA	Internet Assigned Numbers Authority
IBDMA	Ingress Back DMA (in EZchip architecture)
IBGP	Internal BGP
IC	Integrated Circuit
ICFD	Ingress Classify Frame Descriptor (in EZchip architecture)
ICMP	Internet Control Message Protocol
I-DEI	Instance Drop Eligible Indicator
IDS	Intrusion Detection Systems
IETF	Internet Engineering Task Force
IFDMA	Ingress Forward Direct Memory Access (in EZchip architecture)
I-FFT	Inverse Fast Fourier Transform
IGMP	Internet Group Management Protocol
IGP	Interior Gateway Protocol
IHL	Internet Header Length
IID	Independent and Identically Distributed
ILEC	Incumbent Local Exchange Carrier
ILM	Incoming Label Mapping
ILP	Instruction Level Parallelism
iLQF	Iterative Longest Queue First
IMS	IP Multimedia System (or Services)
I-NNI	Internal NNI
IntServ	Integrated Services
iOCF	Iterative Oldest Cell First
IOQS	Ideal Output Queued Switch
IP	Internet Protocol
IP/MPLS	IP over MPLS
IPC	Instructions per Cycle
IPC	Inter-Process Communications
IPLS	IP-only Like Service
IPoA	IP over ATM
IPoS	IP over SONET
IPS	Intrusion Prevention Systems
IPsec	Internet Protocol Security
IPTV	TV service over IP

IPv4	IP version 4
IPv6	IP version 6
IQ	Input Queued
IRRM	Iterative Round-Robin Matching
IS	Instruction Stream
IS-95	Interim Standard 95
ISA	Industry Standard Architecture
ISA	Instruction Set Architecture
iSCSI	SCSI over IP
ISDN	Integrated Services Digital Network
I-SID	Service Instance Identifier
ISO	International Standard Organization
ISO/OSI	ISO/Open System Interconnect
ISOC	Internet Society
ISP	Internet Service Provider
IST	Internal Spanning Tree (in Ethernet bridging)
I-Tag	Instance Service VLAN tag (in Ethernet bridging)
I-Tag TPID	Service Instance Identifier
ITU	International Telecommunication Union
ITU-T	Telecommunication Standardization Sector
IVL	Independent VLAN Learning
IW	Issue Width
IXA	Internet eXchange Architecture
IXP	Internet eXchange Processor
JCA	Joint Coordination Activity
JEDEC	Joint Electron Device Engineering Council
JTACS	Japan Total Access Communications System
Kbps	Kilo (10^3) bits per second
KMEM	Key Memory (in EZchip architecture)
KREG	Key Register (in EZchip architecture)
L1	Layer 1
L2	Layer 2
L2CP	Layer 2 Control Protocol
L2TP	Layer 2 Tunneling Protocol
L2TPext	L2TP extensions
L2VPN	Layer 2 Virtual Private Network
L3	Layer 3
L3VPN	Layer 3 Virtual Private Network
LA	Look-Aside
LAA	Layer 2 Tunneling Protocol (L2TP) Access Aggregation
LAC	L2TP Access Concentrator
LAN	Local Area Network
LB	LoopBack
LBM/LBR	Loopback Message and Reply (in Ethernet)
LCAS	Link Capacity Adjustment Scheme
LCCE	L2TP Control Connection Endpoint
LCO	Local Central Office
LCT	Last Conformance Time (in leaky bucket)
LC-trie	Level Compressed trie
LDP	Label Distribution Protocol
LEC	Local Exchange Company

LER	Label Edge Router
LH	Long Haul
LIB	Label Information Base
LIFO	Last-in, First-out
LLC	Logical Link Control
LLID	Logical Link Identification
LMSC	LAN/MAN Standards Committee
LNS	L2TP Network Server
LOH	Line Overhead (in SDH)
LOS	Line of Sight
LPC-trie	Level and Path Compressed trie
LPM	Longest Prefix Match
LQD	Longest Queue Drop
LRE	Long Reach Ethernet
LSB	Least significant bit
LSI	Large-Scale Integration
LSP	Label Switched Path
LSPID	LSP Identifiers
LSR	Label Switching Router
LT	Link Trace
LTE	Long Term Evolution
LTM/LTR	Link Trace Message and Reply (in Ethernet)
LVDS	Low Voltage Differential Signaling
LX	Long-wave (in Ethernet)
LZ	Lempel-Ziv algorithm
MA	Maintenance Association
MAC	Medium Access Control
MAN	Metropolitan Area Network
MB-OFDM	Multi-Band Orthogonal Frequency Division Multiplexing
MCID	MST Configuration Identifier (in Ethernet bridging)
MCP	Multi-Core Processor
MDE	Microcode Development Environment
MDI	Medium Dependent Interface (in Ethernet)
ME	Maintenance Entity
MEF	Metro Ethernet Forum
MEG	ME Group
MEMS	Micro Electro Mechanical System
MEN	Metro Ethernet Network
MEP	MEG End Point
MF	More Fragments bit
MIB	Management Information Base
MII	Media Independent Interface
MIMO	Multiple Input and Multiple Output
MIN	Multistage Interconnection Network
MIP	MEG Intermediate Point
MIP	Mobile IP
MLD	Multicast Listener Discovery
MM	Maximum Matching
MME	Mobility Management Entity
MMRP	MRP MAC address Registration Protocol
MoCA	Multimedia over Coax Alliance
MOD_ACC	Modification Accelerator Block (in EZchip architecture)

MoE	T-MPLS over Ethernet
MOEMS	Micro Opto-Electro Mechanical System
MoO	T-MPLS over OTH
MoP	T-MPLS over PDH
MoR	T-MPLS over RPR
MoS	T-MPLS over SDH
MOSPF	Multicast Extensions to OSPF
MP	Message Passing
MP2MP	Multipoint-to-Multipoint
MP-BGP	Multiprotocol Extensions for BGP-4
MPCP	MAC Control Protocol
MPHY	Multiple PHY Ports
MPI	Message Passing Interface
MPLS	Multi-Protocol Label Switching
MPMD	Multi Program over Multiple Data streams
MPOA	Multi Protocol over ATM
MPP	Massively Parallel Processors
MPSoC	Multiprocessor System-on-Chip
MRP	Multiple Registration Protocol
MS	Mobile Station
MSB	Most significant bit
MSC	Mobile Switching Center
M-SISD	Multiple SISD
MSOH	Multiplex SOH (in SDH)
MSP	Multi Service Platform
MSPP	Multi Service Provisioning Platform
MSSP	Multi Service Switching Platform
MST	Multiple Spanning Tree (in Ethernet bridging)
MSTI	Multiple Spanning Tree Instance (in Ethernet bridging)
MSTID	Multiple Spanning Tree Instance ID (in Ethernet bridging)
MSTP	Multi Service Transport Platforms
MSTP	Multiple Spanning Tree Protocol (in Ethernet bridging)
MT	Mobile Terminal
MTU	Maximum Transmission Unit
MTU	Multi Tenant Unit
MU	Memory Unit
MVRP	MRP VLAN Registration Protocol
MWM	Maximum Weight Matching
NAMPS	Narrowband Analog Mobile Phone Service
NAP	Network Access Provider
NAT	Network Address Translation
NCA	No Customer Address
NCL	Network Classification Language
NE	Network Elements
NFS	Network File System
NG-SDH/ SONET	Next Generation SONET/SDH
NHLFE	Next Hop Label Forwarding Entry
NIC	Network Interface Card
NLOS	non-Line of Sight
NLRI	Network Layer Reachability Information
NMT	Nordic Mobile Telephone

NNI	Network–Network Interface
NoC	Network on Chip
NOW	Network of Workstations
NP	Network Processor
NPF	Network Processing Forum
NPSI	Network Processing Streaming Interface
nrt-VBR	Non-real-time VBR
NS	Network Service
NSE	Network Search Engines
NSP	Network Service Provider
NSS	Network Subsystem (in RAN)
NT	Network Terminator
NUMA	Non-Uniform Memory Access
OADM	Optical Add/Drop Multiplexer
OAM	Operations, Administration and Maintenance
OBS	Optical Burst Switching
OC	Optical Carrier
OCh	Optical Channel
OCS	Optical Circuit Switching
ODN	Optical Distribution Network
ODU	Optical Data Unit
OE	Optical–Electrical
OFDM	Orthogonal Frequency Division Multiplexing
OFDMA	Orthogonal Frequency Division Multiple Access
OIF	Optical Interfacing Forum
OIF	Optical Internetworking Forum
OLT	Optical Line Terminal
OMEM	Output Memory (in EZchip architecture)
ONT	Optical Network Termination
ONU	Optical Network Unit
ONU-ID	ONU-Identifier
OPS	Optical Packet Switching
OPU	Optical Payload Unit
OQ	Output Queued
OREG	Other Register (in EZchip architecture)
OS	Operating System
OSF	Open Software Foundation
OSPF	Open Shortest Path First
OSPF-TE	Open Shortest Path First–Traffic Engineering
OTH	Optical Transport Hierarchy
OTN	Optical Transport Network
OTU	Optical channel Transport Unit
OUI	Organizationally Unique Identifier
OUT_IF	Output Interface (in EZchip architecture)
OXC	Optical Cross-Connects
P2MP	Point-to-MultiPoint
P2P	Point-to-Point (or Peer to Peer)
PA	Ethernet Preamble
PACT	Parallel Architectures and Compilation Techniques
PAN	Personal Area Network
PATRICIA	Practical Algorithm To Retrieve Information Coded In Alphanumeric

PB	Provider Bridge
PBBN	Provider Backbone Bridged Network
PBB-TE	Provider Backbone Bridges–Traffic Engineering
PBN	Provider Bridged Network
PBPS	Parallel Buffered Packet Switches
PBS	Peak Burst Size
PBT	Provider Backbone Transport (now called PBB-TE)
PBX	Private Branch eXchange
PC	Personal Computer
PC	Program Counter
PCBd	Physical Control Block downstream
PCISIG	PCI Special Interest Group
PCM	Pulse Code Modulation
PCP	Priority Code Point
PCRF	Policy and Charging Rules Function
PCS	Physical Coding Sublayer (in Ethernet)
PDCF	Packet Data Convergence Protocol
PDH	Plesiochronous Digital Hierarchy
PDN	Packet Data Network or Public Data Network
PDR	Peak Data Rate
PDU	Protocol Data Unit
PE	Provider Edge or Equipment
PEB	Provider Edge Bridge
PF	Policy Function
PFQ	Packet Fair Queue
PFQ	Per Flow Queuing
PGPS	Packet Approximation of GPS
P-GW	Packet Data Network (PDN) Gateway
PHB	Per Hop Behavior—a DiffServ term (IP CoS)
PHY	Physical Layer
PI	Proportional-Integral
PICMG	PCI Industrial Computer Manufacturers Group
PID	Protocol Identifier
PIM	Parallel Iterative Matching
PIM-DM	Protocol Independent Multicast–Dense mode
PIM SM	Protocol Independent Multicast–Sparse Mode
PIR	Peak Information Rate
PLC	Power Line Communications
Plend	Physical Length downstream
PLI	GEM's Payload Length Indicator
PLMN	Public Land Mobile Network
PLOAMd	Physical Layer OAM downstream
PLOAMu	Physical Layer OAM upstream
PLOu	Physical Layer Overhead upstream
PLS	Physical Layer Signaling (in Ethernet)
PLSu	Power Leveling Sequence upstream
PMA	Physical Medium Attachment (in Ethernet)
PMD	Physical Medium Dependent (in Ethernet)
PNNI	Private Network–Network Interface
POH	Path Overhead (in SDH)
PON	Passive Optical Network
POP	Point-of-Presence Protocol

POS	Packet over SONET/SDH
POSIX	Portable Operating System Interface for UNIX
POS-PHY	Packet-Over-SONET–Physical Layer
POTS	Plain Old Telephone Service
PPC	Port Path Cost (in Ethernet bridging)
PPD	Partial Packet Discard
PPE	Packet Processor Engine
PPL	Packet Programming Language
PPP	Point-to-Point Protocol
PPPoA	PPP over ATM
PPPoE	PPP over Ethernet
PPPoEoA	PPPoE over ATM
PPPoEoE	PPPoE over Ethernet
PPS	Parallel Packet Switch
PPTP	Point-to-Point Tunneling Protocol
PPV	Pay per View
PPVPN	Provider Provisioned Virtual Private Network
PQ	Priority Queue
PRAM	Parallel Random Access Machines
PRS-1	1 Gbps Packet PHY Reconciliation Sublayer
PRS-10	10 Gbps Packet PHY Reconciliation Sublayer
PS	Processor Sharing
PSM	Parallel Shared Memory
PSN	Packet Switched Network
PSTN	Public Switched Telephone Network
Psync	Physical Synchronization
PT	Payload-Type
PTA	Point-to-Point Protocol Terminated Aggregation
PTE	Path Terminating Equipment
Pthreads	POSIX threads
PTI	GEM's Payload Type Indicator
PTR	Peak Target Rate
PU	Processing Unit
PVC	Permanent Virtual Circuit
PVM	Parallel Virtual Machine
PVR	Personal Video Records
PW	Pseudo-wire
PWE3	Pseudo Wire Emulation Edge-to-Edge
PWMCW	PW MPLS Control Word
PWOT	Pseudowire Over Transport
QDR	Quad Data Rate
QFP	Quantum Flow Processor
QoS	Quality of Service
RADIUS	Remote Authentication Dial In User Service
RAM	Random Access Memory
RAN	Radio Access Network
RAS	Row Address Strobe
RB	Root Bridge (in Ethernet bridging)
RBOC	Regional Bell Operating Company
RCO	Regional Central Office
RCS	Rate-Controlled Service

RCSP	Rate-Controlled Static Priority
RD	Route Distinguisher
RDL	Remote Digital Terminal
RED	Random Early Detection
REM	Random Exponential Marking
RF	Radio Frequency
RFC	Recursive Flow Classification
RFC	Request For Comments
RFD	Receive Frame Descriptor
RG	Residential Gateway
RGMII	Reduced GMII
RIB	Routing Information Base
RIO	RED In/Out
RIP	Routing Information Protocol
RISC	Reduced Instruction Set Computing
RLC	Radio Link Control
RLDRAM	Reduced Latency DRAM
RMEM	Result Memory (in EZchip architecture)
RMI	Remote Method Invocation
RMII	Reduced MII
RNB	Rearrangeable Non-Blocking
RNC	Radio Network Controller (3G)
RND	Random Dropping
ROADM	Reconfigurable OADM
ROM	Read Only Memory
RP	Root Port (in Ethernet bridging)
RPA	Reservation with Preemption and Acknowledgment
RPC	Remote Procedure Call
RPC	Root Path Cost (in Ethernet bridging)
RPR	Resilient Packet Ring
RR	Round-Robin
RR	Route Reflector (in BGP)
RS	Reconciliation Sublayer
RSOH	Regenerator SOH (in SDH)
RSTP	Rapid Spanning Tree Protocol (in Ethernet bridging)
RSVP	Resource ReSerVation Protocol
RSVP-TE	Resource Reservation Protocol—Traffic Engineering
RT	Route Target
RTBI	Reduced TBI
RTOS	Real-Time OS
RTP	Real Time Protocol
RTSP	Real Time Streaming Protocol
rt-VBR	Real-time VBR
SA	Ethernet Source MAC Address
SAN	Storage Area Network
SAP	Service Access Points
SAR	Segmentation and Reassembly
SAS	Shared Address Space
SAToP	Structure-Agnostic Time Division Multiplexing over Packet
SBC	Single Board Computer
SCFQ	Self-Clocked Fair Queueing
SCSI	Small Computer System Interface

SDH	Synchronous Digital Hierarchy
SDMA	Subcarrier Division Multiple Access
SDRAM	Synchronous DRAM
SDSL	Symmetric DSL
SEAL	Simple and Easy Adaptation Layer
SECDED	Single Error Correcting, Double Error Detection
SEFF	Smallest Eligible virtual Finish time First
SERDES	SERializer/DESerializer
SF	Super frame
SFD	Ethernet's Start of Frame Delimiter
SFF	Smallest virtual Finish time First
SFI	SerDes-Framer Interface
SFI-4 or -5	SERDES Framer Interface level 4 or 5
SG	Study Group
SGMII	Serial GMII
SGSN	Serving GPRS Support Node
S-GW	Serving Gateway
SIO	Separated I/O
SIP	Session Initiation Protocol
SIP	Source IP address
SLA	Service Level Agreement
SLB	Simple Leaky Bucket
SMDS	Switched Multimegabit Data Service
SME	Small and Medium Enterprise
SMII	Serial MII
SMP	Symmetrical Multi Processors
SMT	Simultaneous Multithreading
SNAP	Sub Network Access Protocol
SNB	Strict Non-Blocking
SNDCP	Subnetwork Dependent Convergence Protocol
SNMP	Simple Network Management Protocol
SoC	System on Chip
SOH	Section Overhead (in SDH)
SOHO	Small Office Home Office
SONET	Synchronized Optical Networks
SP	Service Provider
SP	Strict Priority
SPB	Shortest Path Bridging
SPE	Synchronous Payload Envelope (in SDH)
SPFQ	Starting Potential-based Fair Queue
SPI	System Packet Interface
SPI	System-Physical Interface
SPI-S	Scalable System Packet Interface
SPMD	Single Program over Multiple Data
SPPA	Smith Predictor-based PI-controller for AQM
SPP	Special Purpose Processor
SRAM	Static RAM
SRC	Source
SRED	Stabilized Random Early Detection
SREG	Specific Register (in EZchip architecture)
SRR	Smoothed Round Robin
SRT	Source-Routing Transparent

srTCM	Single-Rate Three Color Marker
SS	Service specific (in AAL type 5)
SSAP	Source Service Access Point
SSL	Secure Sockets Layer
SST	Single-Spanning-Tree
SStart	Starting time of the allocation
SStop	Stopping time of the allocation
ST	Space-Time
ST	Spanning Tree (in Ethernet bridging)
S-Tag	Service VLAN tag
S-Tag TPID	Service Tag Identifier
STM	Synchronous Transport Module
STP	Shielded Twisted Pairs
STP	Spanning Tree Protocol (in Ethernet bridging)
STS	Synchronous Transport Signal
S-VID	Service VLAN Identifier
S-VLAN	Service VLAN
SVL	Shared VLAN learning
SX	Short-wave (in Ethernet)
TACS	Total Access Communications System
TALS	TDM Access Line Service
TAT	Theoretical Arrival Time (in Virtual scheduling)
TB	Token Bucket
TBI	Ten Bit Interface (in Ethernet bridging)
Tbps	Tera (10^{12}) bps
TBSF	Tandem Banyan Switching Fabric
TC	Toll Center
TCA	Traffic Conditioning Agreement
TCAM	Ternary CAM
TCI	Tag Control Information (in Ethernet)
TCM	Three Color Markers
TCN	Topology Change Notification
T-CONT	Transmission Container
TCP	Transmission Control Protocol
TDM	Time Division Multiplexing
TDMA	Time Division Multiple Access
TE	Traffic Engineering
tHEC	Type HEC
3GPP	Third Generation Partnership Project (for cellular 2G, 3G, and 4G)
TL	Type/Length (in Ethernet)
T-Line	TDM Line
TLP	Thread Level Parallelism
TLS	Transparent LAN Service
TLS	Transport Layer Security
TLV	Type, Length, and Value
TM	Terminal Multiplexer
TM	Traffic Manager
T-MPLS	Transport MPLS
TOH	Transport Overhead (in SDH)
TOP	Task Optimized Processor (in EZchip architecture)
TOS	Type of Service
TP	Twisted Pairs telephone wires

TPID	Type ID, Ethertype (in Ethernet)
TRAN layer	Transport layer
trTCM	Two-Rate Three Color Marker
TS	Time-Space
TSI	Time-Slot Interchange
TSL	Transport Layer Security
TST	Time-Space-Time
tswTCM	Time Sliding Window Three Color Marker
TTL	Time to Live
TU	Tributary Unit (in SDH)
TUG	Tributary Unit Group (in SDH)
UBR	Unspecified Bit Rate
UDP	User Datagram Protocol
UE	User Equipment
UI	Unnumbered Information
UMA	Uniform Memory Access
UMA	Unlicensed Mobile Access
UMB	Ultra Mobile Broadband
UMTS	Universal Mobile Telecommunication System (3G)
UNI	User-Network Interface
UNI-ID	UNI identification
UPC	Unified Parallel C
UREG	User-Defined Register (in EZchip architecture)
USB	Universal Serial Bus
UTP	Unshielded Twisted Pairs
UTRAN	Universal Terrestrial Radio Access Network
UU	User-to-User indication
UWB	Ultra-Wideband
VBR	Variable Bit Rate
VC	Virtual Channel
VC	Virtual Circuit
VC	Virtual Clock
VC	Virtual Container (in SDH)
VCAT	Virtual Concatenation (in NG-SDH)
VCG	Virtual Concatenated Group (in NG-SDH)
VCI	VC identifier
VCI	Virtual Channel Identifier
VDSL	Very high-speed DSL
VID	VLAN ID
VLAN	Virtual Local Area Network
VLIW	Very Long Instruction Word
VLR	Visitor Location Register
VLSM	Variable Length Subnet Mask
VOD	Video on Demand
VoIP	Voice over Internet Protocol
VOQ	Virtual Output Queue
VP	Virtual Path
VPC	Virtual Path Connection
VPI	Virtual Path Identifier
VPI	VP identifier
VPLS	Virtual Private LAN Service

VPN	Virtual Private Network
VPN-L	VPN Label
VPTA	Virtual Path Tunneling Architecture
VPWS	Virtual Private Wire Service
VRF	VPN Routing and Forwarding
VSA	Virtual Scheduling Algorithm
VSR	Very Short Reach (in Ethernet)
VT	Virtual Tributary
WAN	Wide Area Network
W-CDMA	Wideband Code Division Multiple Access
WDM	Wave Division Multiplexing
WDMA	WDM Access
WFI	Worst-case Fair Index
WFQ	Weighted Fair Queueing
Wi-MAX	Worldwide Interoperability for Microwave Access
WIS	WAN Interface Sublayer (in Ethernet)
WLAN	Wireless Local Area Network
WNB	Wide-sense Non-Blocking
WPAN	Wireless Personal Area Network
WRED	Weighted Random Early Detection
WRR	Weighted Round-Robin
XAUI	10 Gbps Attachment Unit Interface (in Ethernet)
XGMII	10 Gigabit Media Independent Interface (in Ethernet)
XGMII	10 Gbps Media Independent Interface (in Ethernet)
XGXS	XGMII Extended Sublayer (in Ethernet)
XSBI	10 Gbps Sixteen Bit Interface (in Ethernet)
ZBT	Zero Bus Turnaround

References

1. 3GPP, General Packet Radio Service (GPRS) Enhancements for Evolved Universal Terrestrial Radio Access Network (E-UTRAN) Access (Release 8), *3GPP TS 23.401 V8.1.0 (2008-03)*, 2008.

2. 3GPP, Evolved Universal Terrestrial Radio Access (E-UTRA) and Evolved Universal Terrestrial Radio Access (E-UTRAN); Overall description; Stage 2, *3GPP TS 36.300 V8.4.0 (2008-03)*, 2008.

3. 3GPP, Evolved Universal Terrestrial Access Network (E-UTRAN), Architecture description, *3GPP TS 36.401 V8.1.0 (2008-03)*, 2008.

4. 3GPP, Evolved Universal Terrestrial Access Network (E-UTRAN), S1 General Aspects and Principles (Release 8), *3GPP TS 36.410 V8.0.0 (2007-12)*, 2007.

5. Aboul-Magd, O., Constraint-Based LSP Setup Using LDP (CR-LDP) Extensions for Automatic Switched Optical Network (ASON), *RFC 3475, IETF*, 2003.

6. Aboul-Magd, O., and Rabie, S., A Differentiated Service Two-Rate, Three-Color Marker with Efficient Handling of In-Profile Traffic, *RFC 4115, IETF*, 2005.

7. Adams, A., Nicholas, J., and Siadak, W., Protocol Independent Multicast–Dense Mode (PIM-DM): Protocol Specification (Revised), *RFC 3973, IETF*, 2005.

8. Adelson-Velsky, G. M., and Landis, E. M., An Algorithm for the Organization of Information, *Dokady Alademiia Nauk SSSR*, 146:263–266, 1962 (translated to English by Ricci, M. K.); also in *Soviet Mathematics, Doklady*, 3:1259–1263, 1962).

9. Afsharian, S., Bertolino, A., De Angelis, G., Iovanna, P., and Mirandola, R., Model Based Approach to Design Applications for Network Processor, *Proceedings of RISE 2004, Lecture Notes in Computer Science*, Vol. 3475:93–101, Springer Verlag, 2005.

10. Agere, The Case for a Classification Language, White Paper, *http://nps.agere.com/support/non-nda/docs/Classification_New.pdf*, 2001.

11. Aggarwal, R., Rosen, E., Morin, T., Rekhter, Y., and Kodeboniya, C., BGP Encodings and Procedures for Multicast in MPLS/BGP IP VPNs, *draft-ietf-l3vpn-2547bis-mcast-bgp, IETF*, 2007.

12. Ahmad, S., and Mahapatra R. N., TCAM Enabled On-Chip Logic Minimization, *Proceedings of the 42nd ACM/IEEE International Design Automation Conference*, pp. 678–683, 2005.

13. Ahmadi, H., and Denzel, W. E., A Survey of Modern High-Performance Switching Techniques, *IEEE Journal on Selected Areas in Communications*, 7(7):1091–1103, 1989.

14. Allan, D., Bragg, N., McGuire A., and Reid, A., Ethernet as Carrier Transport Infrastructure, *IEEE Communications Magazine*, 44(2):134–140, 2006.

15. Almquist, P., Type of Service in the Internet Protocol Suite, *RFC 1349, IETF*, 1992.

16. Alt, H., Blum, N., Mehlhorn, K., and Paul, M., Computing a Maximum Cardinality Matching in a Bipartite Graph in Time $O(n^{3/2} + (m/\log n)^{1/2})$, *Information Processing Letters*, 37(4):237–240, 1991.

17. Anderson, T. E., Owicki, S. S., Saxe, J. B., and Thacker, C. P., High-Speed Switch Scheduling for Local-Area Networks, *ACM Transaction on Computer Systems*, 11(4):319–352, 1993.

18. Andersson, A., and Nilsson, S., Improved Behavior of Tries by Adaptive Branching, *Information Processing Letters*, 46:295–300, 1993.

19. Andersson, L., and Madsen, T., Provider Provisioned Virtual Private Network (VPN) Terminology, *RFC 4026, IETF*, 2005.

20. Andersson, L., and Rosen, E., Framework for Layer 2 Virtual Private Networks (L2VPNs), *RFC 4664, IETF*, 2006.

21. Andersson, L., Doolan, P., Feldman, N., Fredette, A., and Thomas, B., LDP Specifications, *RFC 3036, IETF*, 2001.

22. Andersson, L., Minei, I., and Thomas, B., LDP Specification, *RFC 5036, IETF*, 2007.

23. Anjum, F. M., and Tassiulas, L., Fair Bandwidth Sharing Among Adaptive and Non-Adaptive Flows in the Internet, *Proceedings IEEE INFOCOM*, 1999.

24. Aoe, J., Morimoto, K., Shishibori, M., and Park, K., A Trie Compaction Algorithm for a Large Set of Keys, *IEEE Transactions Knowledge Data Engineering*, 8:476–490, 1996.

25. Asadullah, S., Ahmed, A., Popoviciu, C., Savola, P., and Palet, J., ISP IPv6 Deployment Scenarios in Broadband Access Networks, *RFC 4779, IETF*, 2007.

26. Ashwood-Smith, P., and Berger, L., Generalized Multi-Protocol Label Switching (GMPLS) Signaling Constraint-Based Routed Label Distribution Protocol (CR-LDP) Extensions, *RFC 3472, IETF*, 2003.

27. Athuraliya, S., Low, S. H., Li, V. H., and Yin, Q., REM: Active Queue Management, *IEEE Network Magazine*, 15(3), 2001.

28. ATM Forum, LAN Emulation Over ATM 1.0, *af-lane-0021.000*, The ATM Forum (now the MFA Forum), 1995.

29. ATM Forum, Multi-Protocol Over ATM Specification v1.0, *af-mpoa-0087.0000*, The ATM Forum (now the MFA Forum), 1997.

30. ATM Forum, Private Network-Network Interface Specification v.1.1, *af-pnni-0055.002*, The ATM Forum (now the MFA Forum), 2002.

31. ATM Forum, Utopia, An ATM-PHY Interface Specification, Level 1, Version 2.01, *af-phy-0017.000*, The ATM Forum (now the MFA Forum), 1994.

32. ATM Forum, Utopia Level 2, Version 1.0, *af-phy-0039.000*, The ATM Forum (now the MFA Forum), 1995.

33. ATM Forum, Utopia Level 3, *af-phy-0136.000*, The ATM Forum (now the MFA Forum), 1999.

34. ATM Forum, Utopia Level 4, *af-phy-0144.000*, The ATM Forum (now the MFA Forum), 2000.

35. Attiya, H., and Hay, D., The Inherent Queuing Delay of Parallel Packet Switches, *IEEE Transactions on Parallel and Distributed Systems*, 17(8):1048–1056, 2006.

36. Attiya, H., Hay, D., and Keslassy, I., Packet-Mode Emulation of Output-Queued Switches, *Proceedings of the Eighteenth ACM Symposium on Parallelism in Algorithms and Architectures*, pp. 138–147, 2006.

37. Augustyn, W., and Serbest, Y., Service Requirements for Layer 2 Provider-Provisioned Virtual Private Networks (L2VPNs), *RFC 4665, IETF*, 2006.

38. Awduche, D., Berger, L., Gan, D., Li, T., Srinivasan, V., and Swallow, G., RSVP-TE: Extensions to RSVP for LSP Tunnels, *RFC 3209, IETF*, 2001.

39. Aweya, J., Ouellette, M., and Montuno, D.Y., DRED: A Random Early Detection Algorithm for TCP/IP Networks, *International Journal of Communication Systems*, 15(4):287–307, 2002.

40. Baer, J.-L., Low, D., Crowley, P., and Sidhwaney, N., Memory Hierarchy Design for a Multi-processor Look-Up Engine, *Proceedings of the 12th International IEEE Conference on Parallel Architectures and Compilation Techniques*, pp. 206–216, 2003.

41. Baldi, M., and Risso, F., A Framework for Rapid Development and Portable Execution of Packet-Handling Applications, *Proceedings of 5th IEEE International Symposium on Signal Processing and Information Technology*, 2005.

42. Ballardie, A., Core-Based Trees (CBT) Multicast Routing Architecture, *RFC 2201, IETF*, 1997.

43. Batcher, K. E., Sorting Networks and their Applications, *Proceedings of Spring Joint Computing Conference, AFIPS*, pp. 307–314, 1968.

44. Bates, S., Chen, E., and Chandra, R., BGP Route Reflection: An Alternative to Full Mesh Internal BGP (IBGP), *RFC 4456, IETF*, 2006.

45. Bates, T., Chandra, R., Katz, D., and Rekhter, Y., Multiprotocol Extensions for BGP-4, *RFC 4760, IETF*, 2007.

46. Bates, T., Rekhter, Y., Chandra, R., and Katz, D., Multiprotocol Extensions for BGP-4, *RFC 2858, IETF*, 2000.

47. Benes, V. E., *Mathematical Theory of Connecting Networks and Telephone Traffic*, Academic Press, 1965.

48. Benini, L., and De Micheli, G., Networks on Chips: A New SoC Paradigm, *Computer*, 35(1):70–78, 2002.

49. Bennet, J. C. R., and Zhang, H., Hierarchical Packet Fair Queueing Algorithms, *Proceedings ACM SIGCOMM '96*, 1996.

50. Bennet, J. C. R., and Zhang, H., WF^{2}Q: Worst-Case Fair Weighted Fair Queueing, *Proceedings IEEE INFOCOM*, pp. 120–128, 1996.

51. Bernstein, G., Caviglia, D., Rabbat, R., and van Helvoort, H., VCAT/LCAS in a Clamshell, *IEEE Communications Magazine*, 44(5):34–36, 2006.

52. Birkhoff, G., Tres Observaciones Sobre el Algebra Lineal, *Universidad Nacional de Tucumán Revista Serie A*, 5:147–151, 1946.

53. Black, B., and Kompella, K., Maximum Transmission Unit Signaling Extensions for the Label Distribution Protocol, *RFC 3988, IETF*, 2005.

54. Blake, S., Black, D., Carlson, M., Davies, E., Wang, Z., and Weiss, W., An Architecture for Differentiated Services, *RFC 2475, IETF*, 1998.

55. Bottorff, P., General Discussion of Provider Backbone Transport in 802.1ah Networks, *http://www.ieee802.org/1/files/public/docs2006/ah-bottorff-pbt-v1-0506.pdf*, 2006.

56. Braden, B., Clark, D., Crowcroft, J., Davie, B., Deering, S., Estrin, D., Floyd, S., Jacobson, V., Minshall, G., Partridge, C., Peterson, L., Ramakrishnan, K., Shenker, S., Wroclawski, J., and Zhang, L., Recommendations on Queue Management and Congestion Avoidance in the Internet, *RFC 2309, IETF*, 1998.

57. Braden, B., Zhang, L., Berson, S., Herzog, S., and Jasmin, S., Resource ReSerVation Protocol (RSVP)—Version 1 Functional Specification, *RFC 2205, IETF*, 1997.

58. Braden, R., and Postel, J., Requirements for Internet Gateways, *RFC 1009, IETF*, 1987.

59. Braden, R., Borman, D., and Partridge, C., Computing the Internet Checksum, *RFC 1071, IETF*, 1988.

60. Braden, R., Clark, D., and Shenker, S., Integrated Services in the Internet Architecture: An Overview, *RFC 1633, IETF*, 1994.

61. Bryant, S., and Pate, P., Pseudo-Wire Emulation Edge-to-Edge (PWE3) Architecture, *RFC 3985, IETF*, 2005.

62. Bryant, S., Swallow, G., Martini, L., and McPherson, D., Pseudowire Emulation Edge-to-Edge (PWE3) Control Word for Use over an MPLS PSN, *RFC 4385, IEEE*, 2006.

63. Buddhikot, M. M, Suri, S., and Waldvogel, M., Space Decomposition Techniques for Fast Layer-4 Switching, *Proceedings Conference on Protocols for High Speed Networks*, pp. 25–41, 1999.

64. CableLabs, Cable Modem Termination System Network Side Interface Specification, *SP-CMTS-NSI, http://www.cablelabs.com/specifications/SP_CMTS_NSII01-960702. pdf*, 1996.

65. CableLabs, Cable Modem to Customer Premises Equipment Interface Specification, *SP-CMCI, http://www.cablelabs.com/specifications/SP-CMCI-I10-050408.pdf*, 2005.

66. Cain, B., Deering, S., Kouvelas, I., Fenner, B., and Thyagarajan, A., Internet Group Management Protocol, Version 3, *RFC 3376, IETF*, 2002.

67. Calhoun, P., Loughney, J., Guttman, E., Zorn, G., and Arkko, J., Diameter Base Protocol, *RFC 3588, IETF*, 2003.

68. Callon, R., and Suzuki, M., A Framework for Layer 3 Provider-Provisioned Virtual Private Networks (PPVPNs), *RFC 4110, IETF*, 2005.

69. Camarillo, G., and García-Martín, M.-A., *The 3G IP Multimedia Subsystem (IMS): Merging the Internet and the Cellular Worlds (2nd Edition)*, Wiley, 2005.

70. Campbell, A.T., Chou, S., Kounavis, M. E., Stachtos, V. D., and Vicente, J. B., NetBind: A Binding Tool for Constructing Data Paths in Network Processor-Based Routers, *Proceedings of the 5th IEEE International Conference on Open Architectures and Network Programming*, pp. 91-103, 2002.

71. Carugi, M., and McDysan, D., Service Requirements for Layer 3 Provider Provisioned Virtual Private Networks (PPVPNs), *RFC 4031, IETF*, 2005.

72. Chang, C.-S., Lee, D.-S., and Jou, Y.-S., Load Balanced Birkhoff-von Neumann Switches, part I: One Stage Buffering, *IEEE HPSR*, 2001.

73. Chang, C., Lee, D., and Yue, C., Providing Guaranteed Rate Services in the Load Balanced Birkhoff-von Neumann Switches, *IEEE/ACM Transaction on Networking*, 14(3): 644-656, 2006.

74. Chao, H. J., Saturn: A Terabit Packet Switch Using Dual Round-Robin, *IEEE Global Telecommunications Conference*, 1:487-495, 2000.

75. Chao, H. J., Next Generation Routers, *Proceedings of the IEEE*, 90:1518-1558, 2002.

76. Chao, H. J., and Guo, X., *Quality of Service Control in High-Speed Networks*, Wiley, 2002.

77. Chen, E., Route Refresh Capability for BGP-4, *RFC 2918, IETF*, 2000.

78. Cheung, S., and Pencea, C., BSFQ: Bin Sort Fair Queuing, *Proceedings IEEE INFOCOM*, 2002.

79. Chuang, S.-T., Iyer, S., and McKeown, N., Practical Algorithms for Performance Guarantees in Buffered Crossbars, *Proceedings IEEE INFOCOM*, Vol. 2:981-991, 2005.

80. Chuanxiong, G., SRR, An O(1) Time Complexity Packet Scheduler for Flows in Multi-Service Packet Networks, *Proceedings ACM SIGCOMM*, 2001.

81. Clark, D. D., and Wroclawski, J., An Approach to Service Allocation in the Internet, *draft-clark-diff-svc-alloc-00.txt, IETF*, 1997.

82. Click, *http://www.read.cs.ucla.edu/click/*.

83. Clos, C., A Study of Non-Blocking Switching Networks, *Bell System Technical Journal*, 32(5):406-424, 1953.

84. Coffman, E. G., and Eve, J., File Structures Using Hashing Functions, *Communications of the ACM*, 13(7):427-436, 1970.

85. Coffman, K. G., and Odlyzko, A. M., Internet Growth: Is There a "Moore's Law" for Data Traffic? in *Handbook of Massive Data Sets*, Abello, J., Pardalos, P. M., and Resende, M.G., (Eds.), Kluwer Academic Publishers, pp. 47-93, 2002.

86. Comer, D. E., *Network Systems Design Using Network Processors*, Prentice Hall, 2004.

87. Conta, A., Deering, S., and Gupta, M., Internet Control Message Protocol (ICMPv6) for the Internet Protocol Version 6 (IPv6) Specification, *RFC 4443, IETF*, 2006.

88. Coppo, P., D'Ambrosio, M., and Melen, R., Optimal Cost/Performance Design of ATM Switches, *IEEE/ACM Transactions on Networking*, 1(5):566–575, 1993.

89. Cormen, T. H., Leiserson, C. E., Rivest, R. L., and Stein, C., *Introduction to Algorithms (2nd Edition)*, MIT Press and McGraw-Hill, 2001.

90. Crescenzi, P., Dardini, L., and Grossi, R., IP Address Lookup Made Fast and Simple, *Technical Report TR-99-01*, Dipartmento Di Informatica, Universit a Di Pisa, 1999; also in *Proceedings 7th Annual European Symposium on Algorithms, Lecture Notes in Computer Science*, Vol. 1643:65–76, 1999.

91. Crowley, P., Fiuczynski, M. E., and Baer, J.-L., On the Performance of Multithreaded Architectures for Network Processors, *Technical Report 2000-10-01*, Department of Computer Science, University of Washington, 2000.

92. Crowley, P., Franklin, M., Hadimioglu, H., and Onufryk, P., *Network Processor Design: Issues and Practices*, Vol. 1, Morgan Kaufmann, 2002.

93. Culler, D., Karp, R., Patterson, D., Sahay, A., Schauser, K. E., Santos, E., Subramonian, R., and von Erocken, T., LogP: Towards a Realistic Model of Parallel Computation, *Proceedings of the 4th ACM SIGPLAN Symposium on Principles and Practices of Parallel Computing*, pp. 1–12, 1993.

94. Culler, D. E., Singh, J. P., and Gupta, A., *Parallel Computer Architecture—A Hardware/Software Approach*, Morgan Kaufmann, 1999.

95. Decker, E., Langille, P., Rijsinghani, A., and McCloghrie, K., Definitions of Managed Objects for Bridges, *RFC 1493, IETF*, 1993.

96. Decraene, B., Le Roux, J. L., and Minei, I., LDP extension for Inter-Area LSP, *draft-ietf-mpls-ldp-interarea-03, IETF*, 2008.

97. Deering, S., Host Extensions for IP multicasting, *RFC 1112, IETF*, 1989.

98. Deering, S., and Hinden, R., Internet Protocol, Version 6 (IPv6) Specification, *RFC 2460, IETF*, 1998.

99. Degermark, M., Brodnik, A., Carlsson, S., and Pink, S., Small Forwarding Tables for Fast Routing Lookups, *Proceedings ACM SIGCOMM Computer Communication Review*, Vol. 27:3–14, 1997.

100. Demers, A., Keshav, S., and Shenker, S., Analysis and Simulations of a Fair Queueing Algorithm, *Proceedings ACM SIGCOMM*, Vol. 19:3–12, 1989.

101. Dias, D. M., and Kumar, M., Packet Switching in NlogN Multistage Networks, *Proceedings IEEE GLOBECOM*, pp. 114–120, 1984.

102. Dierks, T., and Rescorla, E., The Transport Layer Security (TLS) Protocol, Version 1.1, *RFC 4346, IETF*, 2006.

103. Dolev, S., and Kesselman, A., Bounded Latency Scheduling Scheme for ATM Cells, *Computer Networks: The International Journal of Computer and Telecommunications Networking*, 32(3):325–331, 2000.

104. Dongarra, J., Foster, I., Fox, G., Gropp, W., Kennedy, K., Torczon, L., and White, A. (Eds.), *Sourcebook of Parallel Computing*, Morgan Kaufmann, 2003.

105. DSL Forum, Core Network Architecture for Access to Legacy Data Network over ADSL, *TR-025, http://www.dslforum.org/techwork/tr/TR-025.pdf*, 1999.

106. DSL Forum, Multi-Service Architecture & Framework Requirements, *TR-058, http://www.dslforum.org/techwork/tr/TR-058.pdf*, 2003.

107. DSL Forum, DSL Evolution—Architecture Requirements for the Support of QoS-Enabled IP Services, *TR-059, http://www.dslforum.org/techwork/tr/TR-059.pdf*, 2003.

108. DSL Forum, Migration to Ethernet Based DSL Aggregation, *TR-101, http://www.dslforum.org/techwork/tr/TR-101.pdf*, 2006.

109. Eatherton, W., Dittia, Z., and Varghese, G., Full Tree Bit Map: Hardware/Software IP Lookup Algorithms with Incremental Updates, *Proceedings IEEE INFOCOM*, 1999.

110. Eatherton, W., Varghese, G., and Dittia, Z., Tree Bitmap: Hardware/Software IP Lookups with Incremental Updates, *Proceedings ACM SIGCOMM Computer Communication Review*, Vol. 34:97–122, 2004.

111. Effenberger, F., Cleary, D., Haran, O., Kramer, G., Ding-Li, R., Oron, M., and Pfeiffer, T., An Introduction to PON Technologies, *IEEE Optical Communications*, in *IEEE Communications Magazine*, 45(3):S17–S25, 2007.

112. Eicker, E., and Lippert, T., Scalable Ethernet Clos-Switches, *Euro-Par 2006 Parallel Processing, Lecture Notes in Computer Science*, Vol. 4128, Springer, pp. 874–883, 2006.

113. Eisler, M., XDR: External Data Representation Standard, *Request for Comments 4506, IETF*, 2006.

114. Ekman, M., Warg, F., and Nilson, J., An In-Depth Look at Computer Performance Growth, *ACM SIGARCH Computer Architecture News*, 33(1):144–147, 2005.

115. Eng, K. Y., Hluchyj, M. G., and Yeh, Y.-S., A Knockout Switch for Variable-Length Packets, *IEEE Journal on Selected Areas in Communications*, 5(9):1426–1435, 1987.

116. Estrin, D., Farinacci, D., Helmy, A., Thaler, D., Deering, S., Handley, M., Jacobson, V., Liu, C., Sharma, P., and Wei, L., Protocol Independent Multicast-Sparse Mode (PIM-SM): Protocol Specification, *RFC 2362, IETF*, 1998.

117. EZchip, *EZdesign—Microcode Development Tools for NP-1 Family: User Manual*, Version 3.4, Document Number: 27-7799-07, EZchip Technologies, Inc., 2003.

118. EZchip, *NP-1 Network Processor: Architecture and Instruction Set: User Manual*, Document Number: 27-7816-02, EZchip Technologies, Inc., 2002.

119. Fang, W., and Seddigh, N., Time Sliding Window Three Color Marker, *RFC 2859, IETF*, 2000.

120. Farinacci, D., Li, T., Hanks, S., Meyer, D., and Traina, P., Generic Routing Encapsulation (GRE), *RFC 2784, IETF*, 2000.

121. Farrel, A., Fault Tolerance for the Label Distribution Protocol (LDP), *RFC 3479, IETF*, 2003.

122. Feldman, A., and Muthukrishnan, S., Tradeoffs for Packet Classification, *Proceedings IEEE INFOCOM*, Vol. 3:1193–1202, 2000.

123. Feng, W.-C., Improving Internet Congestion Control and Queue Management Algorithms, PhD Dissertation, Computer Science and Engineering, University of Michigan, 1999.

124. Feng, W.-C., Kandlur, D. D., Saha, D., and Shin, K. G., BLUE: A New Class of Active Queue Management Algorithms, *Technical Report CSE-TR-387-99*, University of Michigan, 1999.

125. Feng, W.-C., Shin, K. G., Kandlur, D. D., and Saha, D., The BLUE Active Management Algorithms, *IEEE/ACM Transactions on Networking*, 10(4):513–528, 2002.

126. Fenner, W., Internet Group Management Protocol, Version 2, *RFC 2236, IETF*, 1997.

127. Ferrari, D., and Verma, D., A Scheme for Real-Time Channel Establishment in Wide-Area Networks, *IEEE Journal on Selected Areas of Communications*, 8(3):368–379, 1990.

128. Floyd, S., References on RED (Random Early Detection) Queue Management, *http://www.icir.org/floyd/red.html*, 2002.

129. Floyd, S., and Jacobson, V., Random Early Detection Gateways for Congestion Avoidance, *IEEE/ACM Transactions on Networking*, 1(4):397–341, 1993.

130. Floyd, S., and Jacobson, V., Link-Sharing and Resource Management Models for Packet Networks, *IEEE/ACM Transactions on Networking*, 3(4):365–386, 1995.

131. Flynn, M. J., Very High-Speed Computing Systems, *Proceedings of the IEEE*, Vol. 54:1901–1909, 1966.

132. Flynn, M. J., Some Computer Organizations and Their Effectiveness, *IEEE Transactions on Computers*, 21(9):948–960, 1972.

133. Fortune, S., and Wyllie, J., Parallelism in Random Access Machines, *Proceedings of the 10th ACM Symposium on Theory of Computing*, pp. 114–118, 1978.

134. Franklin, M., Crowley, P., and Onufryk, P., *Network Processor Design*, Vol. 2, Morgan Kaufmann, 2003.

135. Franklin, M., Crowley, P., Hadimioglu, H., and Onufryk, P., *Network Processor Design: Issues and Practices*, Vol. 3, Morgan Kaufmann, 2005.

136. Fredkin, E., Trie Memory, *Communications of the ACM*, 3(9):490–500, 1960.

137. Fredman, M. L., and Tarjan, R. E., Fibonacci Heaps and their Uses in Improved Network Optimization Algorithms, *Journal of the Association for Computing Machinery*, 34(3):596–615, 1987.

138. Friend, R., Making the Gigabit IPsec VPN Architecture Secure, *IEEE Computer*, 37(6):54–60, 2004.

139. Frobenius, G., Über zerlegbare Determinanten, *Sitzungsberichte der Koniglich Preußischen Akademie der Wisssenschaften zu Berlin*, XVIII, pp. 274–277, 1917.

140. Fuller, V., Li, T., Yu, J., and Varadhan, K., Classless Inter-Domain Routing (CIDR): An Address Assignment and Aggregation Strategy, *RFC 1519, IETF*, 1993.

141. Ganjali, Y., Keshavarzian, A., and Shah, D., Input-Queued Switches: Cell Switching vs. Packet Switching, *Proceedings IEEE INFOCOM*, Vol. 3:1651–1658, 2003.

142. Ganjali, Y., Keshavarzian, A., and Shah, D., Cell Switching Versus Packet Switching in Input-Queued Switches, *IEEE/ACM Transactions on Networking*, 13(4):782–789, 2005.

143. Gao, Y., and Hou, J., A State Feedback Control Approach to Stabilizing Queues for ECN-Enabled TCP Connections, *Proceedings IEEE INFOCOM*, 2003.

144. Gavrilovska, A., SPLITS Stream Handlers: Deploying Application-Level Services to Attached Network Processors, PhD Dissertation, Georgia Institute of Technology, 2004.

145. Gavrilovska, A., Kumar, S., and Schwan, K., The Execution of Event-Action Rules on Programmable Network Processors, *Proceedings of the 1st Workshop on Operating System and Architectural Support for the On-Demand IT Infrastructure (OASIS)*, 2004.

146. Geist, A., Beguelin, A., Dongara, J. J., Jiang, W., Manchek, R., and Sonderam, V., *PVM: Parallel Virtual Machine—A Users' Guide and Tutorial to Networked Parallel Computing*, MIT Press, 1994.

147. Giordano, S., Procissi, G., Rossi, F., and Vitucci, F., Design of a Multi-Dimensional Packet Classifier for Network Processors, *Proceedings of the IEEE International Conference on Communications*, Vol. 2:503–508, 2006.

148. Gleeson, B., Lin, A., Heinanen, J., Armitage, G., and Malis, A., A Framework for IP Based Virtual Private Networks, *RFC 2764, IETF*, 2000.

149. Goke C., and Lipovsky, G. J., Banyan Networks for Partitioning Multiprocessor Systems, *Proceedings of the 1st Annual Symposium on Computer Architecture*, pp. 21–28, 1973.

150. Golestani S. J., A Framing Strategy for Congestion Management, *IEEE Journal on Selected Areas of Communications*, 9(7):1064–1077, 1991.

151. Golestani, S. J., A Self-Clocked Fair Queuing Scheme for Broadband Applications, *Proceedings IEEE INFOCOM*, pp. 636–646, 1994.

152. Gries, M., Kulkarni, C., Sauer, C., and Keutzer, K., Exploring Trade-Offs in Performance and Programmability of Processing Element Topologies for Network Processors, *2nd Workshop on Network Processors (NP2) at the 9th International Symposium on High Performance Computer Architecture*, pp. 75–87, 2003.

153. Gross, G., Kaycee, M., Li, A., Malis, A., and Stephens, J., PPP Over AAL5, *RFC 2364, IETF*, 1998.

154. Grossman, D., and Heinanen, J., Multiprotocol Encapsulation over ATM Adaptation Layer 5, *RFC 2684, IETF*, 1999.

155. Guerin, R., and Peris, V., Quality-of-Service in Packet Networks: Basic Mechanisms and Directions, *Computer Networks*, 31(3):169–189, 1999.

156. Guez, D., Kesselman, A., and Rosen, A., Packet-Mode Policies for Input-Queued Switches, *Proceedings of the Sixteenth Annual ACM Symposium on Parallelism in Algorithms and Architectures*, pp. 93–103, 2004.

157. Gunning, P., Wilkinson, M., Rragami, L., Semnani, S., and Smith, K., Multiservice Ethernet Access, *BT Technology Journal*, 24(2):72-78, 2006.

158. Gupta, P., and McKeown, N., Designing and Implementing a Fast Crossbar Scheduler, *IEEE Micro*, 19(1):20–28, 1999.

159. Gupta, P., and McKeown, N., Packet Classification on Multiple Fields, *Proceedings SIGCOMM '99, Computer Communication Review*, Vol. 29:147–160, 1999.

160. Gupta, P., and McKeown, N., Classification Using Hierarchical Intelligent Cuttings, *IEEE Micro*, 20(1):34–41, 2000.

161. Gupta, P., and McKeown, N., Algorithms for Packet Classification, *IEEE Network*, 15(2):24–32, 2001.

162. Gupta, P., Lin, S., and McKeown, N., Routing Lookups in Hardware at Memory Access Speeds, *Proceedings of INFOCOM*, pp. 1240–1247, 1998.

163. Gwehenberger, G., Anwendung einer binären Verweiskettenmethode beim Aufbau von Listen, *Elektronische Rechenanlagen*, 10(5):223–226, 1968.

164. Hall, P., On Representatives of Subsets, *Journal of the London Mathematical Society*, (10):26–30, 1935.

165. Hambrusch, E. S., Models of Parallel Computation, *Proceedings of Workshop on Challenges for Parallel Processing, International Conference on Parallel Processing*, pp. 92–95, 1996.

166. Hamzeh, K., Pall, G., Verthein, W., Taarud, J., Little, W., and Zorn, G., Point-to-Point Tunneling Protocol (PPTP), *RFC 2637, IETF*, 1999.

167. Hedrick, C. L., Routing Information Protocol, *RFC 1058, IETF*, 1988.

168. Heinanen, J., and Guerin, R., A Single Rate Three Color Marker, *RFC 2697, IETF*, 1999.

169. Heinanen, J., and Guerin, R., A Two Rate Three Color Marker, *RFC 2698, IETF*, 1999.

170. Heinanen, J., Baker, F., Weiss, W., and Wroclawski, J., Assured Forwarding PHB Group, *RFC 2597, IETF*, 1999.

171. Hennessy, L. J., and Patterson, A. D., *Computer Architecture: A Quantitative Approach (4th Edition)*, Morgan Kaufmann, 2006.

172. Hernandez-Valencia, E., Scholten, M., and Zhu, Z., The Generic Framing Procedure (GFP): An Overview, *IEEE Communications Magazine*, 40(5):63–71, 2002.

173. High Performance Fortran Forum, High Performance Fortran Language Specification, Version 1.1, 1994.

174. Hollot, C., Misra, V., Towsley, D., and Gong, W., On Designing Improved Controllers for AQM Routers Supporting TCP Flows, *Proceedings IEEE INFOCOM*, pp. 1726–1734, 2001.

175. Huang, H., Zhao, S., Pan, J., and Su, C., A Fast IP Routing Lookup Scheme for Gigabit Switching Routers, *Proceedings IEEE INFOCOM*, 1999.

176. Husak, D., Network Processors: A Definition and Comparison, White paper, C-PORT/ FreeScale, *http://www.freescale.com/files/netcomm/doc/white_paper/ COMMPROCWP.pdf*, 2000.

177. Hwang, K., *Advanced Computer Architecture: Parallelism, Scalability, Programmability*, McGraw-Hill, 1993.

178. IEEE Std 802-2001, *IEEE Standard for Local and Metropolitan Area Networks: Overview and Architecture*, 2002.

179. IEEE Std 802.1ad-2005, *IEEE Standard for Local and Metropolitan Area Networks: Virtual Bridged Local Area Networks—Provider Bridges*, 2005.

180. IEEE Std 802.1ag-2007, *IEEE Standard for Local and Metropolitan Area Networks: Connectivity Fault Management*, 2007.

181. IEEE P802.1ah/D4.0, *IEEE Standard for Local and Metropolitan Area Networks, Virtual Bridged Local Area Networks: Provider Backbone Bridges*, 2007.

182. IEEE Std 802.1ak-2007, *IEEE Standard for Local and Metropolitan Area Networks: Multiple Registration Protocol*, 2007.

183. IEEE Std 802.1D-2004, *Standard for Local and Metropolitan Area Networks: Media Access Control (MAC) Bridges*, 2004.

184. IEEE Std 802.1Q-2005, *IEEE Standard for Local and Metropolitan Area Networks, Virtual Bridged Local Area Networks*, 2006.

185. IEEE Std 802.2-1998 Edition (R2003), *IEEE Standard for Local and Metropolitan Area Networks*, Part 2: Logical Link Control, 1998.

186. IEEE Std 802.3-2005, *IEEE Standard for Local and Metropolitan Area Networks*, Part 3: Carrier Sense Multiple Access with Collision Detection (CSMA/CD) Access Method and Physical Layer Specifications, 2005.

187. IEEE Std 802.16-2004, *IEEE Standard for Local and Metropolitan Area Networks*, Part 16: Air Interface for Fixed Broadband Wireless Access Systems, 2004.

188. IEEE Std 802.16e-2005, *IEEE Standard for Local and Metropolitan Area Networks*, Part 16: Air Interface for Fixed Broadband Wireless Access Systems, Amendment 2: Physical and Medium Access Control Layers for Combined Fixed and Mobile Operation in Licensed Bands, 2005.

189. IEEE 802.16m, Project Authorization Request, *http://standards.ieee.org/board/nes/ projects/802-16m.pdf*.

190. IEEE Std 802.17-2004, *IEEE Standard for Local and Metropolitan Area Networks*, Part 17: Resilient Packet Ring (RPR) Access Method and Physical Layer Specifications, 2004.

191. Iliadis, I., and Denzel, W. E., Resequencing Worst-Case Analysis for Parallel Buffered Packet Switches, *IEEE Transactions on Communications*, 55(3):605–617, 2007.

192. Intel, *IXA SDK ACE Programming Framework, SDK 2.01 Reference*, Document Number A46817-002, Revision 3.4, Intel, 2001.

193. International Organization for Standardization, Information Technology Telecommunications and Information Exchange Between Systems—High-Level Data Link Control (HDLC) Procedures—Frame Structure—Amendment 2: Extended Transparency Options For Start/Stop Transmission, *ISO/IEC 3309*, 1992.

194. International Telecommunication Union, Characteristics of Plesiochronous Digital Hierarchy (PDH) Equipment Functional Blocks, *ITU-T Recommendation G.705 (10/00)*, 2000.

195. International Telecommunication Union, Network Node Interface for the Synchronous Digital Hierarchy (SDH), *ITU-T Recommendation G.707/Y.1322 (01/07)*, 2007.

196. International Telecommunication Union, Interfaces for the Optical Transport Network (OTN), *ITU-T Recommendation G.709/Y.1331 (03/03)*, 2003.

197. International Telecommunication Union, Architecture of Transport Networks Based on the Synchronous Digital Hierarchy (SDH), *ITU-T Recommendation G.803 (03/00)*, 2000.

198. International Telecommunication Union, Generic Functional Architecture of Transport Networks, *ITU-T Recommendation G.805 (03/00)*, 2000.

199. International Telecommunication Union, Requirements for Automatic Switched Transport Networks (ASTN), *ITU-T Recommendation G.807/Y.1302 (07/01)*, 2001.

200. International Telecommunication Union, Terms and Definitions for Optical Transport Networks (OTN), *ITU-T Recommendation G.870/Y.1352 (03/08)*, 2008.

201. International Telecommunication Union, Architecture of Optical Transport Networks, *ITU-T Recommendation G.872 (11/01)*, 2001.

202. International Telecommunication Union, Gigabit-Capable Passive Optical Networks (G-PON): Transmission Convergence Layer Specification, *ITU-T Recommendation G.984.3 (03/08)*, 2008.

203. International Telecommunication Union, High Bit Rate Digital Subscriber Line (HDSL) Transceivers, *ITU-T Recommendation G.991.1 (10/98)*, 1998.

204. International Telecommunication Union, Asymmetric Digital Subscriber Line (ADSL) Transceivers, *ITU-T Recommendation G.992.1 (07/99)*, 1999.

205. International Telecommunication Union, Very High Speed Digital Subscriber Line Transceivers, *ITU-T Recommendation G.993.1 (06/04)*, 2004.

206. International Telecommunication Union, Generic Framing Procedure (GFP), *ITU-T Recommendation G.7041/Y.1303 (08/05)*, 2005.

207. International Telecommunication Union, Link Capacity Adjustment Scheme (LCAS) for Virtual Concatenated Signals, *ITU-T Recommendation G.7042/Y.1305 (03/06)*, 2006.

208. International Telecommunication Union, Virtual Concatenation of Plesiochronous Digital Hierarchy (PDH) Signals, *ITU-T Recommendation G.7043/Y.1343 (07/04)*, 2004.

209. International Telecommunication Union, Distributed Call and Connection Management (DCM) Based on PNNI, *ITU-T Recommendation G.7713.1/Y.1704.1 (03/03)*, 2003.

210. International Telecommunication Union, Distributed Call and Connection Management: Signalling Mechanism Using GMPLS RSVP-TE, *ITU-T Recommendation G.7713.2/ Y.1704.2 (03/03)*, 2003.

211. International Telecommunication Union, Distributed Call and Connection Management: Signalling Mechanism Using GMPLS CR-LDP, *ITU-T Recommendation G.7713.3/Y.1704.3 (03/03)*, 2003.

212. International Telecommunication Union, Architecture for the Automatically Switched Optical Network (ASON), *ITU-T Recommendation G.8080/Y.1304 (06/06)*, 2006.

213. International Telecommunication Union, Terms and Definitions for Automatically Switched Optical Networks (ASON), *ITU-T Recommendation G.8081/Y.1353 (03/08)*, 2008.

214. International Telecommunication Union, Terms and Definitions for Transport MPLS, *ITU-T Recommendation G.8101/Y.1355 (12/06)*, 2006.

215. International Telecommunication Union, Architecture of Transport MPLS (T-MPLS) Layer Network, *ITU-T Recommendation G.8110.1/Y.1370.1 (11/06)*, 2006.

216. International Telecommunication Union, Interfaces for the Transport MPLS (T-MPLS) Hierarchy, *ITU-T Recommendation G.8112/Y.1371 (10/06)*, 2006.

217. International Telecommunication Union, Phoneline Networking Transceivers—Foundation (HomePNA 2.0), *ITU-T Recommendation G.9951 (02/01)*, 2001.

218. International Telecommunication Union, Phoneline Networking Transceivers—Payload Format and Link Layer Requirements (HomePNA 2.0), *ITU-T Recommendation G.9952 (11/01)*, 2001.

219. International Telecommunication Union, Phoneline Networking Transceivers—Isolation Function (HomePNA 2.0), *ITU-T Recommendation G.9953 (03/03)*, 2003.

220. International Telecommunication Union, Home Networking Transceivers—Enhanced Physical, Media Access, and Link Layer Specifications (HomePNA 3.0 and 3.1), *ITU-T Recommendation G.9954 (01/07)*, 2007.

221. International Telecommunication Union, B-ISDN General Network Aspects, *ITU-T Recommendation I.311 (08/96)*, 1996.

222. International Telecommunication Union, B-ISDN ATM Layer Specification, *ITU-T Recommendation I.361 (02/99)*, 1999.

223. International Telecommunication Union, Traffic Control and Congestion Control in B-ISDN, *ITU-T, Recommendation I.371*, 1995.

224. International Telecommunication Union, Transmission Systems for Interactive Cable Television Services, *ITU-T Recommendation J.112 (03/98)*, 1998.

225. International Telecommunication Union, Second-Generation Transmission Systems for Interactive Cable Television Services—IP Cable Modems, *ITU-T Recommendation J.122 (12/07)*, 2007.

226. International Telecommunication Union, Overview of Third-Generation Transmission Systems for Interactive Cable Television Services—IP Cable Modems, *ITU-T Recommendation J.222.0 (12/07)*, 2007.

227. International Telecommunication Union, Operation and Maintenance Mechanism for MPLS Networks, *ITU-T Recommendation Y.1711 (02/04)*, 2004.

228. International Telecommunication Union, Requirements for OAM Functions in Ethernet-Based Networks and Ethernet Services, *ITU-T Recommendation Y.1730 (01/04)*, 2004.

229. International Telecommunication Union, OAM Functions and Mechanisms for Ethernet Based Networks, *ITU-T Recommendation Y.1731 (02/08)*, 2008.

230. Iyer, S., and McKeown, N. W., Analysis of the Parallel Packet Switch Architecture, *IEEE/ACM Transaction on Networking*, 11(2):314–324, 2003.

231. Iyer, S., Zhang, R., and McKeown, N., Routers with a Single Stage of Buffering, *ACM SIGCOMM Computer Communication Review*, 32(4):251–264, 2002.

232. Jacobson, V., Nichols, K., and Poduri, K., An Expedited Forwarding PHB, *RFC 2598, IETF*, 1999.

233. Jain, R., Chiu, D. M., and Hawe, W. R., A Quantitative Measure of Fairness and Discrimination for Resource Allocation Shared Computer Systems, *Digital Equipment Corporation Technical Report TR-301*, 1984.

234. Jajszczyk, A., Optical Networks—the Electro-Optic Reality, *Optical Switching and Networking*, 1(1):3–18, 2005.

235. Jamoussi, B., Andersson, L., Callon, R., Dantu, R., Wu, L., Doolan, P., Worster, T., Feldman, N., Fredette, A., Girish, M., Gray, E., Heinanen, J., Kilty, T., and Malis, A., Constraint-Based LSP Setup Using LDP, *RFC 3212, IETF*, 2002.

236. Javidi, T., Magill, R., and Hrabik, T., A High-Throughput Scheduling Algorithm for a Buffered Crossbar Switch Fabric, *Proceedings of the IEEE International Conference on Communications*, Vol. 5:1586–1591, 2001.

237. JEDEC Standard, DDR2 SDRAM Specification, JESD79-2D, *JEDEC Solid State Technology Association, Electronic Industries Alliance*, 2008.

238. JEDEC Standard, DDR3 SDRAM Specification, JESD79-3A, *JEDEC Solid State Technology Association, Electronic Industries Alliance*, 2007.

239. Jeon, H. S., Riegel, M., and Jeong, S. J., Transmission of IP Over Ethernet Over IEEE 802.16 Networks, *draft-ietf-16ng-ip-over-ethernet-over-802.16-04.txt, IETF*, 2007.

240. Jerraya, A. A., and Wolf, W., *Multiprocessor Systems-on-Chip*, Morgan Kaufmann, 2005.

241. Jones, N., and Murton, C., Extending Point-to-Point Protocol (PPP) Over Synchronous Optical NETwork/Synchronous Digital Hierarchy (SONET/SDH) with Virtual Concatenation, High Order and Low Order Payloads, *RFC 3255, IETF*, 2002.

242. Kalmanek, C. R., Kanakia, H., and Keshav, S., Rate Controlled Servers for Very High Speed Networks, *Proceedings GLOBECOM*, pp. 12–20, 1990.

243. Kanhere, S., and Sethu, H., Prioritized Elastic Round Robin: An Efficient and Low-Latency Packet Scheduler with Improved Fairness, *Technical Report DU-CS-03-03*, Department of Computer Science, Drexel University, 2003.

244. Kanhere, S., Sethu, H., and Parekh, A., Fair and Efficient Packet Scheduling Using Elastic Round Robin, *IEEE Transactions on Parallel and Distributed Systems*, 13(3):324–336, 2002.

245. Karlin, S., and Peterson, L., VERA: An Extensible Router Architecture, *Proceedings of the 4th International Conference on Open Architectures and Network Programming (OPENARCH)*, pp. 3–14, 2001; and also in *Computer Networks*, 38(3):277–293, 2002.

246. Karol, M. J., Hluchyj, M. G., and Morgan, S. P., Input Versus Output Queueing on a Space-Division Packet Switch, *IEEE Transactions on Communications*, 35(12):1347–1356, 1987.

247. Kartalopoulos, S. V., *Next Generation SONET/SDH: Voice and Data*, IEEE Press/Wiley, 2004.

248. Katz, D., Kompella, K., and Yeung, D., Traffic Engineering (TE) Extensions to OSPF Version 2, *RFC 3630, IETF*, 2003.

249. Kent, S., and Seo, K., Security Architecture for the Internet Protocol, *RFC 4301, IETF*, 2005.

250. Keshav, S., *An Engineering Approach to Computer Networking: ATM Networks and the Telephone Network*, Addison-Wesley, 1997.

251. Keslassy, I., The Load-Balanced Router, PhD Dissertation, Stanford University, 2004.

252. Keslassy, I., and McKeown, N., Maintaining Packet Order in Two-Stage Switches, *Proceedings of IEEE INFOCOM*, Vol. 2:1032–1041, 2002.

253. Keslassy, I., Chang, C.-S., McKeown, N., and Lee, D.-S., Optimal Load-Balancing, *Proceedings of IEEE INFOCOM*, Vol. 3:1712–1722, 2005.

254. Keutzer, K., Malik, S., and Newton, A. R., From ASIC to ASIP: The Next Design Discontinuity, *Proceedings of the IEEE International Conference on Computer Design: VLSI in Computers and Processors*, 2002.

255. Khosravi, H., and Anderson, T., Requirements for Separation of IP Control and Forwarding, *RFC 3654, IETF*, 2003.

256. Kleene, S. C., *Mathematical Logic*, Wiley, 1967.

257. Kleinrock, L., Time-Shared Systems: A Theoretical Treatment, *Journal of the ACM*, 14(2):242–261, 1967.

258. Kleinrock, L., *Queueing Systems, Volume 2: Computer Applications*, Wiley, 1976.

259. Knight, P., Ould-Brahim, H., and Gleeson, B., Network Based IP VPN Architecture Using Virtual Routers, *draft-ietf-l3vpn-vpn-vr-03.txt, IETF*, 2006.

260. Knuth, D. E., *The Art of Computer Programming, Volume 3: Sorting and Searching (2nd Edition)*, Addison-Wesley, 1998.

261. Koelbel, C. H., Loveman, D. V., Schreiber, R. S., Steele, G. L., and Zosel, M. E., *The High Performance Fortran Handbook*, MIT Press, 1994.

262. Kohler, E., The Click Modular Router, PhD Dissertation, MIT, *http://pdos.csail.mit.edu/papers/click:kohler-phd/thesis.pdf*, 2001.

263. Kohler, E., Morris, R., Chen, B., Jannotti, J., and Kaashoek, M. F., The Click Modular Router, *ACM Transactions on Computer Systems*, 18(3):263–297, 2000.

264. Kompella, K., and Rekhter, Y., Virtual Private LAN Service (VPLS) Using BGP for Auto-Discovery and Signaling, *RFC 4761, IETF*, 2007.

265. Kompella, K., Rekhter, Y., and Kullberg, A., Signaling Unnumbered Links in CR-LDP, *RFC 3480, IETF*, 2003.

266. Kunniyur, S., and Srikant, R., Analysis and Design of an Adaptive Virtual Queue (AVQ) Algorithm for Active Queue Management, *Proceedings ACM SIGCOMM*, 2001.

267. Labovitz, C., Scalability of the Internet Backbone Routing Infrastructure, PhD Dissertation, University of Michigan, 1999.

268. Lakshman, T. V., Neidhardt, A., and Ott, T., The Drop From Front Strategy in TCP Over ATM and Its Interworking with Other Control Features, *Proceedings IEEE INFOCOM*, 1996.

269. Lakshman, T. V., and Stiliadis, D., High-Speed Policy-Based Packet Forwarding Using Efficient Multi-Dimensional Range Matching, *Proceedings of the ACM SIGCOMM Conference on Applications, Technologies, Architectures, and Protocols for Computer Communication*, pp. 203–214, 1998.

270. Lampson, B., Srinivasan, V., and Varghese, G., IP Lookups Using Multiway and Multicolumn Search, *Proceedings IEEE INFOCOM*, pp. 1248–1256, 1998.

271. Lasserre, M., and Kompella, V., Virtual Private LAN Services Using LDP, *RFC 4762, IETF*, 2007.

272. Lau, J., Townsley, M., and Goyret, I., Layer Two Tunneling Protocol—Version 3 (L2TPv3), *RFC 3931, IETF*, 2005.

273. Laubach, M., and Halpern, J., Classical IP and ARP over ATM, *RFC 2225, IETF*, 1998.

274. Lee, C.-H., Sorin, W. V., and Kim, B. Y., Fiber to the Home Using a PON Infrastructure, *Journal of Lightwave Technology*, 24(12):4568–4583, 2006.

275. Lee, E. A., The Problem with Threads, *IEEE Computer*, 39(5):33–42, 2006.

276. Lee, K., Coulson, G., Blair, G., Joolia, A., and Ueyama, K., Towards a Generic Programming Model for Network Processors, *Proceedings of the 12th IEEE International Conference on Networks*, pp. 504–510, 2004.

277. Leighton, F. T., *Introduction to Parallel Algorithms and Architectures: Arrays, Trees, Hypercubes*, Morgan Kaufmann, 1992.

278. Lekkas, P. C., *Network Processors: Architectures, Protocols, and Platforms*, McGraw-Hill, 2003.

279. Lenzini, L., Mingozzi, E., and Stea, G., Eligibility-Based Round Robin for Fair and Efficient Packet Scheduling in Wormhole Switching Networks, *IEEE Transactions on Parallel and Distributed Systems*, 15(3):244–256, 2004.

280. Li, X., Zhou, Z., and Hamdi, M., Space-Memory-Memory Architecture for CLOS-Network Packet Switches, *Proceedings of the IEEE International Conference on Communications*, Vol. 2:1031–1035, 2005.

281. Li, Y., Ko, K.-T., and Chen, G., A Smith Predictor-Based PI-Controller for Active Queue Management, *IEICE Transactions on Communications*, E88-B(11):4293–4300, 2005.

282. Light Reading, *http://www.lightreading.com/*.

283. Lin, D., and Morris, R., Dynamics of Random Early Detection, *Proceedings ACM SIGCOMM*, pp. 127–137, 1997.

284. Lin, M., and McKeown, N., The Throughput of a Buffered Crossbar Switch, *IEEE Communications Letters*, 9(5):465–467, 2005.

285. Lin, Y. S., and Shung, C. B., Quasi-Pushout Cell Discarding, *IEEE Communications Letters*, 1(5):146–148, 1997.

286. Linley Group, *http://www.linleygroup.com/*.

287. Liu, C., and Layland, J., Scheduling Algorithms for Multiprogramming in a Hard Real-Time Environment, *Journal of the ACM*, 20(1):46–61, 1973.

288. Lougheed, K., and Rekhter, Y., Border Gateway Protocol (BGP), *RFC 1105, IETF*, 1989.

289. Lougheed, K., and Rekhter, Y., Border Gateway Protocol (BGP), *RFC 1163, IETF*, 1990.

290. Lougheed, K., and Rekhter, Y., Border Gateway Protocol 3 (BGP-3), *RFC 1267, IETF*, 1991.

291. Magill, R. B., Rohrs, C. E., and Stevenson, R. L., Output-Queued Switch Emulation by Fabrics with Limited Memory, *IEEE Journal on Selected Areas in Communications*, 21(4): 606–615, 2003.

292. Malis, A., Pate, P., Cohen, R., and Zelig, D., Synchronous Optical Network/Synchronous Digital Hierarchy (SONET/SDH) Circuit Emulation over Packet (CEP), *RFC 4842, IETF*, 2007.

293. Malkin, G., RIP Version 2, *RFC 2453, IETF*, 1998.

294. Malkin, G., and Minnear, R., RIPng for IPv6, *RFC 2080, IETF*, 1997.

295. Mamakos, L., Lidl, K., Evarts, J., Carrel, D., Simone, D., and Wheeler, R., A Method for Transmitting PPP over Ethernet (PPPoE), *RFC 2516, IETF*, 1999.

296. Mammoliti, V., Zorn, G., Arberg, P., and Rennison, R., DSL Forum Vendor-Specific RADIUS Attributes, *RFC 4679, IETF*, 2006.

297. Mannie, E., Generalized Multi-Protocol Label Switching (GMPLS) Architecture, *RFC 3945, IETF*, 2004.

298. Marowka, A., Analytic Comparison of Two Advanced C Language-Based Parallel Programming Models, *Proceedings of the Third International Symposium on Parallel and Distributed Computing/Third International Workshop on Algorithms, Models and Tools for Parallel Computing on Heterogeneous Networks*, pp. 284–291, 2004.

299. Marsan, M. A., Bianco, A., Giaccone, P., Leonardi, E., and Neri, F., Packet Scheduling in Input-Queued Cell-Based Switches, *Proceedings of IEEE INFOCOM*, pp. 1085–1094, 2001.

300. Marsan, M. A., Bianco, A., Leonardi, E., and Milia, L., RPA: A Flexible Scheduling Algorithm for Input Buffered Switches, *IEEE Transactions on Communications*, 47(12):1921–1933, 1999.

301. Martini, L., IANA Allocations for Pseudowire Edge to Edge Emulation (PWE3), *RFC 4446, IETF*, 2006.

302. Martini, L., and Sajassi, A., 802.1ah Ethernet Pseudowire, *draft-martini-pwe3-802. 1ah-pw-02.txt, IETF*, 2008.

303. Martini, L., Bocci, M., and Balus, F., Dynamic Placement of Multi Segment Pseudo Wires, *draft-ietf-pwe3-dynamic-ms-pw-06, IETF*, 2007.

304. Martini, L., Jayakumar, J., Bocci, M., El-Aawar, N., Brayley, J., and Koleyni, G., Encapsulation Methods for Transport of Asynchronous Transfer Mode (ATM) Over MPLS Networks, *RFC 4717, IETF*, 2006.

305. Martini, L., Kawa, C., and Malis, A., Encapsulation Methods for Transport of Frame Relay Over Multiprotocol Label Switching (MPLS) Networks, *RFC 4619, IETF*, 2006.

306. Martini, L., Rosen, E., and El-Aawar, N., Encapsulation Methods for Transport of Layer 2 Frames Over MPLS Networks, *RFC 4905, IETF*, 2007.

307. Martini, L., Rosen, E., and El-Aawar, N., Transport of Layer 2 Frames Over MPLS, *RFC 4906, IETF*, 2007.

308. Martini, L., Rosen, E., El-Aawar, N., and Heron, G., Encapsulation Methods for Transport of Ethernet Over MPLS Networks, *RFC 4448, IETF*, 2006.

309. Martini, L., Rosen, E., El-Aawar, N., Smith, T., and Heron, G., Pseudowire Setup and Maintenance Using the Label Distribution Protocol (LDP), *RFC 4447, IETF*, 2006.

310. Martini, L., Rosen, E., Heron, G., and Malis, A., Encapsulation Methods for Transport of PPP/ High-Level Data Link Control (HDLC) Over MPLS Networks, *RFC 4618, IETF*, 2006.

311. McKenney, P. E., Stochastic Fairness Queuing, *Proceedings IEEE INFOCOM*, 1990.

312. McKeown, N., Scheduling Algorithms for Input-Queued Cell Switches, PhD Dissertation, University of California at Berkeley, 1995.

313. McKeown, N., iSLIP: A Scheduling Algorithm for Input-Queued Switches, *IEEE/ ACM Transactions on Networking*, 7(2):188–201, 1999.

314. McKeown, N., Mekkittikul, A., Anantharam, V., and Walrand, J., Achieving 100% Throughput in an Input-Queued Switch, *IEEE Transactions on Communications*, 47(8):1260–1267, 1999.

315. McKeown, N., Varaiya, P., and Warland, J., Scheduling Cells in an Input-Queued Switch, *IEE Electronics Letters*, 29(25):2174–2175, 1993.

316. Message Passing Interface Forum, "MPI: A Message-Passing Interface Standard", 1995.

317. Metro Ethernet Forum, Ethernet Services Attributes Phase 2, *MEF 10.1*, 2006.

318. Metro Ethernet Forum, Ethernet Local Management Interface (E-LMI), *MEF 16*, 2006.

319. Metro Ethernet Forum, Service OAM Requirements and Framework—Phase I, *MEF 17*, 2007.

320. Mogul, J., and Postel, J., Internet Standard Subnetting Procedure, *RFC 950, IETF*, 1985.

321. Mohan, D., and Sajassi, A., L2VPN OAM Requirements and Framework, *draft-ietf-l2vpn-oam-req-frmk-09.txt, IETF*, 2007.

322. Morris, R., Kohler, E., Jannotti, J., and Kaashoek, M. F., The Click Modular Router, *Proceedings of the Seventeenth ACM Symposium on Operating Systems Principles (SOSP '99)*; also in *Operating Systems Review*, 34:217–231, 1999.

323. Morrison, D. R., PATRICIA—Practical Algorithm To Retrieve Information Coded In Alphanumeric, *Journal of the ACM*, 15(4):514–534, 1968.

324. Moy, J., Multicast Extensions to OSPF, *RFC 1584, IETF*, 1994.

325. Moy, J., OSPF Version 2, *RFC 2328, IETF*, 1998.

326. Munteanu, D., and Williamson, C., An FPGA-Based Network Processor for IP Packet Compression, *Proceedings SCS SPECTS*, pp. 599–608, 2005.

327. Myers, G., Overview of IP Fabrics' PPL Language and Virtual Machine, white paper, *http://www.ipfabrics.com/pdf/Overview_of_PPL_and_VM.pdf*, 2006.

328. Nadeau, T. D., Morrow, M., Busschbach, P., Aissaoui, M., and Allan, D., Pseudo Wire (PW) OAM Message Mapping, *draft-ietf-pwe3-oam-msg-map-06.txt, IETF*, 2008.

329. Nagarajan, A., Generic Requirements for Provider Provisioned Virtual Private Networks (PPVPN), *RFC 3809, IETF*, 2004.

330. Nakamura, M., Ueda, H., Makino, S., Yokotani, T., and Oshima, K., Proposal of Networking by PON Technologies for Full and Ethernet Services in FTTx, *Journal of Lightwave Technology*, 22(11):2631–2640, 2004.

331. Network Processing Forum (NPF), *CSIX-L1: Common Switch Interface Specification-L1*, Network Processing Forum Implementation Agreement, 2000.

332. Network Processing Forum (NPF), Network Processing Forum Streaming Interface (NPSI) Implementation Agreement, *NPF2001.121.25*, 2002.

333. Nichols, K., Blake, S., Baker, F., and Black, D., Definition of the Differentiated Services Field (DS Field) in the IPv4 and IPv6 Headers, *RFC 2474, IETF*, 1998.

334. Nilsson, S., and Karlsson, G., Fast Address Look-Up for Internet Routers, *Proceedings of IEEE Broadband Communications: The Future of Telecommunications*, pp. 11–22, 1998.

335. Nilsson, S., and Karlsson, G., IP-Address Lookup Using LC-Tries, *IEEE Journal on Selected Areas in Communications*, 17(6):1083–1092, 1999.

336. Nilsson, S., and Tikkanen, M., Implementing a Dynamic Compressed Trie, *Proceedings 2nd Workshop on Algorithm Engineering*, 1998.

337. Nojima, S., Tsutsui, E., Fukuda, H., and Hashimoto, M., Integrated Services Packet Network Using Bus Matrix Switch, in *Broadband Switching: Architectures, Protocols, Design, and Analysis*, Chas, C., Konangi, V. K., and Sreetharan, M., (Eds.), pp. 296–304, 1991.

338. Ohta, H., Assignment of the "OAM Alert Label" for Multiprotocol Label Switching Architecture (MPLS) Operation and Maintenance (OAM) Functions, *RFC 3429, IETF*, 2002.

339. O'Mahony, M. J., Politi, C., Klonidis, D., Nejabati, R., and Simeonidou, D., Future Optical Networks, *Journal of Lightwave Technology*, 24(12):4684–4696, 2006.

340. Ooghe, S., Voigt, N., Platnic, M., Haag, T., and Wadhwa, S., Framework and Requirements for an Access Node Control Mechanism in Broadband Multi-Service Networks, *draft-ietf-ancp-framework-05.txt, IETF*, 2008.

341. Optical Internetworking Forum (OIF), Scalable System Packet Interface Implementation Agreement: System Packet Interface Capable of Operating as an Adaptation Layer for Serial Data Links, *IA # OIF-SPI-S-01.0*, 2006.

342. Optical Internetworking Forum (OIF), Serdes Framer Interface Level 5 Phase 2 (SFI-5.2): Implementation Agreement for 40Gb/s Interface for Physical Layer Devices, *IA#OIF-SFI-02.0*, 2006.

343. Ott, T. J., Lakshman, T. V., and Wong, L. H., SRED: Stabilized RED, *Proceedings IEEE INFOCOM*, 1999.

344. Overmars, M. H., and van der Stappen, A. F., Range Searching and Point Location Among Fat Objects, *Journal Algorithms*, 21(3):629–656, 1996.

345. Paganini, F., Doyle, J. C., and Low, S. H., Scalable Laws for Stable Network Congestion Control, *Proceedings Conference on Decision and Control*, 2001.

346. Pagiamtzis, K., and Sheikholeslami, A., Content-Addressable Memory (CAM) Circuits and Architectures: A Tutorial and Survey, *IEEE Journal of Solid-State Circuits*, 41(3):712–727, 2006.

347. Parekh, A. K., A Generalized Processor Sharing Approach to Flow Control in Integrated Services Networks, PhD Dissertation, MIT, 1992.

348. Parekh, A. K., and Gallager, R., A Generalization Processor Sharing Approach to Flow Control in Integrated Services Networks: The Single Node Case, *IEEE/ACM Transactions on Networks*, 1(3):344–357, 1993.

349. Patel, J. H., Performance of Processor-Memory Interconnections for Multiprocessors, *IEEE Transactions on Computers*, 30(10):771–780, 1981.

350. Perlman, R., *Interconnections: Bridges, Routers, Switches, and Internetworking Protocols (2nd Edition)*, Addison-Wesley, 1999.

351. Peterson, W. W., and Brown, D. T., Cyclic Codes for Error Detection, *Proceedings of the IRE*, pp. 228–235, 1961.

352. Plummer, D., Ethernet Address Resolution Protocol: Or Converting Network Protocol Addresses to 48-Bit Ethernet Address for Transmission on Ethernet Hardware, *RFC 826, IETF*, 1982.

353. Poikselka, M., Niemi, A., Khartabil, H., and Mayer, G., *The IMS: IP Multimedia Concepts and Services (2nd Edition)*, Wiley, 2006.

354. Postel, J., User Datagram Protocol, *RFC 768, IETF*, 1980.

355. Postel, J., Internet Protocol, *RFC 791, IETF*, 1981.

356. Postel, J., Internet Control Message Protocol, *RFC 792, IETF*, 1981.

357. Postel, J., Transmission Control Protocol, *RFC 793, IETF*, 1981.

358. Postel, J., and Reynolds, J. K., Standard for the Transmission of IP Datagrams Over IEEE 802 Networks, *RFC 1042, IETF*, 1988.

359. PVM: Parallel Virtual Machine, Oak Ridge National Laboratory (ORNL), *http://www.csm.ornl.gov/pvm/*.

360. Rajagopalan, B., Label Distribution Protocol (LDP) and Resource ReSerVation Protocol (RSVP) Extensions for Optical UNI Signaling, *RFC 3476, IETF*, 2003.

361. Ramabhadran, S., and Pasquale, J., Stratified Round Robin: A Low Complexity Packet Scheduler with Bandwidth Fairness and Bounded Delay, *Proceedings ACM SIGCOMM*, pp. 239–250, 2003.

362. Ramakrishnan, K. K., Floyd, S., and Black, D., The Addition of Explicit Congestion Notification (ECN) to IP, *RFC 3168, IETF*, 2001.

363. Rekhter, Y., and Li, T., An Architecture for IP Address Allocation with CIDR, *RFC 1518, IETF*, 1993.

364. Rekhter, Y., and Li, T., A Border Gateway Protocol 4 (BGP-4), *RFC 1771, IETF*, 1995.

365. Rekhter, Y., Li, T., and Hares, S., A Border Gateway Protocol 4 (BGP-4), *RFC 4271, IETF*, 2006.

366. Rigney, C., Willens, S., Rubens, A., and Simpson, W., Remote Authentication Dial-In User Service (RADIUS), *RFC 2865, IETF*, 2000.

367. Rosen, E., and Rekhter, Y., BGP/MPLS VPNs, *RFC 2547, IETF*, 1999.

368. Rosen, E., and Rekhter, Y., BGP/MPLS IP Virtual Private Networks (VPNs), *RFC 4364, IETF*, 2006.

369. Rosen, E., Tappan, D., Fedorkow, G., Rekhter, Y., Farinacci, D., Li, T., and Conta, A., MPLS Label Stack Encoding, *RFC 3032, IETF*, 2001.

370. Rosen, E., Viswanathan, A., and Callon, R., Multiprotocol Label Switching Architecture, *RFC 3031, IETF*, 2001.

371. Rosolen, V., Bonaventure, O., and Leduc, O., Impact of Cell Discard Strategies on TCP/IP in ATM UBR Networks, *Proceedings 6th Workshop on Performance Modeling and Evaluation of ATM Networks*, 1998.

372. Ruf, L., Bossardt, M., Plattner, B., and Stadler, R., A Linux-Based Nose OS for Network Processors, *TIK Report 205*, Computer Engineering and Networks Laboratory (TIK), Swiss Federal Institute of Technology (ETH), 2001.

373. Ruf, L., Farks, K., Hug, H., and Plattner, B., The PromethOS NP Service Programming Interface, *TIK Report 228*, Computer Engineering and Networks Laboratory (TIK), Swiss Federal Institute of Technology (ETH), 2005.

374. Ruiz-Sanchez, M. A., Biersack, E. W., and Dabbous, W., Survey and Taxonomy of IP Address Lookup Algorithms, *IEEE Network*, 15(2):8–23, 2001.

375. Saltsidis, P., IEEE802.1Qay/D0.0, *http://www.ieee802.org/1/files/public/docs2007/ay-saltsidis-initial-draft-0507.pdf*, 2007.

376. Sangli, S. R., Tappan, D., and Rekhter, Y., BGP Extended Communities Attribute, *RFC 4360, IETF*, 2006.

377. Saturn Development Group, SATURN Compatible Packet Over SONET Interface Specification for Physical Layer Devices (Level 3), *PMC-980495*, Issue 3, PMC-Sierra, 1998.

378. Saturn Group, POS-PHY: Saturn Compatible Packet Over SONET, Interface Specification for Physical Layer Devices (Level 2), *PMC-971147*, Issue 5, PMC-Sierra, 1998.

379. Serpanos, D. N., and Antoniadis, P. I., FIRM: A Class of Distributed Scheduling Algorithms for High-Speed ATM Switches with Multiple Input Queues, *Proceedings IEEE INFOCOM*, Vol. 2:548–555, 2000.

380. Serpanos, D. N., Baldi, M., and Giladi, R., Network Systems Architecture, *IEEE Network*, 21(4):6–7, 2007.

381. Shah, N., and Keutzer, K., Network Processors: Origin of Species, *Proceedings of the 17th International Symposium on Computer and Information Sciences, XVII*, 2002.

382. Shah, N., Plishker, W., and Keutzer, K., NP-Click: A Programming Model for the Intel IXP1200, *Proceedings of the Second Workshop on Network Processors (NP-2)*; and in the *Ninth International Symposium on High Performance Computer Architectures (HPCA)*, 2003.

383. Shah, N., Plishker, W., Ravindran, K., and Keutzer, K., NP-Click: A Productive Software Development Approach for Network Processors, *IEEE Micro*, 24(5):45–54, 2004.

384. Shenker, S., Partridge, C., and Guerin, R., Specification of Guaranteed Quality of Service, *RFC 2212, IETF*, 1997.

385. Shreedhar, M., and Varghese, G., Efficient Fair Queuing Using Deficit Round Robin, *Proceedings ACM SIGCOM*, pp. 231–242, 1995.

386. Simpson, W., PPP Over SONET/SDH, *RFC 1619, IETF*, 1994.

387. Simpson, W., The Point-to-Point Protocol (PPP), STD 50, *RFC 1661, IETF*, 1994.

388. Simpson, W., PPP in HDLC-Like Framing, STD 51, *RFC 1662, IEEE*, 1994.

389. Sklower, K., A Tree-Based Packet Routing Table for Berkeley Unix, *Proceedings USENIX Winter*, pp. 93–104, 1991.

390. Spalink, T., Karlin, S., Peterson, L., and Gottlieb, Y., Building a Robust Software-Based Router Using Network Processors, *Proceedings of the Eighteenth ACM Symposium on Operating Systems Principles*, pp. 216–229, 2001.

391. Srinivasan, R., RPC: Remote Procedure Call Protocol Specification Version 2, *RFC 1831, IETF*, 1995.

392. Srinivasan, R., XDR: External Data Representation Standard, *RFC 1832, IETF*, 1995.

393. Srinivasan, V., Suri, S., and Varghese, G., Packet Classification Using Tuple Space Search, *Proceedings ACM SIGCOMM*, pp. 135–146, 1999.

394. Srinivasan, V., Varghese, G., Suri, S., and Waldvogel, M., Fast and Scalable Layer Four Switching, *Proceedings of the ACM SIGCOMM Conference on Applications, Technologies, Architectures, and Protocols for Computer Communication*, pp. 191–202, 1998.

395. Srisuresh, P., and Holdrege, M., IP Network Address Translator (NAT) Terminology and Considerations, *RFC 2663, IETF*, 1999.

396. Stiliadis, D., and Varma, A., A General Methodology for Designing Efficient Traffic Scheduling and Shaping Algorithms, *Proceedings IEEE INFOCOM*, 1997.

397. Stiliadis, D., and Varma, A., Efficient Fair Queuing Algorithms for Packet-Switched Networks, *IEEE/ACM Transactions on Networking*, 6(2):175–185, 1998.

398. Sun, J., Chen, G., Ko, K. T., Chan, S., and Zukerman, M., PD-Controller: A New Active Queue Management Scheme, *Proceedings IEEE GLOBECOM*, Vol. 6:3103–3107, 2003.

399. Suri, S., Varghese, G., and Chandranmenon, G., Leap Forward Virtual Clock: A New Fair Queuing Scheme with Guaranteed Delays and Throughput Fairness, *Proceedings IEEE INFOCOM*, 1997.

400. Suri, S., Varghese, G., and Warkhede, P. R., Multiway Range Trees: Scalable IP Lookup with Fast Updates, *Technical Report 99-28*, Washington University, 1999.

401. Suter, B., Lakshman, T. V., Stiliadis, D., and Choudhury, A. K., Design Considerations for Supporting TCP with Per-Flow Queueing, *Proceedings IEEE INFOCOM*, pp. 299–306, 1998.

402. Tacda, T., Framework and Requirements for Layer 1 Virtual Private Networks, *RFC 4847, IETF*, 2007.

403. Thaler, D., Interoperability Rules for Multicast Routing Protocols, *RFC 2715, IETF*, 1999.

404. Thiele, L., Chakraborty, S., Gries, M., and Knzli, S., Design Space Exploration of Network Processor Architectures, *First Workshop on Network Processors at the 8th International Symposium on High Performance Computer Architecture (HPCA8)*, 2002.

405. Thiele, L., Chakraborty, S., Gries, M., Maxiaguine, A., and Greutert, J., Embedded Software in Network Processors—Models and Algorithms, in *First Workshop on Embedded Software, Lecture Notes in Computer Science*, Vol. 2211, Springer Verlag, pp. 416–434, 2001.

406. Tobagi, F. A., Fast Packet Switch Architectures for Broadband Integrated Services Networks, *Proceedings IEEE*, Vol. 78:133–167, 1990.

407. Tobagi, F. A., Chiussi, F. M., and Kwok, T., Architecture, Performance, and Implementation of the Tandem Banyan Fast Packet Switch, *IEEE Journal on Selected Areas in Communications*, 9(8):1173–1193, 1991.

408. Tomic, S., Statovci-Halimi, B., Halimi, A., Muellner, W., and Fruehwirth, J., ASON and GMPLS— Overview and Comparison, *Photonic Network Communications*, 7(2):111–130, 2004.

409. Townsley, W., Valencia, A., Rubens, A., Pall, G., Zorn, G., and Palter, B., Layer Two Tunneling Protocol "L2TP," *RFC 2661, IETF*, 1999.

410. Traina, P., McPherson, D., and Scudder, J., Autonomous System Confederations for BGP, *RFC 3065, IETF*, 2001.

411. Tse, E. S., Switch Fabric Design for High Performance IP Routers: A Survey, *Journal of Systems Architecture: The EUROMICRO Journal*, 51(10-11):571–601, 2005.

412. Tsuchiya, P., A Search Algorithm for Table Entries with Non-Contiguous Wildcarding, *Unpublished Report, Bellcore, http://citeseer.ist.psu.edu/tsuchiya91search.html*, 1992.

413. Turner, J. S., New Directions in Communications (or Which Way to the Information Age?), *IEEE Communication Magazine*, 24(10):8-15, 1986.

414. Turner, J., Strong Performance Guarantees for Asynchronous Crossbar Schedulers, *Proceedings IEEE INFOCOM, the 25th IEEE International Conference on Computer Communications*, 2006.

415. Vainshtein, A., and Stein, Y. J., Structure-Agnostic Time Division Multiplexing (TDM) Over Packet (SAToP), *RFC 4553, IETF*, 2006.

416. Valiant, L. G., A Bridging Model for Parallel Computation, *Communications of the ACM*, 33(8):103-111, 1990.

417. Valiant, L. G., and Brebner, G. J., Universal Schemes for Parallel Communication, *Proceedings of the Thirteenth Annual ACM Symposium on Theory of Computing*, pp. 263-277, 1981.

418. van Emde Boas, P., Preserving Order in a Forest in Less than Logarithmic Time and Linear Space, *Information Processing Letters*, 6(3):80-82, 1977.

419. van Emde Boas, P., Kaas, R., and Zijlstra, E., Design and Implementation of an Efficient Priority Queue, *Mathematical Systems Theory*, 10(2):99-127, 1977.

420. Verma, D., Zhang, H., and Ferrari, D., Guaranteeing Delay Jitter Bounds in Packet Switching Networks, *Proceedings of TRICOMM*, pp. 35-46, 1991.

421. Villamizar, C., Chandra, R., and Govindan, R., BGP Route Flap Damping, *RFC 2439, IETF*, 1998.

422. Vohra, Q., and Chen, E., BGP Support for Four-Octet AS Number Space, *RFC 4893, IETF*, 2007.

423. von Neumann, J., A Certain Zero-Sum Two-Person Game Equivalent to the Optimal Assignment Problem, in *Contributions to the Theory of Games (Annals of Mathematics Study No. 28)*, Kuhn, H., and Tucker, A., (Eds.), Princeton University Press, 2:5-12, 1953.

424. Waitzman, D., Partridge, C., and Deering, S., Distance Vector Multicast Routing Protocol, *RFC 1075, IETF*, 1988.

425. Waldvogel, M., Fast Longest Prefix Matching: Algorithms, Analysis, and Applications, PhD Dissertation, ETH No. 13266, Swiss Federal Institute of Technology, 2000.

426. Waldvogel, M., Varghese, G., Turner, J., and Plattner, B., Scalable High-Speed IP Routing Lookups, *Proceedings ACM SIGCOMM*, pp. 25-36, 1997.

427. Waldvogel, M., Varghese, G., Turner, J., and Plattner, B., Scalable High-Speed Prefix Matching, *ACM Transactions on Computer Systems*, 19(4):440-482, 2001.

428. Willard, D. E., Log-Logarithmic Worst Case Range Queries are Possible in Space $\Theta(n)$, *Information Processing Letters*, 17(2):81-84, 1983.

429. Willard, D. E., New Trie Data Structures Which Support Very Fast Search Operations, *Journal of Computer and System Sciences*, 28(3):379-394, 1984.

430. WiMAX Forum, WiMAX Forum Network Architecture Stage 2-3: Release 1, Version 1.2, *http://www.wimaxforum.org/technology/documents/WiMAX_End-to-End_Network_Systems_Architecture_Stage_2-3_Release_1.1.2.zip*.

431. Worster, T., Rekhter, Y., and Rosen, E., Encapsulating MPLS in IP or Generic Routing Encapsulation (GRE), *RFC 4023, IETF*, 2005.

432. Wroclawski, J., The Use of RSVP with *IETF* Integrated Services, *RFC 2210, IETF*, 1997.

433. Wroclawski, J., Specification of the Controlled-Load Network Element Service, *RFC 2211, IETF*, 1997.

434. Xiao, H., Zhang, L., and Wu, D., Software Component Model for Network Processor Based System, *Proceedings of the 6th International Conference on Parallel and Distributed Computing Applications and Technologies*, pp. 427–429, 2005.

435. Yang, L., Dantu, R., Anderson, T., and Gopal, R., Forwarding and Control Element Separation (ForCES) Framework, *RFC 3746, IETF*, 2004.

436. Yeh, Y.-S., Hluchyj, M. G., and Acampora, A. S., The Knockout Switch: A Simple, Modular Architecture for High-Performance Packet Switching, *IEEE Journal on Selected Areas in Communications*, 5(8):1274–1283, 1987.

437. Yi, K., and Gaudiot, J.-L., Architectural Support for Network Applications on Simultaneous MultiThreading Processors, *Proceedings of the IEEE International Parallel and Distributed Processing Symposium*, p. 46, 2007.

438. Yoo, S. J. B., Optical Packet and Burst Switching Technologies for the Future Photonic Internet, *Journal of Lightwave Technology*, 24(12):4468–4492, 2006.

439. Zane, F., Narlikar, G., and Basu, A., CoolCAM: Power-Efficient TCAMs for Forwarding Engines, *Proceedings IEEE INFOCOM*, 2003.

440. Zhang, H., Service Disciplines for Guaranteed Performance Service in Packet-Switching Networks, *Proceedings IEEE*, Vol. 83:1374–1396, 1995.

441. Zhang, H., and Ferrari, D., Rate-Controlled Service Disciplines, *Journal of High Speed Networks*, 3(4):389–412, 1994.

442. Zhang, L., Virtual Clock: A New Traffic Control Algorithm for Packet Switching Networks, *Proceedings ACM SIGCOMM*, pp. 19–29, 1990.

443. Ziv, J., and Lempel, A., A Universal Algorithm for Sequential Data Compression, *IEEE Transactions on Information Theory*, 23:337–343, 1977.

444. Ziv, J., and Lempel, A., Compression of Individual Sequences Via Variable-Rate Coding, *IEEE Transactions Information Theory*, 24(5):530–536, 1978.

Index

Printed and bound by CPI Group (UK) Ltd, Croydon, CR0 4YY

03/10/2024

01040320-0005